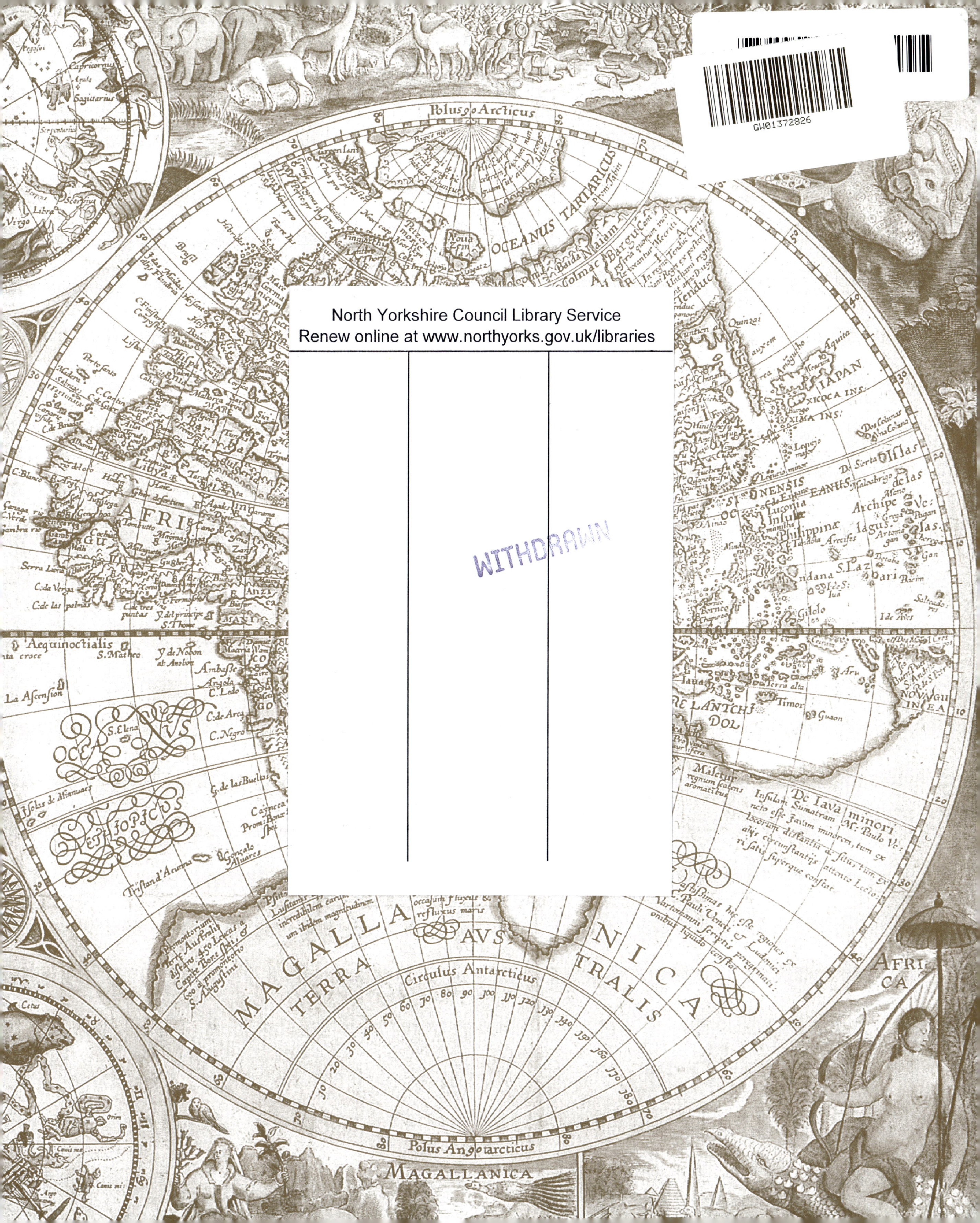

EXPLORATION MAP BY MAP

EXPLORATION MAP BY MAP

CONTENTS

THE ANCIENT WORLD
PREHISTORY–c.500 CE

THE MIDDLE AGES
c.500–1460

THE EARLY MODERN WORLD
1400–1700

CONSULTANTS & CONTRIBUTORS

MAIN CONSULTANT AND CONTRIBUTOR

Philip Parker is an acclaimed author, editor, and historian specializing in the classical and medieval worlds. He studied History at Cambridge University and International Relations at Johns Hopkins University's School of Advanced International Studies.

SPECIALIST CONSULTANTS

Dr Fozia Bora, Dr Pearl K. Brower, Dr Anke Hein, Katherine Hunt, Lauren Keenan, Dr Jagjeet Lally, Seun Matiluko, Rafael Shimabukuro-Cabrera, Timothy K. Topper

CONTRIBUTORS

Simon Adams, John Farndon, Paul Lane, Shafik Meghji

UPHEAVAL AND INDUSTRY 1700–c.1850

THE AGE OF GLOBAL EMPIRES c.1850–1914

THE MODERN WORLD 1910–PRESENT

INTRODUCTION

Humankind is by nature an exploratory species. From the time our human ancestors made the first tentative steps out of Africa almost 2 million years ago, through ancient Phoenician expansion, to the Early Modern circumnavigations of the seas, and the invention of space probes that have travelled far beyond the edges of the Solar System, our instinct has always been to venture into the unknown.

And what better way to celebrate this spirit of adventure than through maps, on which routes can be understood at a glance, traced across land and sea. This book charts landmark journeys throughout history – ancient, medieval, and modern – in fascinating detail. Specially commissioned cartography allows you to follow, step by step, the progress of groundbreaking expeditions, such as those that tried for centuries to find the elusive Northwest Passage, or the famous voyage of HMS *Beagle*, with a young, inquisitive Charles Darwin on board.

Historical maps also illustrate many travels and plot the evolution of our ever-changing worldview. On medieval maps, for example, we see biblical stories and mythical creatures alongside countries and continents, and *terra incognita* or "unknown land" features on maps from the ancient world up until the 19th century.

Taking us from prehistory to the present day, the chronological structure of the book highlights a fact that is often overlooked – how each journey further into the unknown is built on the knowledge of those that have gone before. The early Portuguese navigators, such as Gil Eanes, who pushed steadily down the West African coast in the 14th century, blazed a trail for Vasco da Gama's final voyage around the continent to India in 1497. And in the same way, they both followed in the footsteps of the Phoenician Hanno, who made part of that journey in the 6th century BCE.

The travels and discoveries of Western explorers such as Christopher Columbus, Vasco da Gama, Ferdinand Magellan, and Captain James Cook are well documented, but this book also features lesser-known journeys that are just as noteworthy. The Polynesians, for example, crossed almost the whole width of the Pacific in double-hulled canoes, centuries before the Vikings traversed the Atlantic in around 1000 CE. And during the early 15th century, Chinese admiral Zheng He sailed from eastern China all the way to East Africa, decades before the Portuguese arrived there in 1498. In 1766, the first woman to sail around the world, Jeanne Baret, led the way for many other legendary female travellers, and centuries before any Europeans arrived, the original inhabitants of Australia, the Americas, Africa, and Siberia were already making epic journeys across their lands, preserving their knowledge through oral traditions rather than in written records.

Following each route along these maps reveals so many fascinating details and, in every case, the story expands to take in not only the better-known characters but also other key members of renowned expeditions. By venturing into the unknown, explorers found themselves "strangers in a strange land" and their Indigenous guides, interpreters, and porters were often the people who enabled them to succeed. Without them, many of these endeavours would have ended in failure or even death.

Excepting those very first prehistoric human migrations, which entered previously uninhabited landscapes, every exploration is also the story of an encounter. The explorer, whether consciously or by design, carries a piece of their own culture and transplants it elsewhere. Sometimes this has resulted in dynamic, hybrid societies; yet, all too often it has led instead to conflict and the destruction of Indigenous cultures. As this book illustrates, when the power balance is unequal, exploration rapidly turns into exploitation.

It is to be hoped, as our species ventures beyond our planet, that it will be in the spirit of remembering that those who go in search of the new will often encounter the old – often finding things that were never lost – and that exploration, even if carried out by individuals, has always had shared consequences for all of humanity.

PHILIP PARKER

▷ **Largest early world map**
Made in 1587 from 60 sheets of hand-drawn manuscript, measuring just over 3 m (10 ft) in diameter, this planisphere by Italian gentleman scholar Urbano Monte draws on Renaissance knowledge to visualize the world as a whole, with the North Pole at its centre.

MARE OCEANO

THE ANCIENT WORLD

EARLY HOMININS ORIGINATED IN AFRICA 6 TO 7 MILLION YEARS AGO. AROUND 1.8 MILLION YEARS AGO, THE ANCESTORS OF MODERN HUMANS BEGAN TO TRAVEL, MIGRATING ACROSS AFRICA AND INTO EURASIA. THEY LATER PEOPLED EVERY CONTINENT, JOURNEYING ON IN SEARCH OF KNOWLEDGE, TRADE, AND NEW LANDS.

△ Into the Americas
Bifacial stone points, of a type known as Clovis (see p.17) and dating from around 13,000 years ago, were a new technology – where tools had fluted points – that spread among the early settlers of the Americas.

THE FIRST JOURNEYS

Humans have been travellers from the time our first human ancestors began to migrate from settlements in Africa, around 1.8 million years ago. They travelled in response to changing climate patterns and the lure of better hunting conditions until, by the dawn of the Middle Ages, they had peopled almost every part of the globe.

The first hominins, the ancestors of modern humans, evolved in Africa. One species, *Homo erectus*, became the first known explorers, leaving the continent around 1.8 million years ago to spread across Eurasia. Although *Homo erectus* eventually died out, *Homo sapiens* – modern humans – followed around 200,000 years ago, at first tentatively, in waves that swept across West Asia and then receded. Finally, around 70,000 years ago, a new migration took place out of Africa, which, by the dawn of the Middle Ages, had covered every continent and reached almost every corner of the world.

These early travellers advanced slowly, on foot, or sometimes by boat, pulled onwards by a changing climate and the promise of better hunting. By 60,000 years ago they had reached Australia, and by 54,000 years ago had spread to Europe and across to Siberia. The Americas were peopled about 20,000 years ago. With tools made only of stone, bone, and wood, and building no large permanent constructions, these early travellers left only a light footprint on the land. Living millennia before writing was invented, they remain nameless to us, yet they were true pioneers, venturing to lands on which no human had set foot before.

Trade, exploration, and expansion

The result, by about 10,000 BCE, was a series of relatively isolated islands of humanity, each with little knowledge of the others. The invention of agriculture and the growth of settled villages, and then of towns and cities, changed that. Complex societies emerged, first in Egypt, Mesopotamia, and China, a little later in India, Mesoamerica, and Peru, and then in Europe, Southeast Asia, and sub-Saharan Africa. Those societies began to trade. Luxury items were particularly prized, especially lapis lazuli from Afghanistan, which adorned jewellery as far away as pharaonic Egypt around 3,000 BCE.

Trade led to exploration of a different sort. Phoenician sailors ventured into the Mediterranean from around 1000 BCE, founding settlements on its shores, and also into the Atlantic, becoming the first outsiders to explore the West African coastline, and sailing as far north as Britain. Egyptian expeditions travelled down the Nile to lands they called Punt and Yam, possibly in modern Somalia, in search

△ The Romans in North Africa
The ruins of Timgad, once an impressive Roman colony, lie in the Sahara Desert in modern-day Algeria. The city was founded by the Emperor Trajan around 100 CE, when the Roman Empire was at its height.

EARLY EXPLORATION

It took nearly 3 million years for human ancestors to move definitively out of Africa, but from 70,000 years ago, modern human beings explored and settled almost the entire globe within a 40,000-year period, except for islands such as New Zealand, Hawaii, and Iceland, and the Polar regions. After this initial migration, exploration often took the form of encounters between expeditions from diverse regions which had previously had little contact with each other.

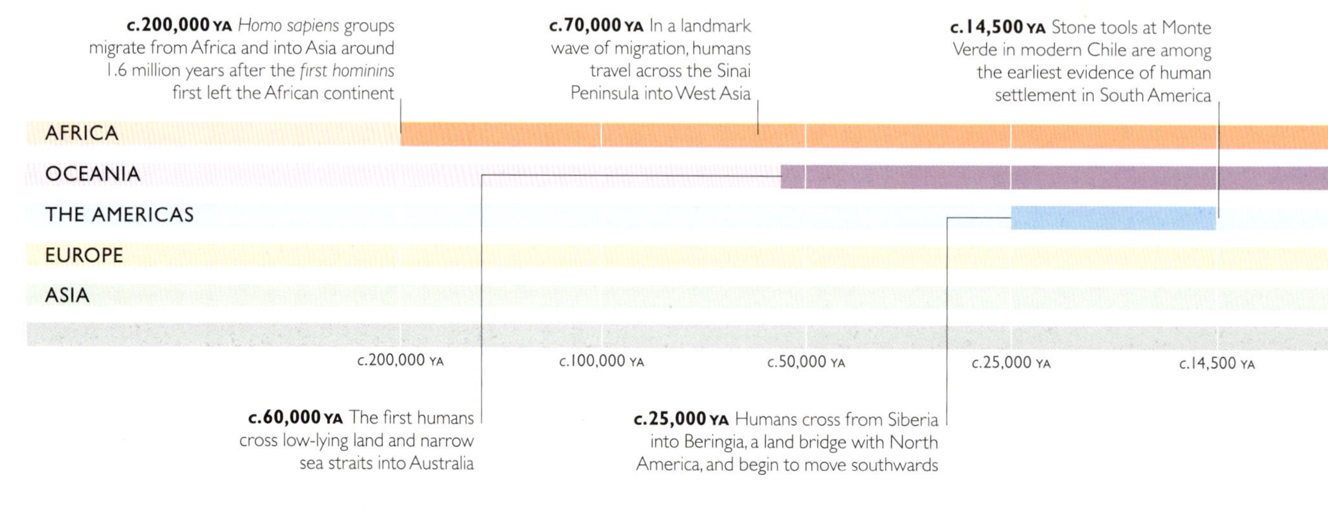

c.200,000 YA *Homo sapiens* groups migrate from Africa and into Asia around 1.6 million years after the *first hominins* first left the African continent
c.70,000 YA In a landmark wave of migration, humans travel across the Sinai Peninsula into West Asia
c.14,500 YA Stone tools at Monte Verde in modern Chile are among the earliest evidence of human settlement in South America
AFRICA
OCEANIA
THE AMERICAS
EUROPE
ASIA
c.200,000 YA
c.100,000 YA
c.50,000 YA
c.25,000 YA
c.14,500 YA
c.60,000 YA The first humans cross low-lying land and narrow sea straits into Australia
c.25,000 YA Humans cross from Siberia into Beringia, a land bridge with North America, and begin to move southwards

◁ **Zhang Qian's expedition departure**
This 7th-century mural on the walls of Mogao Cave 323 at Dunhuang shows the explorer leaving the Imperial Han court to embark on his expedition to "barbarian lands" in the Western Regions.

"It's... an incredible achievement to voyage halfway around the world and carve out life in a new environment."

DR KIRA WESTAWAY, AUSTRALIAN GEOCHRONOLOGIST

of gold and ivory. These expeditions resulted in interactions between diverse cultures. Some created economic networks that bound regions together, while others made a direct and often catastrophic impact through the spread of disease or outright conquest, when armies or colonists followed in the wake of merchants. The emergence of great empires, such as the Achaemenid Persian in the 6th century BCE or the Roman Empire from the early centuries BCE, accelerated this trend. Rome's explorations were far from peaceful as its legions pushed into new land and were met with resistance from the Indigenous populations. The association between trade, empires, and exploration and the differing perspectives of those who "explored" lands new to them, and those who had long lived in them, was already becoming well established.

Continuing migrations

The Polynesians settled strings of previously uninhabited islands in the South Pacific from *c.*800 BCE; the Celts spread throughout Central and Western Europe from their homeland in the middle Danube region from *c.*600 BCE; the Greeks had established colony cities from western Spain to the Kashmir by 300 BCE; and *c.*300 CE Bantu-speaking peoples moved into southern Africa from West and Central Africa.

▷ **Rock art**
Drawn in the X-ray style that shows internal organs, this rock painting from around 20,000 years ago is from Nourlangie Rock Shelter in the Kakadu National Park in northern Australia, home for tens of thousands of years to the Bininj people.

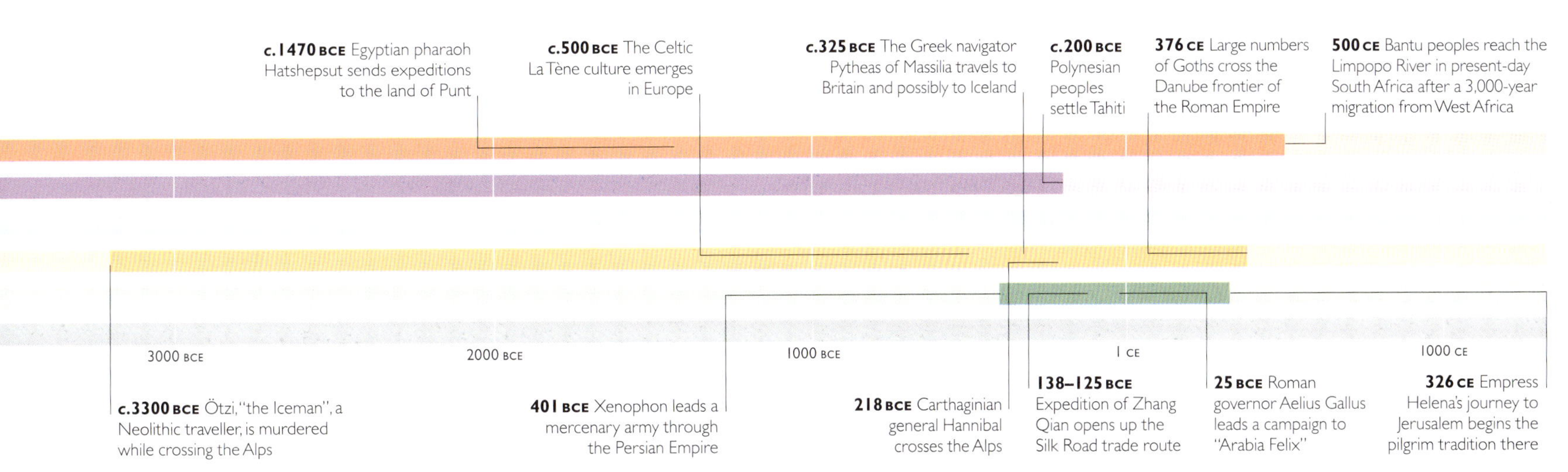

c.3300 BCE Ötzi, "the Iceman", a Neolithic traveller, is murdered while crossing the Alps

c.1470 BCE Egyptian pharaoh Hatshepsut sends expeditions to the land of Punt

c.500 BCE The Celtic La Tène culture emerges in Europe

401 BCE Xenophon leads a mercenary army through the Persian Empire

c.325 BCE The Greek navigator Pytheas of Massilia travels to Britain and possibly to Iceland

218 BCE Carthaginian general Hannibal crosses the Alps

c.200 BCE Polynesian peoples settle Tahiti

138–125 BCE Expedition of Zhang Qian opens up the Silk Road trade route

25 BCE Roman governor Aelius Gallus leads a campaign to "Arabia Felix"

376 CE Large numbers of Goths cross the Danube frontier of the Roman Empire

326 CE Empress Helena's journey to Jerusalem begins the pilgrim tradition there

500 CE Bantu peoples reach the Limpopo River in present-day South Africa after a 3,000-year migration from West Africa

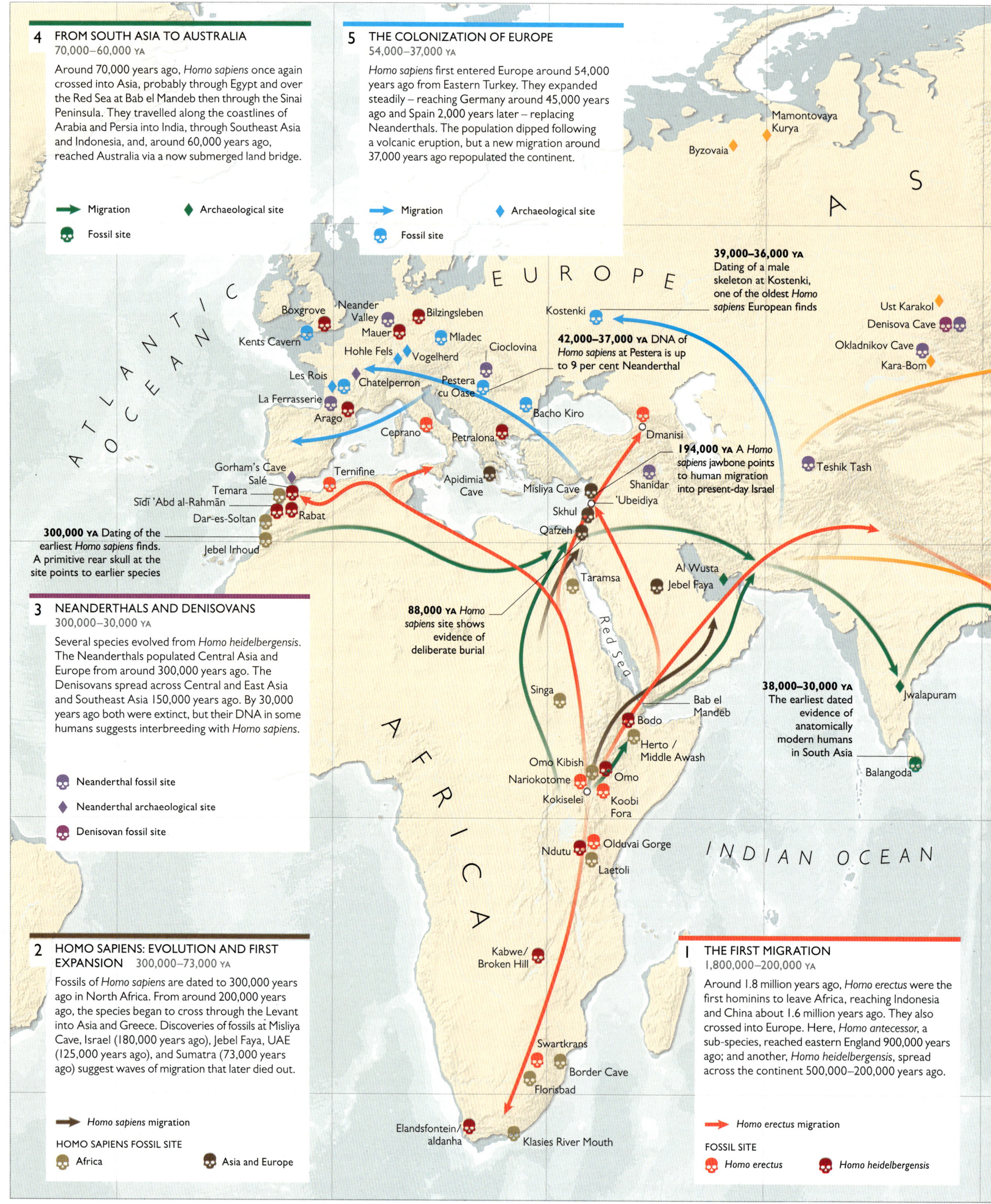

1 THE FIRST MIGRATION
1,800,000–200,000 YA

Around 1.8 million years ago, *Homo erectus* were the first hominins to leave Africa, reaching Indonesia and China about 1.6 million years ago. They also crossed into Europe. Here, *Homo antecessor*, a sub-species, reached eastern England 900,000 years ago; and another, *Homo heidelbergensis*, spread across the continent 500,000–200,000 years ago.

Homo erectus migration

FOSSIL SITE

Homo erectus | *Homo heidelbergensis*

2 HOMO SAPIENS: EVOLUTION AND FIRST EXPANSION 300,000–73,000 YA

Fossils of *Homo sapiens* are dated to 300,000 years ago in North Africa. From around 200,000 years ago, the species began to cross through the Levant into Asia and Greece. Discoveries of fossils at Misliya Cave, Israel (180,000 years ago), Jebel Faya, UAE (125,000 years ago), and Sumatra (73,000 years ago) suggest waves of migration that later died out.

Homo sapiens migration

HOMO SAPIENS FOSSIL SITE

Africa | Asia and Europe

3 NEANDERTHALS AND DENISOVANS
300,000–30,000 YA

Several species evolved from *Homo heidelbergensis*. The Neanderthals populated Central Asia and Europe from around 300,000 years ago. The Denisovans spread across Central and East Asia and Southeast Asia 150,000 years ago. By 30,000 years ago both were extinct, but their DNA in some humans suggests interbreeding with *Homo sapiens*.

Neanderthal fossil site

Neanderthal archaeological site

Denisovan fossil site

4 FROM SOUTH ASIA TO AUSTRALIA
70,000–60,000 YA

Around 70,000 years ago, *Homo sapiens* once again crossed into Asia, probably through Egypt and over the Red Sea at Bab el Mandeb then through the Sinai Peninsula. They travelled along the coastlines of Arabia and Persia into India, through Southeast Asia and Indonesia, and, around 60,000 years ago, reached Australia via a now submerged land bridge.

Migration | Archaeological site

Fossil site

5 THE COLONIZATION OF EUROPE
54,000–37,000 YA

Homo sapiens first entered Europe around 54,000 years ago from Eastern Turkey. They expanded steadily – reaching Germany around 45,000 years ago and Spain 2,000 years later – replacing Neanderthals. The population dipped following a volcanic eruption, but a new migration around 37,000 years ago repopulated the continent.

Migration | Archaeological site

Fossil site

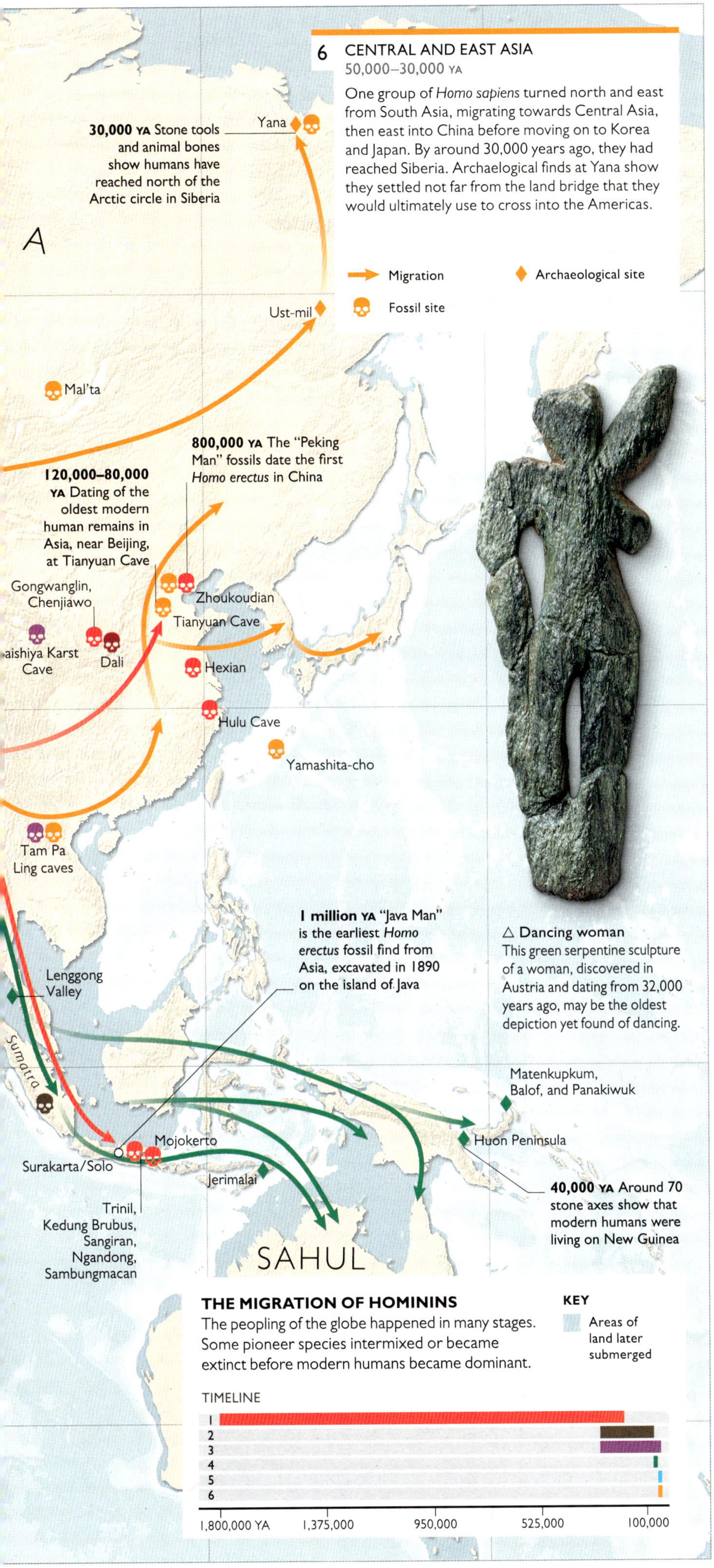

6 CENTRAL AND EAST ASIA
50,000–30,000 YA

One group of *Homo sapiens* turned north and east from South Asia, migrating towards Central Asia, then east into China before moving on to Korea and Japan. By around 30,000 years ago, they had reached Siberia. Archaeological finds at Yana show they settled not far from the land bridge that they would ultimately use to cross into the Americas.

Migration
Archaeological site
Fossil site

30,000 YA Stone tools and animal bones show humans have reached north of the Arctic circle in Siberia

Yana
A
Ust-mil
Mal'ta

800,000 YA The "Peking Man" fossils date the first *Homo erectus* in China

120,000–80,000 YA Dating of the oldest modern human remains in Asia, near Beijing, at Tianyuan Cave

Gongwanglin, Chenjiawo
Zhoukoudian
Tianyuan Cave
aishiya Karst Cave
Dali
Hexian
Hulu Cave
Yamashita-cho
Tam Pa Ling caves

1 million YA "Java Man" is the earliest *Homo erectus* fossil find from Asia, excavated in 1890 on the island of Java

△ **Dancing woman**
This green serpentine sculpture of a woman, discovered in Austria and dating from 32,000 years ago, may be the oldest depiction yet found of dancing.

Lenggong Valley
Sumatra
Matenkupkum, Balof, and Panakiwuk
Huon Peninsula
Mojokerto
Surakarta/Solo
Jerimalai
Trinil, Kedung Brubus, Sangiran, Ngandong, Sambungmacan

40,000 YA Around 70 stone axes show that modern humans were living on New Guinea

SAHUL

THE MIGRATION OF HOMININS
The peopling of the globe happened in many stages. Some pioneer species intermixed or became extinct before modern humans became dominant.

KEY
Areas of land later submerged

TIMELINE
1
2
3
4
5
6
1,800,000 YA 1,375,000 950,000 525,000 100,000

EARLY HOMININ MIGRATIONS

Hominins – humans and their ancestors – have long been explorers. Around 1.8 million years ago, they began to migrate out of Africa in several waves, embarking on journeys that ended only when the present landmasses were settled, around a thousand years ago.

Hominins are a group that include modern humans, extinct humans, and our immediate ancestors. Around 4 million years ago, early hominins diverged from the apes on the African continent. By around 2 million years ago, *Homo erectus* – the first known hominin to walk with an upright gait – had evolved. This pioneering species became the first to migrate out of Africa, around 1.8 million years ago, travelling through the Sinai into modern Israel, Syria, and Lebanon, across a land bridge to Arabia, before reaching China. Several species – *Homo heidelbergensis*, Neanderthals, and Denisovans – followed in their wake. *Homo sapiens* – modern humans – evolved about 300,000 years ago in Africa; they also migrated out of the continent in several waves. The last of these early *Homo sapiens* left Africa about 70,000 years ago and are the ancestors of all modern-day populations.

Homo sapiens began travelling along coastal routes from around 200,000 years ago. Their move from Africa was probably driven by climate changes that made food scarcer, leading these early explorers to previously unpopulated lands. By around 35,000 years ago, humans had populated most of Europe, Asia (including Japan), Siberia, and Australia, leaving only New Zealand, Polynesia, the Americas, and isolated islands left to discover.

PREHISTORIC CLUES

As well as fossils, the first migrants left evidence of their travels in stone tools, cooking hearths, and, from around 50,000 years ago, in cave art. Part of the roof of the Altamira cave, near Santander in Spain, is covered in engravings of bison, vividly painted over in red and black pigments. The carbon present in the black paint has been radiocarbon dated to c. 14,820–13,130 years ago.

A depiction of a bison, Altamira cave, Spain

ROUTES INTO EUROPE

Migration was a key factor in shaping prehistoric Europe. Between 54,000 and 42,000 years ago, three significant migratory waves took place, first from Africa, then from Southwest Asia, and finally from Northeast Asia.

△ **The Sleeping Lady**
This clay figure of a reclining woman, found in a Neolithic tomb dating to c.4000–2500 BCE on Malta, is thought to depict a goddess or represent death.

Around 54,000 years ago, *Homo sapiens* hunter-gatherers, who had originated in Africa, arrived in the southeast of Europe. Some 45,000 years ago, these migrants reached the northern European plain. Within two millennia, they had become technologically adept, creating bone and antler tools, basketry, and projectile weapons.

The warmer climate that came with the end of the Ice Age (*c.*9500 BCE) paved the way for an agricultural revolution. Bringing domesticated animals and seed crops, Neolithic farmers from Southwest Asia journeyed across and around the Mediterranean. They had arrived in Southern Europe by *c.*6000 BCE, then migrated north, reaching Scandinavia, Britain, and Ireland by 4000 BCE.

Some 2,000 years later, a group of ancient herders, known as the Yamnaya, began moving into Europe from the eastern steppe region. They had domesticated horses by around 5000 BCE, and by 3500 BCE had learned how to construct wagons, enabling them to explore far from their homeland. They also developed metallurgical skills and made weapons and tools that were superior to stone implements.

Scientists have found many indications that regional cultures were being assimilated across Europe. For example, analysis of the remains of the so-called "Amesbury Archer" – who died *c.*2300 BCE, during the early Bronze Age, and was buried three miles from Stonehenge in the UK – suggests that he had grown up in the Alps. His grave contained around 100 objects, including copper knives and bell-shaped Beaker pottery, a style that originated in Iberia some 4,500 years ago.

ÖTZI THE ICEMAN

In 1991, hikers in the Ötzal Alps between Austria and Italy chanced upon the remains of a late Neolithic-era man. Dating back some 5,300 years, the body and its clothing had been well preserved in glacial ice. The corpse was found alongside stone tools and a copper axe. The knowledge and skill required to process metalware arrived in Europe from Asia Minor in around 3000 BCE, facilitated by the migration of Yamnaya steppe peoples.

Recent studies suggest Ötzi had a darker skin tone than this reconstruction shows

Painting of a horse
Lascaux in France is home to cave paintings that date back around 17,000 years. Prehistoric humans used pigments made from minerals to portray animals, including horses and cattle, as well as human figures and abstract signs.

PEOPLING THE AMERICAS

The Americas were the last major land mass to be populated, as *Homo sapiens* most likely travelled from Siberia and gradually, over hundreds of years, across North America to reach the tip of South America. The route they took across America and the precise timing of when they reached the south continues to be debated.

Until recently, the long-held view was that the ancestors of all Indigenous Americans reached America through Beringia during the last ice age, around 13,500 years ago, by crossing a land bridge that existed between Siberia and Alaska. As the ice sheets melted on land, these ancestors moved south into Central America, and by 10,500 years ago had reached South America.

Yet, the discovery of older sites with evidence of human activity, such as Monte Verde in Chile and Meadowcroft in Pennsylvania, suggests *Homo sapiens* crossed into North America at least 2,000 years earlier, and reached modern-day Chile around 14,500 years ago. Genetic studies of maternal DNA, as well as analysis of the evolution of gut bacteria and linguistic studies, indicate even earlier settlements, up to 20,000 years ago. A further theory contends that humans used the frozen ice-age seas to cross the North Atlantic and settled the eastern seaboard of North America first.

Whatever the route or timing, the earliest Americans, who probably came in several waves, settled across the continent, subsisting on seafood along the coasts and hunting megafauna, such as mastodons, in the interior. A final migration from Siberia and Alaska populated the high Arctic and Greenland around 5,000 years ago.

> *"They made history, those latter-day Asians, who by jumping continents became the first Americans."*
>
> DAVID J. MELTZER, *FIRST PEOPLES IN A NEW AMERICA*, 2009

ANCIENT FOOTPRINTS

The discovery in 2009 of a set of more than 60 fossilized human footprints at White Sands, New Mexico, pushed the possible date of human migration into North America back even further than 19,000 years ago. The footprints were preserved in layers of gypsum, along with ancient grass seeds radiocarbon-dated to around 23,000 years ago, suggesting the region was once lush grassland rather than arid desert. Most of the footprints can be attributed to young adults, and one pair has a toddler's prints beside them.

Fossilized footprints discovered in New Mexico

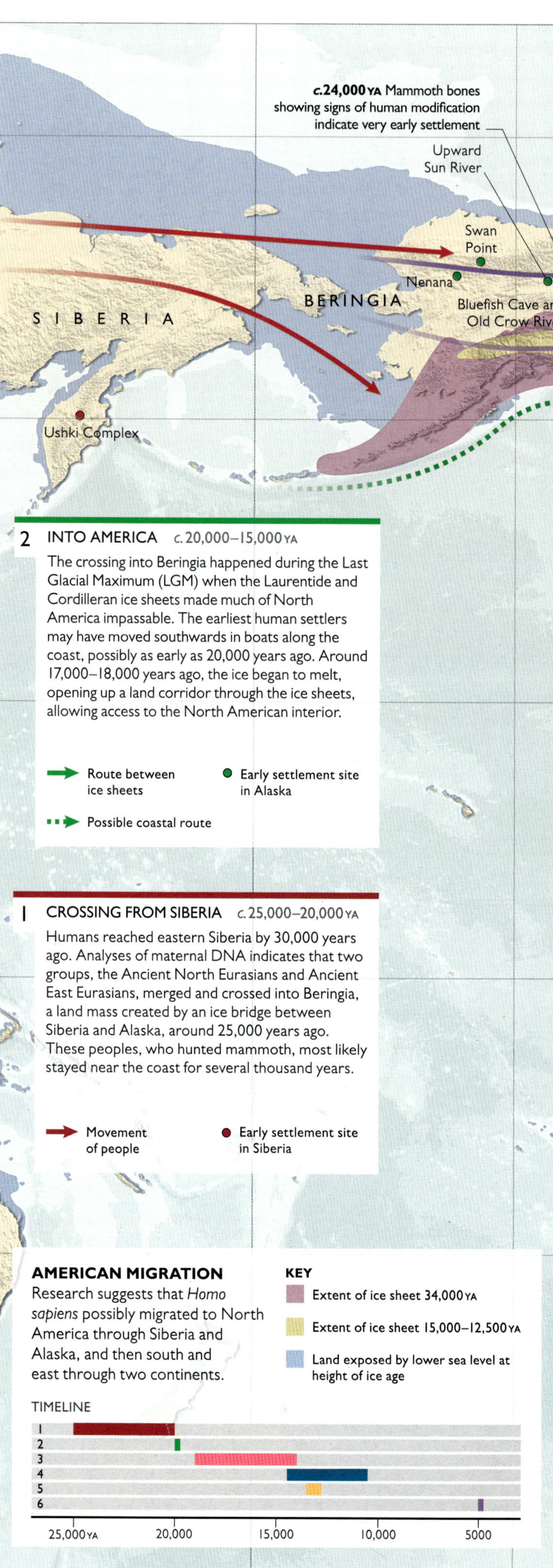

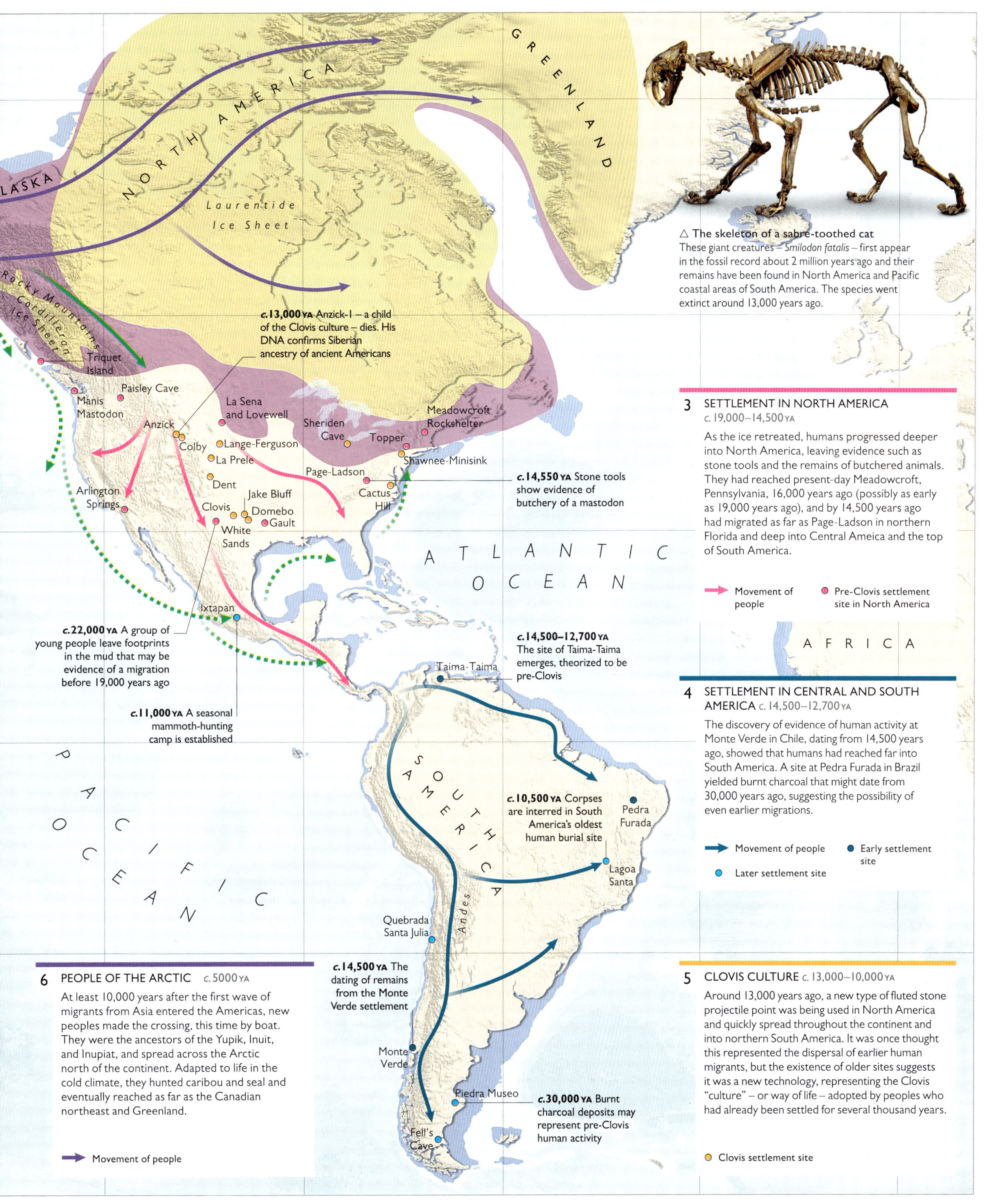

△ **The skeleton of a sabre-toothed cat**
These giant creatures – *Smilodon fatalis* – first appear in the fossil record about 2 million years ago and their remains have been found in North America and Pacific coastal areas of South America. The species went extinct around 13,000 years ago.

3 SETTLEMENT IN NORTH AMERICA
c. 19,000–14,500 YA

As the ice retreated, humans progressed deeper into North America, leaving evidence such as stone tools and the remains of butchered animals. They had reached present-day Meadowcroft, Pennsylvania, 16,000 years ago (possibly as early as 19,000 years ago), and by 14,500 years ago had migrated as far as Page-Ladson in northern Florida and deep into Central Ameica and the top of South America.

Movement of people

Pre-Clovis settlement site in North America

4 SETTLEMENT IN CENTRAL AND SOUTH AMERICA c. 14,500–12,700 YA

The discovery of evidence of human activity at Monte Verde in Chile, dating from 14,500 years ago, showed that humans had reached far into South America. A site at Pedra Furada in Brazil yielded burnt charcoal that might date from 30,000 years ago, suggesting the possibility of even earlier migrations.

Movement of people

Early settlement site

Later settlement site

5 CLOVIS CULTURE c. 13,000–10,000 YA

Around 13,000 years ago, a new type of fluted stone projectile point was being used in North America and quickly spread throughout the continent and into northern South America. It was once thought this represented the dispersal of earlier human migrants, but the existence of older sites suggests it was a new technology, representing the Clovis "culture" – or way of life – adopted by peoples who had already been settled for several thousand years.

Clovis settlement site

6 PEOPLE OF THE ARCTIC c. 5000 YA

At least 10,000 years after the first wave of migrants from Asia entered the Americas, new peoples made the crossing, this time by boat. They were the ancestors of the Yupik, Inuit, and Inupiat, and spread across the Arctic north of the continent. Adapted to life in the cold climate, they hunted caribou and seal and eventually reached as far as the Canadian northeast and Greenland.

Movement of people

AUSTRONESIAN MIGRATIONS

In the prehistoric era, two waves of different peoples – separated by many millennia – slowly migrated into southeast Asia, Australasia, and the eastern islands of the Pacific Ocean. Walking first across land bridges when sea levels were low, and then sailing across oceans, they reached almost every island in the region.

The colonization of southeast Asia and Oceania took place over many millennia. The earliest arrivals were the ancestors of the First Australians, who moved south from Asia and then crossed the sea to northern Australia around 60,000 BCE, making them the first people in history to complete successful sea voyages.

Aboriginal people stayed on Sahul – a supercontinent comprising Australia, Tasmania, and New Guinea – which was not broken up by rising sea levels until the end of the last Ice Age and the start of the current Holocene geological epoch *c.*9,700 BCE. They were largely isolated from other cultures until the arrival of Europeans in the 17th century. As a result, they have the oldest continuous culture in existence.

Four thousand years ago, a second group of people migrated south in search of new lands. The Austronesians originated in Taiwan. After they were forced out by overpopulation, they sailed south, first settling in the islands of southeast Asia and Sunda (the ancient landmass that included Borneo, Sumatra, and Java), and then spreading out across the isles within Melanesia and remote, distant Micronesia.

The Austronesians' development of outrigger and double-hulled canoes enabled them to sail between islands. Spectacularly, they traversed the Indian Ocean to settle on the African island of Madagascar. Theirs was a maritime culture of adventure and daring, but one that also produced finely decorated pottery and obsidian tools and artefacts.

> *"Their ships... sail like birds, while ours are like lead."*
>
> FRANCISCO COMBÉS, ON THE FILIPINO KARAKOA WARSHIP, *HISTORIA DE LAS ISLAS DE MINDANAO, IOLO Y SUS ADYACENTESS*, 1667

OUTRIGGER CANOE

The Austronesians' development of outrigger canoes and double-hulled sailing boats after 1500 BCE enabled them to cross the Pacific and Indian oceans and colonize the far-flung islands. Such craft are double-ended: they can sail in either direction without turning round. Austronesian voyagers carried their future on such boats, taking everything with them that they needed for settling on a new island.

A double-hulled craft or *drua*

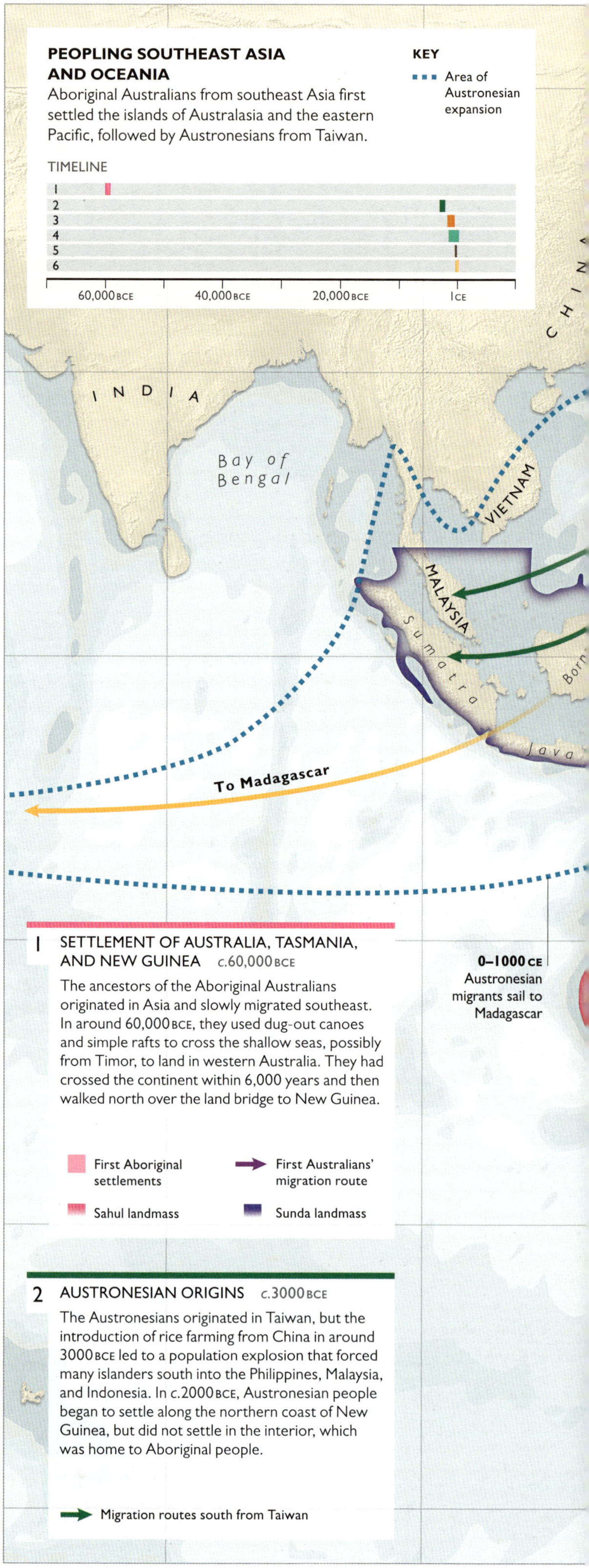

PEOPLING SOUTHEAST ASIA AND OCEANIA

Aboriginal Australians from southeast Asia first settled the islands of Australasia and the eastern Pacific, followed by Austronesians from Taiwan.

KEY

Area of Austronesian expansion

TIMELINE

1 2 3 4 5 6

60,000 BCE 40,000 BCE 20,000 BCE ICE

INDIA

Bay of Bengal

VIETNAM

MALAYSIA

Sumatra

Java

To Madagascar

0–1000 CE Austronesian migrants sail to Madagascar

1 SETTLEMENT OF AUSTRALIA, TASMANIA, AND NEW GUINEA *c.*60,000 BCE

The ancestors of the Aboriginal Australians originated in Asia and slowly migrated southeast. In around 60,000 BCE, they used dug-out canoes and simple rafts to cross the shallow seas, possibly from Timor, to land in western Australia. They had crossed the continent within 6,000 years and then walked north over the land bridge to New Guinea.

First Aboriginal settlements

First Australians' migration route

Sahul landmass

Sunda landmass

2 AUSTRONESIAN ORIGINS *c.*3000 BCE

The Austronesians originated in Taiwan, but the introduction of rice farming from China in around 3000 BCE led to a population explosion that forced many islanders south into the Philippines, Malaysia, and Indonesia. In *c.*2000 BCE, Austronesian people began to settle along the northern coast of New Guinea, but did not settle in the interior, which was home to Aboriginal people.

Migration routes south from Taiwan

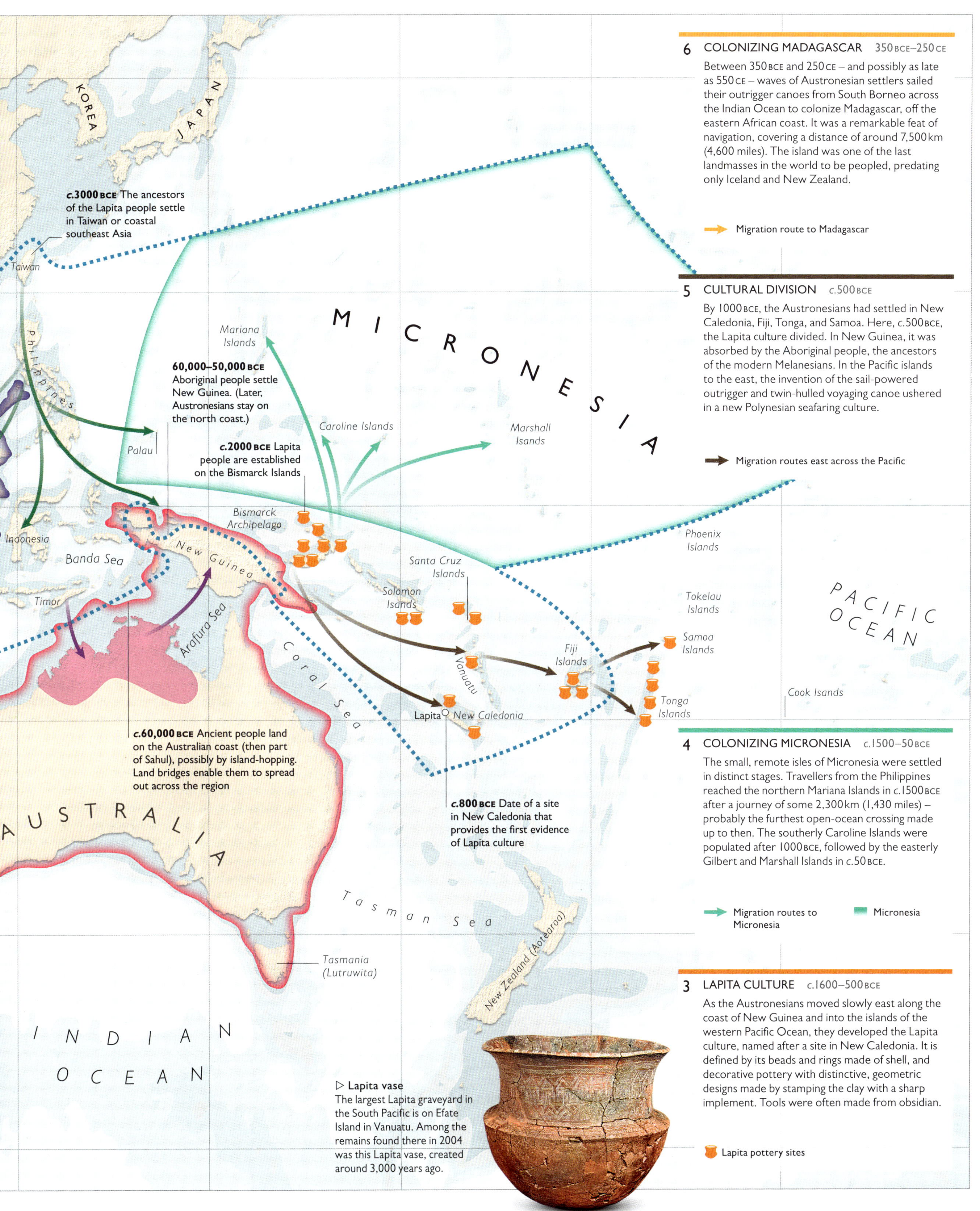

6 COLONIZING MADAGASCAR 350 BCE–250 CE

Between 350 BCE and 250 CE – and possibly as late as 550 CE – waves of Austronesian settlers sailed their outrigger canoes from South Borneo across the Indian Ocean to colonize Madagascar, off the eastern African coast. It was a remarkable feat of navigation, covering a distance of around 7,500 km (4,600 miles). The island was one of the last landmasses in the world to be peopled, predating only Iceland and New Zealand.

Migration route to Madagascar

5 CULTURAL DIVISION c.500 BCE

By 1000 BCE, the Austronesians had settled in New Caledonia, Fiji, Tonga, and Samoa. Here, c.500 BCE, the Lapita culture divided. In New Guinea, it was absorbed by the Aboriginal people, the ancestors of the modern Melanesians. In the Pacific islands to the east, the invention of the sail-powered outrigger and twin-hulled voyaging canoe ushered in a new Polynesian seafaring culture.

Migration routes east across the Pacific

4 COLONIZING MICRONESIA c.1500–50 BCE

The small, remote isles of Micronesia were settled in distinct stages. Travellers from the Philippines reached the northern Mariana Islands in c.1500 BCE after a journey of some 2,300 km (1,430 miles) – probably the furthest open-ocean crossing made up to then. The southerly Caroline Islands were populated after 1000 BCE, followed by the easterly Gilbert and Marshall Islands in c.50 BCE.

Migration routes to Micronesia

Micronesia

3 LAPITA CULTURE c.1600–500 BCE

As the Austronesians moved slowly east along the coast of New Guinea and into the islands of the western Pacific Ocean, they developed the Lapita culture, named after a site in New Caledonia. It is defined by its beads and rings made of shell, and decorative pottery with distinctive, geometric designs made by stamping the clay with a sharp implement. Tools were often made from obsidian.

Lapita pottery sites

▷ **Lapita vase**
The largest Lapita graveyard in the South Pacific is on Efate Island in Vanuatu. Among the remains found there in 2004 was this Lapita vase, created around 3,000 years ago.

PHOENICIAN EXPANSION

The Phoenicians were among the greatest long-distance traders of ancient times. From the 11th century BCE, they founded Mediterranean colonies and trade routes and were the first outsiders to explore the coast of West Africa.

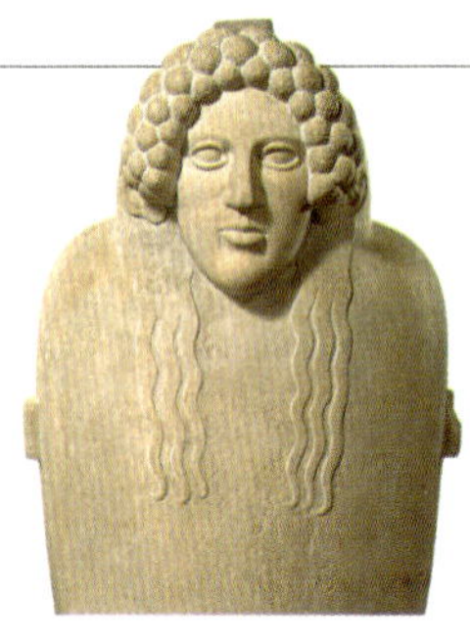

△ **Marble sarcophagus**
This 5th-century BCE coffin carved in the shape of a long-haired woman is Phoenician and shows Greek artistic influences.

Around 1100 BCE, the Phoenicians began migrating westwards from their homeland in modern Lebanon. Under pressure from the expansion of the neighbouring Assyrian Empire, and overpopulation in growing port cities such as Byblos and Tyre, Phoenician ships sailed in search of new prospects. Expert navigators and skilled merchants, Phoenicians traded along their sea routes, notably in a purple dye made from the murex sea snail. They were also leading exporters of cedar wood.

The Phoenicians founded a series of colonies: Utica in modern Tunisia in 1100 BCE; an enclave on an island off Athens; Palermo on Sicily in the western Mediterranean; in North Africa as far west as Mogador (Essasouira) in Morocco; and along the east coast of Spain, where they founded Gadir (Cádiz) before 800 BCE. Often sited on defensible promontories, with double harbours, these colonies served as commercial bases and as points to dominate the extraction of minerals, such as silver from Spain. Only later did Carthage, founded in North Africa by Phoenicians in 813 BCE, grow into a land-based empire, coming into conflict with the Romans and ultimately leading to the city's destruction in 146 BCE.

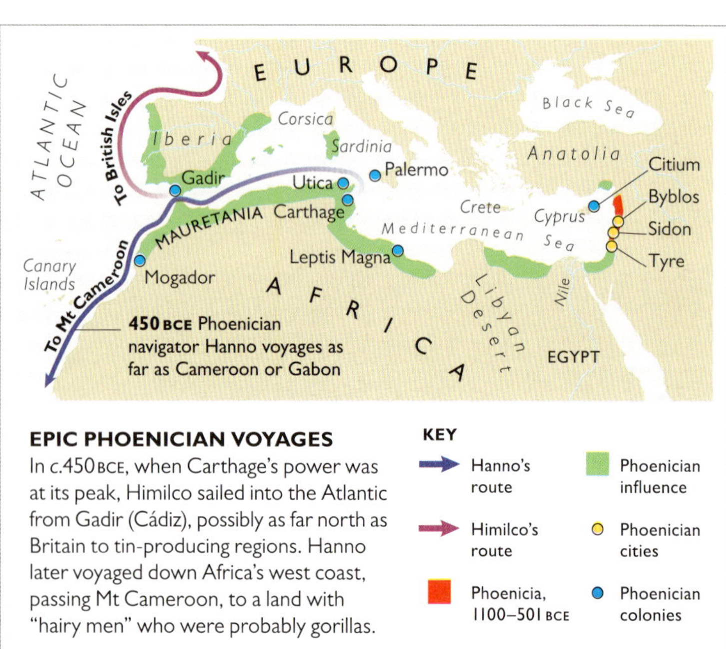

EPIC PHOENICIAN VOYAGES
In c.450 BCE, when Carthage's power was at its peak, Himilco sailed into the Atlantic from Gadir (Cádiz), possibly as far north as Britain to tin-producing regions. Hanno later voyaged down Africa's west coast, passing Mt Cameroon, to a land with "hairy men" who were probably gorillas.

KEY
- Hanno's route
- Himilco's route
- Phoenicia, 1100–501 BCE
- Phoenician influence
- Phoenician cities
- Phoenician colonies

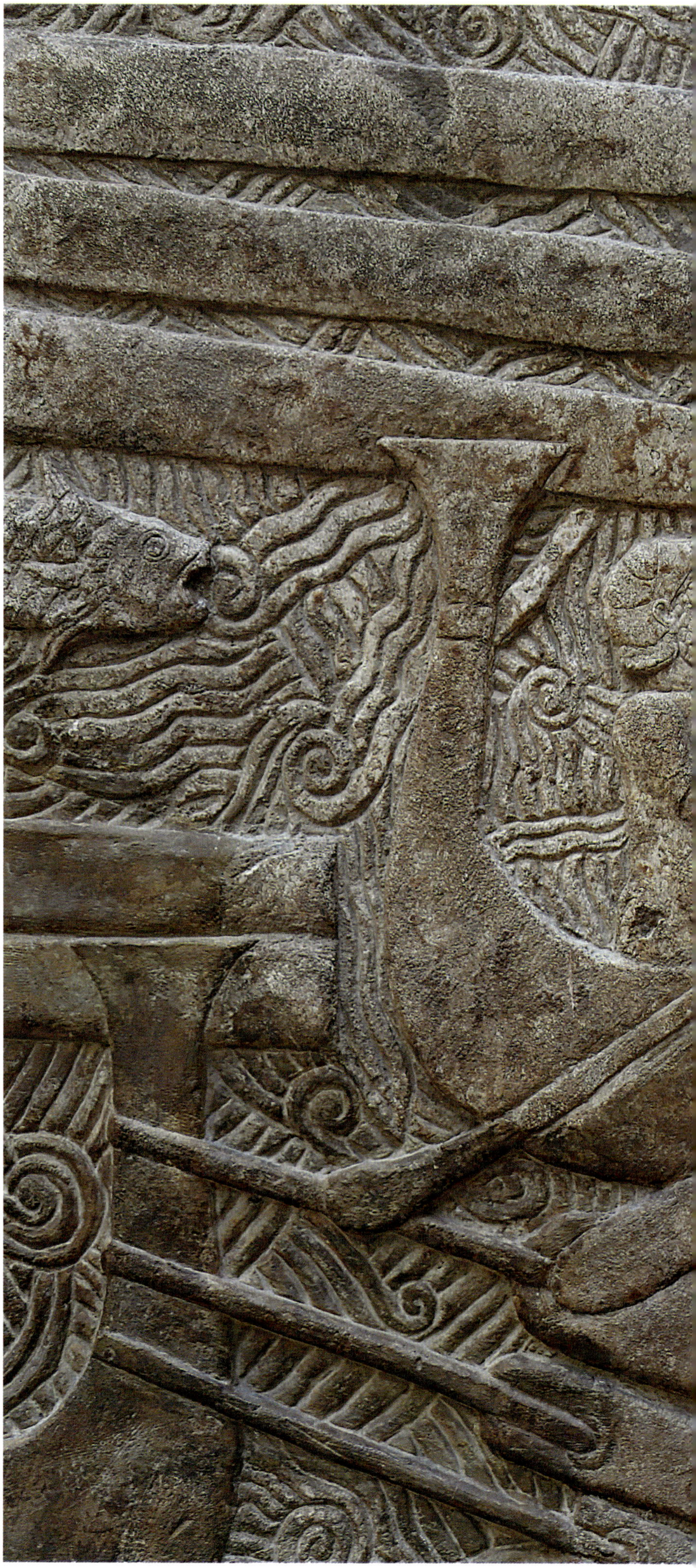

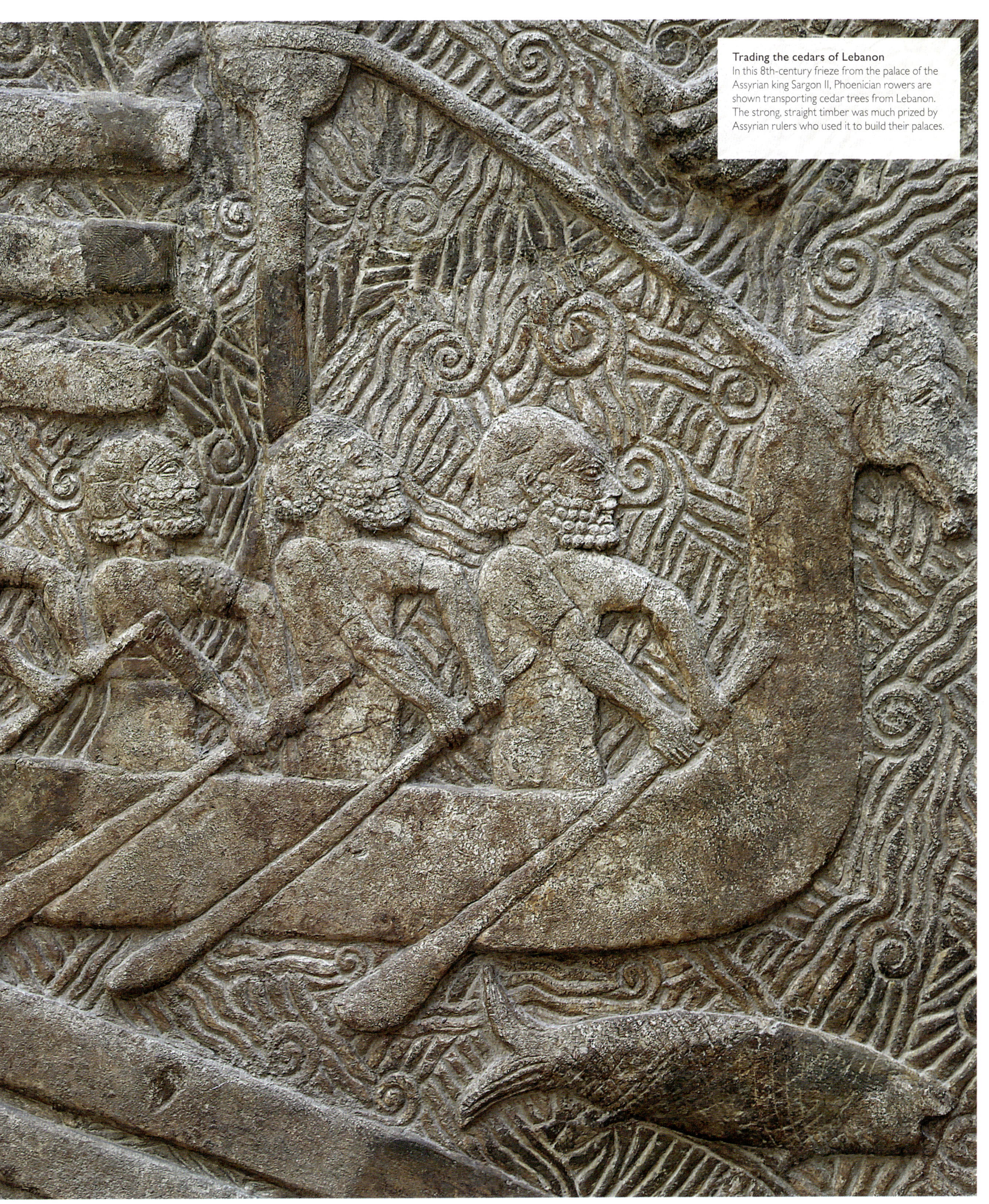

Trading the cedars of Lebanon
In this 8th-century frieze from the palace of the Assyrian king Sargon II, Phoenician rowers are shown transporting cedar trees from Lebanon. The strong, straight timber was much prized by Assyrian rulers who used it to build their palaces.

To the land of Punt
On the return of Hatshepsut's chancellor, Nehsy, from the 1480 BCE expedition to Punt, Hatshepsut decreed that the mission be recorded on wall friezes in her funerary temple at Deir el-Bahri, by Luxor. Axe-bearing soldiers, who would have been part of the company, appear in this section.

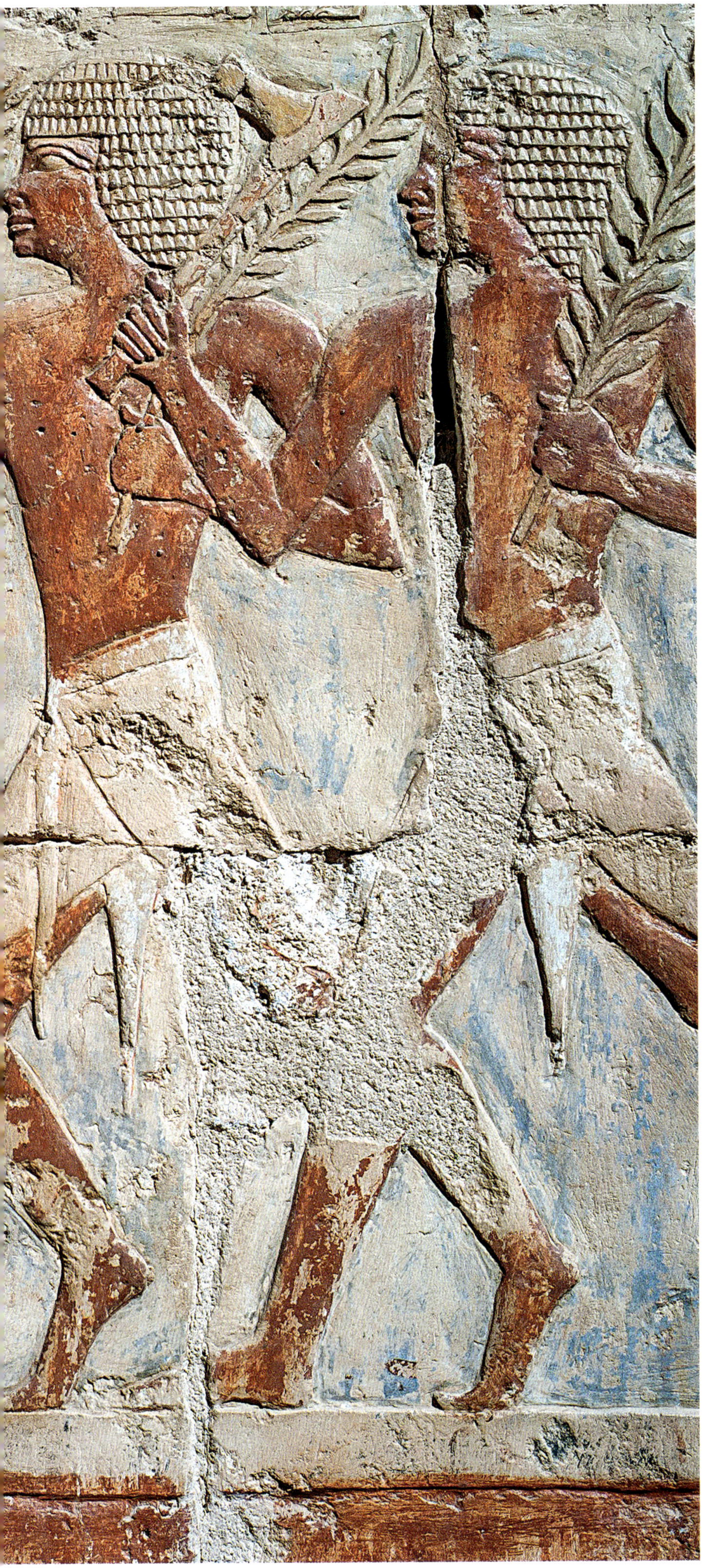

EXPLORING ANCIENT EGYPT

South of the territory of the Egyptian pharaohs lay kingdoms such as Kush, with which they traded and fought. Even further afield were rumoured lands, to which the pharaohs sent expeditions in search of the source of the gold, ivory, and tropical goods they produced.

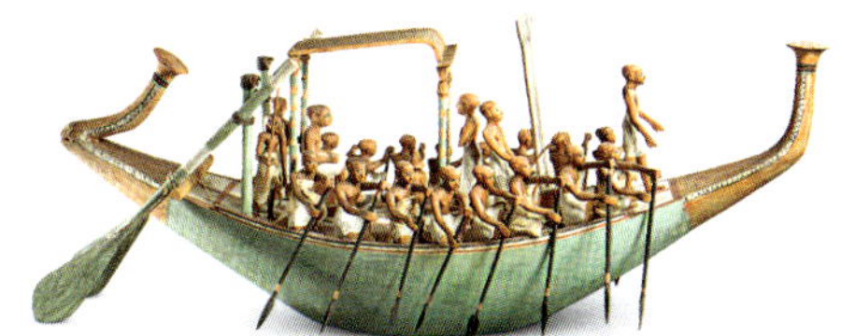

△ **Egyptian paddling boat**
This model vessel was uncovered in a 20th-century BCE tomb in Thebes, Upper Egypt. With two large steering oars situated at the stern, it is typical of the boats used on the Nile during this period.

In around 3100 BCE, after Egypt's unification, the sway of its pharaohs ended just south of Elephantine (modern Aswan). Expansion halted in the following millennium, then during the Old Kingdom – *c.*2686–2181 BCE – the first southerly exploration took place. Harkhuf, governor of Upper Egypt under Merenre (r.2255–2246 BCE) and Pepi II (*c.*2275–2181 BCE), was sent on four successive expeditions beyond Nubia to prepare for Egyptian expansion towards a land called Iyam (which may be in modern Sudan).

Over the following centuries, Egyptian merchants established sea routes from Red Sea ports to access the Horn of Africa. Around 1990 BCE, Mentuhotep III sent Henenu, his high steward, in search of Punt (which may have been in present-day Somalia), from which he returned with exotic spices. In the reign of Hatshepsut (1479–1458 BCE), an expedition led by Nehsy, her chancellor, repeated this journey, travelling to Punt with five ships and more than 1,000 merchants, sailors, and marines. They extracted tribute from local rulers and brought back exotic goods. There were no further voyages to Punt.

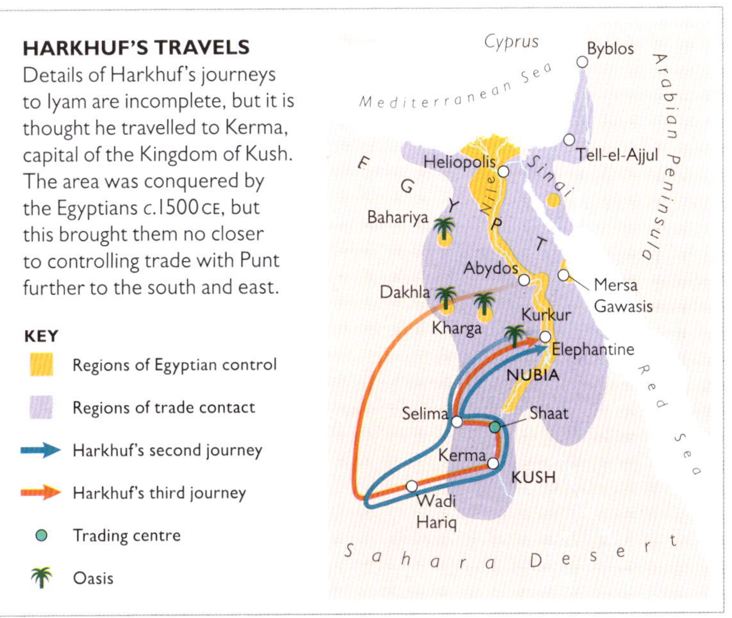

HARKHUF'S TRAVELS
Details of Harkhuf's journeys to Iyam are incomplete, but it is thought he travelled to Kerma, capital of the Kingdom of Kush. The area was conquered by the Egyptians c.1500 CE, but this brought them no closer to controlling trade with Punt further to the south and east.

KEY
- Regions of Egyptian control
- Regions of trade contact
- Harkhuf's second journey
- Harkhuf's third journey
- Trading centre
- Oasis

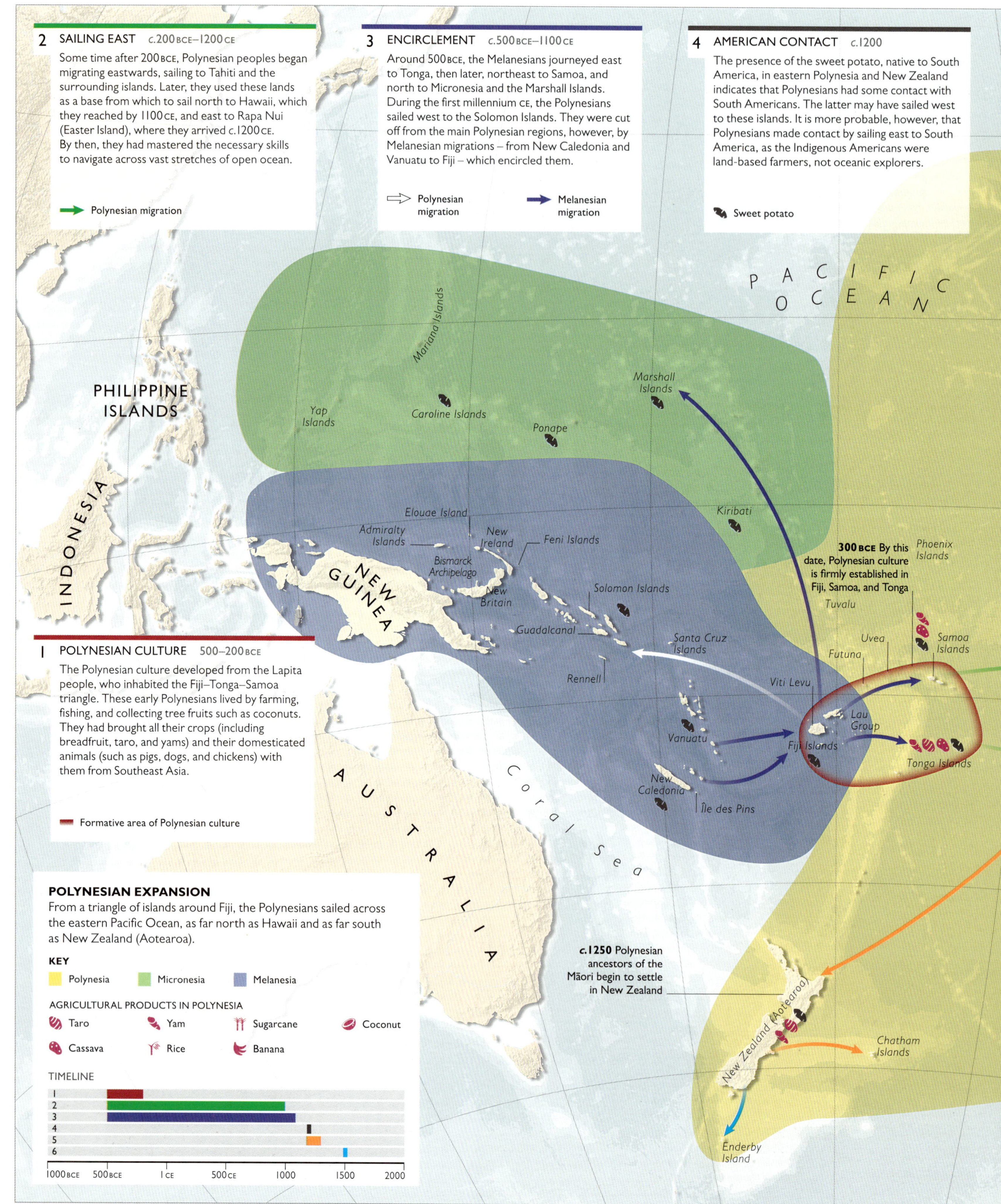
2 SAILING EAST c.200 BCE–1200 CE
Some time after 200 BCE, Polynesian peoples began migrating eastwards, sailing to Tahiti and the surrounding islands. Later, they used these lands as a base from which to sail north to Hawaii, which they reached by 1100 CE, and east to Rapa Nui (Easter Island), where they arrived c.1200 CE. By then, they had mastered the necessary skills to navigate across vast stretches of open ocean.
Polynesian migration
3 ENCIRCLEMENT c.500 BCE–1100 CE
Around 500 BCE, the Melanesians journeyed east to Tonga, then later, northeast to Samoa, and north to Micronesia and the Marshall Islands. During the first millennium CE, the Polynesians sailed west to the Solomon Islands. They were cut off from the main Polynesian regions, however, by Melanesian migrations – from New Caledonia and Vanuatu to Fiji – which encircled them.
Polynesian migration
Melanesian migration
4 AMERICAN CONTACT c.1200
The presence of the sweet potato, native to South America, in eastern Polynesia and New Zealand indicates that Polynesians had some contact with South Americans. The latter may have sailed west to these islands. It is more probable, however, that Polynesians made contact by sailing east to South America, as the Indigenous Americans were land-based farmers, not oceanic explorers.
Sweet potato
PACIFIC OCEAN
PHILIPPINE ISLANDS
INDONESIA
Mariana Islands
Yap Islands
Caroline Islands
Ponape
Marshall Islands
Kiribati
Elouae Island
Admiralty Islands
New Ireland
Feni Islands
Bismarck Archipelago
NEW GUINEA
New Britain
Solomon Islands
Guadalcanal
Santa Cruz Islands
Rennell
Vanuatu
New Caledonia
Île des Pins
Coral Sea
AUSTRALIA
300 BCE By this date, Polynesian culture is firmly established in Fiji, Samoa, and Tonga
Phoenix Islands
Tuvalu
Uvea
Futuna
Samoa Islands
Viti Levu
Lau Group
Fiji Islands
Tonga Islands
1 POLYNESIAN CULTURE 500–200 BCE
The Polynesian culture developed from the Lapita people, who inhabited the Fiji–Tonga–Samoa triangle. These early Polynesians lived by farming, fishing, and collecting tree fruits such as coconuts. They had brought all their crops (including breadfruit, taro, and yams) and their domesticated animals (such as pigs, dogs, and chickens) with them from Southeast Asia.
Formative area of Polynesian culture
c.1250 Polynesian ancestors of the Māori begin to settle in New Zealand
New Zealand (Aotearoa)
Chatham Islands
Enderby Island
POLYNESIAN EXPANSION
From a triangle of islands around Fiji, the Polynesians sailed across the eastern Pacific Ocean, as far north as Hawaii and as far south as New Zealand (Aotearoa).
KEY
Polynesia
Micronesia
Melanesia
AGRICULTURAL PRODUCTS IN POLYNESIA
Taro
Yam
Sugarcane
Coconut
Cassava
Rice
Banana
TIMELINE
1
2
3
4
5
6
1000 BCE
500 BCE
1 CE
500 CE
1000
1500
2000

POLYNESIAN NAVIGATORS

Originally migrating from Southeast Asia, the Polynesians had reached Samoa in the Pacific by around 1000 BCE. From around 200 BCE, these immensely skilled navigators sailed east across the entire Pacific Ocean in one of the most remarkable known feats of human exploration.

The expansion of the Polynesians from the islands of Fiji, Samoa, and Tonga that began in around 200 BCE saw them sail east to Tahiti, north to Hawaii, and, around 1280 CE, reach the islands of New Zealand. This exploration was prompted by a lack of resources on the Pacific islands they had first inhabited some 3,000 years ago.

Their migration to previously uninhabited islands was possible because of their maritime and navigational skills. They sailed in ocean-going, wooden, double-hulled and outrigger canoes, powered by reed sails and paddles. These were large enough to carry people, food, tools, and weapons. Some craft were vast: when Captain Cook measured a Māori canoe (*waka*) – based on the vessels of Polynesian ancestors – in New Zealand in 1770, he found it was 1 m (3 ft) longer than the ship in which he was circling the globe.

The islands they reached – even beyond the horizon – were located by the cloud cover the land threw up, wind changes, wave patterns, and the migration of birds. They also navigated by the sun and stars. Each island had its own "zenith" star – for example, Sirius for Tahiti. When Sirius was overhead, they knew they were in the latitude of Tahiti. Such skills were passed down orally through the generations.

> *"How shall we account for this nation's [spread across]… every quarter of the Pacific Ocean?"*
>
> CAPTAIN JAMES COOK, JOURNAL, FEBRUARY 1778

NAVIGATIONAL KNOWLEDGE

Polynesian navigators used palm-stick charts held together with coconut fibre. The framework represented the journey distances; the shells or coral pebbles threaded on the sticks marked the islands they encountered. It is unclear when these were first made, or how widely they were used, but they were very common in the Marshall Islands.

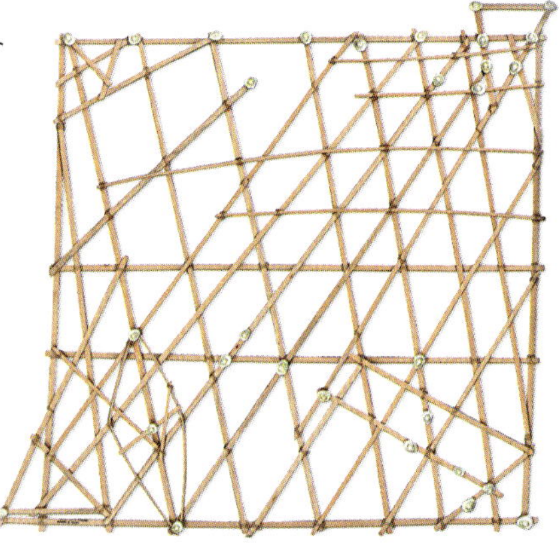

Polynesian navigational stick chart

△ **The *moai***
Between 1250 and 1500, more than 1,040 *moai* – monolithic human figures – were carved from volcanic tuff (solidified ash) and erected on Rapa Nui (Easter Island). The statues were symbols of power and also marked the location of sacred spirits.

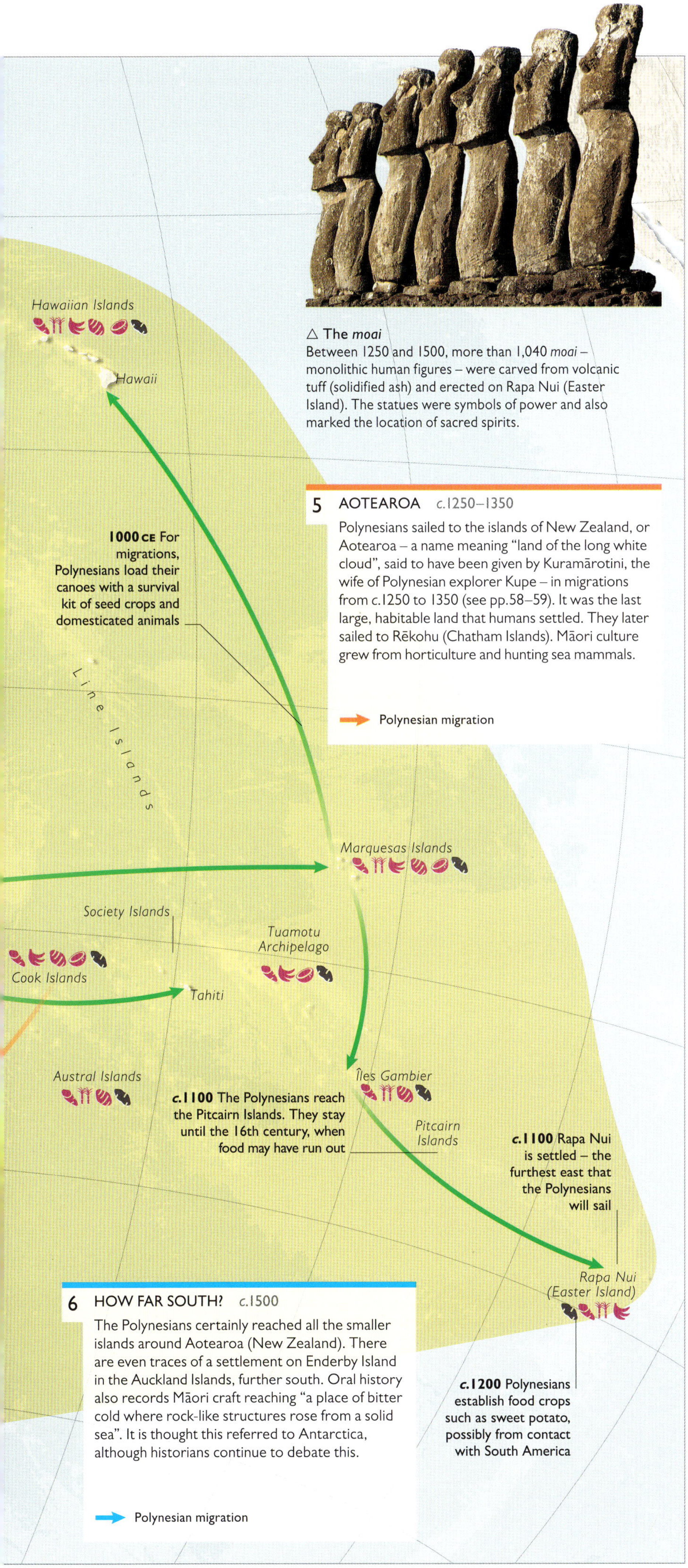

5 AOTEAROA *c.*1250–1350

Polynesians sailed to the islands of New Zealand, or Aotearoa – a name meaning "land of the long white cloud", said to have been given by Kuramārotini, the wife of Polynesian explorer Kupe – in migrations from c.1250 to 1350 (see pp.58–59). It was the last large, habitable land that humans settled. They later sailed to Rēkohu (Chatham Islands). Māori culture grew from horticulture and hunting sea mammals.

Polynesian migration

Hawaiian Islands
Hawaii
Line Islands
Marquesas Islands
Society Islands
Tuamotu Archipelago
Cook Islands
Tahiti
Austral Islands
Îles Gambier
Pitcairn Islands
Rapa Nui (Easter Island)

1000 CE For migrations, Polynesians load their canoes with a survival kit of seed crops and domesticated animals

***c.*1100** The Polynesians reach the Pitcairn Islands. They stay until the 16th century, when food may have run out

***c.*1100** Rapa Nui is settled – the furthest east that the Polynesians will sail

***c.*1200** Polynesians establish food crops such as sweet potato, possibly from contact with South America

6 HOW FAR SOUTH? *c.*1500

The Polynesians certainly reached all the smaller islands around Aotearoa (New Zealand). There are even traces of a settlement on Enderby Island in the Auckland Islands, further south. Oral history also records Māori craft reaching "a place of bitter cold where rock-like structures rose from a solid sea". It is thought this referred to Antarctica, although historians continue to debate this.

Polynesian migration

MACEDONIA
THRACE
Byzantium
Black Sea
Sinop
Trapezus
Caspian Sea
Athens
Ephesus
Sardis
LYDIA
Halys
Pteria
CAPPADOCIA
ARMENIA
Centrites
Zabatas
Tigris
MESOPOTAMIA
Euphrates
MEDIA
PARTHIA
Ecbatana
Great Salt Desert
ASSYRIA
Mediterranean Sea
401 BCE Cunaxa
Babylon
PERSIS
Zagros Mountains
Susa
SAGARTIA
Iranian Plateau
BABYLONIA
Syrian Desert
ARABIA
Memphis
Nile
EGYPT
Sahara Desert
Pasargadae
Persepolis
Persian Gulf
Red Sea
399 BCE Thrace, although technically within the Persian Empire, is sufficiently populated by Greeks to put it beyond the reach of Artaxerxes
Summer 400 BCE At first, only the sick and women accompanying the army sail, but at Sinop the Greeks hire more vessels
Spring 400 BCE From the summit of Mount Theches, the Greek soldiers cry out "Thalassa, Thalassa" ("the Sea, the Sea") as they spy the coast near Trapezus
402–401 BCE From his base in Sardis, Cyrus the Younger seeks Spartan help and recruits Greek mercenaries
401 BCE Attacked by Persian satraps from the front and rear, the Greeks escape across the freezing waters of the River Centrites
401 BCE Clearchus and other Greek generals are imprisoned and executed by Tissaphernes. The Greek mercenaries elect new leaders, including Xenophon

▷ **Corinthian helmet**
Most of the 10,000 mercenaries would have worn headgear in this style. Named after the Greek city state where it originated, it offered good protection, but also restricted the wearer's sightlines.

1 THE ROYAL ROAD 522–486 BCE

Darius I consolidated and reconstructed an existing series of routes to create a high-quality transport network that spanned the Persian Empire. The Royal Road ran 2,500 km (1,550 miles) from Susa to Sardis, a distance that took 90 days to travel on foot. It also incorporated way stations where royal *pirradazis* (messengers) could change horses and make the journey in just nine days, enabling the Persian ruler to communicate with and govern his empire more effectively.

The Royal Road from Sardis to Susa
Other Persian routes

2 THE ADVANCE OF THE 10,000 401 BCE

Cyrus the Younger, satrap (governor) of Lydia, revolted against his brother Artaxerxes II in an attempt to seize the Persian throne. As he moved eastwards, he gathered 20,000 Greek mercenaries under the command of the Spartan general Clearchus, trying to reach the Persian heartland before Artaxerxes had a chance to react. Some of the Greek hoplites deserted when they heard that Cyrus was trying to overthrow the king.

Advance of the 10,000 Greek mercenaries
Area ruled by Cyrus the Younger

3 THE BATTLE OF CUNAXA 401 BCE

Cyrus met his brother's army north of Babylon. In a tight battle, the Greek mercenaries pushed back the royal army's left flank while Cyrus charged the centre, wounding Artaxerxes. But then Cyrus himself was killed; although the Greeks fought on, they were now stranded in Persian territory. Artaxerxes' satrap Tissaphernes invited the Greek generals to negotiate, but it was a ruse: he arrested and executed them. The remaining Greeks elected new leaders, including Xenophon, and began a gruelling march north to the sea.

Battle of Cunaxa
Generals killed

4 THE RETREAT TO THE SEA 401–400 BCE

The Greeks crossed the river Zapatas, then headed, without guides, into the mountains, pursued by Tissaphernes. Sometimes they found local allies such as the Molossians, but otherwise they were under almost constant attack from Persians, Carduchians, Armenians, Taochians, and Chalybes. Xenophon's strong leadership enabled the army to remain a cohesive force and also to survive the bitter cold. Along the way he took notes that later formed the basis of the *Anabasis*. Finally, near Trapezus, the Greeks sighted the sea.

Retreat to the Black Sea
Mount Theches

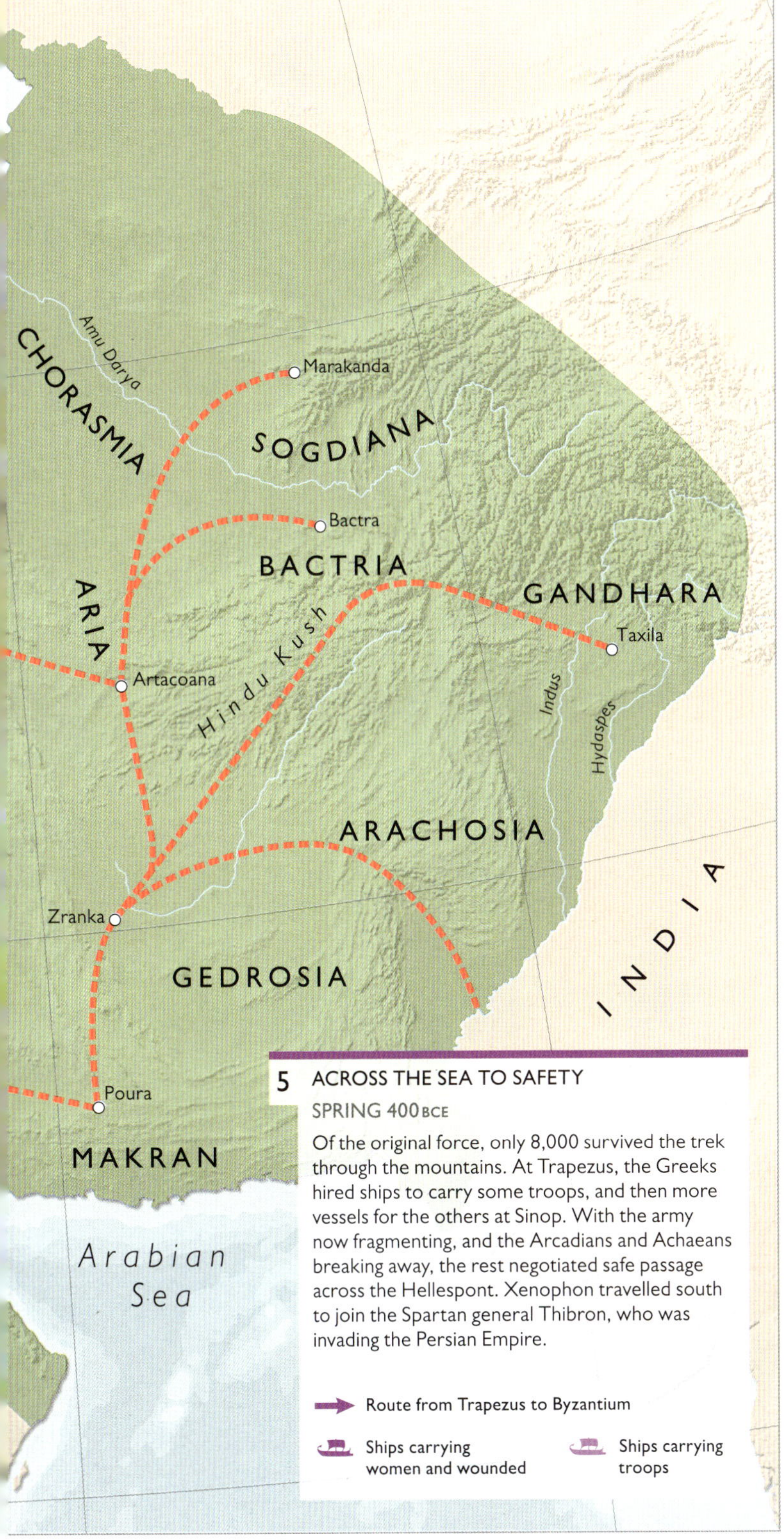

CROSSING THE EMPIRE
Eighty years after two unsuccessful invasions of Greece by Darius I, and the construction of the Royal Road under his command, the Persian Empire still controlled widespread territories.

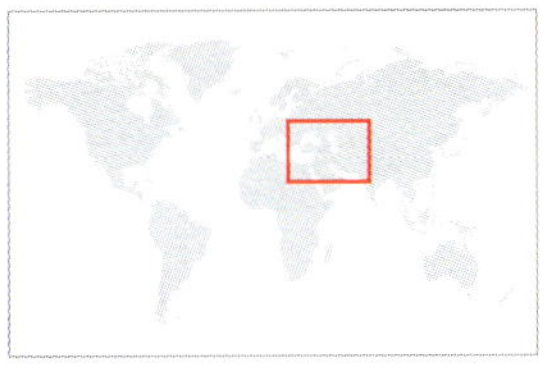

KEY
Persian Empire
Greek-held territories

TIMELINE
1 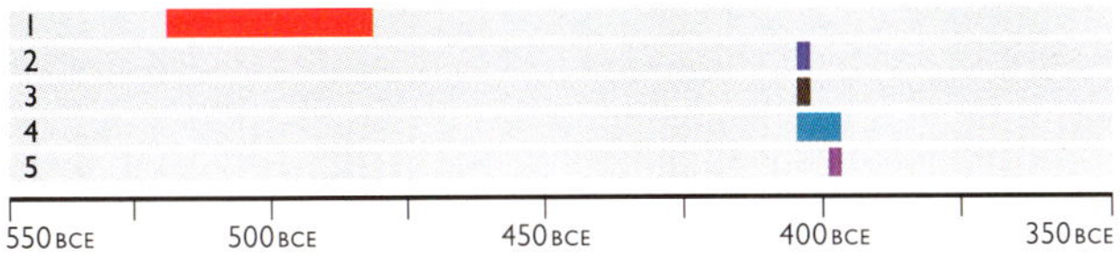
2
3
4
5
550 BCE 500 BCE 450 BCE 400 BCE 350 BCE

PERSIAN COURIERS

The vast Persian Empire (then ruled by the Achaemenid dynasty) spanned more than 5,500 km (3,400 miles) from north Greece to India's borders, with a road network for fast communication. Few enemies infiltrated far – one exception being Alexander the Great – though Greek commander Xenophon led an epic escape from within.

The 20 satrapies (or provinces) that made up the Persian Empire at its height under Darius I (r. 522-486 BCE) covered a huge diversity of landscapes and peoples. Ruling this from the royal capitals of Susa and Persepolis in the empire's centre presented enormous difficulties. The construction of a Royal Road linking Susa to Sardis in Ionia (now the west coast of Turkey) allowed both messengers and armies to reach its western satrapies more quickly.

As Persian royal power waned in the later 5th century BCE following two failed invasions of Greece, more foreigners entered the nation's armed forces. Among them were Greek hoplites, heavily armed mercenaries on whom the Persians increasingly came to rely. In 401 BCE, one group, under the leadership of Xenophon, became involved in a civil war in the empire. They marched more than 2,500 km (1,550 miles) from Sardis towards Babylon, and then a similar distance to escape after their defeat.

Xenophon subsequently wrote an account of their withdrawal, the *Anabasis*. It provides precious details about the Persian Empire and of one of the greatest journeys undertaken within its borders.

"It is to the enemy that I should wish to have all roads seem easy – for their retreat."

XENOPHON, *ANABASIS*, BOOK VI V. 18, c.370 BCE

5 ACROSS THE SEA TO SAFETY
SPRING 400 BCE
Of the original force, only 8,000 survived the trek through the mountains. At Trapezus, the Greeks hired ships to carry some troops, and then more vessels for the others at Sinop. With the army now fragmenting, and the Arcadians and Achaeans breaking away, the rest negotiated safe passage across the Hellespont. Xenophon travelled south to join the Spartan general Thibron, who was invading the Persian Empire.

Route from Trapezus to Byzantium
Ships carrying women and wounded
Ships carrying troops

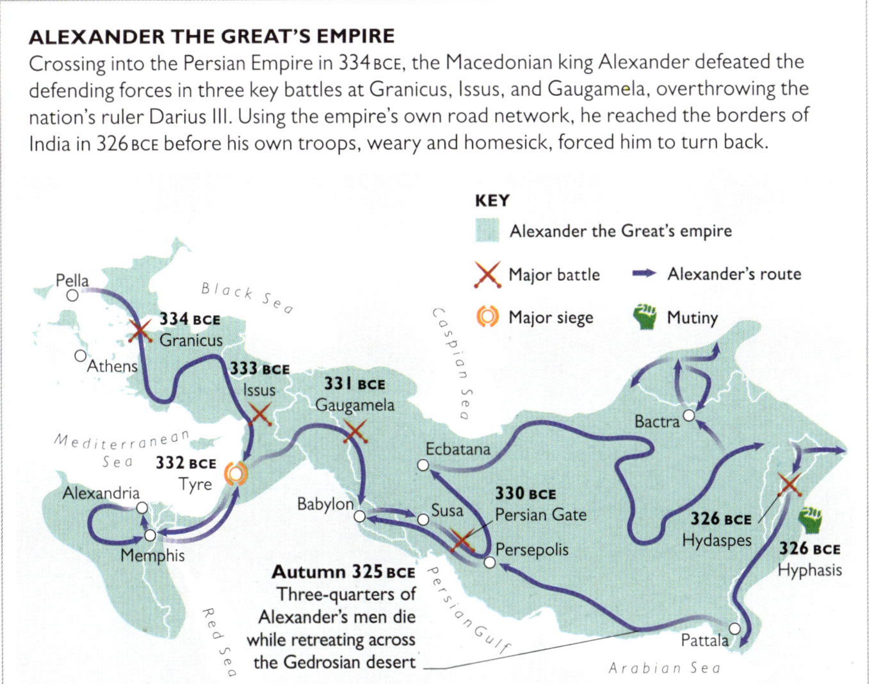

ALEXANDER THE GREAT'S EMPIRE
Crossing into the Persian Empire in 334 BCE, the Macedonian king Alexander defeated the defending forces in three key battles at Granicus, Issus, and Gaugamela, overthrowing the nation's ruler Darius III. Using the empire's own road network, he reached the borders of India in 326 BCE before his own troops, weary and homesick, forced him to turn back.

THE CELTS

The first northern Europeans to appear in historical records around the 7th century BCE, the Celts migrated from the middle Danube region to reach Spain, France, the Balkans, Britain, and even Asia Minor, in an expansion that lasted more than 500 years. Although ultimately conquered by other groups, their language, legends, and culture have survived to the present day.

Never politically united, the Celts were linked by a closely related set of languages and a culture centred on a warrior aristocracy, farmers, and druids – learned priests and teachers. They clustered around *oppida*, large hill forts that acted as refuges for the surrounding population. The Celts had probably been living along the Middle Danube – located in Croatia, Hungary, and Serbia – for over three centuries by the time they were mentioned in records by the Greeks and Romans around the 7th century BCE. Hallstatt metal-workers, who shared elements of Celtic culture, emerged in this area around 800 BCE, controlling trade routes into the Mediterranean region, and gradually migrating west into Bohemia (present-day Czech Republic), Bavaria, and eastern Gaul

△ **Celtic torc**
This brass neck ring, from c.100 CE and found in Dumfriesshire, Scotland, is decorated with scroll patterns typical of La Tène craftsmanship.

The Battersea Shield
This detail is from a bronze shield cover found in Battersea, Britain, dating to c.350–50 BCE. The La Tène-style decoration of circles and spirals features red glass studs. The metal cover would have once been attached to a wooden shield.

(France). Around 500 BCE, the Hallstatt culture was replaced by La Tène culture – named after the site in Switzerland. Culturally, this was also connected with the Celts on the Middle Rhone, characterized by pottery and metalwork with geometric patterns.

The La Tène Celts expanded even further, into almost all of Gaul, the Iberian Peninsula – where they merged with existing peoples to form a hybrid Celtiberian culture – and Italy, where in 390 BCE a Gaulish Celtic tribe briefly captured Rome. By 279 BCE, Celts had settled in Britain and Ireland. Several tribes also crossed into Asia Minor (Turkey), where they established an outlying Celtic kingdom known as Galatia.

> *"[Celtic warriors] amass a great quantity of gold which is used for ornament."*
>
> GREEK HISTORIAN DIODORUS SICULUS, 1ST CENTURY BCE

CELTIC BELIEFS

Celtic religion focused on sacred groves and springs, where rituals were performed by priests known as druids. There were many different local deities, but some, such as Cernunnos, a horned god representing nature, and the three mysterious hooded female goddesses (right) known as the Matrones ("mothers"), were worshipped widely throughout the Celtic world.

Hooded deities from the 3rd century

"Along the shore are the Fish-Eaters... and beyond them the Wild Flesh-Eaters... and behind them, further inland, there lies a city called Meroë."

THE PERIPLUS OF THE ERYTHRAEAN SEA, c.50 CE

△ *Mapping the Erythraean Sea*
Dutch cartographer Jan Jansson's 1658 version of Abraham Ortelius's 1597 map shows places included in *The Periplus of the Erythraean Sea*. However, he added two smaller maps, showing the voyage of the Carthaginian seafarer Hanno down West Africa (above upper left) and the mythical islands of Atlantis and Hyperborea (above upper right).

GREEK EXPLORATION

The expansion of the Greek Empire from around 800 BCE, along with the establishment of colonies throughout the Mediterranean and along the shores of the Black Sea, fuelled a demand for accounts of the new regions that had been explored and for practical navigation guides for Greek traders.

The foundation of the colony city of Pithecusae (near present-day Naples) around 800 BCE marked the start of the establishment of Greek colonies along the shores of the Mediterranean and Black Sea. Gradually, ancient Greek scholars and historians such as Herodotus compiled information about these areas. Although Herodotus travelled down the Nile, to the region far west beyond the Pillars of Hercules (present-day Straits of Gibraltar), he had to rely on the accounts of others. While Herodotus states that he did not know of any tin-producing islands called Cassiterides, this partly mythologized name refers to north Atlantic lands, such as Cornwall and the Scillies. Intrepid travellers went in search of tin, and one explorer, Pytheas, reached the British Isles and reported evidence of tin-mining around 320 BCE (see below).

To the east, from around 100 BCE, Greek merchants used the monsoon winds to sail across the Indian Ocean, setting out from Red Sea ports in Egypt such as Myos Hormos (present-day Quseir al-Qadim) and travelling as far as India. Some compiled manuals called *periploi* (voyages), which described the ports merchants called in to and the best trade goods to be acquired in each of them. The best-known of these, *The Periplus of the Erythraean Sea*, was likely compiled in *c.*50 CE. The Erythraean Sea refers to the present-day Red Sea, Arabian Sea, Persian Gulf, and Indian Ocean.

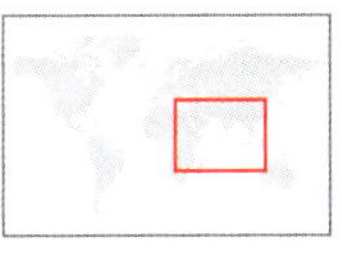
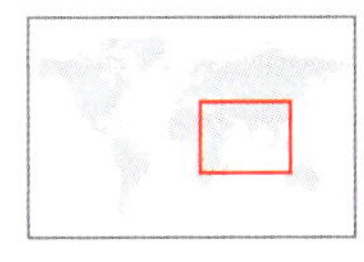

LOCATOR

KEY

1 Muziris (India) – Periplus said this "abounds in ships sent there with cargoes from Arabia".

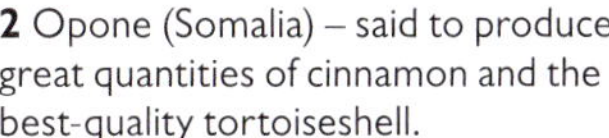
2 Opone (Somalia) – said to produce great quantities of cinnamon and the best-quality tortoiseshell.

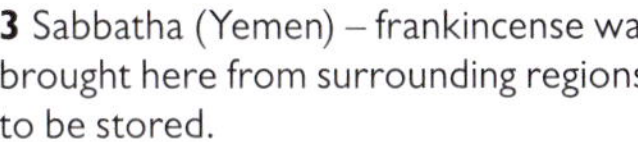
3 Sabbatha (Yemen) – frankincense was brought here from surrounding regions to be stored.

THE VOYAGE OF PYTHEAS

Greek sailors from Massalia (present-day Marseille, France) had a tradition of sailing west beyond the Pillars of Hercules into the Atlantic, then up the coastline of what is now Portugal and France. Around 320 BCE, one explorer, Pytheas, went even further. His account mentions tin trading in Cornwall, Britain, and his voyages to the Orcas (Orkney Islands). It also describes sailing six days north of Britain, where he saw "frozen seas" and as far as Thule (possibly Scandinavia or Iceland), where it was almost always dark in winter.

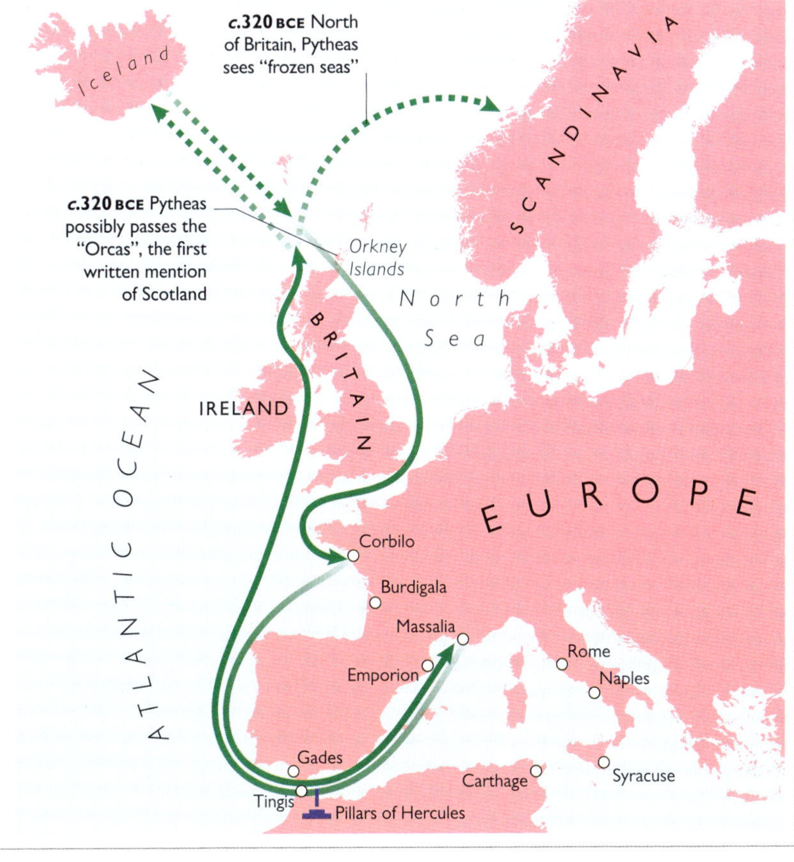

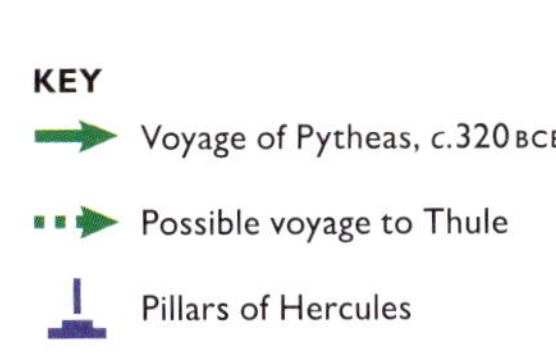

BANTU EXPANSIONS IN AFRICA

Today, about a third of Africa's population speak a Bantu language. With Bantu-language speakers distributed across 9 million sq km (3 million sq miles) – approximately one-third of Africa's landmass – there has been much debate about when, how, and why the languages of the Bantu peoples expanded across such vast distances.

Bantu languages are the largest subgroup of the Niger-Congo language group – one of Africa's four major language families. They probably originated some 6,000–7,000 years ago in the Mambilla and Bamenda highlands of present-day southeastern Nigeria and western Cameroon. The expansion of Bantu-language speakers across Africa over *c.*4,000 years represents one of the largest migrations in human history, traceable via genetics, new material culture traditions (see box), and the introduction of crops and livestock.

The timing, routes, and consequences of this expansion have been the subject of much debate for over a century, and many issues remain unresolved. Two major directions of spread have been identified: a Western stream associated with southward expansion through the Central African rainforest into west-central and southwestern Africa, and an Eastern stream near the southern rainforest boundary into eastern Africa, and then into southeastern Africa.

The expansion of Bantu-language speakers is linked with the introduction of crops, including yam, pearl and finger millet, and sorghum. Archaeologically, the arrival of speakers of Bantu languages is often marked by evidence of ironworking, new styles of pottery, and more permanent settlements.

"LATE WHITE" ROCK ART

Speakers of Bantu languages created a diverse range of figurative art in clay, wood and metal, and they also made rock art. In eastern and parts of south-central Africa, this often took the form of finger-painted abstract, geometric forms, as well as occasional figurative images (mostly stylized humans), all painted using a pigment made from white kaolin clay. These white artworks can be found alongside earlier, predominantly figurative, images of humans and wild animals in a range of colours (probably produced by foragers) and may have even been superimposed on these. The style is often known as the "late white" tradition. Some of this art was probably associated with initiation ceremonies, while other forms and imagery had symbolic significance for secret societies, hunters, and other specialist groups.

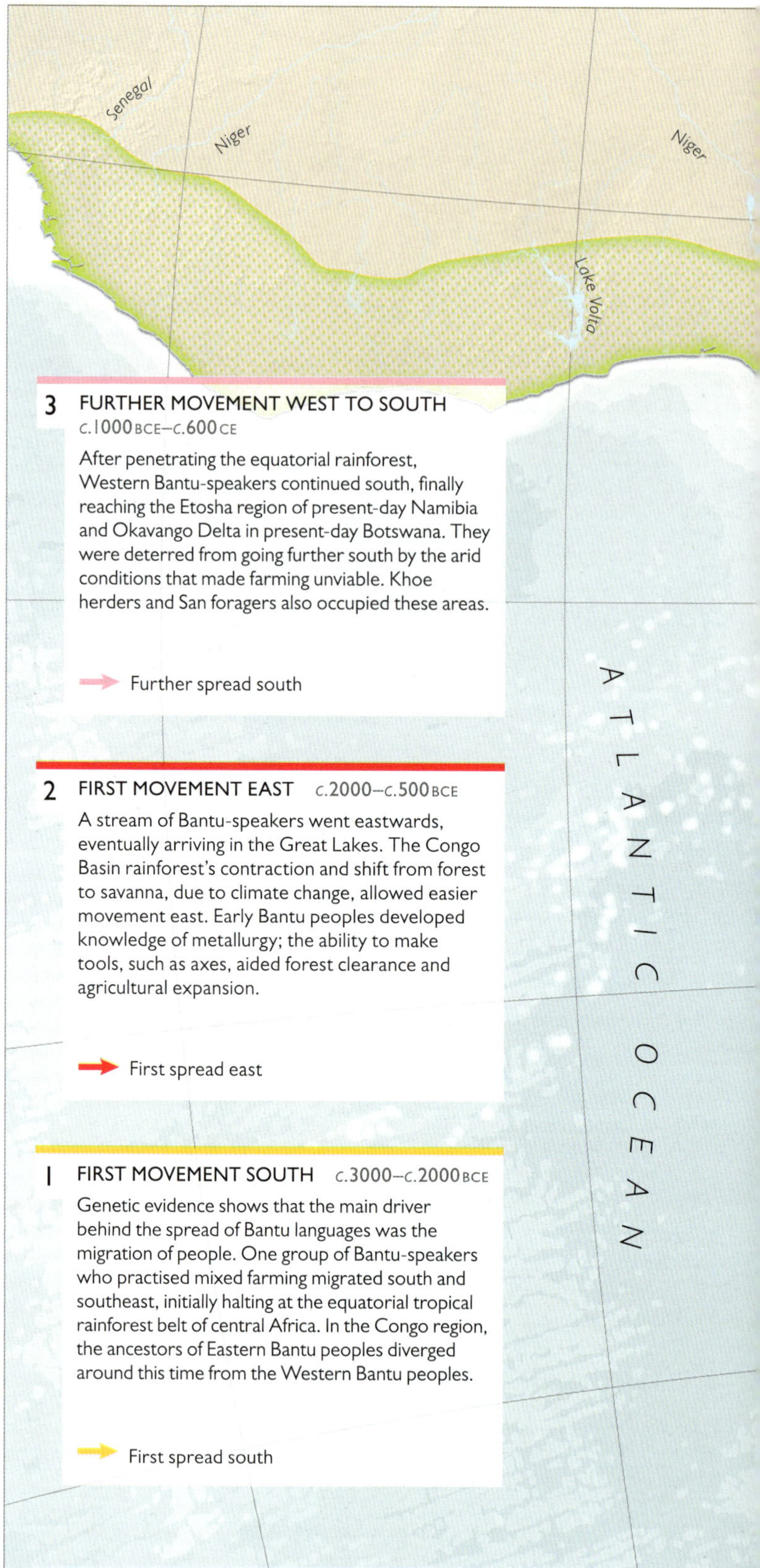

Senegal

Niger

Niger

Lake Volta

ATLANTIC OCEAN

3 FURTHER MOVEMENT WEST TO SOUTH *c.*1000 BCE–*c.*600 CE

After penetrating the equatorial rainforest, Western Bantu-speakers continued south, finally reaching the Etosha region of present-day Namibia and Okavango Delta in present-day Botswana. They were deterred from going further south by the arid conditions that made farming unviable. Khoe herders and San foragers also occupied these areas.

Further spread south

2 FIRST MOVEMENT EAST *c.*2000–*c.*500 BCE

A stream of Bantu-speakers went eastwards, eventually arriving in the Great Lakes. The Congo Basin rainforest's contraction and shift from forest to savanna, due to climate change, allowed easier movement east. Early Bantu peoples developed knowledge of metallurgy; the ability to make tools, such as axes, aided forest clearance and agricultural expansion.

First spread east

1 FIRST MOVEMENT SOUTH *c.*3000–*c.*2000 BCE

Genetic evidence shows that the main driver behind the spread of Bantu languages was the migration of people. One group of Bantu-speakers who practised mixed farming migrated south and southeast, initially halting at the equatorial tropical rainforest belt of central Africa. In the Congo region, the ancestors of Eastern Bantu peoples diverged around this time from the Western Bantu peoples.

First spread south

LANGUAGE SPREAD

Bantu-language speakers were not a homogenous ethnic group, but the movement of Bantu-speaking communities left distinctive cultural, linguistic, and genetic traces.

KEY

- Bantu homeland 3000 BCE
- Northwestern Bantu 500 CE
- Western Bantu 500 CE
- Eastern Bantu 500 CE
- Equatorial rainforest
- Desert

TIMELINE

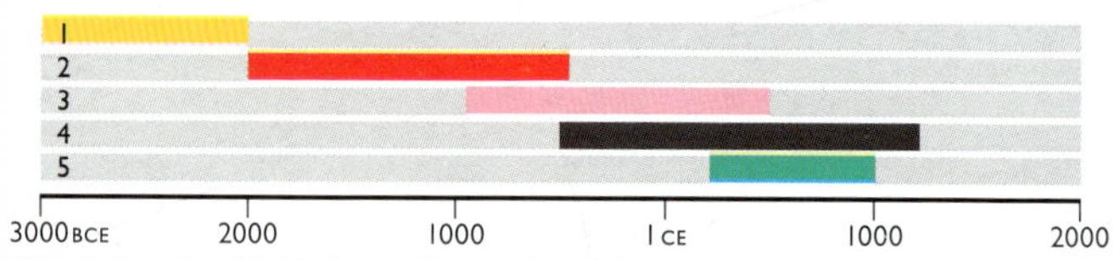

1
2
3
4
5

3000 BCE 2000 1000 1 CE 1000 2000

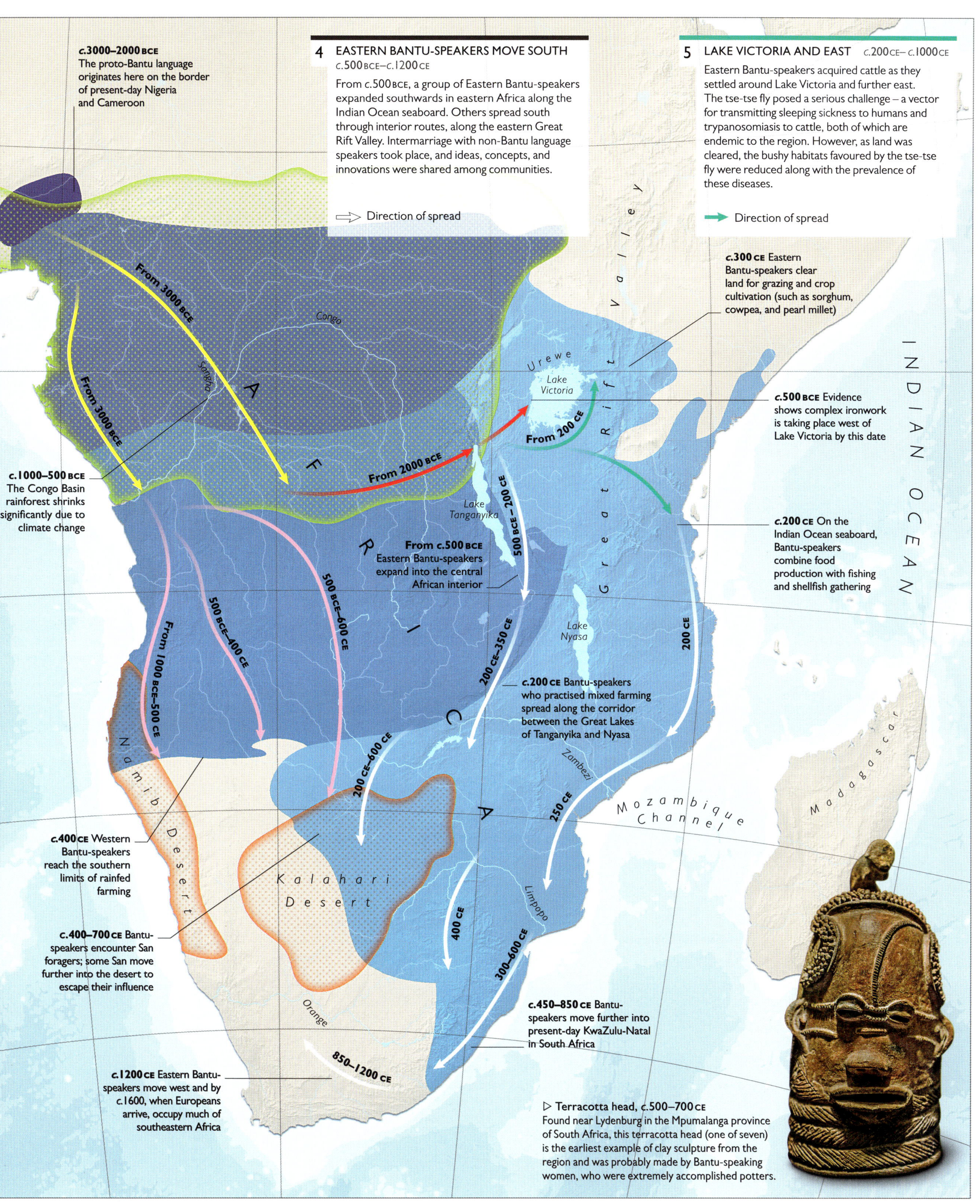

▷ **Terracotta head, c.500–700 CE**
Found near Lydenburg in the Mpumalanga province of South Africa, this terracotta head (one of seven) is the earliest example of clay sculpture from the region and was probably made by Bantu-speaking women, who were extremely accomplished potters.

Han riders
This 8th-century CE mural from Chang'an (now Xi'an) shows Han-dynasty hunters on horseback. The Han acquired many horses through trade and war; Emperor Wudi even fought a war over "heavenly horses" (see pp.38–39) in 104 BCE.

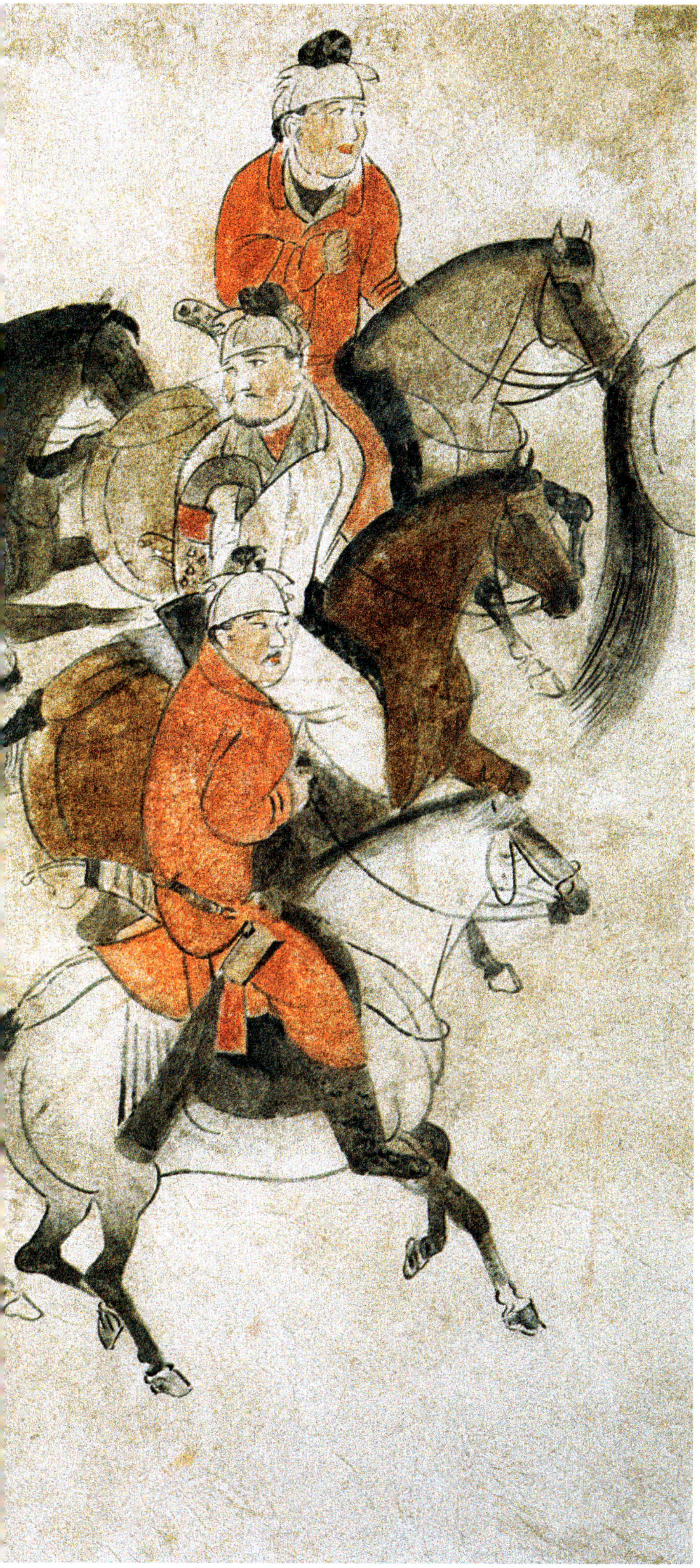

CHINESE MIGRATIONS

In 221 BCE, the Qin state defeated six rival kingdoms and unified China for the first time. The Qin dynasty laid the foundations for huge expansion under the later Han dynasty, whose territory eclipsed even the Roman Empire in size.

△ **Terracotta warrior**
In 1974, an army of life-size tomb guardians, including this kneeling archer, was found at the tomb of Qin Shi Huangdi.

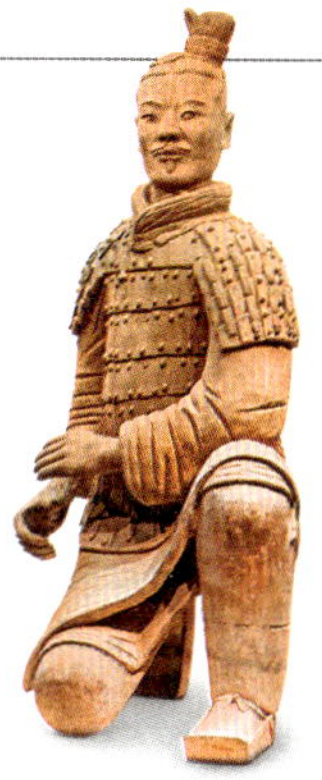

In the 3rd century BCE, Qin Shi Huangdi, the first emperor of Qin, reinforced his new empire's borders by sending millions of people from interior counties to settle on the new frontiers. The Qin migration push included the creation of new farmland to lure migrants, and the state-induced migration of 120,000 wealthy families from the conquered land in the east to Xiangyang, the Qin capital. Reforms standardized currency, language, and weights and measures. The Qin dynasty, however, did not survive long after the emperor's death.

China flourished under its next rulers, the Han (202 BCE–220 CE). Military successes under Emperor Wudi (r.141–87 BCE) doubled the size of the empire, annexing present-day northern Korea and northern Vietnam. As the empire expanded, migrants displaced or assimilated Indigenous peoples such as the Dian and Yue. The Han government encouraged this migration by creating new agricultural colonies – as in the Hexi Corridor (present-day Gansu Province) – or new towns built around mausoleums, where the government granted migrants money to build homes. They also established military garrisons in the new territories gained.

At its height, the Han empire had a population of 50 million people. However, by the 2nd century CE, their central authority had dwindled. In 220 CE, the dynasty collapsed once more into rival kingdoms.

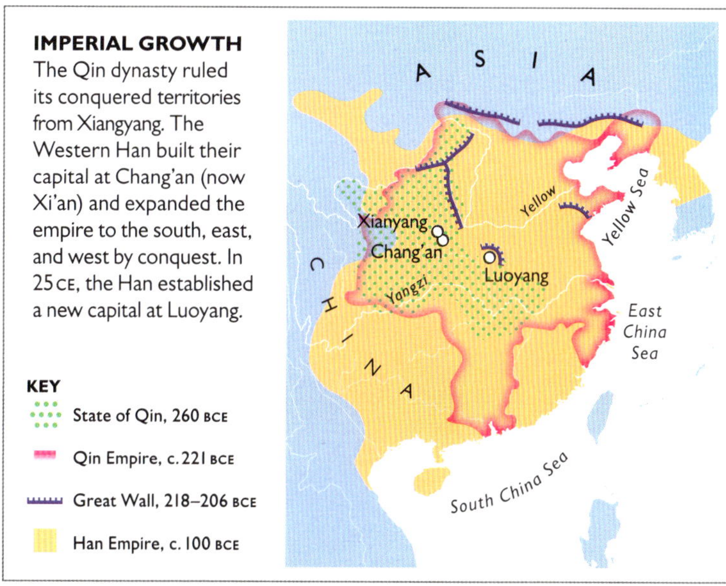

IMPERIAL GROWTH
The Qin dynasty ruled its conquered territories from Xiangyang. The Western Han built their capital at Chang'an (now Xi'an) and expanded the empire to the south, east, and west by conquest. In 25 CE, the Han established a new capital at Luoyang.

KEY
- State of Qin, 260 BCE
- Qin Empire, c. 221 BCE
- Great Wall, 218–206 BCE
- Han Empire, c. 100 BCE

CROSSING THE ALPS

When the leading Carthaginian general Hannibal Barca crossed the Alps with tens of thousands of troops as winter closed in, it was one of the greatest exploits in military history, pioneering a route that his enemies, the Romans, thought impassable.

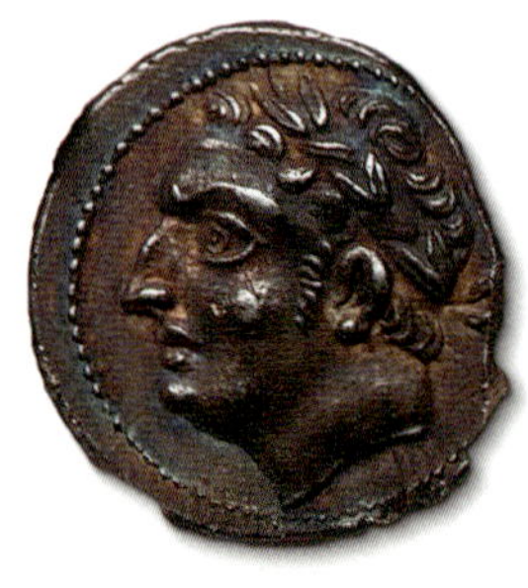

△ **Military genius**
Hannibal, shown on this coin, was a leading supporter of war with Rome, and used his Alpine crossing to thwart peace moves.

When war broke out between Carthage (in present-day Tunisia) and Rome in 218 BCE, Hannibal was in Spain. Rather than attack Italy by sea, as Carthage had previously, he boldly decided to do so by land. Gathering a huge force, he crossed the Pyrenees in August, reaching the Rhône River the next month, where he used 60-m (200-ft) long rafts to ferry across the 37 African war elephants he had with him. Hannibal hoped to use the elephants to overawe the Romans and their cavalry horses in battle.

Winter was coming when he reached the foothills of the Alps in October. Aided by local guides, he began an arduous crossing, vexed by the Allobroges, a local Celtic tribe, who sent rocks crashing down from mountaintops onto the Carthaginians. Hannibal used elite Spanish troops to drive the Celts off and took their fort, allowing him to cross an Alpine pass and begin his descent. Heavy snow made the going hazardous and when he reached the plain of the River Po, Hannibal had lost half his men. However, the epic crossing had caught the Romans completely unawares, leading to their defeat by the Carthaginians at Trebia.

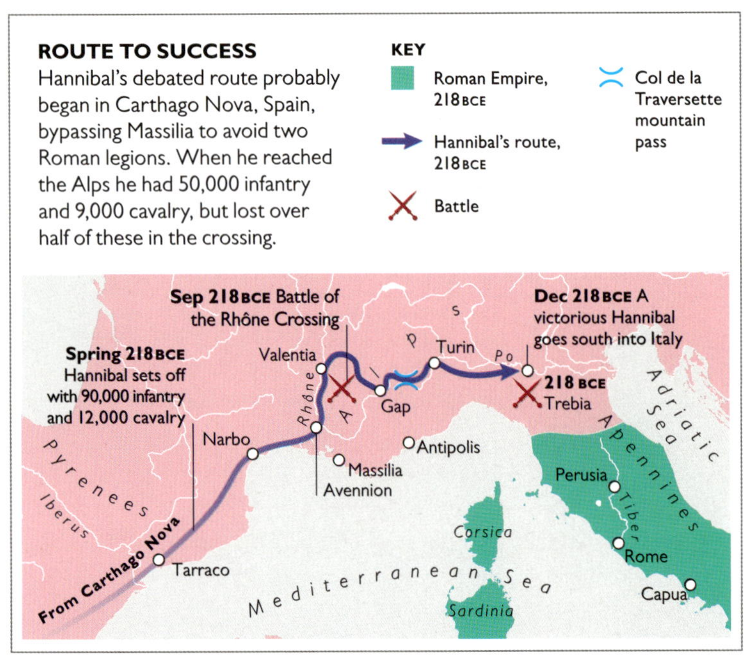

ROUTE TO SUCCESS
Hannibal's debated route probably began in Carthago Nova, Spain, bypassing Massilia to avoid two Roman legions. When he reached the Alps he had 50,000 infantry and 9,000 cavalry, but lost over half of these in the crossing.

KEY
- Roman Empire, 218 BCE
- Hannibal's route, 218 BCE
- Battle
- Col de la Traversette mountain pass

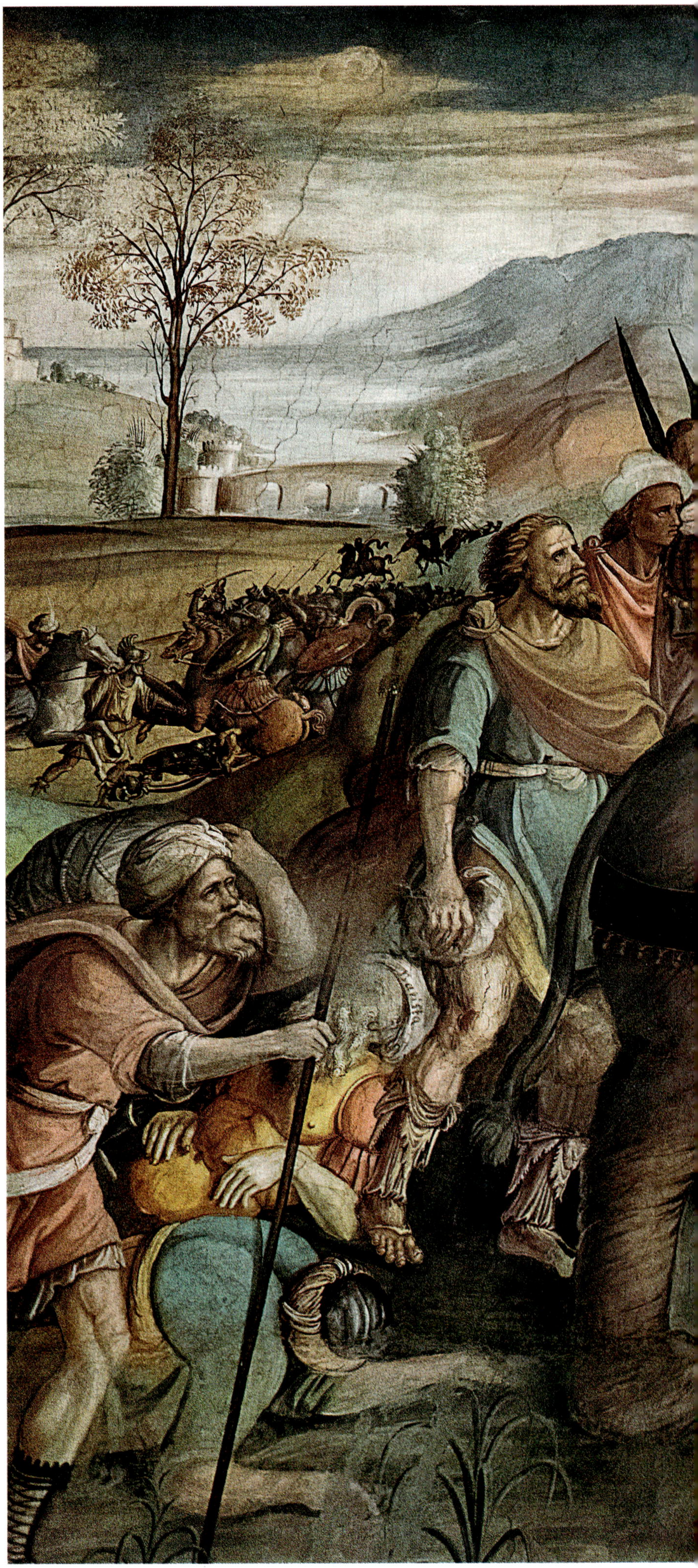

Audacious Alpine journey
A detail from a 16th-century Italian fresco attributed to Jacopo Ripanda depicts Hannibal riding one of his war elephants. Although all but one of his elephants were dead within a year, none of them is recorded as dying on the Alpine slopes.

THE SILK ROAD

This network of trade routes was established in the 2nd century BCE and linked China to Central Asia, Iran, and the Mediterranean. It carried goods, ideas, and travellers in either direction until political instability and new maritime routes caused its demise.

The Silk Road – named by 19th-century historians for the main commodity traded along it – began with the attempts by the Han Chinese emperor Wudi (r.141–87 BCE) to find allies against the nomadic Xiongnu. The 17-year-long expedition of Zhang Qian brought intelligence about the regions west of China, which led the Han to establish outposts in oases as far as Dunhuang and the Tarim Basin, south of the Tian Shan mountains. This Chinese-controlled corridor became a conduit for trade, and a way of projecting Chinese control into Central Asia. At first trade was dominated by Parthian and then Kushan merchants; by the 4th century CE, Sogdians from Bactria were the largest trading group. Along with silk came a range of goods: gunpowder and paper, both invented by the Han, had lasting impacts on the West. Ideologies, too, were exchanged, including Buddhism, which reached China in the 2nd century CE, while Parthian diplomats visited the Chinese court at Chang'an around 100 CE. Later periods when China regained control of the Western Regions, such as the Tang (618–907 CE) or under Mongol rule from 1269, saw the Silk Road flourish. However, the Ottoman capture of Constantinople in 1453 cut off land routes to China, and the rise of direct maritime routes from Europe in the 16th century led to the road's decline.

"If Han will not attack us, we will bring out all the fine horses."

HISTORIAN BAN GU IN THE *HAN SHU*, *c.*62 CE

EXCHANGE OF GOODS

There was a Chinese monopoly on silk production until silkworms were smuggled west into the Byzantine Empire in the 6th century. Demand for the luxury fabric generated huge profits for merchants. The Silk Road carried other merchandise, too, with porcelain and paper going westwards, and glass (which the Chinese, in turn, could not produce), saffron, and even spinach, east. The silver that was used to pay for the trade caused Roman writers to complain it was bankrupting the empire.

An 11th-century spinning wheel for silk

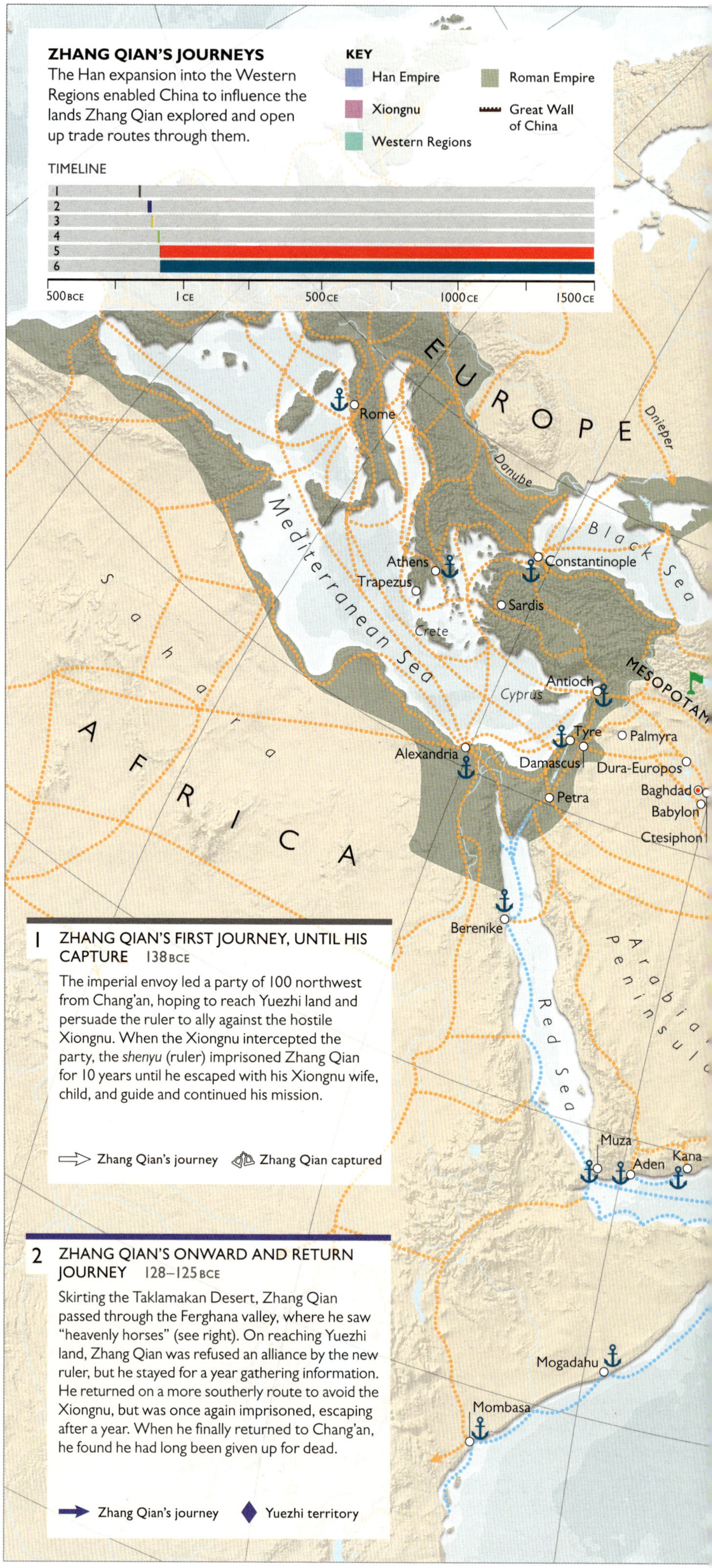

ZHANG QIAN'S JOURNEYS

The Han expansion into the Western Regions enabled China to influence the lands Zhang Qian explored and open up trade routes through them.

KEY

- Han Empire
- Xiongnu
- Western Regions
- Roman Empire
- Great Wall of China

TIMELINE

1, 2, 3, 4, 5, 6

500 BCE | 1 CE | 500 CE | 1000 CE | 1500 CE

1 ZHANG QIAN'S FIRST JOURNEY, UNTIL HIS CAPTURE 138 BCE

The imperial envoy led a party of 100 northwest from Chang'an, hoping to reach Yuezhi land and persuade the ruler to ally against the hostile Xiongnu. When the Xiongnu intercepted the party, the *shenyu* (ruler) imprisoned Zhang Qian for 10 years until he escaped with his Xiongnu wife, child, and guide and continued his mission.

Zhang Qian's journey — Zhang Qian captured

2 ZHANG QIAN'S ONWARD AND RETURN JOURNEY 128–125 BCE

Skirting the Taklamakan Desert, Zhang Qian passed through the Ferghana valley, where he saw "heavenly horses" (see right). On reaching Yuezhi land, Zhang Qian was refused an alliance by the new ruler, but he stayed for a year gathering information. He returned on a more southerly route to avoid the Xiongnu, but was once again imprisoned, escaping after a year. When he finally returned to Chang'an, he found he had long been given up for dead.

Zhang Qian's journey — Yuezhi territory

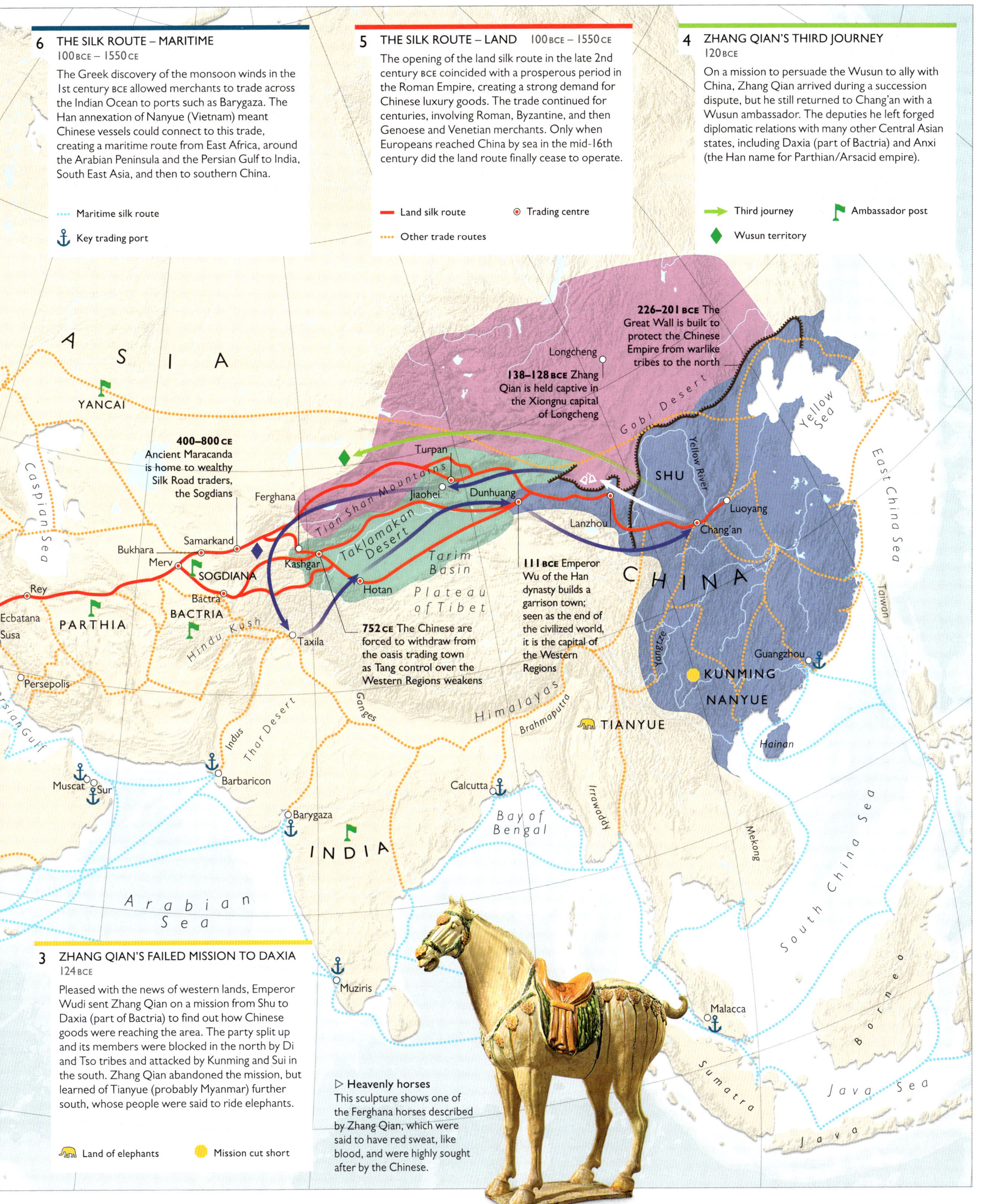

6 THE SILK ROUTE – MARITIME

100 BCE – 1550 CE

The Greek discovery of the monsoon winds in the 1st century BCE allowed merchants to trade across the Indian Ocean to ports such as Barygaza. The Han annexation of Nanyue (Vietnam) meant Chinese vessels could connect to this trade, creating a maritime route from East Africa, around the Arabian Peninsula and the Persian Gulf to India, South East Asia, and then to southern China.

- Maritime silk route
- Key trading port

5 THE SILK ROUTE – LAND

100 BCE – 1550 CE

The opening of the land silk route in the late 2nd century BCE coincided with a prosperous period in the Roman Empire, creating a strong demand for Chinese luxury goods. The trade continued for centuries, involving Roman, Byzantine, and then Genoese and Venetian merchants. Only when Europeans reached China by sea in the mid-16th century did the land route finally cease to operate.

- Land silk route
- Trading centre
- Other trade routes

4 ZHANG QIAN'S THIRD JOURNEY

120 BCE

On a mission to persuade the Wusun to ally with China, Zhang Qian arrived during a succession dispute, but he still returned to Chang'an with a Wusun ambassador. The deputies he left forged diplomatic relations with many other Central Asian states, including Daxia (part of Bactria) and Anxi (the Han name for Parthian/Arsacid empire).

- Third journey
- Ambassador post
- Wusun territory

3 ZHANG QIAN'S FAILED MISSION TO DAXIA

124 BCE

Pleased with the news of western lands, Emperor Wudi sent Zhang Qian on a mission from Shu to Daxia (part of Bactria) to find out how Chinese goods were reaching the area. The party split up and its members were blocked in the north by Di and Tso tribes and attacked by Kunming and Sui in the south. Zhang Qian abandoned the mission, but learned of Tianyue (probably Myanmar) further south, whose people were said to ride elephants.

- Land of elephants
- Mission cut short

▷ **Heavenly horses**
This sculpture shows one of the Ferghana horses described by Zhang Qian, which were said to have red sweat, like blood, and were highly sought after by the Chinese.

BEYOND THE BORDERLANDS

The Roman Empire (27 BCE–476 CE), stretching from modern Iraq in the east to England in the west, provided a vast, if slow, space for travel. Yet some military, trade, and exploratory expeditions penetrated across its borders.

Hadrian was the most travelled of Roman emperors (see box, below) but, during the time of the Empire, a number of journeys ventured beyond the frontier. A Roman fleet explored the West African coast in 146 BCE, and in 19 BCE the proconsul Cornelius Balbus marched over 640 km (400 miles) south from the African coast to attack oases in Garamantia (in present-day Libya). The general Suetonius Paulinus crossed the Atlas Mountains of Morocco in 41 CE in pursuit of Berber rebels and, in 90 CE, Julius Maternus led a more peaceful expedition south from Garamantia, reaching a place called Agisymba (possibly modern Chad), where he captured a rhinoceros.

Further east, a Roman army under Petronius attacked the Nubian kingdom of Meroë in 23 BCE, sacking its capital, Napata. The Romans next reached the region around 61 CE, when Emperor Nero sent a group of legionaries to search for the Nile's source. They passed through Meröe but, further south, their boats became stuck in a swampland full of crocodiles. They claimed to have seen the river's source in two huge rocks, but it is unclear quite how far south they had in fact reached.

A Roman merchant, sent by Nero in the mid-1st century CE, explored along the Baltic coastline, searching for sources of amber. Ancient Chinese texts also contain an account of a group of Roman merchants who travelled to China around 166 CE and who were escorted by Han soldiers to the imperial court at Luoyang.

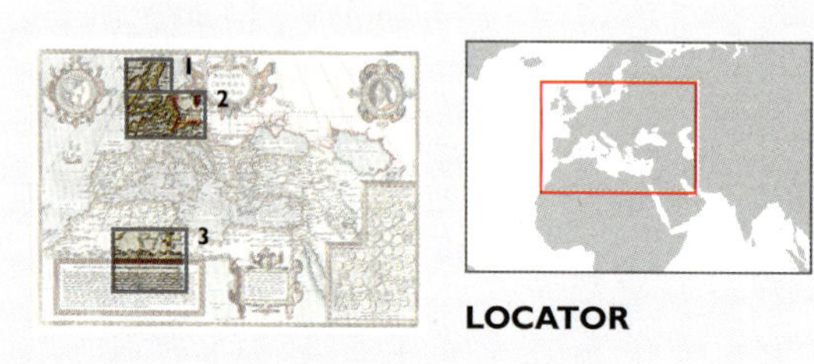

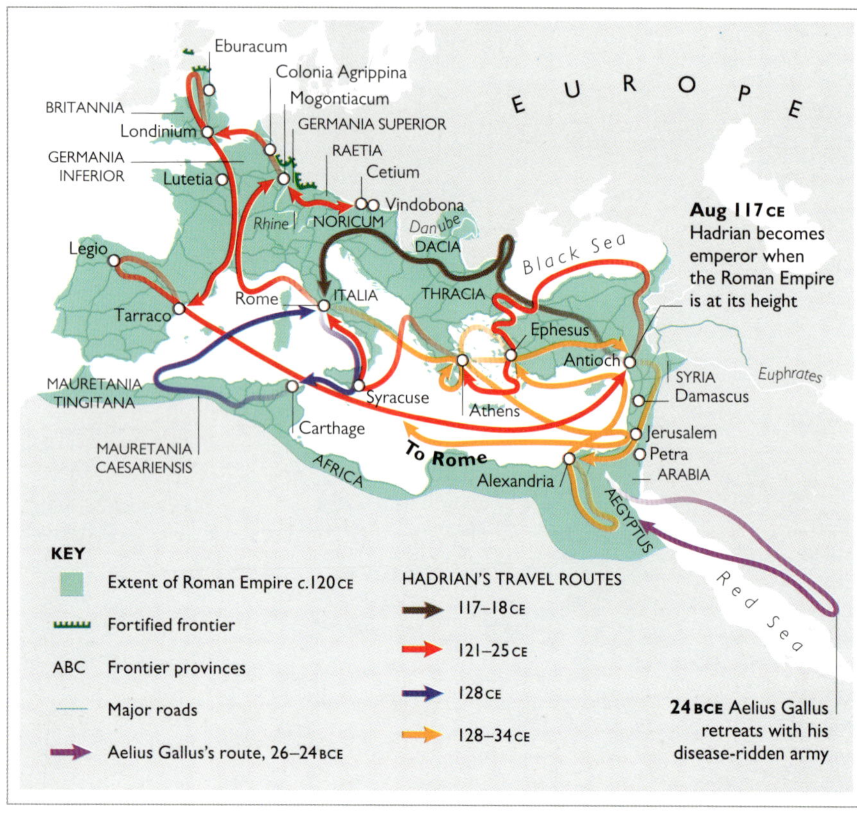

JOURNEYS IN THE ROMAN EMPIRE

Long-distance travel, within and outside the Roman Empire, was officially sponsored. The most southerly point reached by a Roman force was as far as present-day Yemen on the southern Arabian peninsula in 26–24 BCE. Aelius Gallus led 10,000 men in search of frankincense. Their Nabataean guide (from Petra) deliberately led them astray and, after besieging Marsiaba, which was a two-day diversion from the army's objective, Aelius was forced to retreat due to illness.

The Emperor Hadrian visited the far corners of his empire between 117 and 134 CE, building a wall in northern England to mark his empire's boundary.

◁ **Hadrian's Wall**
Hadrian visited Britain in 122 CE to construct a wall on the northern frontier, a defence against unconquered Caledonia (present-day Scotland).

▷ **A fortified frontier**
The gap between the Rhine and Danube frontiers was heavily fortified to prevent incursions by Germanic tribes.

▽ **Atlas mountain border**
The desert kingdom of Garamantia was on the other side of the Atlas mountains. Its oasis strongholds were often attacked by Roman soldiers.

△ **The Roman Empire**
Flemish cartographer Abraham Ortelius's 1598 map shows the empire at its height (c. 120 CE), stretching from England to Southwest Asia (Middle East).

△ **Visigothic metalwork** Dating from 550–600, this belt buckle from the Black Sea area is inset with glass and lapis lazuli. It is typical of high-status female Visigothic dress.

1 THE HUNS 370–453 CE

Moving west from southern Russia, the Huns crossed the Carpathian Mountains in the early 5th century. From the 430s, their leader Attila welded them into a skilled fighting force, building an empire that threatened to overwhelm the Roman frontier. His death in 453 was followed by the rapid dissolution of the Hunnish state, as peoples Attila had conquered broke away.

2 THE GOTHS 376–493 CE

The Goths crossed into the Roman Empire from the Hun Empire in 376, defeated the Romans at Adrianople (Edirne), present-day Turkey, in 378, and headed west under Alaric. They split into two groups: Visigoths, who travelled via Italy to southwest France and Spain; and the Ostrogoths, who stayed in modern-day Hungary then in 489 invaded Italy, establishing a kingdom in 493.

3 THE VANDALS 406–455 CE

Crossing the Rhine with a group of Alans and Sueves, the Vandals quickly surged through Gaul (France). The Alans and Sueves settled in present-day northern Spain, while the Vandals crossed into Africa, gradually moving east until they captured Carthage (near present-day Tunis), the capital of Roman North Africa in 439. They then established a kingdom that lasted almost a century before its reconquest by Byzantium in 533.

4 MIGRATIONS TO BRITAIN 367–500 CE

The withdrawal of the Roman army from Britain in 410 left it vulnerable. Picts raided south from Scotland and Irish Celts went east. Saxons and Angles from northwest Germany and the Jutes from Denmark began crossing the North Sea in small boats. These groups gradually took more land, founding kingdoms across England by the early 6th century.

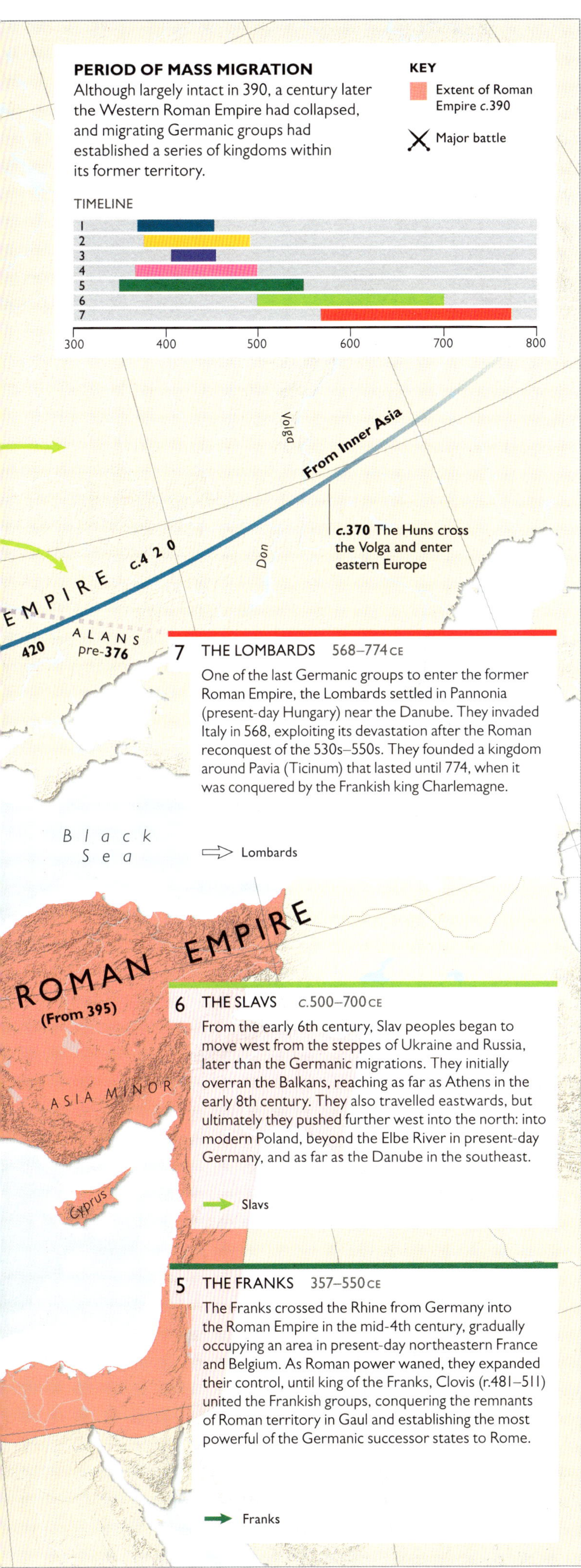

PERIOD OF MASS MIGRATION
Although largely intact in 390, a century later the Western Roman Empire had collapsed, and migrating Germanic groups had established a series of kingdoms within its former territory.
KEY
Extent of Roman Empire c.390
Major battle
TIMELINE
1
2
3
4
5
6
7
300
400
500
600
700
800
Volga
From Inner Asia
c.370 The Huns cross the Volga and enter eastern Europe
Don
EMPIRE c.420
420
ALANS pre-376
7 THE LOMBARDS 568–774 CE
One of the last Germanic groups to enter the former Roman Empire, the Lombards settled in Pannonia (present-day Hungary) near the Danube. They invaded Italy in 568, exploiting its devastation after the Roman reconquest of the 530s–550s. They founded a kingdom around Pavia (Ticinum) that lasted until 774, when it was conquered by the Frankish king Charlemagne.
Black Sea
Lombards
ROMAN EMPIRE
(From 395)
ASIA MINOR
6 THE SLAVS c.500–700 CE
From the early 6th century, Slav peoples began to move west from the steppes of Ukraine and Russia, later than the Germanic migrations. They initially overran the Balkans, reaching as far as Athens in the early 8th century. They also travelled eastwards, but ultimately they pushed further west into the north: into modern Poland, beyond the Elbe River in present-day Germany, and as far as the Danube in the southeast.
Slavs
Cyprus
5 THE FRANKS 357–550 CE
The Franks crossed the Rhine from Germany into the Roman Empire in the mid-4th century, gradually occupying an area in present-day northeastern France and Belgium. As Roman power waned, they expanded their control, until king of the Franks, Clovis (r.481–511) united the Frankish groups, conquering the remnants of Roman territory in Gaul and establishing the most powerful of the Germanic successor states to Rome.
Franks

THE GREAT MIGRATIONS

The movement of Germanic tribes westwards into the Roman Empire from the 3rd century CE constituted one of the largest-scale migrations in European history. Taking control of former Roman provinces, these peoples established kingdoms that formed the nuclei of future European nations.

Although the Roman Empire had suffered previous incursions from peoples such as the Celts and had fought Germans across the Rhine, a far greater threat to its borders emerged in the 3rd century CE. Pressed from the east by other groups migrating from the steppes of Russia and Ukraine, Germanic tribes pushed west in search of security, creating a great domino effect.

From the early 300s to the 700s, Germanic tribes breached the Roman frontier, as Roman emperors struggled to contain their expansion. Sometimes used as mercenaries by the Romans or as a buffer against others, the tribes began to transform the border regions. They themselves also changed, absorbing elements of Roman culture.

The Goths burst into the Balkans in 376 CE, led by their chieftain Alaric, and ravaged the region before attacking Italy. In 406 CE a mass of nomadic peoples – Alans, Sueves, and Vandals – crossed the frozen Rhine near present-day Mainz, travelling on to France and then Spain. They were followed by Franks and, as the newcomers advanced, they conquered former Roman provinces one by one. In 410 CE, the city of Rome itself was sacked by Visigoths, while the Huns, led by Attila, a chieftain with a fearsome reputation, inflicted devastating defeats on the Romans in the 430s. By 476 CE Rome's empire in the west had fallen, but migrations continued, with the Lombards entering Italy in the 550s to establish their own Germanic kingdom.

> *"Attila was a man born into the world to shake the nations, the scourge of all lands."*
>
> JORDANES, *GOTHIC HISTORIES*, C.550

THE SPREAD OF LITERACY

The migrating Germanic peoples were non-literate and relied on oral traditions of their history. As they settled within the Roman Empire and many of them converted to Christianity, a few among their elite – mainly churchmen – learnt to read. The Bible was translated from Greek to Gothic in the late 4th century. Later, Anglo-Saxon (the language of the 8th-century epic poem *Beowulf*), Old Norse, and German acquired their own written literature.

Extract from the first Gothic Bible, translated by Wulfila in the 4th century

THE MIDDLE AGES

FROM c.500 CE, WHEN TRADE FLOURISHED ACROSS THE EURASIAN LANDMASS, DIPLOMATS, SCHOLARS, AND PILGRIMS FOLLOWED THE TRADE ROUTES TO FAR-FLUNG PLACES. ICELAND AND AOTEAROA (NEW ZEALAND) WERE SETTLED AFTER DARING LONG-DISTANCE VOYAGES, AND EUROPEANS BEGAN TO NAVIGATE THE OPEN SEAS.

A CHANGING WORLD

The Middle Ages (c.500–1460) were a culturally vibrant and politically dynamic period. New empires arose to replace the collapsed states of the classical world and fresh links were forged between different regions of the Earth.

By the 6th century, the Roman, Han Chinese, and Indian Gupta empires, which had dominated large areas of the world for centuries, had all collapsed. In their wake, trade routes dried up, longstanding connections were broken, and travel became more difficult. But as new political formations established themselves, travellers began once more to explore. Christian pilgrims, Jewish scholars, Buddhist monks, and Muslim travellers journeyed widely, and in Europe and Asia, seaborne travel was undertaken to venture into new regions for exploration and to forge trading links.

◁ **Reliquary box**
Painted with scenes from the life of Christ, this wooden box contained relics brought back to Rome by a 6th-century pilgrim as a memento of a journey to the Holy Land.

New lands by sea

Europe's first maritime ventures into the unknown were tentative. For centuries, the Atlantic Ocean had been an almost impassable barrier, setting the boundaries of the ancient worlds of Africa and Europe. Ships lacked the technology to venture across the deep ocean and so hugged the coastlines. Then, in the 9th century, the Vikings travelled from Norway across the North Atlantic in light, strong vessels, using the Faroes, then Iceland and Greenland as stepping stones to reach the coast of Newfoundland by 1000 CE. There, they came into contact with Indigenous peoples, with their own rich cultures and traditions of travel across the continent. And a few hundreds of years later, in the 14th century, Portuguese explorers, inspired by the patronage of Prince Henry the Navigator, began to venture into Atlantic waters, reaching the Azores, and the Canaries off the coast of West Africa.

◁ **Pilgrim badge**
This copper alloy badge was probably collected by a pilgrim visiting the shrine of St Thomas á Becket at Canterbury Cathedral, a popular pilgrimage destination in medieval England.

In contrast, seaborne travel had continued unabated in East Asia and the Pacific, linking islands, archipelagoes, and maritime states such as Srivijaya on Sumatra with China and India, and bringing rich exchanges in trade, political customs, and religious traditions. In the late Middle Ages, these voyages extended further, with the Māori reaching and settling Aotearoa (New Zealand) in the 13th century, and the Chinese admiral Zheng He travelling as far south as Tanzania on the East African coast in the early 15th century.

Diplomats, pilgrims, and scholars

With the Islamic Empire in its golden age from the 8th to the 13th century, Muslim travellers began traversing it, making truly epic journeys – from Ibn Fadlan's diplomatic mission to the Volga region of Russia and Ibn Wahhab's quest to

A TIME OF NEW BEGINNINGS

The eight centuries of the Middle Ages saw the world begin to open up in a new way. As time went on, officially sponsored missions of exploration began to cross huge areas of the globe. The Atlantic, which had been a barrier to travel for centuries, was crossed, and the last large, unpopulated land masses in Iceland and New Zealand were settled after daring long-distance voyages.

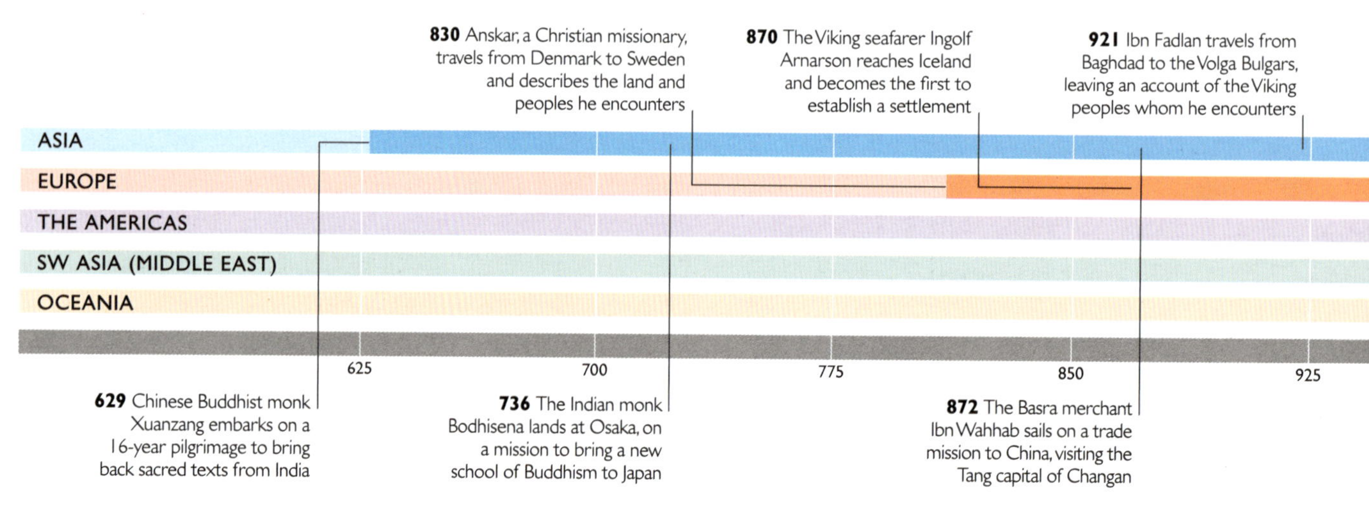

830 Anskar, a Christian missionary, travels from Denmark to Sweden and describes the land and peoples he encounters
870 The Viking seafarer Ingolf Arnarson reaches Iceland and becomes the first to establish a settlement
921 Ibn Fadlan travels from Baghdad to the Volga Bulgars, leaving an account of the Viking peoples whom he encounters
ASIA
EUROPE
THE AMERICAS
SW ASIA (MIDDLE EAST)
OCEANIA
625
700
775
850
925
629 Chinese Buddhist monk Xuanzang embarks on a 16-year pilgrimage to bring back sacred texts from India
736 The Indian monk Bodhisena lands at Osaka, on a mission to bring a new school of Buddhism to Japan
872 The Basra merchant Ibn Wahhab sails on a trade mission to China, visiting the Tang capital of Changan

◁ **Qigexing Buddha, 8th century**
The 7th-century Chinese Buddhist monk Xuanzang visited the Seven Stars Buddhist Temple in Qigexing on his epic 16-year journey to India in search of ancient sacred Buddhist texts.

> *"Do there exist many worlds, or is there but a single world?"*
>
> ALBERTUS MAGNUS, 13TH-CENTURY DOMINICAN FRIAR AND SCIENTIST

trade with China, to Ibn Battuta's odyssey in the 14th century. And from 632 CE, the Hajj became the fifth pillar of Islam, requiring every Muslim who was able, to make the pilgrimage to Mecca at least once in their lifetime.

From around 1050, Christian pilgrims made long and often dangerous journeys to sacred places, such as Santiago in Spain and the city of Rome or, most prized of all, the Holy Land – sites in Southwest Asia (the Middle East) associated with the life of Christ. Jewish scholars and pilgrims also embarked on epic journeys, such as Benjamin of Tudela in the 12th century, who travelled to Europe, Asia, and North Africa. Buddhist pilgrims from China journeyed to India in search of sacred texts. And in 736 CE, an Indian monk, Bodhisena, travelled to Japan as a missionary. In the Americas, climate changes pushed Ancestral Puebloans southwest from Chaco Canyon in New Mexico to Arizona.

A meeting of cultures

The expansion of the Mongol Empire at the start of the 13th century brought new regions in contact with each other, initiating new forms of exchange. Within 50 years, its territories reached from northern China to the borders of Poland and Hungary. Travellers who crossed this vast landmass included the Italian monk Giovanni da Pian del Carpini and Flemish missionary William of Rubruck, who separately travelled to Central Asia in the 1240s and around 1250. The travelogue of Venetian merchant Marco Polo (1270–95) gave Europeans their first insight into Chinese culture.

A time when letters could be transported from the Great Khan's court in China to the Pope in Rome, when Polynesian canoes could navigate by the stars, winds, and ocean swells to travel thousands of miles across the Pacific to Aotearoa (New Zealand), and Arab travellers wrote with wonder of burial customs deep in the Russian steppe, the medieval age saw signficant cross-cultural connections.

▽ **Map of the world**
Taken from the *Book of Marvels of Science*, a 12–13th century Arabic manuscript, this map contains the earliest known example of a scale bar (at the top) and draws on a rich tradition of medieval Arabic cartography.

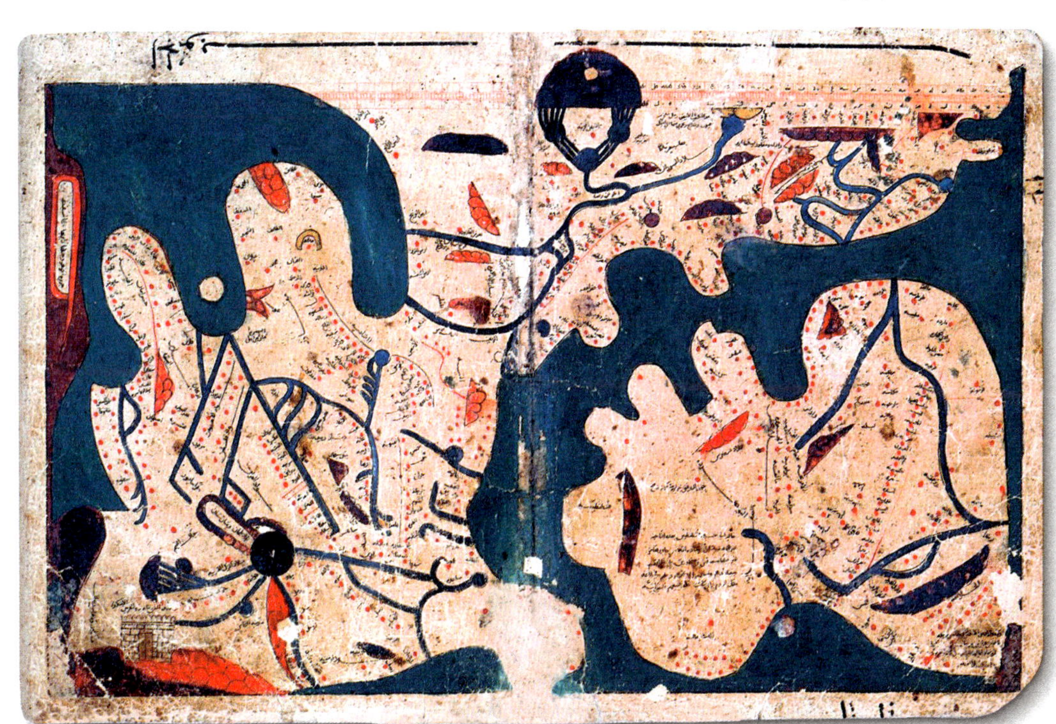

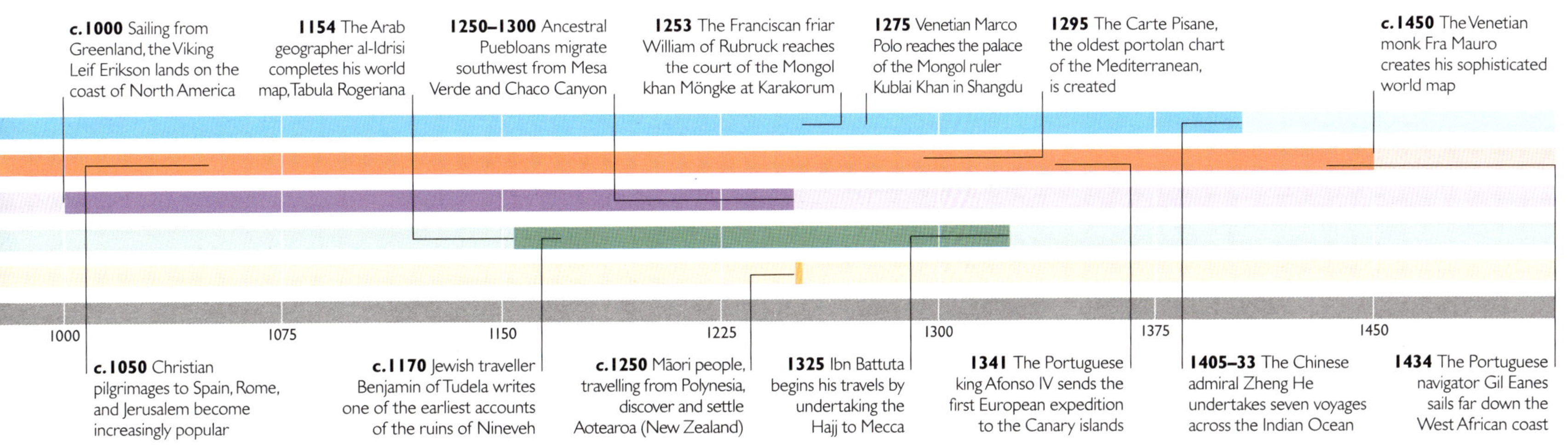

c. 1000 Sailing from Greenland, the Viking Leif Erikson lands on the coast of North America

c. 1050 Christian pilgrimages to Spain, Rome, and Jerusalem become increasingly popular

1154 The Arab geographer al-Idrisi completes his world map, Tabula Rogeriana

c. 1170 Jewish traveller Benjamin of Tudela writes one of the earliest accounts of the ruins of Nineveh

1250–1300 Ancestral Puebloans migrate southwest from Mesa Verde and Chaco Canyon

c. 1250 Māori people, travelling from Polynesia, discover and settle Aotearoa (New Zealand)

1253 The Franciscan friar William of Rubruck reaches the court of the Mongol khan Möngke at Karakorum

1275 Venetian Marco Polo reaches the palace of the Mongol ruler Kublai Khan in Shangdu

1325 Ibn Battuta begins his travels by undertaking the Hajj to Mecca

1295 The Carte Pisane, the oldest portolan chart of the Mediterranean, is created

1341 The Portuguese king Afonso IV sends the first European expedition to the Canary islands

1405–33 The Chinese admiral Zheng He undertakes seven voyages across the Indian Ocean

c. 1450 The Venetian monk Fra Mauro creates his sophisticated world map

1434 The Portuguese navigator Gil Eanes sails far down the West African coast

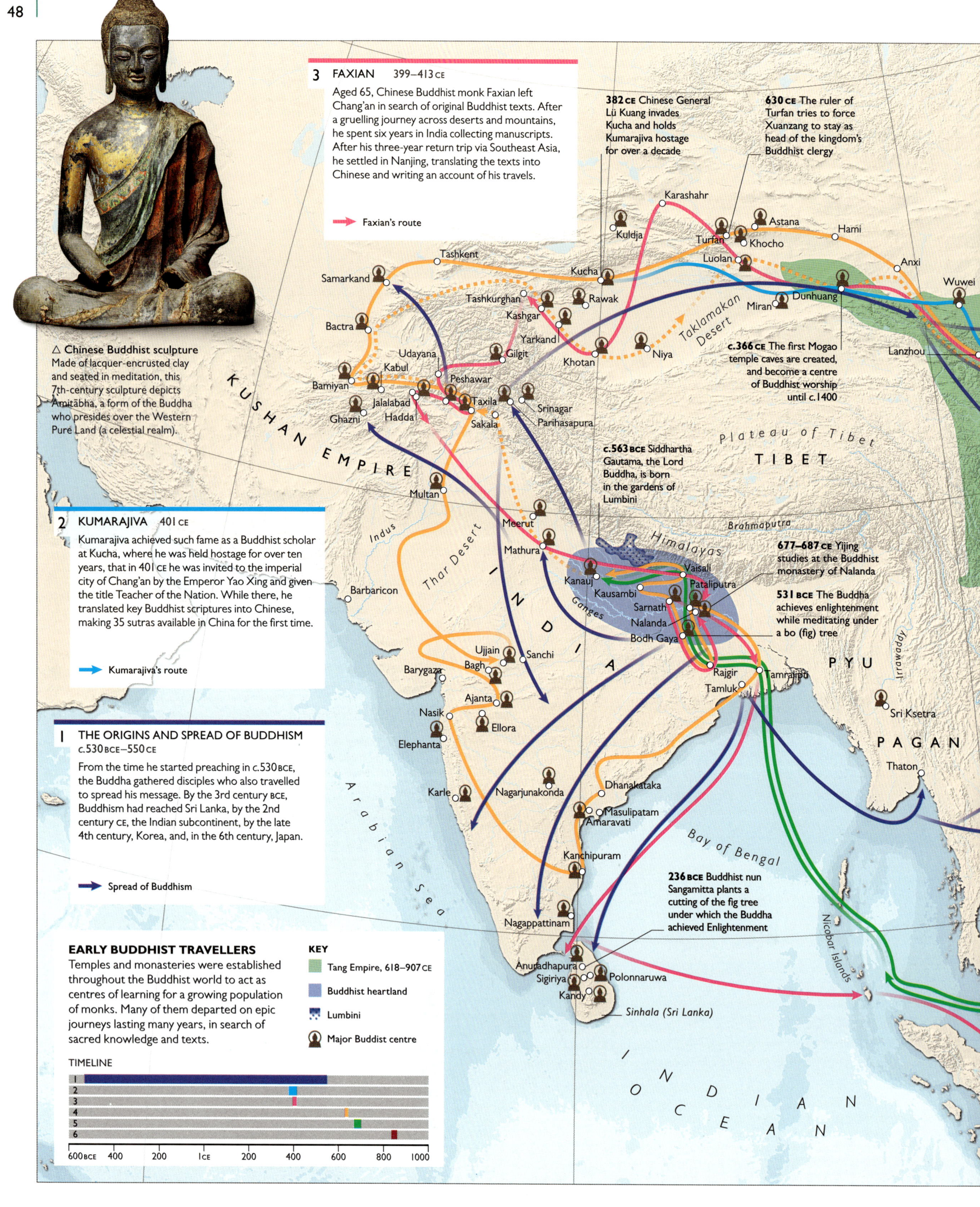

3 **FAXIAN** 399–413 CE

Aged 65, Chinese Buddhist monk Faxian left Chang'an in search of original Buddhist texts. After a gruelling journey across deserts and mountains, he spent six years in India collecting manuscripts. After his three-year return trip via Southeast Asia, he settled in Nanjing, translating the texts into Chinese and writing an account of his travels.

Faxian's route

△ **Chinese Buddhist sculpture**
Made of lacquer-encrusted clay and seated in meditation, this 7th-century sculpture depicts Amitābha, a form of the Buddha who presides over the Western Pure Land (a celestial realm).

2 **KUMARAJIVA** 401 CE

Kumarajiva achieved such fame as a Buddhist scholar at Kucha, where he was held hostage for over ten years, that in 401 CE he was invited to the imperial city of Chang'an by the Emperor Yao Xing and given the title Teacher of the Nation. While there, he translated key Buddhist scriptures into Chinese, making 35 sutras available in China for the first time.

Kumarajiva's route

1 **THE ORIGINS AND SPREAD OF BUDDHISM** c.530 BCE–550 CE

From the time he started preaching in c.530 BCE, the Buddha gathered disciples who also travelled to spread his message. By the 3rd century BCE, Buddhism had reached Sri Lanka, by the 2nd century CE, the Indian subcontinent, by the late 4th century, Korea, and, in the 6th century, Japan.

Spread of Buddhism

EARLY BUDDHIST TRAVELLERS

Temples and monasteries were established throughout the Buddhist world to act as centres of learning for a growing population of monks. Many of them departed on epic journeys lasting many years, in search of sacred knowledge and texts.

KEY

- Tang Empire, 618–907 CE
- Buddhist heartland
- Lumbini
- Major Buddhist centre

TIMELINE

1, 2, 3, 4, 5, 6

600 BCE 400 200 1 CE 200 400 600 800 1000

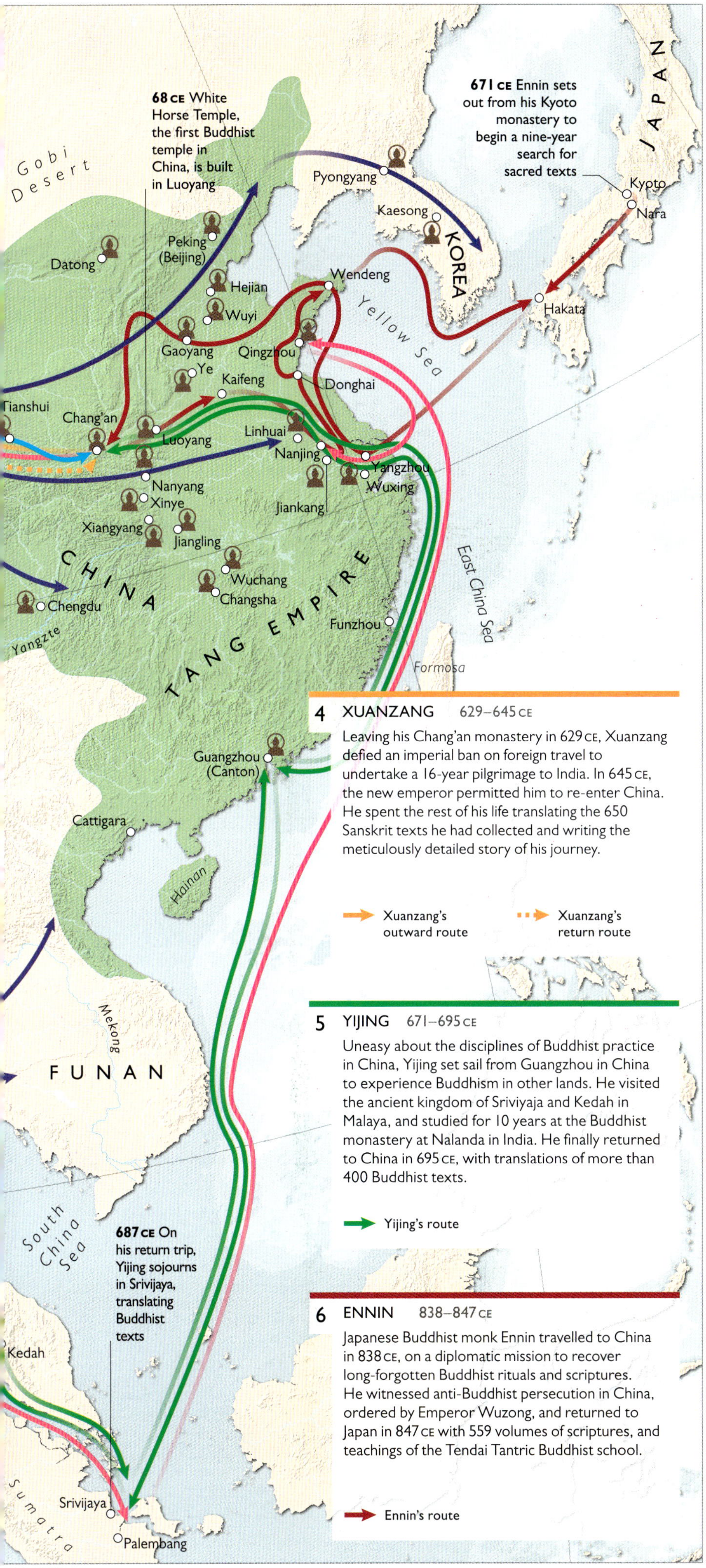

68 CE White Horse Temple, the first Buddhist temple in China, is built in Luoyang

671 CE Ennin sets out from his Kyoto monastery to begin a nine-year search for sacred texts

687 CE On his return trip, Yijing sojourns in Srivijaya, translating Buddhist texts

4 XUANZANG 629–645 CE

Leaving his Chang'an monastery in 629 CE, Xuanzang defied an imperial ban on foreign travel to undertake a 16-year pilgrimage to India. In 645 CE, the new emperor permitted him to re-enter China. He spent the rest of his life translating the 650 Sanskrit texts he had collected and writing the meticulously detailed story of his journey.

Xuanzang's outward route

Xuanzang's return route

5 YIJING 671–695 CE

Uneasy about the disciplines of Buddhist practice in China, Yijing set sail from Guangzhou in China to experience Buddhism in other lands. He visited the ancient kingdom of Sriviyaja and Kedah in Malaya, and studied for 10 years at the Buddhist monastery at Nalanda in India. He finally returned to China in 695 CE, with translations of more than 400 Buddhist texts.

Yijing's route

6 ENNIN 838–847 CE

Japanese Buddhist monk Ennin travelled to China in 838 CE, on a diplomatic mission to recover long-forgotten Buddhist rituals and scriptures. He witnessed anti-Buddhist persecution in China, ordered by Emperor Wuzong, and returned to Japan in 847 CE with 559 volumes of scriptures, and teachings of the Tendai Tantric Buddhist school.

Ennin's route

BUDDHIST TRAVELLERS

The spread of Buddhism from its Indian homeland inspired a vast network of missionaries and pilgrims to travel. Promoting their faith and searching for sacred texts, they traversed Central Asia and China, and reached as far as Southeast Asia and Japan.

From its origins in North India in the mid-6th century BCE, Buddhism was a missionary faith. According to foundational texts, from the time Siddhartha Gautama became the Buddha (enlightened one) at the age of 35, he journeyed around India preaching, and his followers were instructed to do the same. In this way, the faith spread within India and then to bordering regions, such as Sihala (modern-day Sri Lanka) and Nepal. From the 2nd century BCE it passed onward along the Silk Road (see pp. 38–39) to Central Asia and China, and on to Southeast Asia, the Kingdom of Goryeo (modern-day Korea), and Japan.

Buddhist communities sprang up in these far-flung territories over the 1st millennium CE, with temples and monasteries as their focal points. Over time, many Buddhists became dissatisfied with the sacred texts in their possession, which were either written in a language they didn't understand or had scriptural omissions. To fill the gaps in their knowledge, pilgrims set out for India, often enduring enormous hardships as they crossed the mountain ranges of the Pamirs and Himalayas. They visited sacred places in the life of the Buddha, and collected authentic manuscripts of his teachings, which they took back home with them for translation.

Missionaries travelled as well as pilgrims, seeking to spread the faith yet wider or to visit Buddhist communities in other lands. As a result, even though Islamic control of North India had suppressed Buddhism in the subcontinent, by the end of the 1st millennium CE it remained a vibrant faith in much of East and Southeast Asia.

BUDDHIST SACRED TEXTS

After the Buddha achieved enlightenment by renouncing dependence on material things, he taught his message to a growing band of followers for over 50 years, until his death around 483 BCE. At first, the Buddha's teachings were passed on orally but, over time, written records were created and formed an extensive canon of sacred Buddhist texts or sutras.

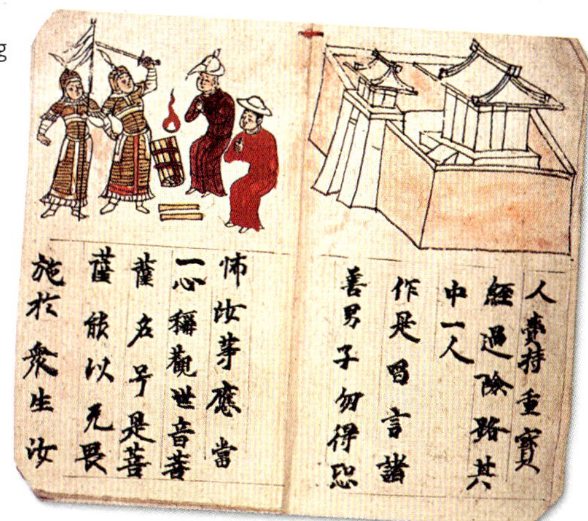

The Guanshiyin Sutra, c. 9th century

EARLY CHRISTIAN PILGRIMS

Christianity had been forbidden by the Romans, but in 313 CE, the Edict of Milan reversed this policy and Christians were allowed to travel openly. A new group of travellers emerged, with Christian pilgrims undertaking journeys to holy Christian sites.

Jerusalem and the Holy Land – the parts of Southwest Asia (the Middle East) associated with the life of Christ – were prime destinations for Christian pilgrimages, with the journey itself considered as sacred as the site visited. The Roman St Helena was the first Christian pilgrim when, in the early 4th century, together with her son Emperor Constantine, she founded the Church of the Nativity in Bethlehem in present-day Israel.

The tradition of Christian pilgrimage was well established by the 7th century, when the Muslim conquest of Palestine made access to Southwest Asia harder for Christian travellers. Consequently, new pilgrimage sites emerged in Europe – notably Rome, with its association with St Peter, but also churches with holy relics. Pilgrims from all over Western Europe converged on St James' at Santiago de Compostela in northern Spain. Other sacred destinations included the shrine of the martyred archbishop Thomas Becket in Canterbury, UK, after his death in 1170; and Maastricht in the Netherlands, where, from the 14th century, a seven-yearly exposition of relics at the grave of St Severus drew thousands of pilgrims.

Some Christian travellers evangelized, including the English saints Willibrord in the 690s and Boniface in the 710s, both journeying to Frisia, northwest Europe; while in the 850s, French St Anskar was the first Christian missionary in Sweden. Irish monks, such as St Brendan, began a tradition of Immrama (voyages) in the 6th century, travelling into the wilderness and far out to sea in search of solitude. Theories that they reached as far as the Americas are unlikely, but it is possible these monastic explorers arrived in Iceland.

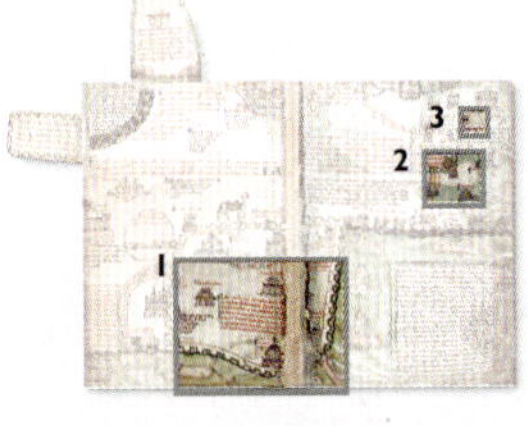

KEY

1 Acre, a major Holy Land port, had 40 churches pilgrims could visit, listed in *Pardouns d'Acre* – a pilgrimage manual.

2 The Temple Mount and tomb of Jesus are two of Jerusalem's principal sites for Christian pilgrims.

3 Bethlehem's Church of the Nativity, built c.330 CE, is believed to be on the site of Jesus's birth.

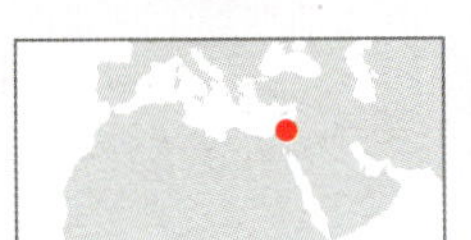

LOCATOR

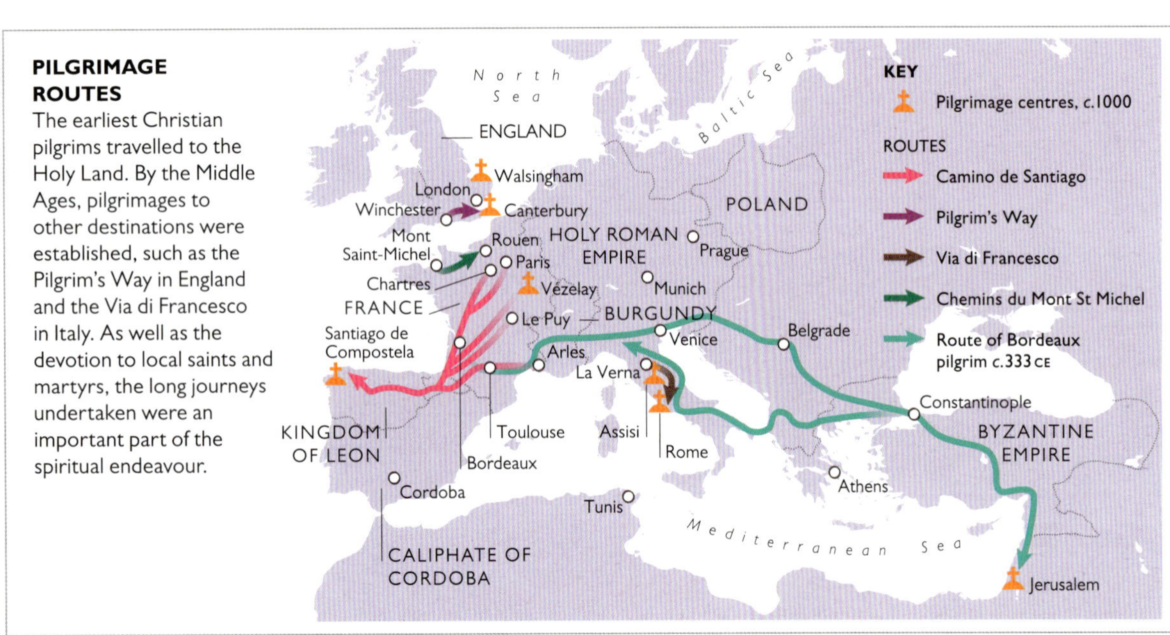

PILGRIMAGE ROUTES
The earliest Christian pilgrims travelled to the Holy Land. By the Middle Ages, pilgrimages to other destinations were established, such as the Pilgrim's Way in England and the Via di Francesco in Italy. As well as the devotion to local saints and martyrs, the long journeys undertaken were an important part of the spiritual endeavour.

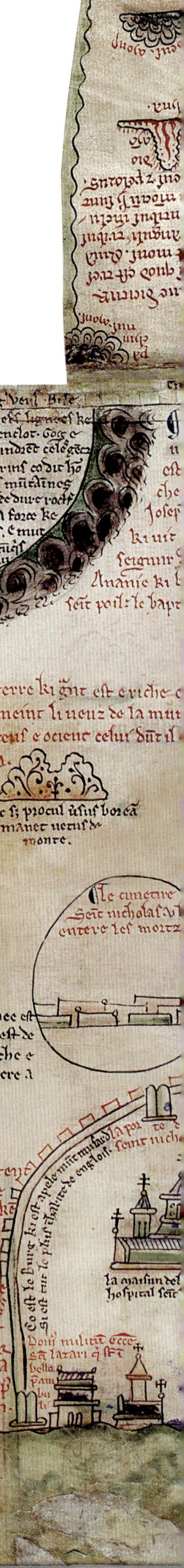

> *"From there two miles on the left-hand side is Bethlehem, where... Jesus Christ was born;... a basilica was made by Constantine's command."*
>
> ITINERARIO BURDIGALIS (THE BORDEAUX ITINERARY), ANONYMOUS, c.333 CE

▽ Pilgrim itinerary
Drawn around 1255 by English Benedictine monk Matthew Paris, this map illustrates a Christian pilgrim route from London to Jerusalem. The exaggerated size of the port of Acre in present-day Israel shows it was an important disembarkation point for pilgrims.

MEDIEVAL ARAB JOURNEYS

The enormous expansion of the Arab Muslim world in the 7th and 8th centuries resulted in its reach stretching from the Atlantic coast of Africa and Spain in the west to Central Asia in the east. While scholars produced maps of these regions, travellers embarked on pilgrimages and journeys in search of knowledge.

A key motivation for Arab travellers from the 7th century was the Hajj pilgrimage to Mecca in present-day Saudi Arabia. Pilgrims such as Ibn Jubayr (see box) and Ibn Battuta (see pp. 66–67) left detailed accounts of their journeys. Other Arab travellers went as envoys, including Ibn Fadlan, who in 921 accompanied an embassy from the Abbasid Caliph to the Volga Bulgars in present-day Russia, and witnessed a ship burial by Rus Vikings.

Some travellers left *Risala*, accounts of journeys motivated by a desire for knowledge. *Tuhfat al-Albab* ("The Gift of Hearts"), by Abu Hamid al-Garnati from al-Andalus (Muslim Spain), recounts over six decades of travels from 1106, including visits to Central Europe and Muslim cities in Southwest Asia (the Middle East), such as Baghdad and Mosul in Iraq, and Damascus, Syria.

Supplementing these writings were geographical compendia such as the *Kitab al-Buldan* ("Book of Countries"), written around 890 by the Baghdad geographer ibn Ya'qubi, and a vigorous tradition of map-making, which culminated in the world map compiled by al-Idrisi in 1154.

▽ **Holy city**
Mecca ("Makka" on the map), the holiest city in Islam, was the focal point of the Hajj pilgrimage and occupied the central position in most maps in the Arab cartographic tradition. Just to the right of it is Gidda, now Jeddah, the capital of Saudia Arabia, which was the main port where pilgrims disembarked.

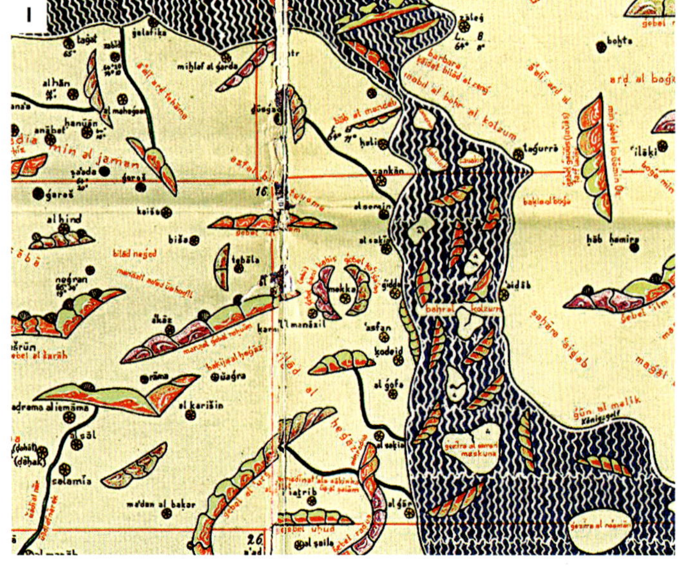

TRAVELS IN EGYPT

Some Arab travellers performing Hajj left accounts of the lands they travelled through. Ibn Jubayr, from Granada in al-Andalus, travelled across North Africa during his pilgrimage from 1183 to 1185. His account of Egypt remarks approvingly of such sights as the hostels for students erected by the Sultan Saladin in Alexandria and a hospital for the mentally unwell in Cairo. He also describes the Pyramids and the Sphinx, but he was most impressed by the lighthouse on the island of Pharos at Alexandria, still in operation almost 1,500 years after its construction. Ibn Jubayr's return journey from Mecca took him through Sicily, where he witnessed an eruption of the volcano on the island of Stromboli.

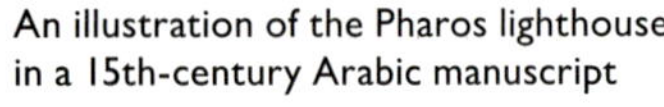
An illustration of the Pharos lighthouse in a 15th-century Arabic manuscript

△ World map of 1154
Born in Ceuta in North Africa, the geographer Muhammad al-Idrisi became court cartographer of King Roger II of Sicily in 1145. Over 18 years, he produced the *Tabula Rogeriana*, a world map inscribed on a 2-m- (6.5-ft-) diameter silver disc. Only the more conventional paper version, oriented with South at the top (as was usual with Arab maps), survives.

◁ Iraq
Containing the Abbasid capital Baghdad (founded in 762), the area of modern Iraq was at the heart of the Islamic world. Al-Idrisi shows the Tigris river snaking through it and, nearby, other important cities such as Kufa, the original Abbasid capital from 750 to 762.

△ Muslim Spain
Al-Andalus was a vibrant centre of Islamic culture and home to travellers such as Ibn Jubayr (see box). Its most important cities were Seville, Córdoba, and Granada, all Muslim-held when al-Idrisi compiled his map, though Córdoba fell to Christian armies in 1236 and Granada in 1248.

THE VIKING VOYAGES

The Viking Age, from the 8th to the 11th century, saw a sea-borne people spread from their Scandinavian homelands, settle across Europe and Asia – from Ireland to Russia – and embark on epic journeys that took them as far as North America.

In the late 8th century, the peoples of Scandinavia began crossing the Baltic and North Seas in fast-running longships to raid coastal regions – voyages driven by climatic deterioration, population growth, and political instability. For the next 300 years, Vikings advanced a reign of terror, seizing hoards of treasure from monasteries and enslaving people to take back and sell. Others, though, were traders, although the margin between the two could be narrow, as peaceful groups of merchants transformed into warrior bands when opportunities arose. Gradually, some Vikings settled in lands they had seized in England, Scotland, Ireland, northern France, southern Italy, Ukraine, and Russia. Others pressed on, sailing down the great rivers of Ukraine and Russia to Constantinople and the Caspian Sea. In their greatest venture of all, Viking voyagers struck out into the North Atlantic, reaching Iceland in the 870s CE, and then, around a century later, Greenland and the east coast of northern Canada.

The story of these intrepid travellers was shared in oral and written sagas, though by the 12th century, their descendants had ceased raiding and adopted a more settled way of life.

> *"People will be attracted thither [to Greenland], if the land has a good name."*
>
> SAGA OF ERIK THE RED, 13TH CENTURY

REACHING NORTH AMERICA

Voyaging west of Greenland in c.1000, Leif Eriksson came across a rocky coast he called Helluland, then a section he called Markland ("forest land"), and finally a land with vines that he named Vinland ("wine land"). The Vikings set up base at a place called Leifbudir ("Leif's booths"). The only evidence of these landings is a Viking settlement at L'Anse aux Meadows in Newfoundland, excavated in the 1960s, which is thought to be Leifbudir.

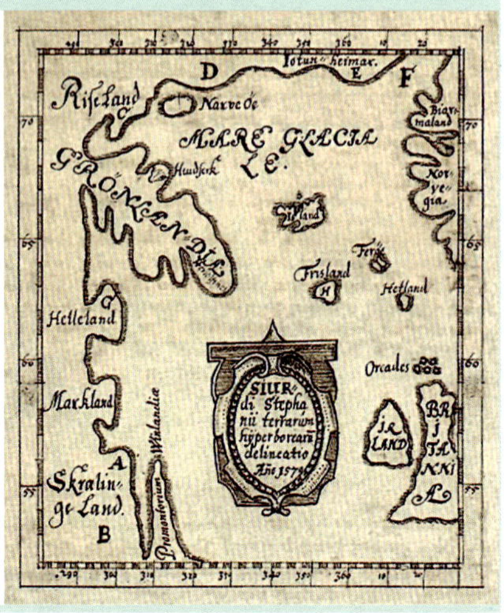

Norse map showing Vinland, 1690

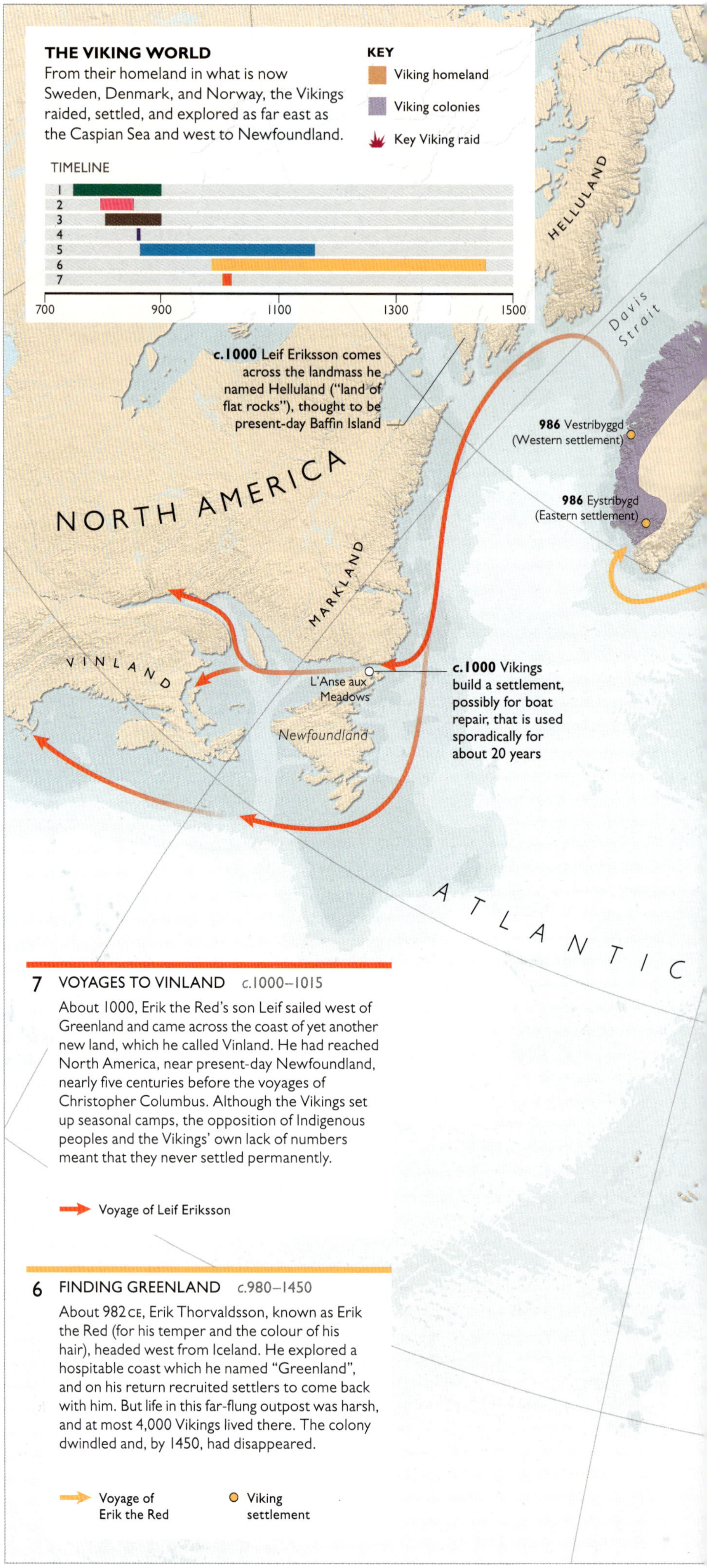

THE VIKING WORLD
From their homeland in what is now Sweden, Denmark, and Norway, the Vikings raided, settled, and explored as far east as the Caspian Sea and west to Newfoundland.

KEY
- Viking homeland
- Viking colonies
- Key Viking raid

TIMELINE: 1–7 (700, 900, 1100, 1300, 1500)

HELLULAND

Davis Strait

c.1000 Leif Eriksson comes across the landmass he named Helluland ("land of flat rocks"), thought to be present-day Baffin Island

986 Vestribyggd (Western settlement)

986 Eystribygd (Eastern settlement)

NORTH AMERICA

MARKLAND

VINLAND

L'Anse aux Meadows

Newfoundland

c.1000 Vikings build a settlement, possibly for boat repair, that is used sporadically for about 20 years

ATLANTIC

7 VOYAGES TO VINLAND c.1000–1015

About 1000, Erik the Red's son Leif sailed west of Greenland and came across the coast of yet another new land, which he called Vinland. He had reached North America, near present-day Newfoundland, nearly five centuries before the voyages of Christopher Columbus. Although the Vikings set up seasonal camps, the opposition of Indigenous peoples and the Vikings' own lack of numbers meant that they never settled permanently.

Voyage of Leif Eriksson

6 FINDING GREENLAND c.980–1450

About 982 CE, Erik Thorvaldsson, known as Erik the Red (for his temper and the colour of his hair), headed west from Iceland. He explored a hospitable coast which he named "Greenland", and on his return recruited settlers to come back with him. But life in this far-flung outpost was harsh, and at most 4,000 Vikings lived there. The colony dwindled and, by 1450, had disappeared.

Voyage of Erik the Red

Viking settlement

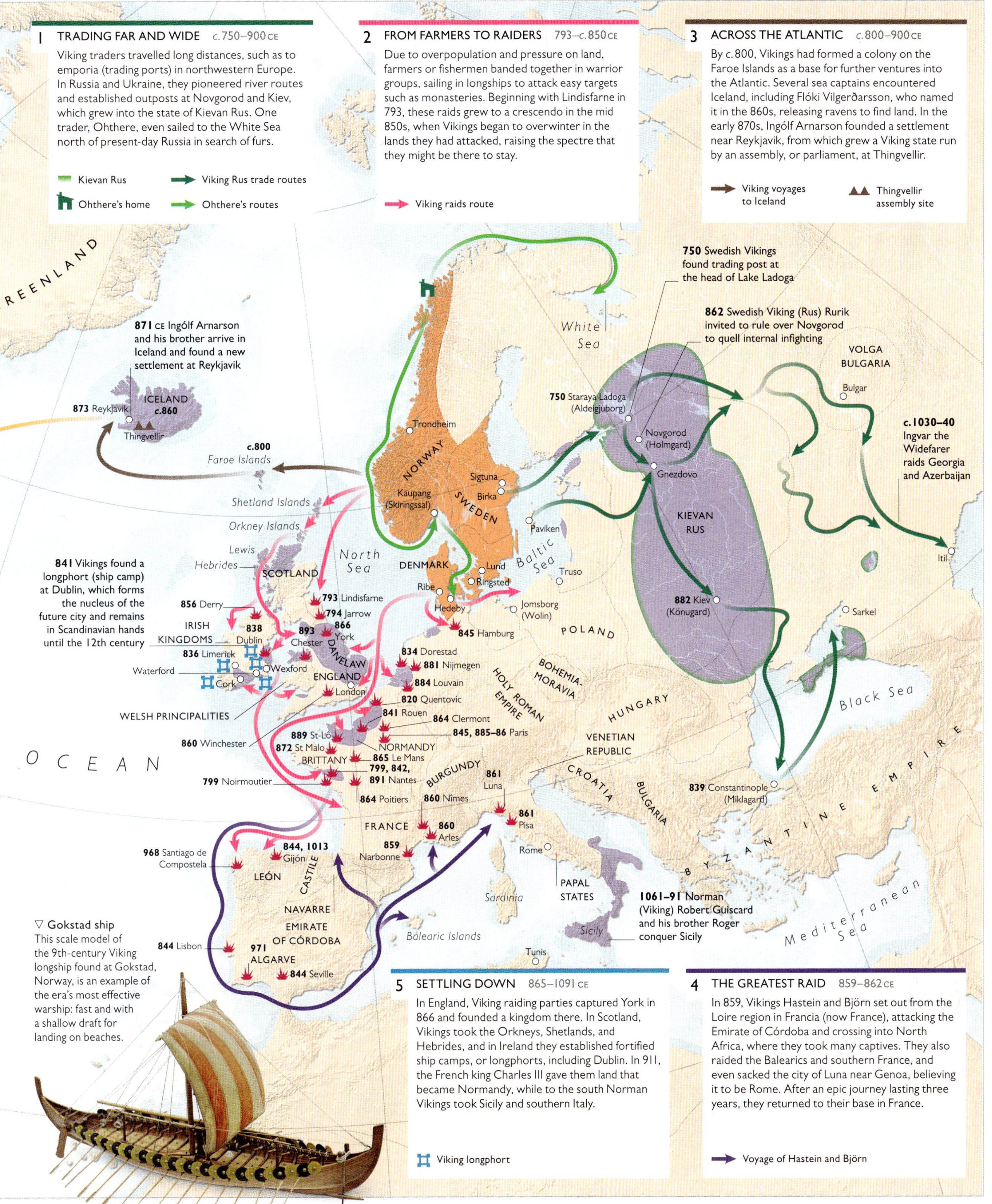

1 TRADING FAR AND WIDE c.750–900 CE

Viking traders travelled long distances, such as to emporia (trading ports) in northwestern Europe. In Russia and Ukraine, they pioneered river routes and established outposts at Novgorod and Kiev, which grew into the state of Kievan Rus. One trader, Ohthere, even sailed to the White Sea north of present-day Russia in search of furs.

Kievan Rus
Viking Rus trade routes
Ohthere's home
Ohthere's routes

2 FROM FARMERS TO RAIDERS 793–c.850 CE

Due to overpopulation and pressure on land, farmers or fishermen banded together in warrior groups, sailing in longships to attack easy targets such as monasteries. Beginning with Lindisfarne in 793, these raids grew to a crescendo in the mid 850s, when Vikings began to overwinter in the lands they had attacked, raising the spectre that they might be there to stay.

Viking raids route

3 ACROSS THE ATLANTIC c.800–900 CE

By c.800, Vikings had formed a colony on the Faroe Islands as a base for further ventures into the Atlantic. Several sea captains encountered Iceland, including Flóki Vilgerðarsson, who named it in the 860s, releasing ravens to find land. In the early 870s, Ingólf Arnarson founded a settlement near Reykjavik, from which grew a Viking state run by an assembly, or parliament, at Thingvellir.

Viking voyages to Iceland
Thingvellir assembly site

▽ **Gokstad ship**
This scale model of the 9th-century Viking longship found at Gokstad, Norway, is an example of the era's most effective warship: fast and with a shallow draft for landing on beaches.

5 SETTLING DOWN 865–1091 CE

In England, Viking raiding parties captured York in 866 and founded a kingdom there. In Scotland, Vikings took the Orkneys, Shetlands, and Hebrides, and in Ireland they established fortified ship camps, or longphorts, including Dublin. In 911, the French king Charles III gave them land that became Normandy, while to the south Norman Vikings took Sicily and southern Italy.

Viking longphort

4 THE GREATEST RAID 859–862 CE

In 859, Vikings Hastein and Björn set out from the Loire region in Francia (now France), attacking the Emirate of Córdoba and crossing into North Africa, where they took many captives. They also raided the Balearics and southern France, and even sacked the city of Luna near Genoa, believing it to be Rome. After an epic journey lasting three years, they returned to their base in France.

Voyage of Hastein and Björn

MEDIEVAL JEWISH TRAVELLERS

In the Middle Ages (c.500–1500 CE), Jewish communities existed across the Mediterranean region and beyond. Jewish travellers, including scholars, merchants, and students, bridged Islamic and Christian lands and brought unique perspectives as they visited religious sites and cities with large Jewish populations.

The Jewish communities in the cities of the former Roman and Sasanian Persian empires were linked by networks that had survived the fall of both states in the 5th to 7th centuries CE. Jewish merchants, or *radaniya*, travelled throughout the Christian kingdoms that emerged in the west and the Muslim caliphate in the east, and were recorded as far afield as China in the 9th century CE. The Jewish merchant Ishaq (or Isaac) bin Yahuda, from Oman, made his fortune trading with China, where he lived for 30 years from *c.*880 to 912 CE.

Other Jewish travellers acted as envoys: another man also named Isaac accompanied a mission from the Frankish ruler Charlemagne in western Germany to Baghdad in 797 CE, and returned with an elephant named Abul-abbas, which the Abbasid caliph Harun

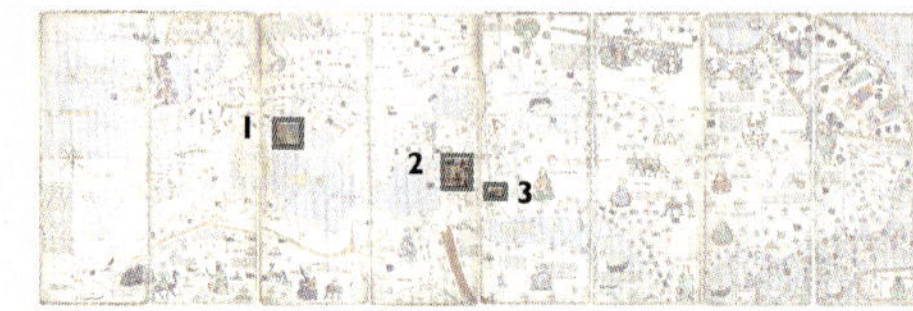

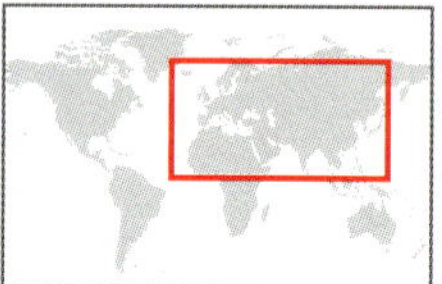

KEY

1 Rome – home to 200 Jews, according to 12th-century Benjamin of Tudela

2 Damascus – a city of 3,000 Jews, where, Benjamin writes, "no district richer in fruit can be seen in all the world"

3 Baghdad – Benjamin notes its 28 Jewish synagogues and the Caliph's palace

LOCATOR

> *"The two hundred Jews [in Rome] are very much respected and pay tribute to no one."*
>
> BENJAMIN OF TUDELA, *BOOK OF TRAVELS*, 12TH CENTURY

al-Rashid had sent as a gift. Ibrahim ibn Yaqub, a Jewish merchant from Tortosa in Andalusia, travelled in the 960s CE through France, Germany, Bohemia, and Italy, where he had an audience with Holy Roman Emperor Otto I, and wrote one of the earliest surviving descriptions of Poland.

Visits by Jewish travellers to Jerusalem were rare – Christian and Muslim rulers, who were antisemitic, discouraged them – but Jewish scholars travelled widely to exchange knowledge. Rabbi Petachiah of Regensburg (in present-day southeast Germany) made the journey from Bohemia (present-day Czechia) to Nineveh in Persia around 1175, risking a rule that if a Jewish person died there, the doctors who treated them took half of their property.

Jewish travellers faced further discrimination: they had to gain permission from their community to leave and pay a fine to go beyond their home country. Yet the tradition of Jewish travel thrived when, in 1585–1601, Pedro Teixeira from Lisbon became the first Jewish explorer to sail around the globe and the first European to travel up the Amazon.

BENJAMIN OF TUDELA 12TH CENTURY

One of the greatest Jewish travellers, Benjamin set off from his native Spain in the early 1160s and embarked on a 12-year journey through France, Italy, and the Balkans to Egypt and North Africa, before returning via Sicily. He visited the great cities of the Muslim world – including Baghdad, Damascus, and Cairo – and wrote about his travels, with an emphasis on Jewish populations he encountered. At Constantinople (present-day Istanbul), he notes that none among the 2,500-strong Jewish community were permitted to ride on horseback, except the imperial physician.

A 19th-century imagining of Benjamin in the Sahara

▽ **Medieval world map**
One of the finest late medieval maps, the Catalan Atlas was produced in 1375 by the cartographers Abraham Cresques and his son Jehuda, both members of the Jewish community in Palma, Majorca.

"Ko te pae tawhiti, whāia kia tata. Ko te pae tata, whakamaua kia tina."
(Seek out distant horizons. And cherish those you attain.)

MĀORI PROVERB

MĀORI NAVIGATORS

In the 13th century, during the last phase of the great Polynesian expansion, the ancestors of the Māori explored the oceans in canoes, discovering the islands of present-day New Zealand, which they called Aotearoa.

Tav: 66

Rosaspina inc:

a roccia

△ **A Māori war canoe**
This 1840s engraving by a European artist shows a *waka taua* or war canoe approaching a *pā*, set on top of a coastal rock formation. These long vessels were designed specifically for combat and often carved from a single hollowed-out log – the trunk of a native Tōtara tree.

According to Māori oral history, the great Polynesian navigator Kupe was the first to discover the islands of Aotearoa (present-day New Zealand). The legend goes that, after leaving a land called Hawaiki and pursuing a giant octopus in the Pacific Ocean, he stumbled across the shores of the East Coast. His wife, Kuramārōtini, named it Aotearoa ("the land of the long white cloud"), and they were soon followed by seven great *waka*, double-hulled canoes, whose leaders each became the founder of an *iwi*, or tribe, from whom all later Māori trace their descent.

△ **First explorer**
This carved *pou* or post at Meretoto marks Kupe's arrival on the East Coast of North Island.

Archaeology confirms the general outlines of these stories, with early settlement beginning on Te Ika-a-Māui (North Island) and reaching, a decade later, Te Waipounamu (South Island). The early settlers stayed on the coast, hunting giant moa (a large flightless bird), which soon became extinct, and gathering shellfish. They also grew the sweet potato, taro, and yam they had brought with them.

From around 1500, the Māori began to build increasing numbers of *pā* (fo≠rtified settlements), which operated as strongholds during the bouts of warfare that erupted. Around the same time, there was a final migration from Te Waipounamu east to Rēkohu (Chatham Islands), where the settlers developed a unique Moriori culture.

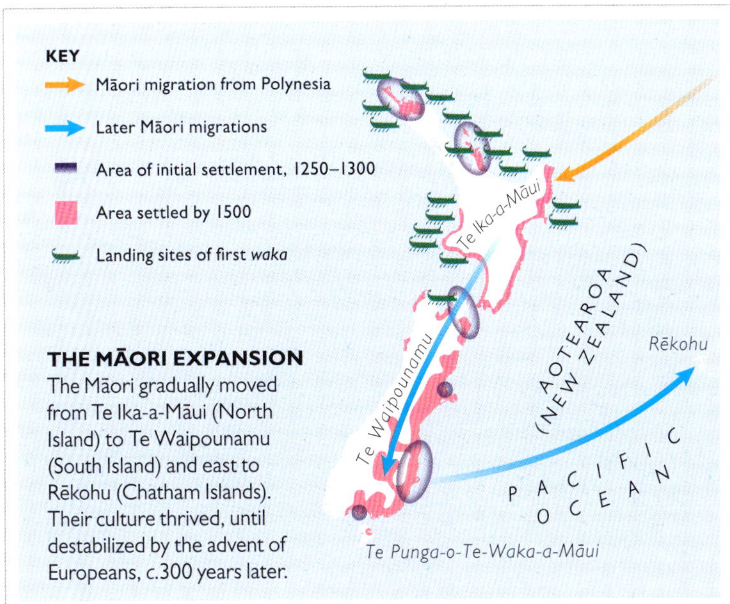

THE MĀORI EXPANSION
The Māori gradually moved from Te Ika-a-Māui (North Island) to Te Waipounamu (South Island) and east to Rēkohu (Chatham Islands). Their culture thrived, until destabilized by the advent of Europeans, c.300 years later.

ANCESTRAL PUEBLOANS

From the 7th century, the Ancestral Puebloans of the arid American southwest began to create large, multi-storeyed settlements called pueblos, sometimes built into cliff faces. However, shifts in climate led to repeated migrations and eventual abandonment of dwellings.

△ **Decorated pitcher**
Ancestral Puebloans made high-quality pottery with raw clay dug from the ground. Traditionally, most potters were women.

Formerly nomadic hunter-gatherers, the Ancestral Puebloans turned to maize farming in the American southwest from around 300 BCE. To support their crops in the near-desert conditions, they developed irrigation systems. Around 750 CE, at Chaco Canyon, New Mexico, they began to build pueblos, groups of masonry dwellings clustered around subterranean chambers called "kivas", used for rituals. In addition to their architecture, the Ancestral Puebloans were skilled basket-makers and potters.

Around a century later, a cooling climate brought newcomers from Colorado, who built complexes known as "great houses" in Chaco Canyon, the most notable being Pueblo Bonito, a three-acre site with over 600 rooms. By the 11th century, the Chaco Canyon complex was home to more than 5,000 people, linked by roads to over 70 other settlements. Intense drought led to the abandonment of Chaco Canyon around 1200.

By now the main Ancestral Puebloan settlements were in the San Juan Basin region of Colorado, at Mesa Verde, where pueblos were built or enlarged. The Yellow Jacket settlement had over 180 kivas, while the Mesa Verde Cliff Palace in southwestern Colorado had 150 rooms.

From around 1250, further droughts struck, and by 1300 Mesa Verde was abandoned. Its inhabitants moved southwest into Arizona, merging with the ancestors of the Hopi and Zuñi peoples.

PAINTED PICTOGRAPHS

The Ancestral Puebloans used caves and sheltered canyon walls as seasonal homes before they settled as farmers in the southwest. They painted pictographs depicting animals, human figures, and other objects onto the walls. The pictographs at Horseshoe Canyon in Utah, shown here, are between 2,000 and 5,000 years old.

Ancient rock markings

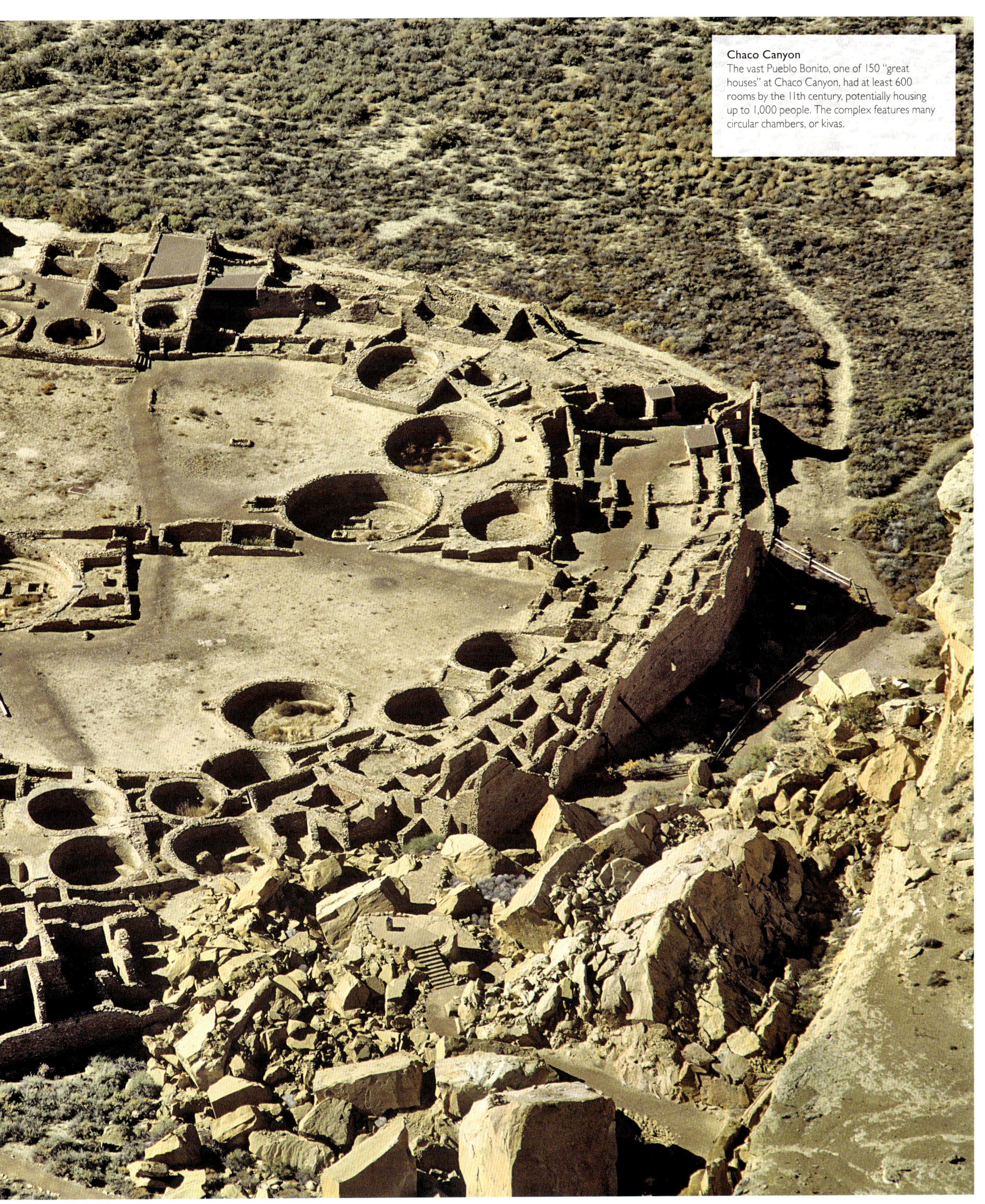

Chaco Canyon
The vast Pueblo Bonito, one of 150 "great houses" at Chaco Canyon, had at least 600 rooms by the 11th century, potentially housing up to 1,000 people. The complex features many circular chambers, or kivas.

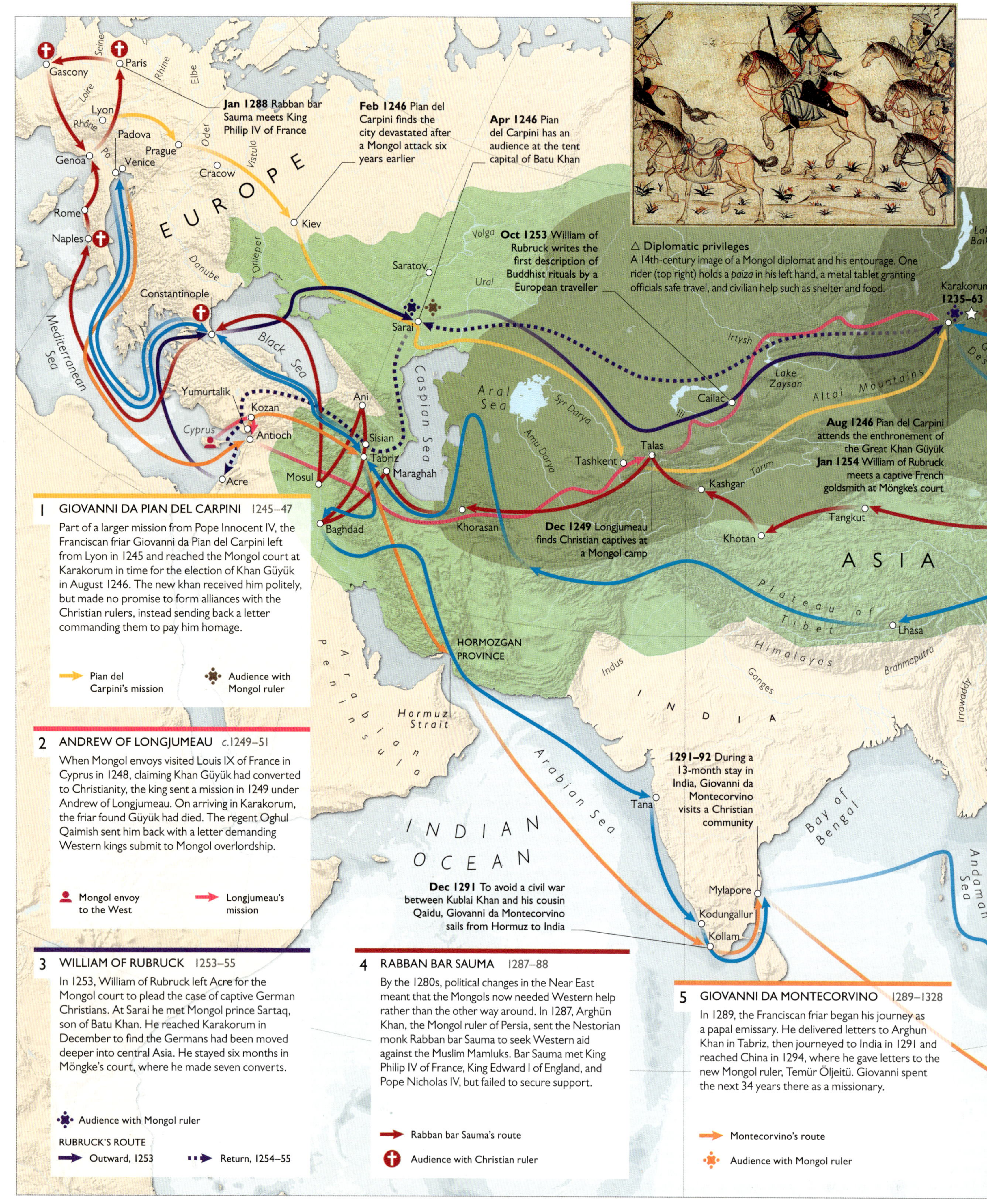
Jan 1288 Rabban bar Sauma meets King Philip IV of France
Feb 1246 Pian del Carpini finds the city devastated after a Mongol attack six years earlier
Apr 1246 Pian del Carpini has an audience at the tent capital of Batu Khan
Oct 1253 William of Rubruck writes the first description of Buddhist rituals by a European traveller
Aug 1246 Pian del Carpini attends the enthronement of the Great Khan Güyük
Jan 1254 William of Rubruck meets a captive French goldsmith at Möngke's court
Dec 1249 Longjumeau finds Christian captives at a Mongol camp
1291–92 During a 13-month stay in India, Giovanni da Montecorvino visits a Christian community
Dec 1291 To avoid a civil war between Kublai Khan and his cousin Qaidu, Giovanni da Montecorvino sails from Hormuz to India
EUROPE
ASIA
INDIAN OCEAN
INDIA
Gascony
Paris
Seine
Rhine
Elbe
Loire
Lyon
Rhône
Padova
Prague
Oder
Vistula
Genoa
Po
Venice
Cracow
Kiev
Rome
Naples
Danube
Dnieper
Volga
Saratov
Ural
Constantinople
Sarai
Black Sea
Caspian Sea
Mediterranean Sea
Yumurtalik
Kozan
Cyprus
Antioch
Acre
Ani
Sisian
Tabriz
Maraghah
Mosul
Baghdad
Aral Sea
Syr Darya
Amu Darya
Irtysh
Lake Zaysan
Cailac
Altai Mountains
Karakorum 1235–63
Lake Baikal
Tashkent
Talas
Kashgar
Tarim
Khorasan
Khotan
Tangkut
Plateau of Tibet
Lhasa
Himalayas
Brahmaputra
Irrawaddy
Indus
Ganges
HORMOZGAN PROVINCE
Hormuz Strait
Arabian Peninsula
Arabian Sea
Tana
Bay of Bengal
Andaman Sea
Mylapore
Kodungallur
Kollam

△ **Diplomatic privileges**
A 14th-century image of a Mongol diplomat and his entourage. One rider (top right) holds a *paiza* in his left hand, a metal tablet granting officials safe travel, and civilian help such as shelter and food.

1 GIOVANNI DA PIAN DEL CARPINI 1245–47

Part of a larger mission from Pope Innocent IV, the Franciscan friar Giovanni da Pian del Carpini left from Lyon in 1245 and reached the Mongol court at Karakorum in time for the election of Khan Güyük in August 1246. The new khan received him politely, but made no promise to form alliances with the Christian rulers, instead sending back a letter commanding them to pay him homage.

Pian del Carpini's mission
Audience with Mongol ruler

2 ANDREW OF LONGJUMEAU c.1249–51

When Mongol envoys visited Louis IX of France in Cyprus in 1248, claiming Khan Güyük had converted to Christianity, the king sent a mission in 1249 under Andrew of Longjumeau. On arriving in Karakorum, the friar found Güyük had died. The regent Oghul Qaimish sent him back with a letter demanding Western kings submit to Mongol overlordship.

Mongol envoy to the West
Longjumeau's mission

3 WILLIAM OF RUBRUCK 1253–55

In 1253, William of Rubruck left Acre for the Mongol court to plead the case of captive German Christians. At Sarai he met Mongol prince Sartaq, son of Batu Khan. He reached Karakorum in December to find the Germans had been moved deeper into central Asia. He stayed six months in Möngke's court, where he made seven converts.

Audience with Mongol ruler
RUBRUCK'S ROUTE
Outward, 1253
Return, 1254–55

4 RABBAN BAR SAUMA 1287–88

By the 1280s, political changes in the Near East meant that the Mongols now needed Western help rather than the other way around. In 1287, Arghūn Khan, the Mongol ruler of Persia, sent the Nestorian monk Rabban bar Sauma to seek Western aid against the Muslim Mamluks. Bar Sauma met King Philip IV of France, King Edward I of England, and Pope Nicholas IV, but failed to secure support.

Rabban bar Sauma's route
Audience with Christian ruler

5 GIOVANNI DA MONTECORVINO 1289–1328

In 1289, the Franciscan friar began his journey as a papal emissary. He delivered letters to Arghun Khan in Tabriz, then journeyed to India in 1291 and reached China in 1294, where he gave letters to the new Mongol ruler, Temür Öljeitü. Giovanni spent the next 34 years there as a missionary.

Montecorvino's route
Audience with Mongol ruler

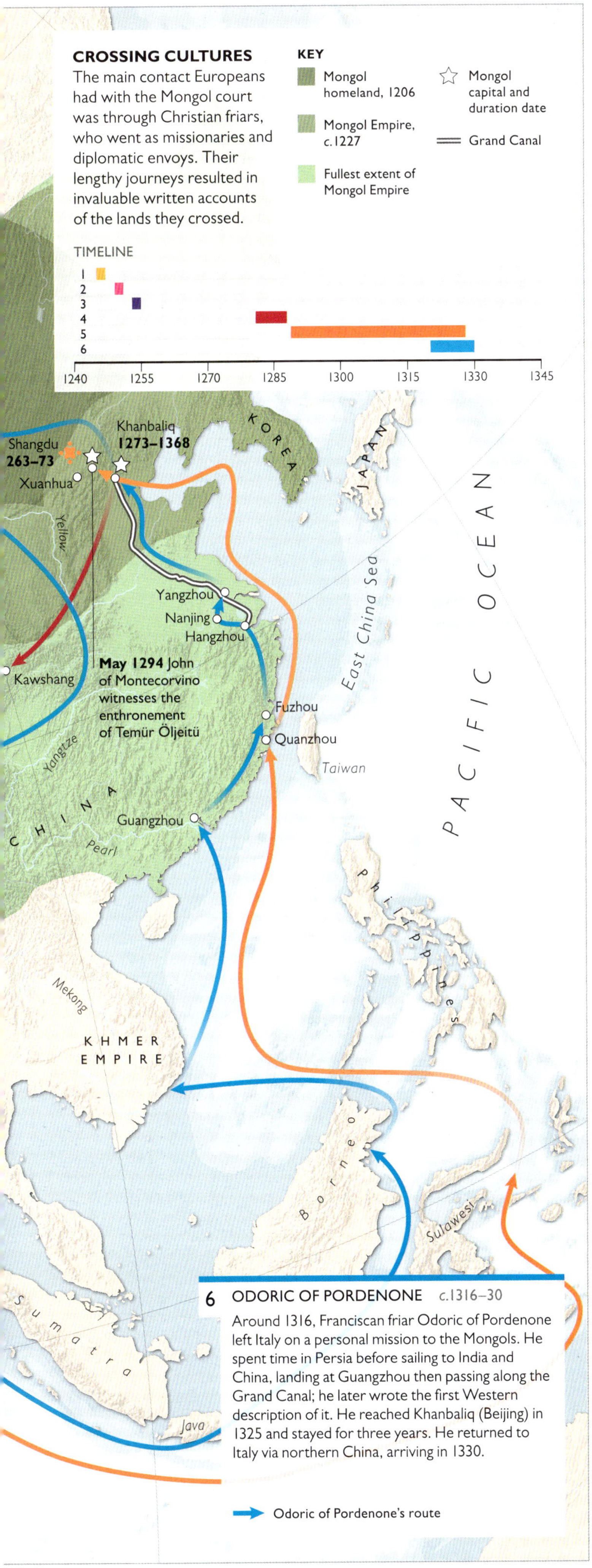

THE MONGOL EXPANSION

In the 13th century, the Mongols – nomadic people from the steppes of Central Asia – forged an empire that covered present-day China, Iran, Eastern Europe, and Southeast Asia. The era of stability that followed saw trade flourish and diplomatic envoys travel between western Europe and the Mongol court.

After Genghis Khan united the Mongol tribes in 1206, he embarked on a series of invasions that engulfed the Khwarazmian Empire of Central Asia and Persia (1221), Kievan Rus (1240), and China (1279). The Great Khan ruled all this from capitals at Karakorum (Harhorin) and later Khanbaliq (Beijing), and a summer capital at Shangdu, popularly known as Xanadu. The territory was bound together by the Mongol military, an elite, brutally effective mounted force; an efficient bureaucracy based on Chinese models; and a messenger system, the *Yam*, that allowed swift travel and communication throughout the vast Mongol lands.

European envoys to the Mongol court sought to make Christian converts and to gain allies to attack the Islamic Empire from the rear. A smaller number of diplomats from the Mongol Empire also went in the opposite direction, to explore the possibility of alliances with Western powers, or simply to make contact with states on the fringes of the Mongol world.

When Möngke Khan died in 1259 with no declared successor, infighting fractured the empire into smaller khanates. By the late 14th century, the Mongols had lost significant territory (present-day China, Iran, and Russia) and the flow of diplomats to and from their courts ceased.

> *"When I found myself among [the Mongols], it seemed... that I had been transported into another world."*
>
> WILLIAM OF RUBRUCK, *THE JOURNEY OF WILLIAM OF RUBRUCK TO THE EASTERN PARTS OF THE WORLD*, 1253–55

CROSSING CULTURES

The main contact Europeans had with the Mongol court was through Christian friars, who went as missionaries and diplomatic envoys. Their lengthy journeys resulted in invaluable written accounts of the lands they crossed.

KEY

- Mongol homeland, 1206
- Mongol Empire, c.1227
- Fullest extent of Mongol Empire
- ☆ Mongol capital and duration date
- Grand Canal

TIMELINE

1 2 3 4 5 6

1240 1255 1270 1285 1300 1315 1330 1345

Shangdu 1263–73
Khanbaliq 1273–1368
Xuanhua
KOREA
JAPAN
Yellow
Yangzhou
Nanjing
Hangzhou
East China Sea
PACIFIC OCEAN
Kawshang
May 1294 John of Montecorvino witnesses the enthronement of Temür Öljeitü
Fuzhou
Quanzhou
Taiwan
Yangtze
CHINA
Guangzhou
Pearl
Philippines
Mekong
KHMER EMPIRE
Borneo
Sulawesi
Sumatra
Java

6 ODORIC OF PORDENONE c.1316–30

Around 1316, Franciscan friar Odoric of Pordenone left Italy on a personal mission to the Mongols. He spent time in Persia before sailing to India and China, landing at Guangzhou then passing along the Grand Canal; he later wrote the first Western description of it. He reached Khanbaliq (Beijing) in 1325 and stayed for three years. He returned to Italy via northern China, arriving in 1330.

→ Odoric of Pordenone's route

ZHOU DAGUAN c.1266–1346

The only first-hand account of life in the Khmer Empire comes from the Chinese diplomat Zhou Daguan. The Mongol Yuan Emperor Temür Khan sent him on a mission to Southeast Asia (1295–96). He visited the kingdom of Champa in the coastal region of what is now Vietnam, and then sailed to Cambodia. Zhou lived in the Khmer capital of Angkor Thom for 11 months, eventually writing a detailed account of late 13th-century life under Khmer ruler Indravarman III, which included descriptions of the capital's magnificent temples.

A 12th-century bas relief of a market scene from a temple at Angkor Thom, Cambodia.

Arrival at Hormuz
This illustration of Polo's travels shows him landing at Hormuz, Persia (now Iran). He commented on the high quality of the goods found there, including gold and spices, but also on the searing winds, hot enough to kill armies.

MARCO POLO'S JOURNEYS

Venetian merchant Marco Polo stayed for 17 years at the court of the Mongol ruler Kublai Khan, providing Europeans with compelling first-hand accounts of China and the Silk Road lands through which he travelled.

In 1271, Marco Polo left Venice with his father and uncle. They passed through Baghdad, then through Central Asia and Samarkand, before reaching the capital of Mongol leader Kublai Khan at Shangdu (Xanadu). Kublai Khan had requested they bring missionaries for his court, but when they arrived in 1275, the Polos had brought only gifts; the missionaries had turned back.

Marco Polo impressed Kublai Khan and was invited to stay at his court; the Polos spent the next 17 years learning Mongol customs and travelling widely, with Marco acting as a diplomatic envoy. When they left in 1291, the Polos sailed from south China to Sumatra and Ceylon (present-day Sri Lanka), accompanying Mongol princess Kököchin to the Il-Khanate (now Iran), before returning to Venice via Constantinople.

In 1298, Marco was taken captive in a naval battle against the Genoese. He recounted his years of travels to his cellmate Rustichello, who collected them in the *Description of the World* (or *The Travels of Marco Polo*), which became a seminal late-medieval travel narrative.

△ **Mongol ruler**
A grandson of Genghis Khan, Kublai Khan conquered China, establishing the Yuan dynasty. He had many foreign advisors at his court, including Marco Polo.

MARCO POLO'S ROUTE
The Polos travelled along established Silk Road routes made more secure through their common control by Mongol rulers: the Il-Khan, the Chagatai, and Kublai Khan himself. Marco's exact routes in China are less clear, but he visited several major cities and made a side trip to Myanmar (Burma).

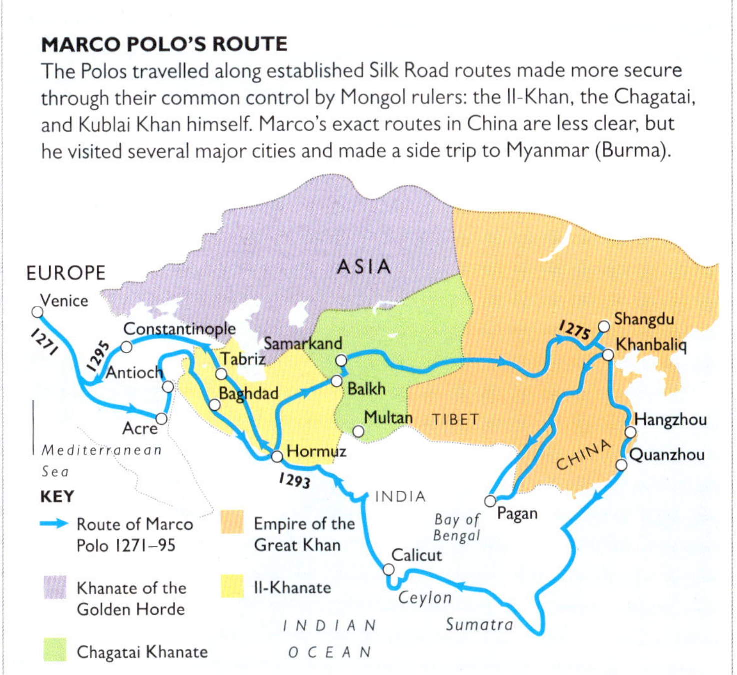

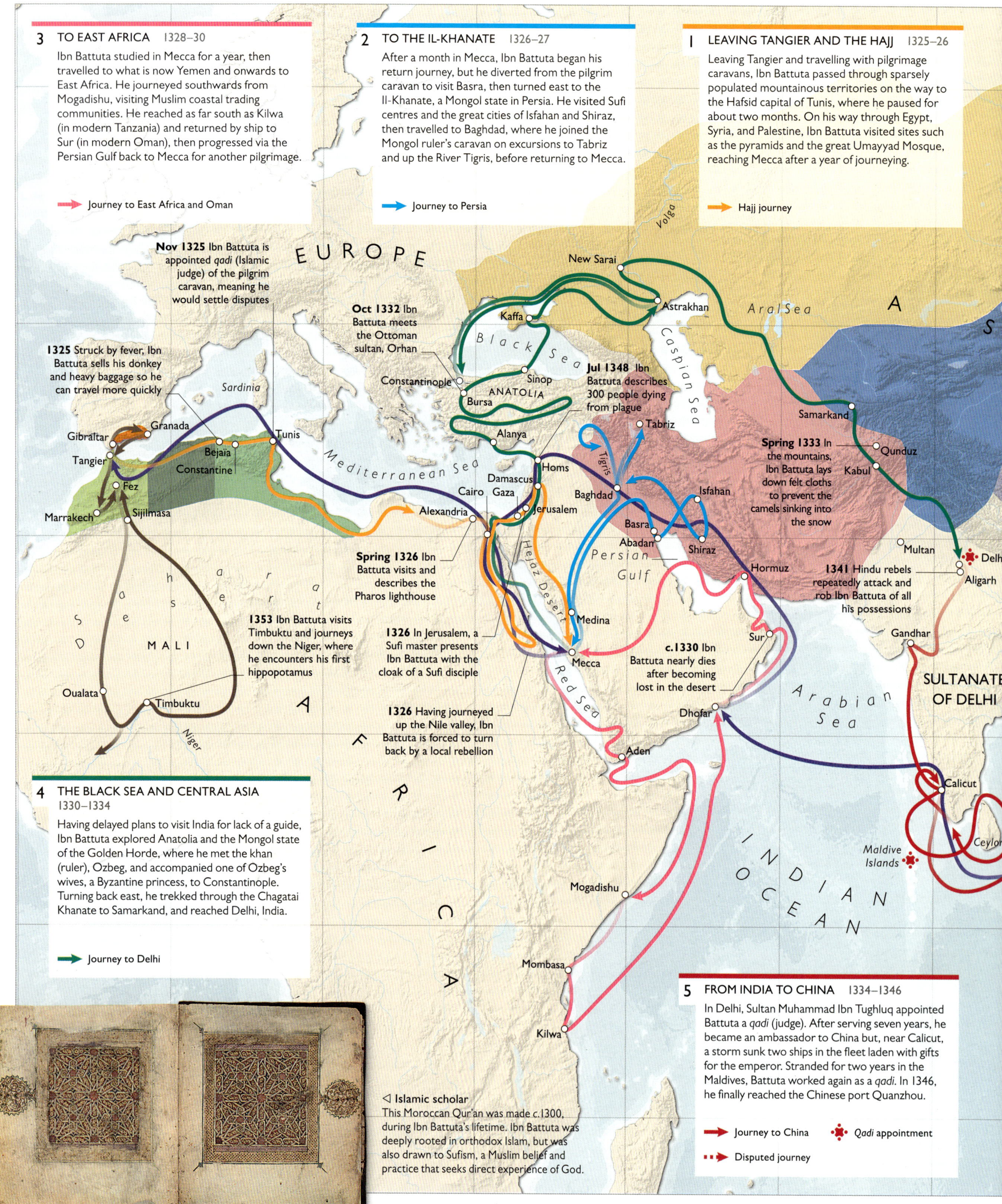

1 LEAVING TANGIER AND THE HAJJ 1325–26

Leaving Tangier and travelling with pilgrimage caravans, Ibn Battuta passed through sparsely populated mountainous territories on the way to the Hafsid capital of Tunis, where he paused for about two months. On his way through Egypt, Syria, and Palestine, Ibn Battuta visited sites such as the pyramids and the great Umayyad Mosque, reaching Mecca after a year of journeying.

2 TO THE IL-KHANATE 1326–27

After a month in Mecca, Ibn Battuta began his return journey, but he diverted from the pilgrim caravan to visit Basra, then turned east to the Il-Khanate, a Mongol state in Persia. He visited Sufi centres and the great cities of Isfahan and Shiraz, then travelled to Baghdad, where he joined the Mongol ruler's caravan on excursions to Tabriz and up the River Tigris, before returning to Mecca.

3 TO EAST AFRICA 1328–30

Ibn Battuta studied in Mecca for a year, then travelled to what is now Yemen and onwards to East Africa. He journeyed southwards from Mogadishu, visiting Muslim coastal trading communities. He reached as far south as Kilwa (in modern Tanzania) and returned by ship to Sur (in modern Oman), then progressed via the Persian Gulf back to Mecca for another pilgrimage.

4 THE BLACK SEA AND CENTRAL ASIA 1330–1334

Having delayed plans to visit India for lack of a guide, Ibn Battuta explored Anatolia and the Mongol state of the Golden Horde, where he met the khan (ruler), Ozbeg, and accompanied one of Ozbeg's wives, a Byzantine princess, to Constantinople. Turning back east, he trekked through the Chagatai Khanate to Samarkand, and reached Delhi, India.

5 FROM INDIA TO CHINA 1334–1346

In Delhi, Sultan Muhammad Ibn Tughluq appointed Battuta a *qadi* (judge). After serving seven years, he became an ambassador to China but, near Calicut, a storm sunk two ships in the fleet laden with gifts for the emperor. Stranded for two years in the Maldives, Battuta worked again as a *qadi*. In 1346, he finally reached the Chinese port Quanzhou.

◁ **Islamic scholar**
This Moroccan Qur'an was made c.1300, during Ibn Battuta's lifetime. Ibn Battuta was deeply rooted in orthodox Islam, but was also drawn to Sufism, a Muslim belief and practice that seeks direct experience of God.

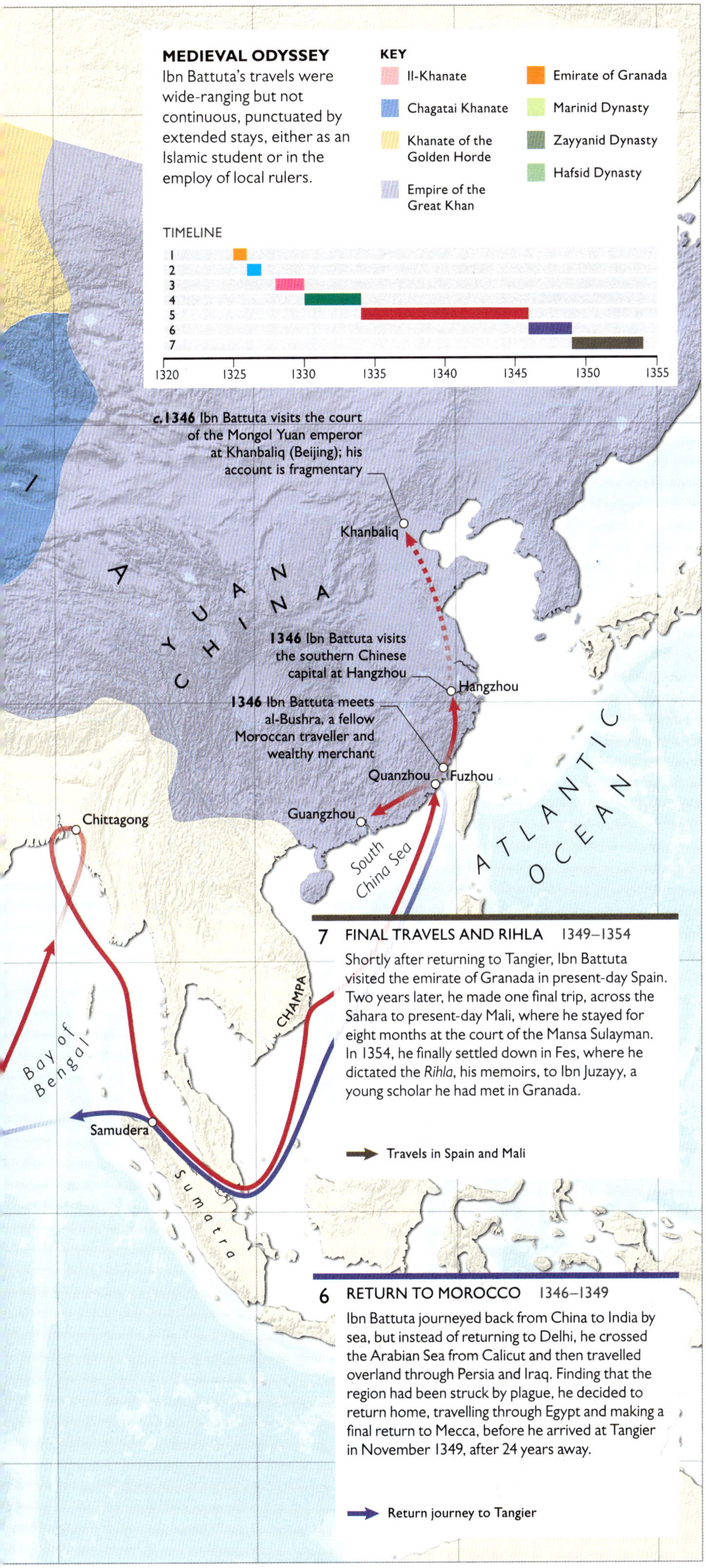

MEDIEVAL ODYSSEY
Ibn Battuta's travels were wide-ranging but not continuous, punctuated by extended stays, either as an Islamic student or in the employ of local rulers.

KEY
- Il-Khanate
- Chagatai Khanate
- Khanate of the Golden Horde
- Empire of the Great Khan
- Emirate of Granada
- Marinid Dynasty
- Zayyanid Dynasty
- Hafsid Dynasty

TIMELINE
1
2
3
4
5
6
7
1320 1325 1330 1335 1340 1345 1350 1355

c.1346 Ibn Battuta visits the court of the Mongol Yuan emperor at Khanbaliq (Beijing); his account is fragmentary

1346 Ibn Battuta visits the southern Chinese capital at Hangzhou

1346 Ibn Battuta meets al-Bushra, a fellow Moroccan traveller and wealthy merchant

Khanbaliq
Hangzhou
Quanzhou
Fuzhou
Guangzhou
Chittagong
Samudera
YUAN CHINA
CHAMPA
South China Sea
ATLANTIC OCEAN
Bay of Bengal
Sumatra

7 FINAL TRAVELS AND RIHLA 1349–1354

Shortly after returning to Tangier, Ibn Battuta visited the emirate of Granada in present-day Spain. Two years later, he made one final trip, across the Sahara to present-day Mali, where he stayed for eight months at the court of the Mansa Sulayman. In 1354, he finally settled down in Fes, where he dictated the *Rihla*, his memoirs, to Ibn Juzayy, a young scholar he had met in Granada.

Travels in Spain and Mali

6 RETURN TO MOROCCO 1346–1349

Ibn Battuta journeyed back from China to India by sea, but instead of returning to Delhi, he crossed the Arabian Sea from Calicut and then travelled overland through Persia and Iraq. Finding that the region had been struck by plague, he decided to return home, travelling through Egypt and making a final return to Mecca, before he arrived at Tangier in November 1349, after 24 years away.

Return journey to Tangier

THE TRAVELS OF IBN BATTUTA

The remarkable journeys of the Moroccan traveller Ibn Battuta began in 1325 with a pilgrimage to Mecca, and then took him to almost every part of the Islamic world. By the time he returned home 29 years later, he had covered around 110,000 km (70,000 miles).

Born in 1304 to a prominent family of judges in the Moroccan city of Tangier, Abu Abdullah Muhammad ibn Battuta was only 21 years old when he set out to perform the Hajj (Islamic pilgrimage) to Mecca. His journey through North Africa mirrored the life he would lead for the next three decades: meeting emirs, sultans, and city governors, receiving gifts customarily offered to pilgrims, being attacked by bandits, and entering into short-lived marriages. Everywhere he went, Ibn Battuta took notes, which would form the basis of his book, *Rihla* (*Travels*), one of the most precious accounts of the Islamic world in the medieval period.

Having visited sacred sites in Syria and Palestine and performed the Hajj, Ibn Battuta chose not to return home. Instead, he embarked upon two decades of travels that took him to East Africa, Anatolia (present-day Türkiye), the Byzantine capital of Constantinople, the Mongol-ruled Il-Khanate and Chagatai Khanate, and, in 1334, the sultanate of Delhi, whose ruler he served for several years as a judge. Sent on an embassy to China in 1342, Ibn Battuta became marooned in the Maldives, and finally resumed his journey to China in 1345, possibly reaching as far north as the Mongol court in Beijing. He then began a slow journey back to Tangier, passing through Syria just as it was being struck by the Black Death in July 1348. The inveterate traveller made further trips north to Muslim-ruled al-Andalus and south across the Sahara to the Empire of Mali. His account is a vital source about that region's history.

CARAVANSERAIS

The trade and pilgrimage routes of the Islamic world – most notably the Silk Routes between China and the Mediterranean – were dotted with caravanserais, inns where travellers could rest overnight. Called *hans* or *funduqs* in some regions, caravanserais were often situated just outside towns and usually consisted of rectangular buildings with an open inner courtyard, where horses, camels, or donkeys could be kept.

A thief steals from sleeping guests at a caravanserai, 13th-century manuscript

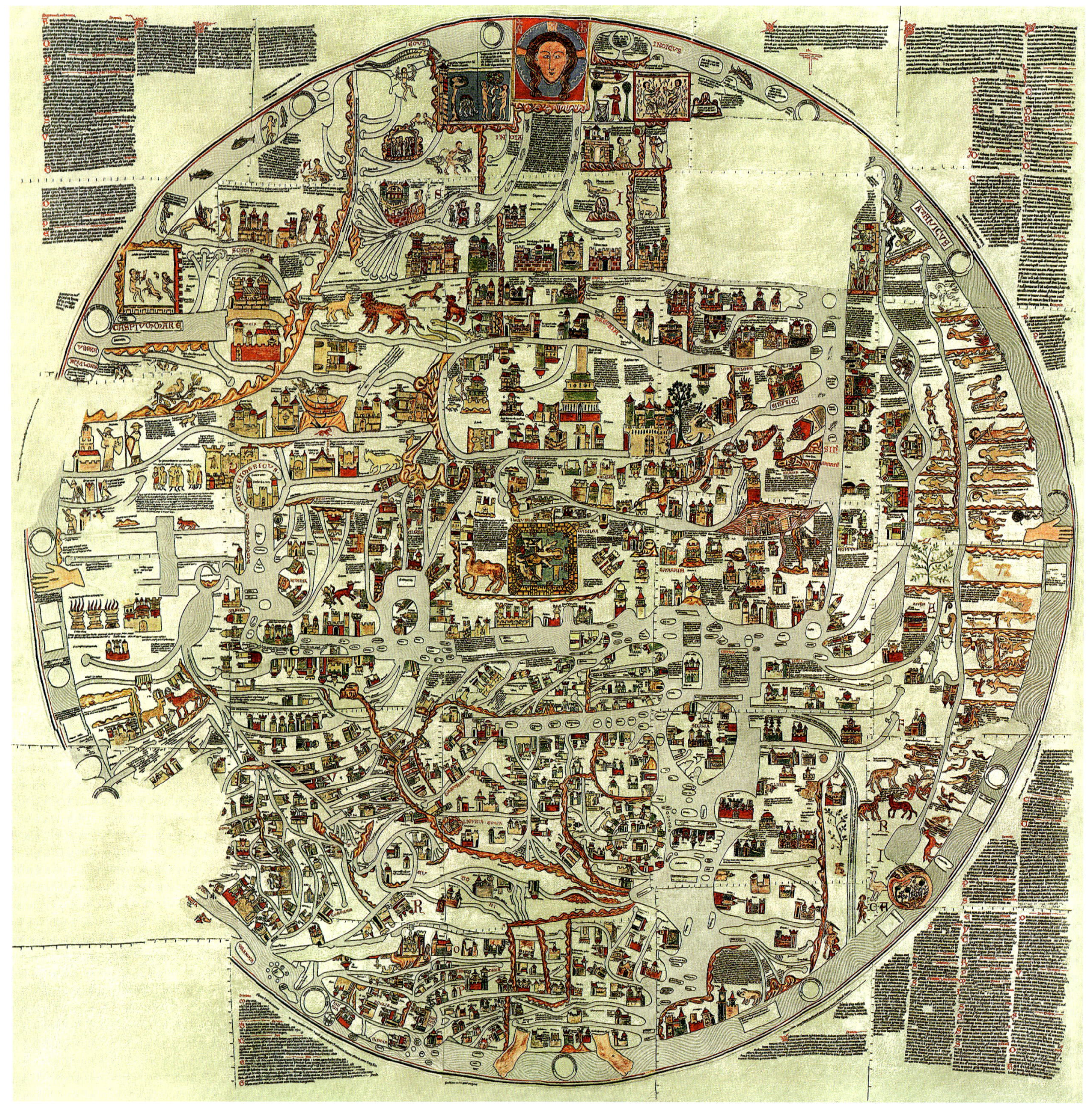

"The pelican is an Egyptian bird which does, if it is true, kill his young ones and mourn them for three days, and then revives them with his own blood."

ANNOTATION FROM THE EBSTORF MAP, 13TH CENTURY

MEDIEVAL MAPS

The cartographic tradition that emerged in the Middle Ages created elaborate maps to tell religious stories. These focused on a sacred geography that laid out a path to salvation, rather than showing navigable routes around the physical world.

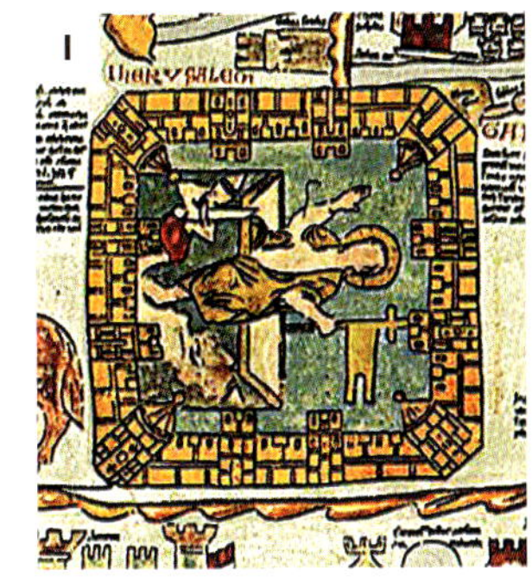

◁ **The Ebstorf map**
The largest known *mappa mundi* at 12 sq m (137 sq ft), the Ebstorf map was drawn in a German Benedictine monastery in the 13th century. It survives only in a facsimile.

▷ **Jerusalem**
As the most important location in Christian belief, Jerusalem was placed in the centre of the map, with Jesus Christ shown emerging from his tomb after the Resurrection.

◁ **Tower of Babel**
The soaring tower was said to have been built in an attempt to reach heaven, in punishment for which God cursed humanity by making it speak hundreds of mutually unintelligible languages.

▽ **Alexander the Great**
The story of the 4th-century BCE Macedonian king Alexander was a common classical reference on *mappae mundi*. Here he is shown consulting the Oracle of Jupiter Ammon in Egypt.

△ **Amazon warriors**
In Asia, not far from the Caspian Sea, two Amazons stand guard outside a fortified town. The mythical female warriors are shown wearing medieval-style helmets.

Ancient geographers, such as Ptolemy of Alexandria in the 2nd century CE, made maps that represented the world as they knew it. Some of this knowledge survived the fall of the Roman Empire in the 5th century, but the maps produced in medieval Europe were very different. While medieval mapmakers drew the three known continents of Asia, Africa, and Europe, they were more interested in places linked with Christ's life. These world maps or *mappae mundi* were centred on Jerusalem. The simplest, called "T–O" maps, show the continents separated by the Mediterranean, the Nile, and the Don, formerly called the Tanais (forming the "T"), encircled by ocean ("O"). Lavish examples, such as the Ebstorf map (left), depict details from the Bible, mythology, and classical tradition.

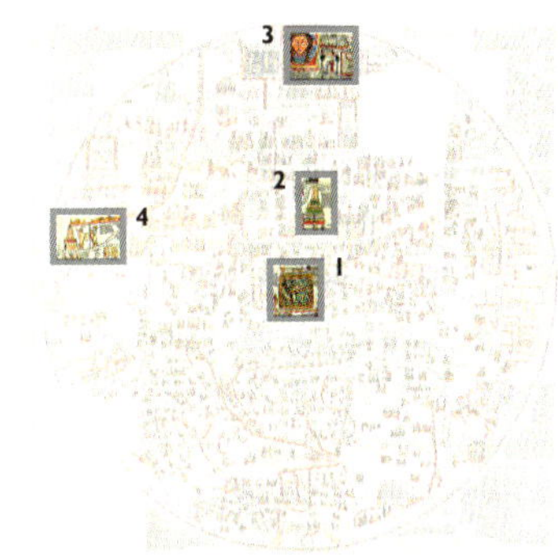

By the 13th century, however, new types of maps appeared, as travellers such as Marco Polo brought back first-hand accounts of East Asia (see pp.64–65), and navigators sailed further into the Mediterranean and beyond (see pp.72–73). These portolan charts, showing accurate coastlines, ports, and practical navigation routes, represented a new desire to explore the world on a physical, rather than spiritual, level.

MONSTERS ON LAND AND SEA

Medieval mapmakers populated regions for which they had little or no concrete information with monsters and fantastical creatures, sometimes illustrating popular myths or legends. For example, sea monsters were a common motif, as shown in this 16th-century map by Abraham Ortelius. The phoenix, a bird that rises from the burnt ashes of its predecessor, was also often portrayed, as well as the half-lion, half-eagle griffon; monopods – humanoid creatures with a single huge foot they used for shade; and the Blemmyes, headless beings with a single gigantic eye in their stomachs.

THE VOYAGES OF ZHENG HE

For nearly three decades in the early 15th century, Admiral Zheng He commanded a fleet of ships, sent by the Ming Chinese emperor, to explore the coasts of the Indian Ocean. On these voyages, he exacted tribute, brought foreign envoys back to the imperial court, intervened in military and political matters, and became the first agent of the Chinese government to visit Africa.

The Yongle Emperor, the third Ming ruler (1402–24), had great ambitions for wielding China's power on the world stage. Focused on proving Chinese supremacy around the Indian Ocean, the emperor commissioned maritime missions, the most famous of which were led by Zheng He, an imperial eunuch.

Zheng He commanded a huge fleet of 317 vessels, including 62 vast *baochuan* or "treasure ships", some more than 91 m (300 ft) long. In 1405, he set out on his first voyage with 20,000–30,000 men, sailing as far as Calicut, India. The next voyage (1408–09) followed a similar route, the third (1409–11) sailed to Silam (Sri Lanka) and the fourth (1413–15) sailed to the Persian port of Hormuz and brought envoys from 18 nations back to China. The fifth and furthest voyage (1417–19) returned the envoys and reached Malindi on the East African coast, while the sixth (1421–22) and seventh (1431–33) visited Mogadishu. Zheng He died in 1433, towards the end or soon after his final voyage.

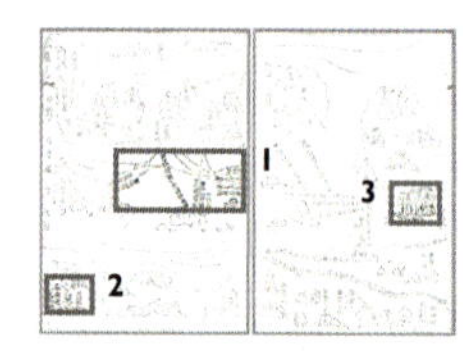

LOCATOR

KEY

1 Dashed lines and text show the routes taken by the fleet's ships.

2 Labels give the stellar altitudes of locations – here at Hafun, Somalia, Polaris is four fingers above the horizon.

3 A large Buddhist temple is depicted at Dondra Head, Silam.

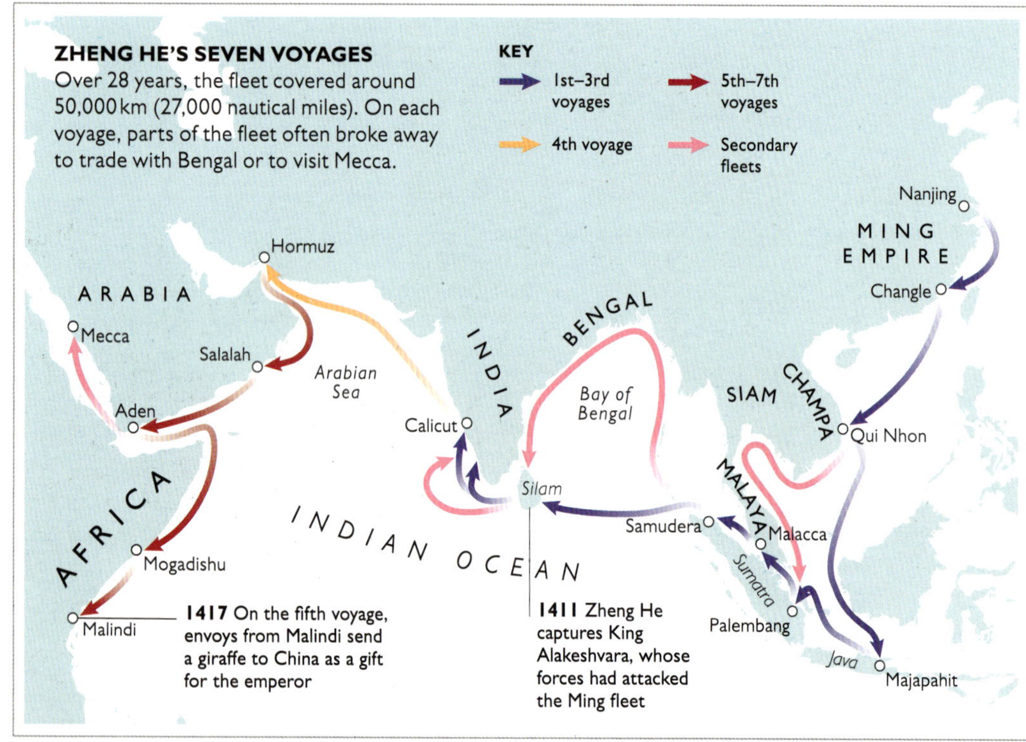
ZHENG HE'S SEVEN VOYAGES
Over 28 years, the fleet covered around 50,000 km (27,000 nautical miles). On each voyage, parts of the fleet often broke away to trade with Bengal or to visit Mecca.
KEY
1st–3rd voyages
5th–7th voyages
4th voyage
Secondary fleets
Nanjing
MING EMPIRE
Changle
Hormuz
ARABIA
Mecca
Salalah
Arabian Sea
Aden
Calicut
INDIA
BENGAL
Bay of Bengal
SIAM
CHAMPA
Qui Nhon
MALAYA
Malacca
Silam
Samudera
Sumatra
Palembang
Java
Majapahit
AFRICA
Mogadishu
Malindi
INDIAN OCEAN
1417 On the fifth voyage, envoys from Malindi send a giraffe to China as a gift for the emperor
1411 Zheng He captures King Alakeshvara, whose forces had attacked the Ming fleet

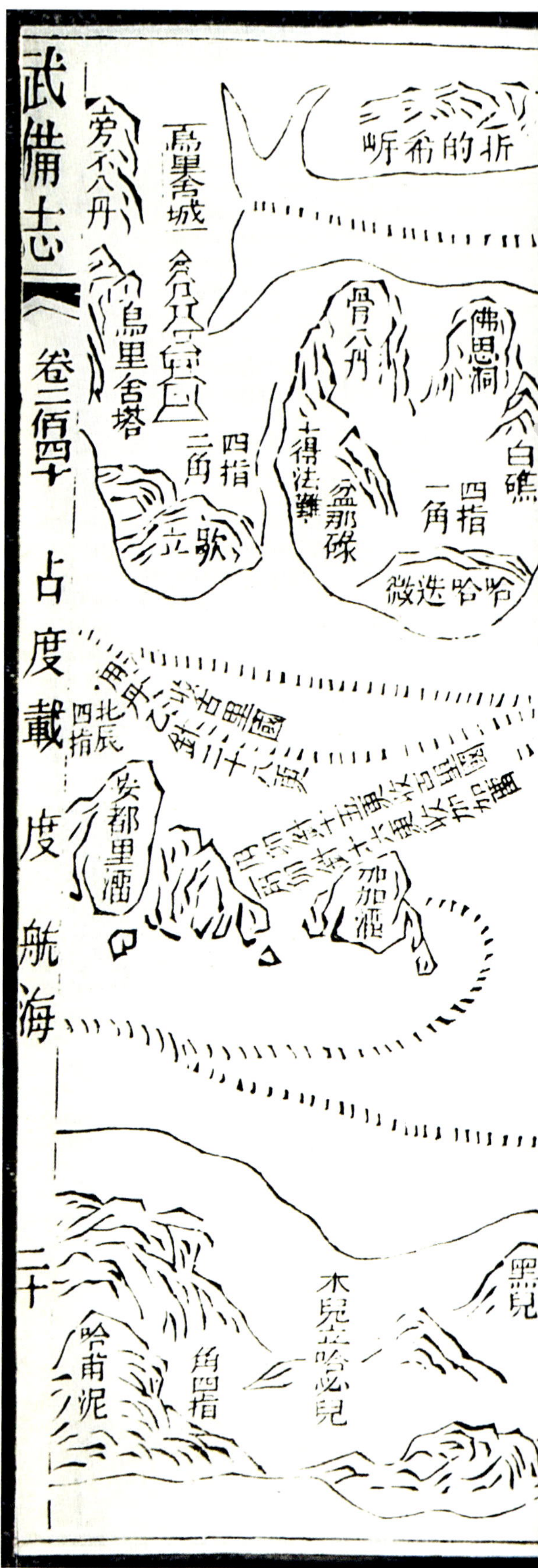

△ **Schematic mapping**
The Mao Kun map, probably created during Zheng He's sixth voyage (1421–22), is a schematic representation of the sea route. The orientation of the map changes in each section. Here the Indian shoreline is at the top; the African shoreline is at the bottom.

"... since the front part is tall and the hind part low, men cannot ride it..."

DESCRIPTION OF THE AFRICAN QILIN/GIRAFFE, MA HUN'S *YING-YAI SHENG-LAN*, 1433

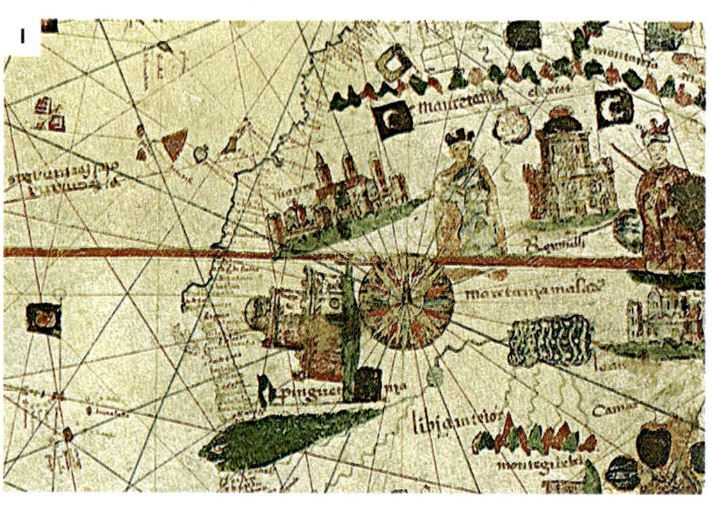

◁ **Cape Bojador**
Strong winds, violent storms, and intense heat made this stretch of the West African coast "the point of no return" for early European navigators, until Gil Eanes sailed beyond it in 1434.

△ **Elmina Castle, in present-day Ghana**
In 1482, the Portuguese built a fortress – *Castelo de São Jorge da Mina* – to protect their trading interests in the region. Soon after, the port began to operate as a centre in the trafficking of enslaved people.

EARLY PORTUGUESE EXPLORATION

In the 15th century, Portugal was a small European kingdom with a long Atlantic coastline and a strong seafaring tradition. Gradually, Portuguese navigators began to embark on long-distance sea voyages, discovering routes to Asia and West Africa, and launching a transformative new era of maritime exploration.

From the 13th century, Portuguese fishing vessels had undergone a series of design adaptations, until a new kind of ship had emerged that, by the 15th century, was known as a caravel. A light vessel, with square sails for sailing into strong winds and lateen sails for agility in shallow, coastal waters, these ships had the potential to transform maritime navigation, and the Portuguese were quick to exploit it.

In 1415, King John I of Portugal, together with his sons, attacked the North African port of Ceuta, a key staging post in the trans-Saharan gold trade. Motivated by a desire to bypass the Muslim states that controlled the trade in North Africa and access the sources of gold directly, the search for sea routes began.

The first voyages took place in the late 1420s, under the patronage of Prince Henry (see box). In 1434, Gil Eanes rounded Cape Bojador in West Africa, and subsequent expeditions nudged further, establishing a fort at Arguin (off Mauritania) in 1443. The pace of exploration then quickened: Alvise Ca' da Mosto sailed to Cape Verde in 1456; Fernão Gomes explored present-day Ghana in 1471, and in 1482 Diogo Cão entered the Congo River and reached modern Angola, setting up *padrãoes* (stone pillars) to mark his route.

Finally, in 1488, Bartolomeu Dias rounded the tip of Africa and, following the coast northeast, obtained proof that a sea route to India was a practical possibility, and that the West could directly access the resources of the East.

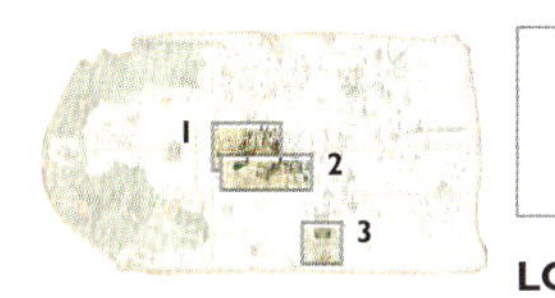

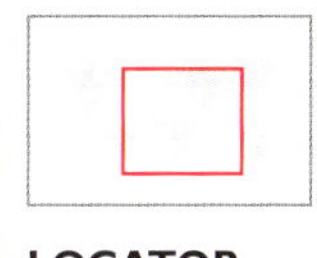

LOCATOR

3

△ World map of Juan de la Cosa, 1500
A navigator on at least two of Columbus's Atlantic voyages (see pp.80–81), de la Cosa drew the first European map to feature the territories that were officially named "America" in 1507.

◁ Cape of Good Hope
Bartolomeu Dias sighted the Cape on his return journey to Portugal, in May 1488. His outward voyage carved a wide arc around the tip of Africa, showing this to be the most successful route to the East.

HENRY THE NAVIGATOR 1394–1460

Born in 1394, Henry was a son of King John I of Portugal. A devout Christian raised on tales of the crusades, he participated in the conquest of the Muslim city of Ceuta in 1415. Overseas travel introduced him to Asia and Africa's trade networks, sparking an interest in exploration. From the 1420s, he sponsored Portuguese voyages and, in 1441, a ship returned with gold dust and enslaved African people. Henry did not go on many expeditions, but his home at Sagres in southern Portugal became a centre for navigators. Henry's legacy is complex. He opened up trade routes, but his voyages began the process of European colonization and the transatlantic slave trade.

A 19th-century portrait of Prince Henry

the names of places & other things
The Hill of
Grasshoppers

THE EARLY MODERN WORLD

IN THE 15TH CENTURY, LONG-DISTANCE SEA VOYAGES BY EUROPEANS BEGAN IN THE ATLANTIC OCEAN AND ALONG THE AFRICAN COAST. THESE ROUTES SOON EXPANDED AND TRAVEL INTENSIFIED. IN THE NEWLY CONNECTED WORLD, FIRST ENCOUNTERS BETWEEN CULTURES WERE SOMETIMES HARMONIOUS, BUT MORE OFTEN HOSTILE.

CONNECTING THE WORLD

From 1492, European long-distance sea voyages ushered in a period of profound transformation as exploration became key to controlling lucrative trade routes, accessing the wealth of the Americas, and reshaping the global balance of power.

The Ottoman capture of Constantinople (present-day Istanbul) in 1453 was hugely significant for Europe. The Byzantine Empire's capital had been both a key trading post between Europe and the East and a buffer against Ottoman expansion into Eastern Europe. With its fall, Ottomans gained control of land trade routes, cutting off Europeans from spices and silk and other sought-after luxuries from India, China, the Pacific Islands, and Africa. To set up new trade routes, Europeans ventured further across the seas.

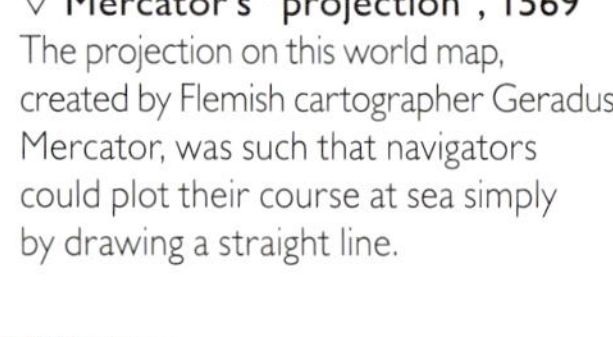

▽ **Mercator's "projection", 1569**
The projection on this world map, created by Flemish cartographer Geradus Mercator, was such that navigators could plot their course at sea simply by drawing a straight line.

The open ocean

Even before Constantinople fell, the Portuguese and Spanish had begun to sail out into the open seas, rather than staying close to the coastlines. In the early 15th century, Portuguese mariners – encouraged by the Portuguese prince, Henry the Navigator – explored south along the west coast of Africa, pushing out into the Atlantic.

The Mediterranean cargo ships of the day were cumbersome for long voyages, so Henry encouraged the development of "caravels". These small ships were not only fast and light, but also had triangular, or "lateen" sails, which enabled them to sail almost directly into the wind. This new technology, combined with that of the magnetic compass, a Chinese invention that had reached Europe by the 14th century, created the potential for long-distance voyages.

While the Portuguese set their sights on a route to Asia around Africa, the Spanish looked west across the Atlantic. The Italian explorer Columbus believed he could reach the Indies by setting out west from Spain, but in 1492 his ships reached the Americas instead. The kingdoms of Spain and Portugal attempted to divide the "new" lands outside of Europe between them with the Treaty of Tordesillas (1494), which drew a line north to south through the Atlantic. It granted Spain the rights to all land to the west, and Portugal the rights to all land to the east, overruling the rights of Indigenous peoples. Other European monarchs challenged this supremacy with voyages of their own and established trade dominance and colonies. By 1700, the naval powers of Portugal, Spain, Britain, and Holland had empires stretching across the globe. Many attempts were made to search out a route to Asia around the tip of North America, but a "northwest passage" continued to prove elusive.

△ **A Song dynasty compass**
The magnetic compass originated in China in the 12th century and revolutionized maritime navigation. This new technology travelled along trade routes to the West.

OCEAN CONNECTIONS

Before 1400, the world's main centres for trade and cultural exchange were in Eurasia, and travel was conducted almost entirely by land. However, once the long-distance sea voyages of European navigators began to connect lands and cultures across vast distances, a more global world was created. These sea paths began with voyages across the Atlantic from Europe, but they soon expanded across the Indian and Pacific oceans.

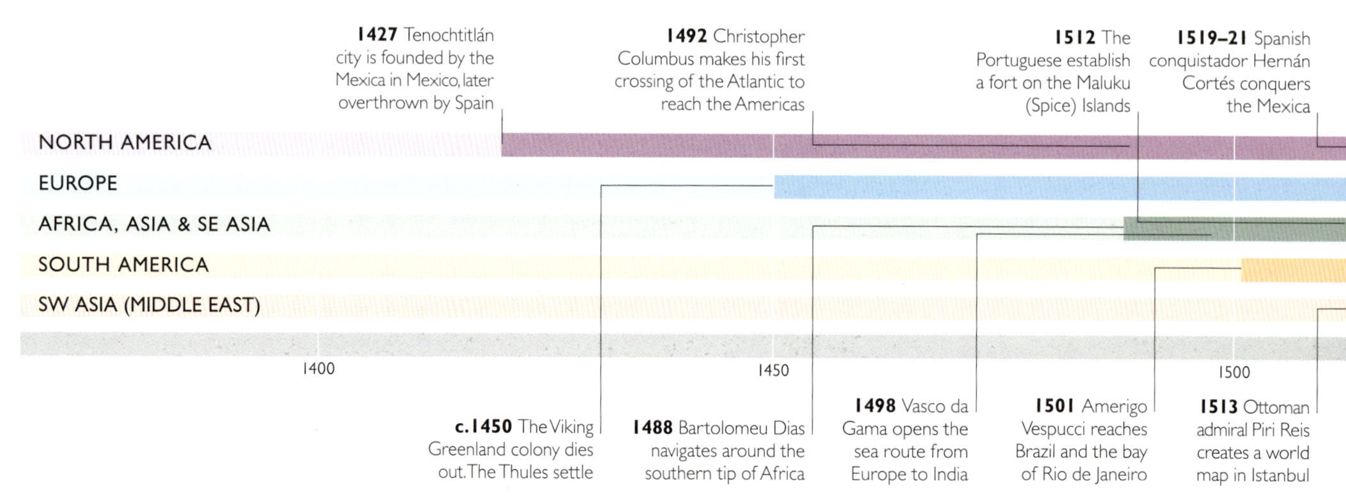

1427 Tenochtitlán city is founded by the Mexica in Mexico, later overthrown by Spain
1492 Christopher Columbus makes his first crossing of the Atlantic to reach the Americas
1512 The Portuguese establish a fort on the Maluku (Spice) Islands
1519–21 Spanish conquistador Hernán Cortés conquers the Mexica
NORTH AMERICA
EUROPE
AFRICA, ASIA & SE ASIA
SOUTH AMERICA
SW ASIA (MIDDLE EAST)
1400
1450
1500
c.1450 The Viking Greenland colony dies out. The Thules settle
1488 Bartolomeu Dias navigates around the southern tip of Africa
1498 Vasco da Gama opens the sea route from Europe to India
1501 Amerigo Vespucci reaches Brazil and the bay of Rio de Janeiro
1513 Ottoman admiral Piri Reis creates a world map in Istanbul

◁ **The nanban trade**
In the late 1500s, Portuguese ships arrived in Japan and established trading. The Japanese described these exotic-looking foreigners, who had arrived from the south, as "nanban" – or southern barbarians.

Expansionist aims

The fall of Constantinople not only brought about the disruption of trade routes but also acted as a catalyst for the Renaissance in Europe. Byzantine scholars fled to the West with ancient texts, contributing to a cultural and intellectual revival that spurred on the spirit of exploration from the mid-15th century. European historians soon began to call this era "The Age of Discovery", though the territories the explorers "discovered" were not uninhabited. In addition, the Europeans were rarely solely seeking knowledge; they were often searching for riches or territory, and their encounters with Indigenous peoples were often brutal and sometimes devastating.

Territorial expansion was a theme of the age. The Ottoman Empire reached its height during the reigns (1512–66) of sultans Selim I and Süleyman I. Selim's conquest of Egypt in 1516 gave the Ottomans direct access to the Indian Ocean, where they combatted the Spanish and Portuguese with naval and diplomatic offensives and leveraged their advantage with the Muslim traders of the region. In Russia, Tsar Ivan the Terrible sent armies of mercenaries into Siberia, until they had reached all the way to the Pacific and dominated the lucrative fur trade.

The exchange of goods and the wealth that was created through trade led to increasing numbers of voyages of "discovery". Activity was especially focused in the Pacific, the hub of the spice trade. The first Europeans to arrive in Japan were shipwrecked Portuguese and, initially, the Japanese were happy to trade with these "barbarians" from the south. However, by 1600, the shogun Tokugawa, who had reunited the country after a long civil war, foresaw the dangers posed by foreign powers and gradually excluded them, allowing only the Dutch to remain in a tiny trading enclave off Nagasaki.

▽ **The Florentine Codex, 1540–85**
Written by a Spanish priest, the Codex documented Mexican culture, and here shows the Nahua Mexica (Aztecs) using local Indigenous maps.

> *"This boundary… shall be drawn… at a distance of three hundred and seventy leagues west of the Cape Verde Islands."*
>
> TREATY OF TORDESILLAS, 1494

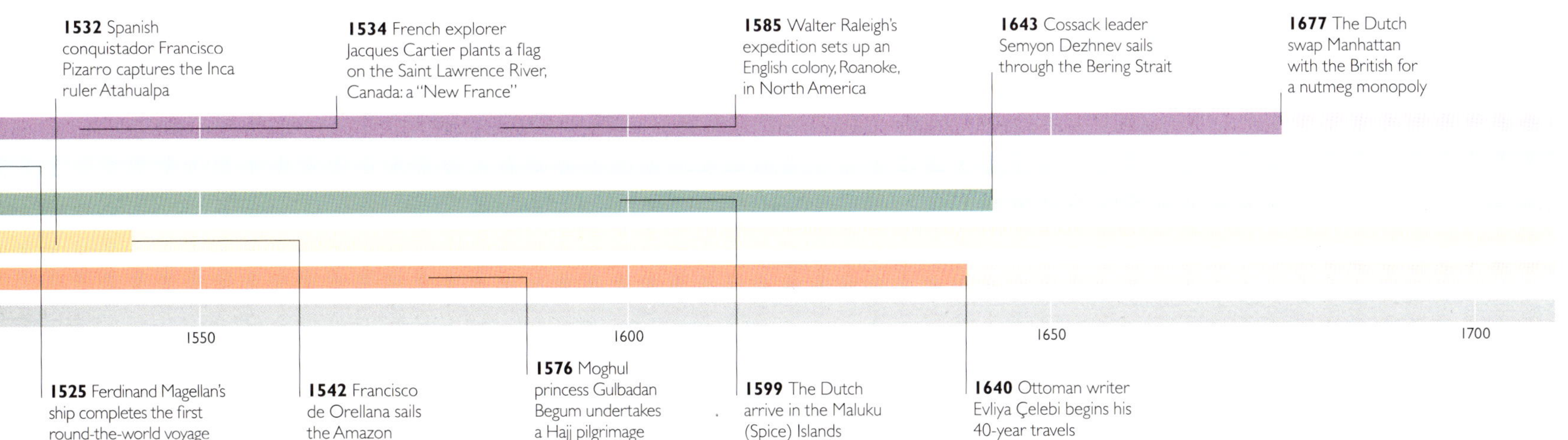

1532 Spanish conquistador Francisco Pizarro captures the Inca ruler Atahualpa
1534 French explorer Jacques Cartier plants a flag on the Saint Lawrence River, Canada: a "New France"
1585 Walter Raleigh's expedition sets up an English colony, Roanoke, in North America
1643 Cossack leader Semyon Dezhnev sails through the Bering Strait
1677 The Dutch swap Manhattan with the British for a nutmeg monopoly
1550
1600
1650
1700
1525 Ferdinand Magellan's ship completes the first round-the-world voyage
1542 Francisco de Orellana sails the Amazon
1576 Moghul princess Gulbadan Begum undertakes a Hajj pilgrimage
1599 The Dutch arrive in the Maluku (Spice) Islands
1640 Ottoman writer Evliya Çelebi begins his 40-year travels

Cornwall Island
Smith Sound
Bathurst Island
Devon Island
Melville Sound
Baffin Bay
Lancaster Sound
Summer 1586 After being blocked by ice, Davis turns south and explores the east coast of Baffin Island
Somerset Island
Bylot Island
Prince of Wales Island
Brodeur Peninsula
To England
Gulf of Boothia
Boothia Peninsula
Baffin Island
Melville Peninsula
26 Apr 1615 Baffin and Bylot reach Southampton Island at the mouth of Hudson Bay but discover that there is no way through to the west
Foxe Basin
Back
Thelon
Foxe Channel
18 Aug 1576 Frobisher wrongly believes he has found the passage
Kazan
CANADA
2 Aug 1610 Hudson's ship enters Hudson Bay
Hudson Strait
Churchill
CHIPEWYAN
Hudson Bay
Labrador
NASKAPI
Nelson
16 Jul 1620 Munk sails for home with the two other surviving crew members
CREE
James Bay
29 Nov 1631 James's men sink their ship to save it from being crushed by ice
MONTAGNAIS
La Grande
INNU
OJIBWA
Albany
EASTERN CREE
Jun 1611 Hudson's crew mutiny. He is set adrift with his son and only a handful of men
ATTIKAMEK

7 FOXE AND JAMES 1631–32

Leaving England in 1631, Luke Foxe led the first ship to sail right around Hudson Bay, and reached what is now Foxe Basin off Baffin Island before being forced back by ice. In Hudson Bay, he met Thomas James, who had sailed from England a few days after him. Foxe sailed home but James overwintered on Charlton Island, sinking his ship to stop it being crushed by ice. The crew refloated the ship and sailed it home in summer 1632.

Foxe's outward voyage in 1631 | Sunken ship
James's outward voyage in 1631–32

6 JENS MUNK 1619–1620

Danish explorer Jens Munk and his crew sailed to what is now Churchill on Hudson Bay, and made the first detailed map of this inland sea. But as winter set in, they became trapped by ice. Most of the crew died in the brutal conditions, leaving Munk alone on the ship. When the ice opened in June 1620, he found that two other crew members had survived on land. The three sailed their ship home in the summer.

Munk's outward voyage | Trapped in ice

5 BAFFIN AND BYLOT 1615–16

After participating in earlier voyages to the north of Canada, English navigators William Baffin and Robert Bylot believed that they might find a Northwest Passage by heading north of Hudson Bay. They made two voyages of their own on the ship *Discovery*, but the first, in 1615, showed the route north of the bay to be impassable. Their second, in 1616, took them as far as what is now Baffin Bay before ice forced them back.

BAFFIN AND BYLOT OUTWARD VOYAGES
1615 | 1616

4 HENRY HUDSON 1610–11

Englishman Hudson made several attempts to find a northern route to Asia via the Arctic. After his second failed attempt, he searched for the Northwest Passage, reaching what is now the Hudson River. In 1610, he arrived in what became known as Hudson Bay, but after finding no way through to the Pacific and spending the winter trapped in ice, his starving crew mutinied and set him adrift in a boat, never to be seen again.

Hudson's outward voyage | Mutiny

◁ **Munk's diaries**
Jens Munk wrote exceptionally detailed accounts of his voyage that were published in 1624. His journal reveals the story of his devastating winter trapped alone in the ice near the Churchill River on Hudson Bay.

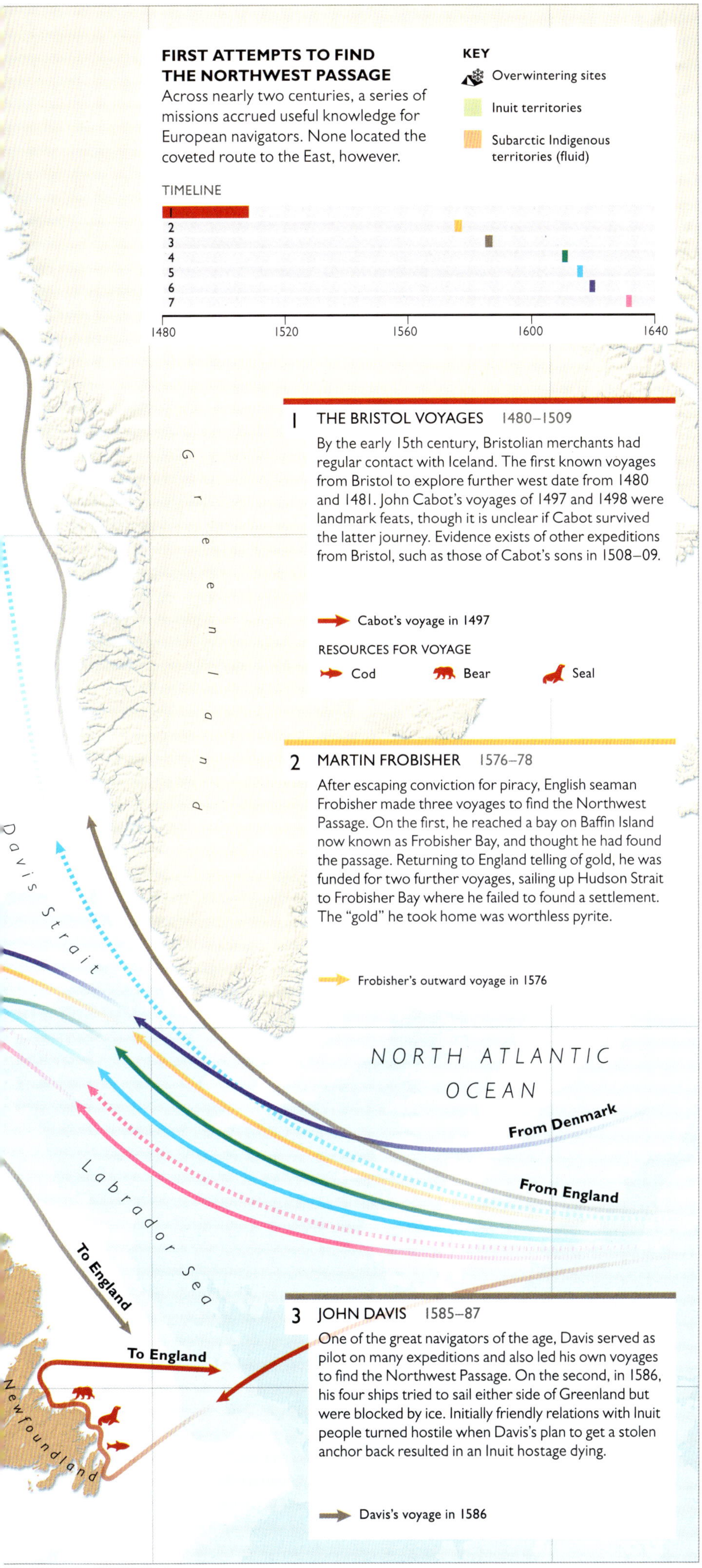

FIRST ATTEMPTS TO FIND THE NORTHWEST PASSAGE
Across nearly two centuries, a series of missions accrued useful knowledge for European navigators. None located the coveted route to the East, however.

KEY
Overwintering sites
Inuit territories
Subarctic Indigenous territories (fluid)

TIMELINE
1
2
3
4
5
6
7
1480 1520 1560 1600 1640

1 THE BRISTOL VOYAGES 1480–1509

By the early 15th century, Bristolian merchants had regular contact with Iceland. The first known voyages from Bristol to explore further west date from 1480 and 1481. John Cabot's voyages of 1497 and 1498 were landmark feats, though it is unclear if Cabot survived the latter journey. Evidence exists of other expeditions from Bristol, such as those of Cabot's sons in 1508–09.

Cabot's voyage in 1497

RESOURCES FOR VOYAGE
Cod Bear Seal

2 MARTIN FROBISHER 1576–78

After escaping conviction for piracy, English seaman Frobisher made three voyages to find the Northwest Passage. On the first, he reached a bay on Baffin Island now known as Frobisher Bay, and thought he had found the passage. Returning to England telling of gold, he was funded for two further voyages, sailing up Hudson Strait to Frobisher Bay where he failed to found a settlement. The "gold" he took home was worthless pyrite.

Frobisher's outward voyage in 1576

3 JOHN DAVIS 1585–87

One of the great navigators of the age, Davis served as pilot on many expeditions and also led his own voyages to find the Northwest Passage. On the second, in 1586, his four ships tried to sail either side of Greenland but were blocked by ice. Initially friendly relations with Inuit people turned hostile when Davis's plan to get a stolen anchor back resulted in an Inuit hostage dying.

Davis's voyage in 1586

Greenland
Davis Strait
NORTH ATLANTIC OCEAN
From Denmark
From England
Labrador Sea
To England
To England
Newfoundland

A NORTHWEST PASSAGE

For 400 years from the late 15th century, Europeans sought a direct sea route to Asia. After John Cabot sailed from England to North America in 1497, others tried to find a way around the northernmost tip of the continent. They did not succeed until 1906 (see pp.196–97).

Ships began venturing west across the North Atlantic from Bristol, England, as early as 1480. That was probably the motivation for Genoese-born Venetian John Cabot to go to the city with his three sons, Ludovico, Sebastiano, and Sancto, to launch his own expedition. With the backing of King Henry VII, Cabot sailed from Bristol in May 1497, on the ship *Matthew*, with 18 men.

For 35 days, they sailed west across the open ocean. On 24 June 1497, they made landfall, perhaps in Labrador, Newfoundland, or on Cape Breton Island. This marked the first European landing in North America since that of the Vikings in the late 10th century – although Cabot mistakenly believed that he had reached Asia's northeast coast.

Cabot detected signs that the land was inhabited, but saw no people. He unfurled English and Venetian flags and claimed the territory for Henry VII, then sailed along the coast before heading for home. Back in England, he reported that the land was good and the sea rich with codfish, and promised that if he could get backing for a second, bigger expedition, he would find a route to Japan. Cabot's second mission left Bristol in May 1498 with five ships and 300 men. Their fate remains a mystery, however.

> *"We sighted the land on the American side, but could not reach the shore for the quantity of ice."*
>
> JENS MUNK, JOURNAL ENTRY, 9 JULY 1619

LATITUDE AND LONGITUDE

In Cabot's day, European navigators had no way of determining their longitude – how far they were to the east or west. But they could calculate their latitude – the distance north or south of the Equator – using a backstaff to measure the angle of the North Star over the horizon. In 1594, English seaman John Davis devised his version of the backstaff – the Davis quadrant – which enabled navigators to measure the angle of the sun above the horizon.

The Seamans Secrets.
How is the vse of this staffe?

The vse of this staffe is altogether contrary to the other, for the center of this staffe where the brasse plate is fastned, must be turned to that part of the Horizon which is from the Sunne, & with your backe toward the Sunne, by the lower edge of the halfe crosse, & through the slitte of the plate you must direct your sight onely to the Horizon, & then mouing the transuersary as occasion requireth, vntill the shadow of your vpper edge of the transuersary doe fall directly vpon the said slitte or long hole, and also at the same instant you see the horizon through the slitte, and then the transuersary sheweth the height desired.

Finding by practise the excellency of the Crosse staffe aboue all other instruments, to satisfie the Seamans expectation, and also knowing that those instruments whose degrees are of largest capacitie, are instruments of most certainty. I haue very carefully laboured to search a good and demonstrable meane how a Crosse staffe might be protected, not only to contaiine

A page from Davis's *The Seaman's Secrets* (1626) showing a backstaff in use

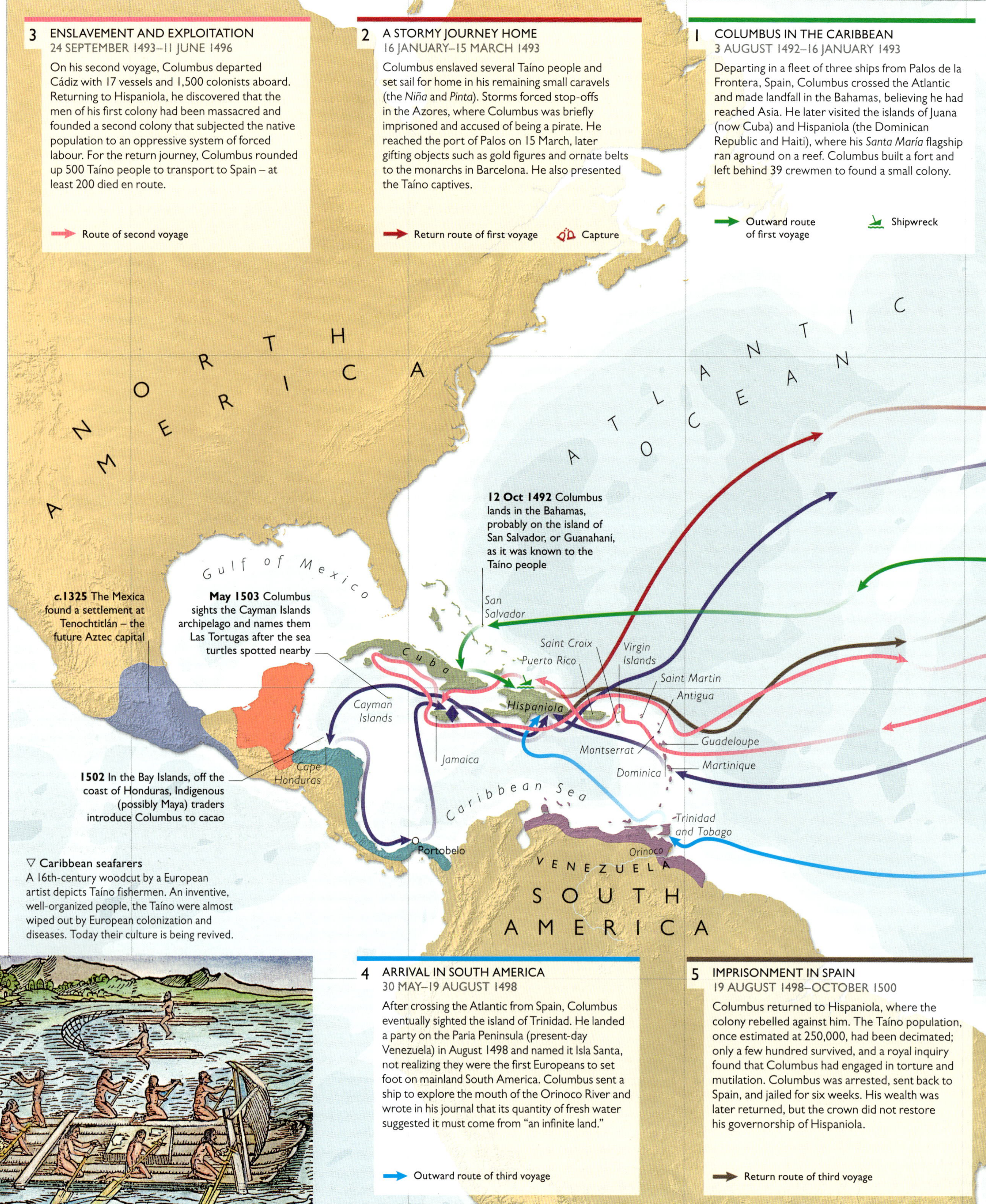

1 COLUMBUS IN THE CARIBBEAN

3 AUGUST 1492–16 JANUARY 1493

Departing in a fleet of three ships from Palos de la Frontera, Spain, Columbus crossed the Atlantic and made landfall in the Bahamas, believing he had reached Asia. He later visited the islands of Juana (now Cuba) and Hispaniola (the Dominican Republic and Haiti), where his *Santa María* flagship ran aground on a reef. Columbus built a fort and left behind 39 crewmen to found a small colony.

2 A STORMY JOURNEY HOME

16 JANUARY–15 MARCH 1493

Columbus enslaved several Taíno people and set sail for home in his remaining small caravels (the *Niña* and *Pinta*). Storms forced stop-offs in the Azores, where Columbus was briefly imprisoned and accused of being a pirate. He reached the port of Palos on 15 March, later gifting objects such as gold figures and ornate belts to the monarchs in Barcelona. He also presented the Taíno captives.

3 ENSLAVEMENT AND EXPLOITATION

24 SEPTEMBER 1493–11 JUNE 1496

On his second voyage, Columbus departed Cádiz with 17 vessels and 1,500 colonists aboard. Returning to Hispaniola, he discovered that the men of his first colony had been massacred and founded a second colony that subjected the native population to an oppressive system of forced labour. For the return journey, Columbus rounded up 500 Taíno people to transport to Spain – at least 200 died en route.

4 ARRIVAL IN SOUTH AMERICA

30 MAY–19 AUGUST 1498

After crossing the Atlantic from Spain, Columbus eventually sighted the island of Trinidad. He landed a party on the Paria Peninsula (present-day Venezuela) in August 1498 and named it Isla Santa, not realizing they were the first Europeans to set foot on mainland South America. Columbus sent a ship to explore the mouth of the Orinoco River and wrote in his journal that its quantity of fresh water suggested it must come from "an infinite land."

5 IMPRISONMENT IN SPAIN

19 AUGUST 1498–OCTOBER 1500

Columbus returned to Hispaniola, where the colony rebelled against him. The Taíno population, once estimated at 250,000, had been decimated; only a few hundred survived, and a royal inquiry found that Columbus had engaged in torture and mutilation. Columbus was arrested, sent back to Spain, and jailed for six weeks. His wealth was later returned, but the crown did not restore his governorship of Hispaniola.

▽ **Caribbean seafarers**
A 16th-century woodcut by a European artist depicts Taíno fishermen. An inventive, well-organized people, the Taíno were almost wiped out by European colonization and diseases. Today their culture is being revived.

THE VOYAGES OF COLUMBUS

Christopher Columbus commanded the first European expeditions to the Caribbean, Central America, and South America. These voyages had a seismic impact, leading to the colonization of the Americas and the devastation of Indigenous peoples.

Born in 1451 into a family of Genoese weavers, Christopher Columbus gained experience at sea by working on merchant ships, and fostered an ambition to find a new western sea route to Asia in order to gain access to the lucrative trade in silk and spices. He petitioned many patrons for sponsorship and eventually succeeded with the Spanish monarchs, Ferdinand of Aragon and Isabella of Castille.

Columbus set off across the Atlantic in 1492. He landed on an island in the present-day Bahamas and, influenced by Ptolemaic geography that misjudged the distance to Asia, Columbus thought he had achieved his goal of reaching the continent. This error led him to describe the Indigenous people he met as "Indians" (Europeans at the time often referred to East and South-East Asia as the "the Indies").

Columbus crossed the Atlantic three more times, reaching Cuba (which he mistook for Japan), Trinidad, northern South America, and the Caribbean coast of Central America. These journeys started the European colonization of the Americas, with lethal consequences for the Indigenous population: many were enslaved, subjected to forced labour, or died from diseases introduced by the colonizers.

While Columbus is often described as the man who "discovered" the Americas, people had lived on these continents for millenia. He was not the first European to cross the Atlantic either: in the 11th century, Vikings founded a settlement on what is now the island of Labrador and used it as a base to explore along the coast. When he died in 1506, Columbus still mistakenly believed he had reached Asia.

VOYAGES TO THE AMERICAS

Columbus set off across the Atlantic in 1492 in search of a direct route to Asia, stumbling upon the Americas in the process and opening the way for centuries of European colonization.

KEY

- Kalinago
- Taíno
- Maya
- Mexica
- Other Indigenous American culture areas
- Indigenous groups off Caribbean coast

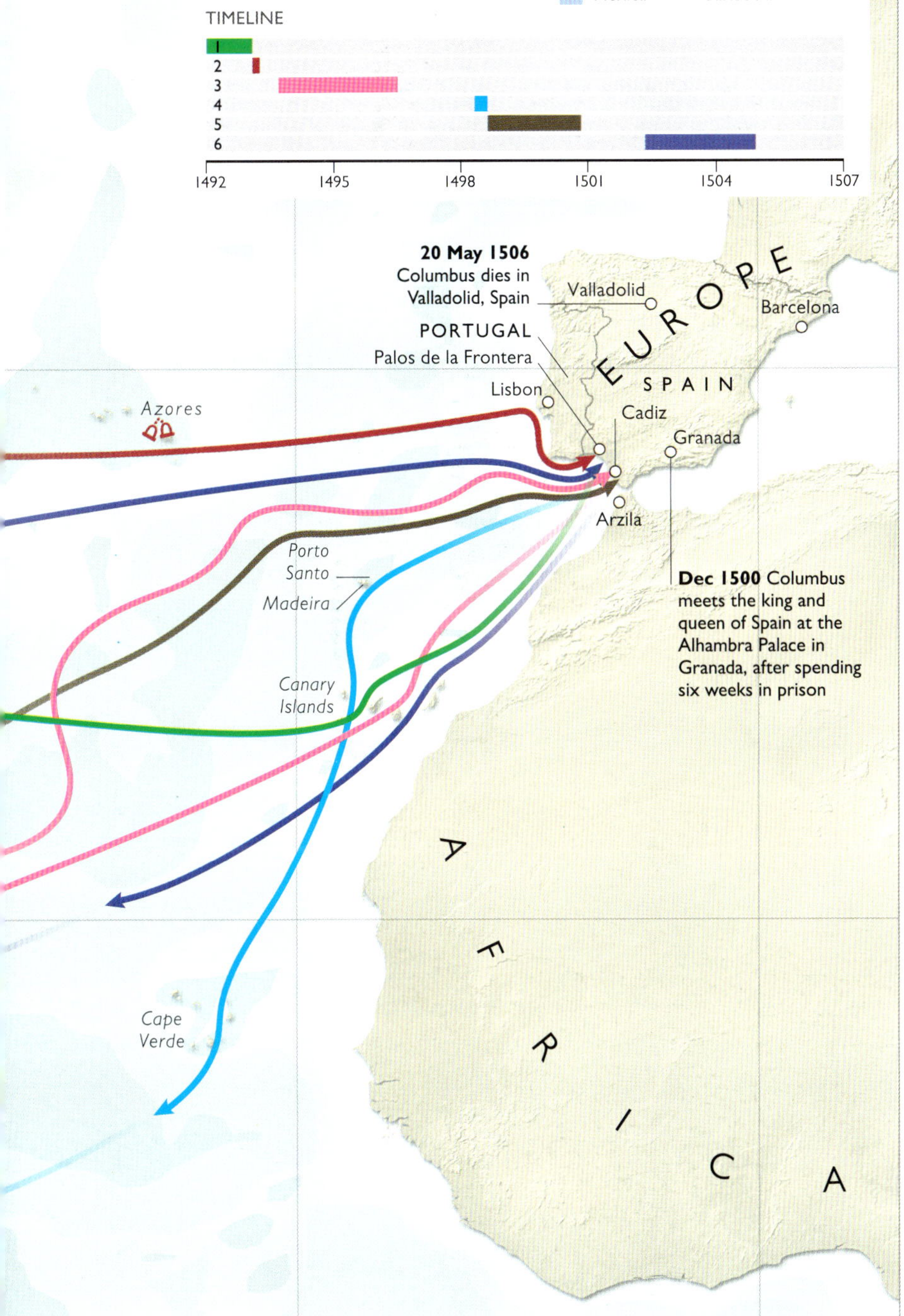

TIMELINE

1, 2, 3, 4, 5, 6

1492 | 1495 | 1498 | 1501 | 1504 | 1507

20 May 1506 Columbus dies in Valladolid, Spain

Dec 1500 Columbus meets the king and queen of Spain at the Alhambra Palace in Granada, after spending six weeks in prison

EUROPE

PORTUGAL

SPAIN

Valladolid

Barcelona

Palos de la Frontera

Lisbon

Cadiz

Granada

Arzila

Azores

Porto Santo

Madeira

Canary Islands

Cape Verde

AFRICA

6 FINAL VOYAGE

9 MAY 1502–7 NOVEMBER 1504

On his fourth voyage, Columbus reached Central America; from July to September 1502, he searched the coast of present-day Honduras, Nicaragua, Costa Rica, and Panama for gold and a strait to India. In February 1503, an attempt to set up a trading post in Panama failed and Columbus's shipworm-ridden fleet became beached on present-day Jamaica. Stranded for a whole year, Columbus then returned to Spain via Hispaniola.

Route of fourth voyage

Marooned

ORIGINAL MAP OF AMERICA

Although Columbus left no known maps of his explorations, Castilian navigator Juan de la Cosa did. Cosa owned Columbus's flagship, the *Santa María*, and sailed on at least two of Columbus's Atlantic voyages. His is the oldest surviving European map to depict the Americas – shown here as a large green area. Made around 1500, the map – or nautical chart – was lost for centuries (see pp.72–73). Today it is on display in Madrid's Naval Museum, Spain. Older Indigenous maps exist but the originals have been lost.

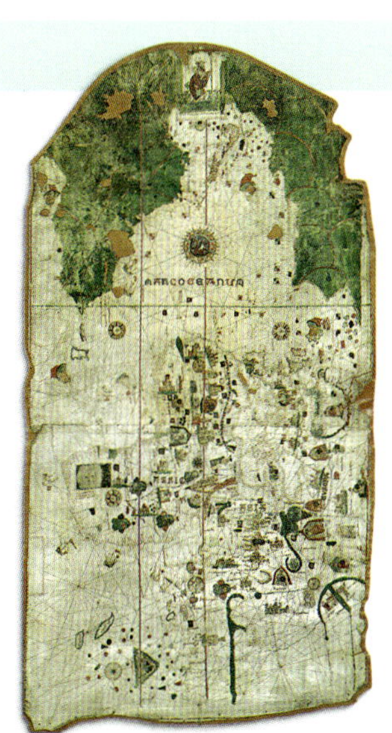

World map by Juan de la Cosa, 1500

VESPUCCI AND THE AMERICAS

An air of mystery and conjecture continues to surround the voyages of Amerigo Vespucci during the late 15th and early 16th centuries. Yet his name, and a memorable turn of phrase that he coined, have entered the popular lexicon.

Born in 1451 in the Italian city-state of Florence, Amerigo Vespucci made at least two voyages across the Atlantic to the Americas. During the first journey (1499–1500), on behalf of Spain, he served as a navigator with Alonso de Ojeda. He undertook his second mission, for Portugal (1501–02), as a pilot under Gonçalo Coelho.

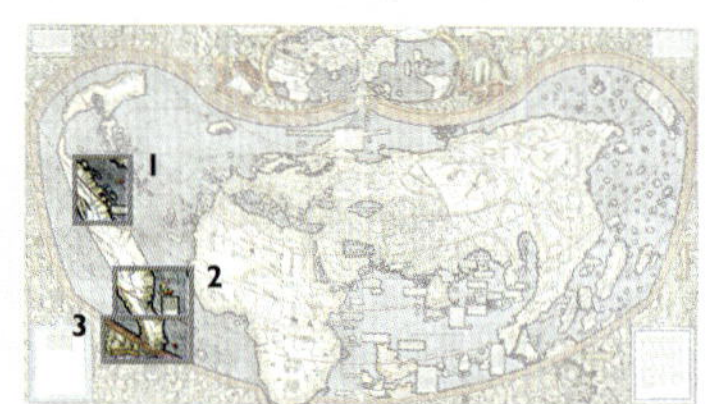

Documents published under Vespucci's name in the first decade of the 16th century mention a further two expeditions (1497–98 and 1503–04). The authenticity of their accounts is contested, however; it is possible that they were concocted by other writers.

But Vespucci's wider impact is unquestioned. He is credited with coining the phrase "New World" to describe the Americas (even though Indigenous peoples had lived there for millennia), which Europeans had previously considered to be part of Asia. In response, a leading German cartographer, Martin Waldseemüller, created a map that used a Latinized version of Vespucci's first name, "America", for the southern half of the "New World". Other mapmakers followed suit; in due course, the word was applied to its northern half too.

Vespucci later settled in Seville, where he was declared a Spanish citizen by royal proclamation. He died in 1512.

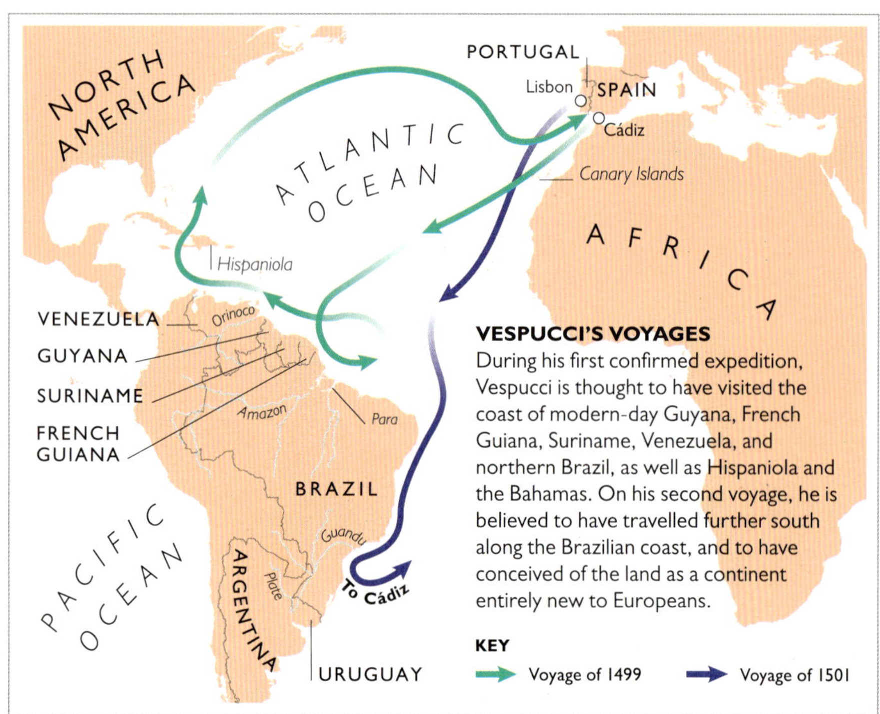

VESPUCCI'S VOYAGES
During his first confirmed expedition, Vespucci is thought to have visited the coast of modern-day Guyana, French Guiana, Suriname, Venezuela, and northern Brazil, as well as Hispaniola and the Bahamas. On his second voyage, he is believed to have travelled further south along the Brazilian coast, and to have conceived of the land as a continent entirely new to Europeans.

△ **The Waldseemüller map**
In 1507, cartographer Martin Waldseemüller created a global map showing the "New World" as a separate continent from Asia. He also used the name "America" to refer to its southern part.

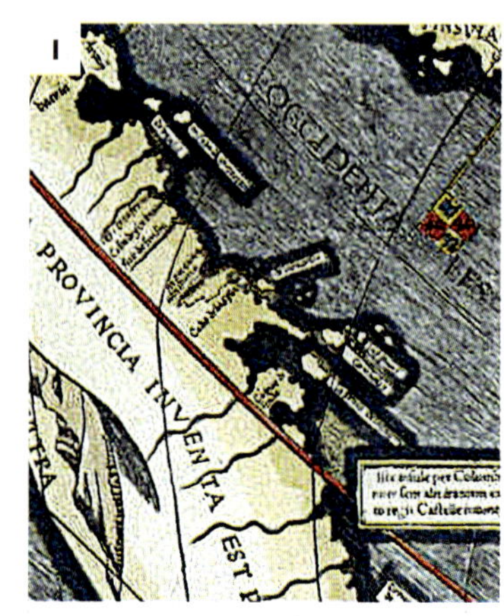

▷ **First arrival in South America**
Vespucci reached the "New World" in mid-1499. He then sailed south, while his compatriots de Ojeda and Juan de la Cosa headed west, towards present-day Venezuela. Later, Vespucci reputedly sighted the mouth of the Amazon River.

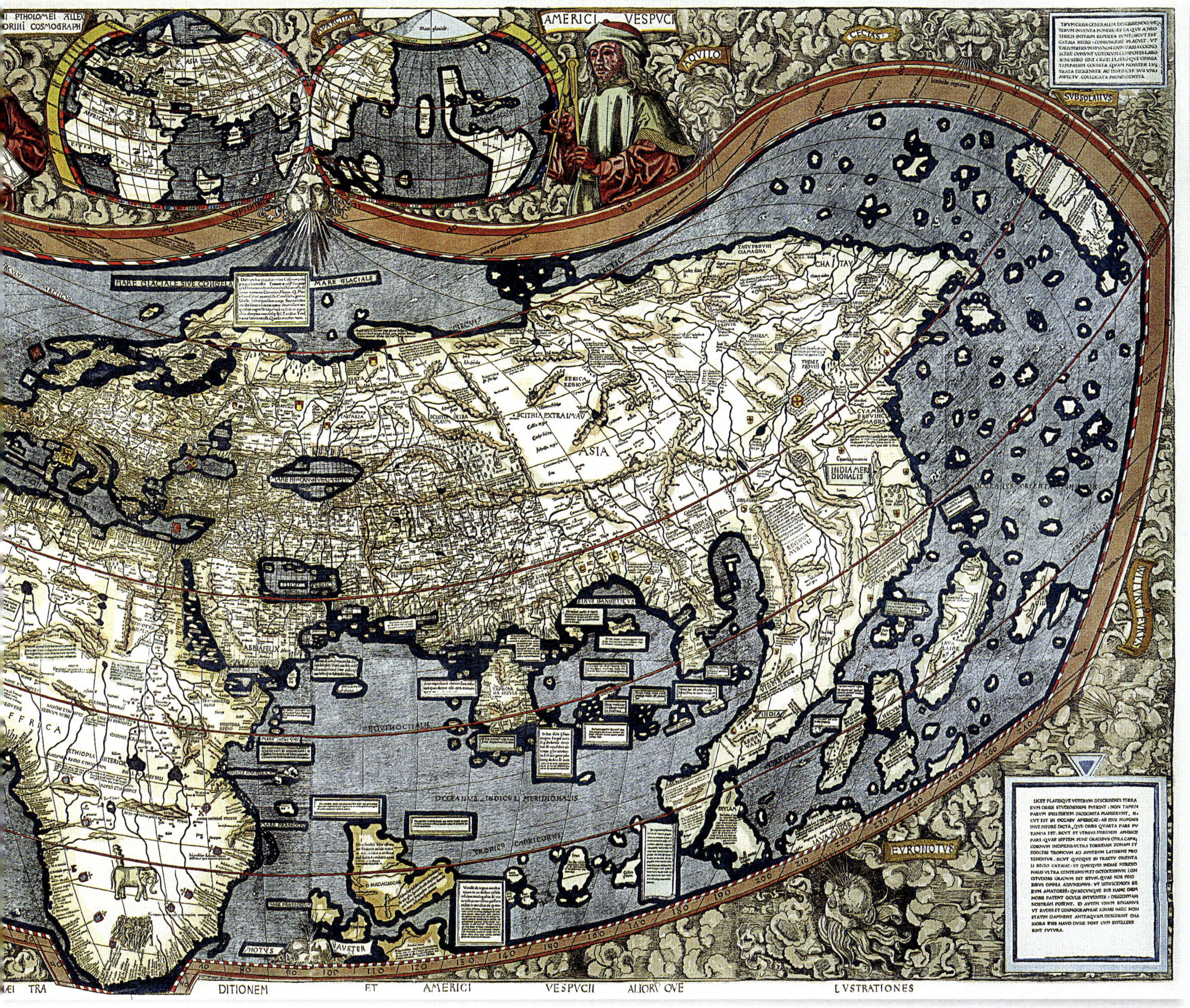

2

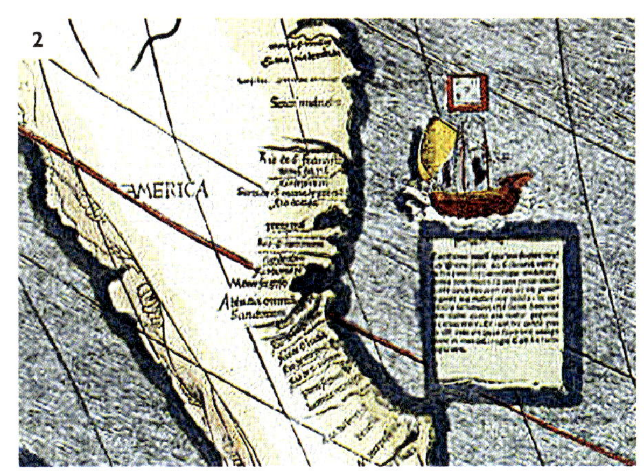

3

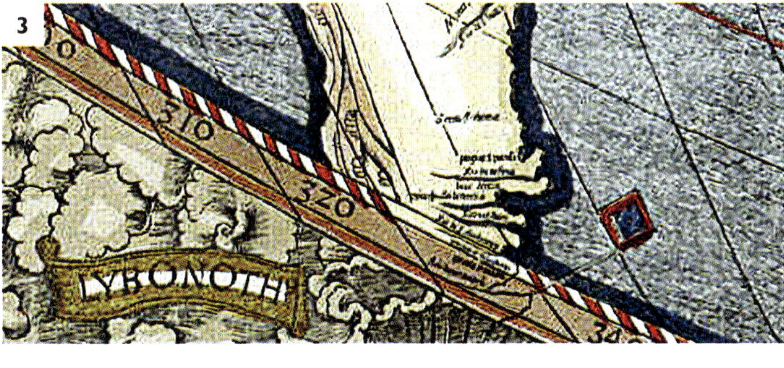

◁ Guanabara Bay, Brazil
On 1 January 1502, during his second confirmed voyage, Vespucci claimed to have sighted the bay now known as Guanabara, which he named Rio de Janeiro ("January River" in Portuguese).

◁ Río de la Plata, Argentina/Uruguay
Vespucci maintained that he reached the estuary of the Río de la Plata ("River Plate") – subsequently the site of Buenos Aires and Montevideo – later during his second voyage. His account is disputed, however.

DA GAMA'S VOYAGES

Portugal's Vasco da Gama opened up a trade route around Africa with three voyages to India in 1497–99, 1502–03, and 1524. European nations could now trade directly with the East, and set up their own trading stations, rather than routing goods through Türkiye and Iran.

◁ **Portuguese caravel**
Nimble, durable, and with ample cargo capacity, these light ships were widely used by Iberian explorers. One such, the *Berrio*, was the first of da Gama's ships to return to Lisbon in July 1499.

For centuries, Europeans had been desperate to gain access to the riches of Asia, particularly its treasured spices. The expeditions of Vasco da Gama changed all that, linking the West and East by an ocean route, and launching the modern era of global trade.

Da Gama set off from Belém, Portugal, on 8 July 1497, his four ships led by the *São Gabriel*. After sailing close to Africa as far as Sierra Leone, he headed far out into the open ocean, sailing for three months and 5,420 km (3,370 miles) before swinging back to reach southeast

Africa beyond the Cape of Good Hope. It was the longest sea journey made out of sight of land up to that time. In Malindi (in present-day Kenya), he hired a pilot, who may have been an Indian from Gujarat named Kanji Malam, to guide him over the Indian Ocean to reach Kozhikode (Calicut), India, on 20 May 1498.

However, the Portuguese had underestimated the quality of goods that were traded in the region and, although initially welcoming, the city's ruler was unimpressed with da Gama's gifts of clothing and food. Defying his pilot, da Gama set sail for Africa without waiting for favourable winds; on the return journey, many of the crew died from scurvy. In 1499, after a 37,000km (23,000 miles) voyage, only 54 of the 170 crew made it home, but with enough spices, silks, and goods to show the route's potential. Da Gama returned to India and was ruthless in his efforts to break into the Indian-African trade network, through which Portugal would build a vast, profitable trading empire – one that, as with all major colonial nations, relied heavily on the slave trade.

LEAVING A MARK

During the course of their voyages, Portuguese explorers marked landing points by erecting pillars called *padrãoes* in prominent places. Early *padrãoes* were simply wooden columns or crosses carried in the ship's hold, but later versions were built in stone and emblazoned with the Portuguese royal coat of arms. They marked Portugal's imperial aims and "christened" foreign shores.

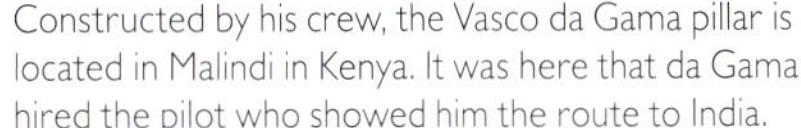

Constructed by his crew, the Vasco da Gama pillar is located in Malindi in Kenya. It was here that da Gama hired the pilot who showed him the route to India.

Glorifying da Gama
The great propaganda value of da Gama's voyage was recognized in the stunning Tournai Tapestry commissioned by Portugal's King Manuel I. It depicts the explorer arriving in Calicut, complete with extravagant displays of luxury goods.

CABRAL REACHES SOUTH AMERICA

On 9 March 1500, a Portuguese fleet under Pedro Álvares Cabral set off from Lisbon to open up the sea route to India. Blown off course, the ships arrived by accident at the land that is now Brazil and – as was customary colonial practice – claimed it for Portugal. They then sailed around Africa to India – the second European expedition there.

Vasco da Gama's voyage around Africa (see pp.84–85) showed Europe the massive potential for ocean trade with the East. Two years later, King Manuel I sent young Pedro Álvares Cabral to establish the route with a substantial fleet of 13 vessels and 1,200 men.

Following da Gama, Cabral sailed to the Cape Verde islands off West Africa, then steered out into the Atlantic and made good progress south. But strong winds took him much further southwest than intended and he eventually made landfall in what is now Brazil. He erected a wooden cross on the shore, claiming the land for Portugal, and sent a ship back to Portugal to tell the king.

Crossing the Atlantic to reach the Cape of Good Hope, the fleet was caught in storms. Four ships were lost, including the one carrying Bartolomeu Dias – the first European to navigate the Cape. Cabral sailed on to India with six ships and moved along the coast buying and selling goods, sometimes using force. He headed for home laden with spices, arriving on 23 June 1501.

However, the violent tactics Cabral employed with Arab traders had tarnished trade relations in India and his voyage was widely seen as a failure.

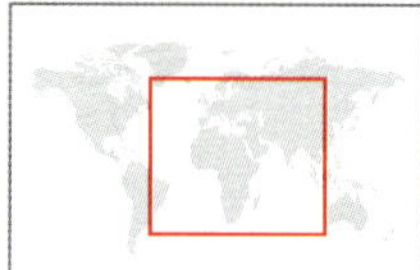

LOCATOR

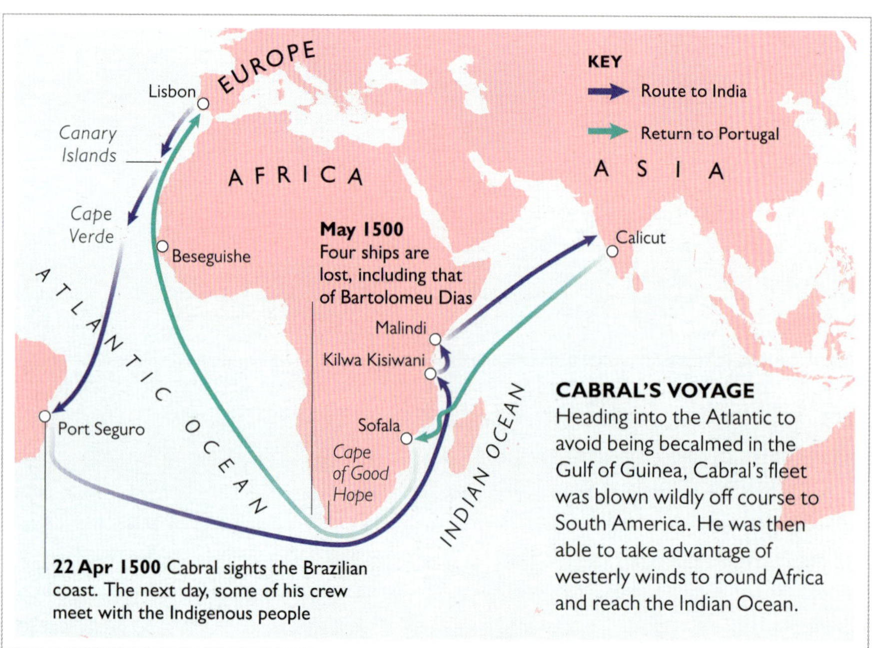

CABRAL'S VOYAGE
Heading into the Atlantic to avoid being becalmed in the Gulf of Guinea, Cabral's fleet was blown wildly off course to South America. He was then able to take advantage of westerly winds to round Africa and reach the Indian Ocean.

△ **Cantino planisphere**
Drawn in Portugal in 1502, this map was informed by knowledge gathered on Cabral's expedition. The chart is named for Alberto Cantino, who smuggled it to Italy: this new information was a precious commodity.

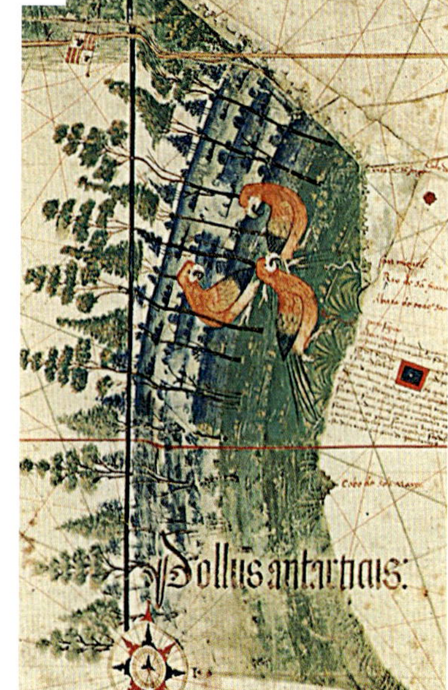

▷ **Brazil to Portugal**
The chart shows some of the Brazilian coast explored by Cabral in 1500. It emphasizes Portugal's asserted claim to Brazil east of the Line of Tordesillas and counters Spanish claims to the west.

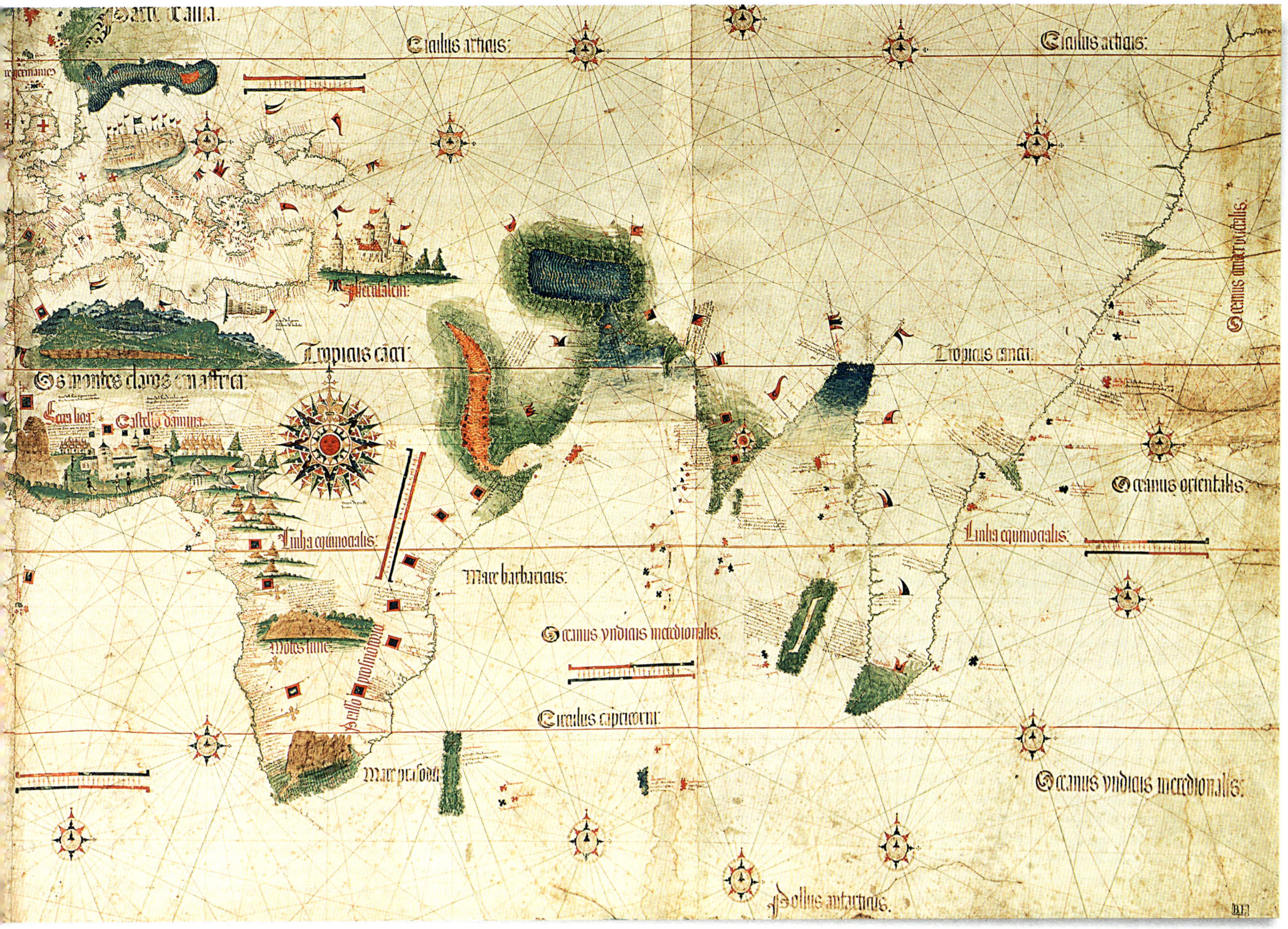

△ **A Portuguese presence in Africa**
Cabral avoided the West African coast as the Portuguese were already established there. They had built the Castelo de São Jorge da Mina, later a key trading post on the Atlantic slave route, in 1482.

◁ **Lines of latitude**
This is one of the first maps to show the key latitudes – the Arctic Circle, Tropic of Cancer, Equator, and Tropic of Capricorn.

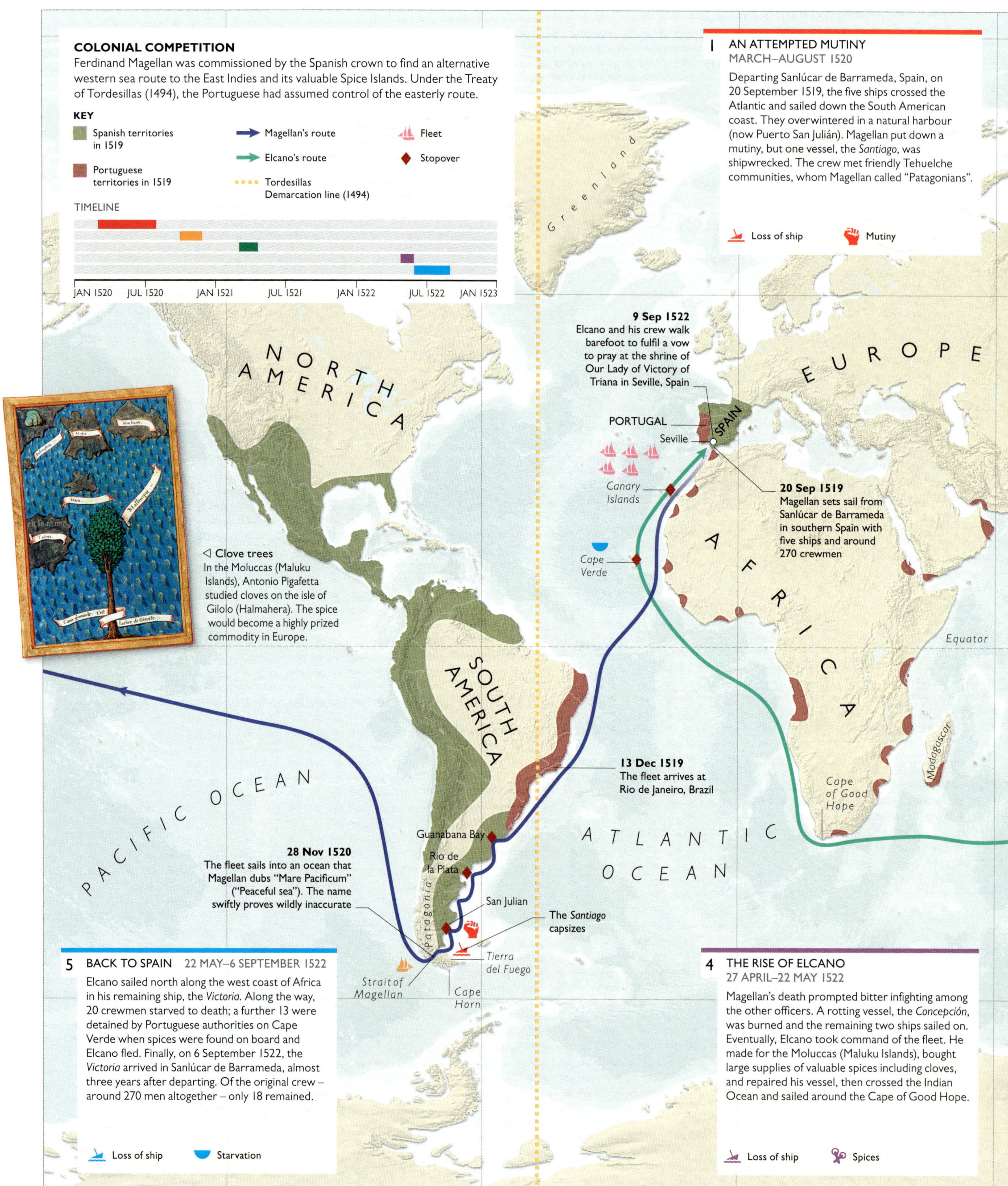

COLONIAL COMPETITION

Ferdinand Magellan was commissioned by the Spanish crown to find an alternative western sea route to the East Indies and its valuable Spice Islands. Under the Treaty of Tordesillas (1494), the Portuguese had assumed control of the easterly route.

◁ **Clove trees**
In the Moluccas (Maluku Islands), Antonio Pigafetta studied cloves on the isle of Gilolo (Halmahera). The spice would become a highly prized commodity in Europe.

1 AN ATTEMPTED MUTINY
MARCH–AUGUST 1520

Departing Sanlúcar de Barrameda, Spain, on 20 September 1519, the five ships crossed the Atlantic and sailed down the South American coast. They overwintered in a natural harbour (now Puerto San Julián). Magellan put down a mutiny, but one vessel, the *Santiago*, was shipwrecked. The crew met friendly Tehuelche communities, whom Magellan called "Patagonians".

4 THE RISE OF ELCANO
27 APRIL–22 MAY 1522

Magellan's death prompted bitter infighting among the other officers. A rotting vessel, the *Concepción*, was burned and the remaining two ships sailed on. Eventually, Elcano took command of the fleet. He made for the Moluccas (Maluku Islands), bought large supplies of valuable spices including cloves, and repaired his vessel, then crossed the Indian Ocean and sailed around the Cape of Good Hope.

5 BACK TO SPAIN 22 MAY–6 SEPTEMBER 1522

Elcano sailed north along the west coast of Africa in his remaining ship, the *Victoria*. Along the way, 20 crewmen starved to death; a further 13 were detained by Portuguese authorities on Cape Verde when spices were found on board and Elcano fled. Finally, on 6 September 1522, the *Victoria* arrived in Sanlúcar de Barrameda, almost three years after departing. Of the original crew – around 270 men altogether – only 18 remained.

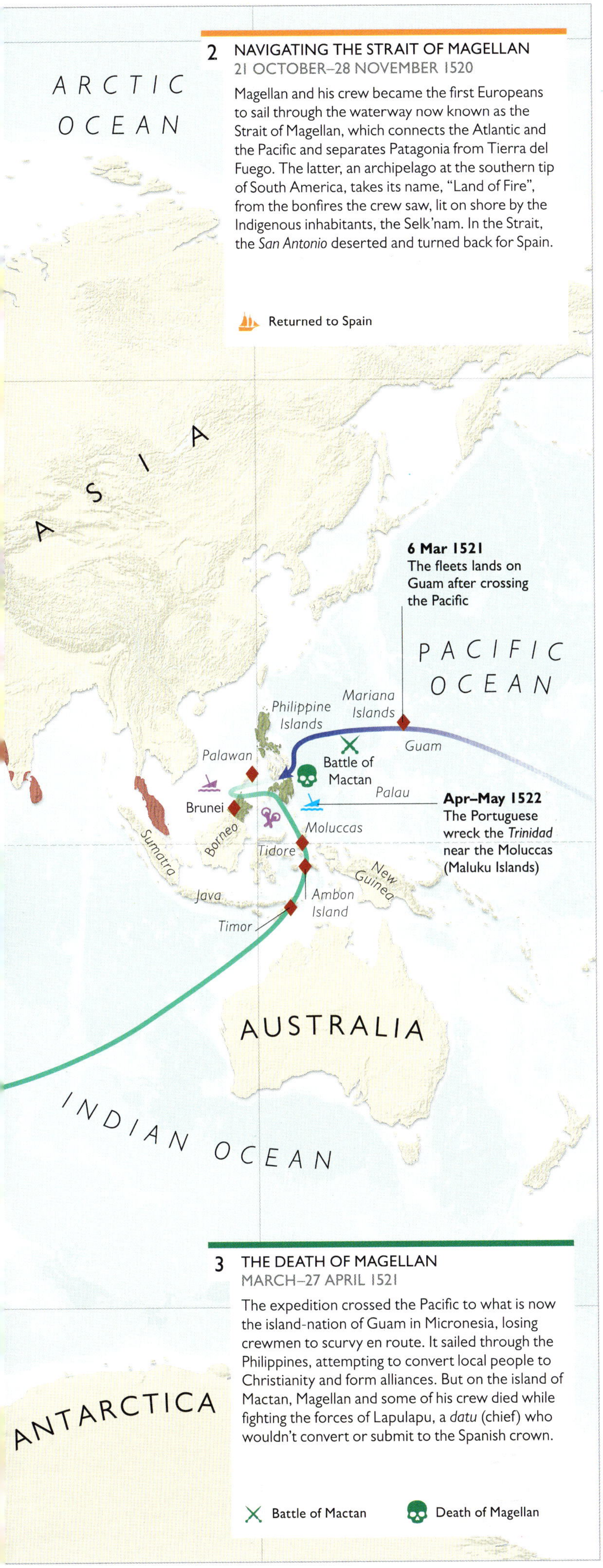

MAGELLAN AND ELCANO'S VOYAGE

In the 16th century, the Magellan–Elcano expedition completed the first circumnavigation of the globe. This monumental feat played a pivotal role in hastening the spread of European colonialism around the world and transforming the societies of South America and Southeast Asia.

Portuguese navigator Ferdinand Magellan was the first commander of an epic three-year voyage around the globe (1519–22). It was funded by the Spanish crown after Magellan fell out with Portugal's king, Manuel I. Although Magellan set out from Spain seeking wealth and fame, his journey had far-reaching effects. It transformed European conceptions of the world, sparked colonial expansion, opened up new trade routes, and drove advances in geography and cartography.

He is the best-known member of the expedition and has geographical features named after him – such as the Strait of Magellan, which connects the Atlantic and Pacific oceans. However, he did not actually complete the circumnavigation, having been killed midway in the Philippines. The fleet returned to Spain under captain Juan Sebastián Elcano, part of the original crew, arriving in September 1522.

Other crew members played key roles in the journey, notably Malay-speaking interpreter Enrique de Malaca. He had been enslaved by Magellan as a 14-year-old during the 1511 Portuguese invasion of Malacca in present-day Malaysia. His fate is unclear, but some historians argue that he could have returned to Spain ahead of Elcano, making him the first person to circle the globe.

> *"We had sailed 14,460 leagues and… had completed the circumnavigation of the world from east to west."*
>
> ANTONIO PIGAFETTA, *MAGELLAN'S VOYAGE*, 1550–59

CONFLICT WITH INDIGENOUS PEOPLES

In Patagonia, Magellan's crewmen kidnapped two of the unusually tall Tehuelche (or Aónikenk) men to take back to Spain, but they fell ill and died on board. Expedition accounts subsequently spread the myth, among others, of the presence of a race of "giants" in Patagonia. Magellan himself perished on Mactan in April 1521, fighting for a local chief who had become an ally. In the Philippines, the Europeans tried to convert locals to Christianity and to seize their lands for Spain.

Magellan (with shield) in fatal combat on Mactan

2 NAVIGATING THE STRAIT OF MAGELLAN

21 OCTOBER–28 NOVEMBER 1520

Magellan and his crew became the first Europeans to sail through the waterway now known as the Strait of Magellan, which connects the Atlantic and the Pacific and separates Patagonia from Tierra del Fuego. The latter, an archipelago at the southern tip of South America, takes its name, "Land of Fire", from the bonfires the crew saw, lit on shore by the Indigenous inhabitants, the Selk'nam. In the Strait, the *San Antonio* deserted and turned back for Spain.

Returned to Spain

ARCTIC OCEAN

ASIA

6 Mar 1521
The fleets lands on Guam after crossing the Pacific

PACIFIC OCEAN

Mariana Islands

Philippine Islands

Guam

Palawan

Battle of Mactan

Palau

Apr–May 1522
The Portuguese wreck the *Trinidad* near the Moluccas (Maluku Islands)

Brunei

Moluccas

Sumatra

Borneo

Tidore

New Guinea

Java

Ambon Island

Timor

AUSTRALIA

INDIAN OCEAN

3 THE DEATH OF MAGELLAN

MARCH–27 APRIL 1521

The expedition crossed the Pacific to what is now the island-nation of Guam in Micronesia, losing crewmen to scurvy en route. It sailed through the Philippines, attempting to convert local people to Christianity and form alliances. But on the island of Mactan, Magellan and some of his crew died while fighting the forces of Lapulapu, a *datu* (chief) who wouldn't convert or submit to the Spanish crown.

ANTARCTICA

Battle of Mactan

Death of Magellan

THE SPANISH IN MESOAMERICA

In the 16th and 17th centuries, Spain colonized much of present-day Mexico and Central America, a brutal process that decimated Indigenous societies. Local alliances and epidemics of European diseases both played key roles.

△ **Razor-sharp weaponry** Mexica warriors fought with obsidian-edged wooden clubs (*macuahuitl*), as shown here in the 16th-century Florentine Codex.

In the early 16th century, Mesoamerica was home to two of the world's most sophisticated and wealthy civilizations. As displayed in the majestic Mexica (Aztec) city-states, such as Tenochtitlán (see p.100), and Maya cities, such as Palenque and Chichén Itzá, their power flourished across a vast region, including what is now southern Mexico, Guatemala, and Belize.

The arrival of Spanish conquistador Hernán Cortés in 1519 was the catalyst for the Spanish colonization of Mesoamerica, which saw Spain begin systematically to occupy the region. Resistance efforts by the Maya and Mexica were suppressed with brute force. European contact also brought diseases such as smallpox, and studies suggest that these, by the early 1600s, had decimated around 90 per cent of the Indigenous population.

Over the following centuries, colonial mines and plantations used the forced labour of Indigenous people and enslaved African people to produce enormous amounts of gold, silver, tobacco, and chocolate, which were then traded across the globe. Meanwhile, horses, cattle, and wheat were transported from Europe to Mesoamerica.

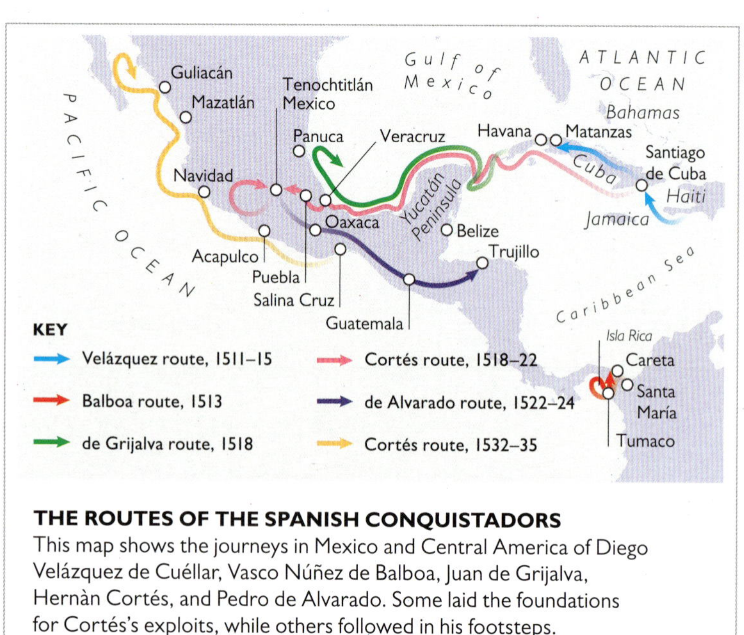

THE ROUTES OF THE SPANISH CONQUISTADORS
This map shows the journeys in Mexico and Central America of Diego Velázquez de Cuéllar, Vasco Núñez de Balboa, Juan de Grijalva, Hernàn Cortés, and Pedro de Alvarado. Some laid the foundations for Cortés's exploits, while others followed in his footsteps.

March into Tenochtitlán
The 16th-century Codex Azcatitlán depicts Cortés (the bearded figure) arriving at the Mexica city. The group, led by his interpreter Malinche, includes an African-born Spaniard (holding the horse's reins) and three Indigenous porters.

1 PIZARRO'S FIRST TWO EXPEDITIONS 1524–28

Pizarro's first expedition (1524–25) along the Pacific coast, with soldier Diego de Almagro and a priest, Hernando de Luque, failed due to difficult winds and a battle with Quito people in which Pizarro was wounded. The second expedition (1526–28) travelled further and in the region of Tumbes seized an Inca raft loaded with treasure, including gold. The port of Tumbes provided more proof of Inca wealth.

First expedition — Second expedition

2 INCA CIVIL WAR 1529–32

Before Pizarro's arrival in 1526, the death of Sapa Inca Huayna Cápac and his successor in a probable smallpox epidemic had led to infighting and later a civil war in 1529. His sons, Huáscar and Atahualpa, fought for control. Although the latter prevailed, it was a Pyrrhic victory: the three-year conflict left the Inca vulnerable, which the Spanish exploited.

Battle

GROWTH OF INCA EMPIRE

Huayna Cápac (r.1493–1525)

Huascar (r.1525–32)

3 PIZARRO'S THIRD EXPEDITION 1530–33

Pizarro set sail from Panama in December 1530 with three ships, around 180 men, and 27 horses. Other conquistadors, such as Hernando de Soto, bolstered this force with their own troops and the army reached Cajamarca in November 1532, just as Atahualpa was celebrating his civil war victory. The Spanish took the new Sapa Inca hostage before executing him in 1533, and then headed south with their Indigenous allies to seize the capital, Cusco.

Third expedition — Execution of Atahualpa

▽ Cerro Rico, Potosí

"Rich Mountain" in Spanish, this was dubbed "the mountain that eats men": from the 1570s, many forced Indigenous and enslaved African workers died in the Imperial Spanish silver mine.

CONQUERING THE INCA
On his third expedition in 1530, Spanish conquistador Francisco Pizarro took control of the Inca Empire, starting the colonization of South America. The map plots the dates cities were colonized.

6 INCA RESISTANCE AND DOWNFALL 1536–72

Initially installed by Pizarro in 1533 as a puppet Sapa Inca, in 1536 Manco Inca II led an uprising against the conquistadors. His army laid siege to Cusco but was eventually driven back. In 1539, Manco Inca II went on to found the Neo-Inca State in the jungles of Vilcabamba, beyond Spanish control. However, this state ended in 1572, when Túpac Amaru, the last Inca leader, was executed by the Spanish.

5 ALMAGRO AND VALDIVIA TRAVEL SOUTH 1535–53

Leaving Cusco in 1535, Diego de Almagro led the first Spanish expedition south into what is now Chile, but was repelled by the Mapuche people. Pedro de Valdivia led a second expedition south, extending Spanish colonial power and founding Santiago in 1541. On a later Chile campaign in 1550, he established forts at Arauco, Tucapel, and Purén. He was eventually killed by the Mapuche in 1553.

4 BELALCÁZAR TRAVELS NORTH 1534–39

One of Pizarro's conquistadors, Sebastián de Belalcázar, had travelled to the city of Quito to plunder treasure and suppress the resistance led by the Inca military leader Rumiñahui. Belalcázar founded a Spanish settlement in Quito in 1534, as well as the colonial cities of Cali and Pasto. Despite encountering resistance near the Cauca River, led by Misak Chief Payán, Belalcázar went on to found Popayán in 1537.

THE SPANISH IN SOUTH AMERICA

In the 15th century, the Inca civilization grew to become the largest empire in the Americas, with a sophisticated bureaucracy and road network. But, by the early 1500s, a series of epidemics and civil war had weakened its power. When a group of Spanish conquistadors arrived, they waged war and seized the riches of South America.

At its height, the Inca Empire spanned a vast area, encompassing over 12 million people and much of western South America. Named Tawantinsuyu (Quechua for "Realm of the Four Parts") by the Inca, the empire originated in the Cusco region in the 13th century and gave rise to cities, fortresses, religious complexes, such as Machu Picchu, and an advanced political government.

In 1519, Spaniard Francisco Pizarro, who was a soldier in Vasco Nunez de Balboa's 1513 Pacific Ocean expedition, was appointed mayor of Panama. He heard rumours of a ruler in the south with riches in the mountains. Eager to find wealth on a par with Cortés's looted Mexica treasure, he gained the Spanish crown's backing, and in 1532 arrived in Tawantinsuyu with a small military.

The conquistadors' arrival was preceded by civil war and epidemics of European diseases from Central America, introduced by Columbus. The Spanish joined forces with allies – from groups subjugated by the Inca, as well as Inca and Andean elites keen to maintain their influence – to overthrow the Sapa Inca (emperor) Atahualpa in 1533. This era saw the start of centuries of Spanish colonialism in South America, destroying the Incas and other Indigenous peoples, such as the Mapuche in what is now central and southern Chile.

"... after Christian Spaniards have passed through... [towns] are left in such a state... as if a fire had consumed them."

PEDRO CIEZA DE LEÓN, CHRONICLE OF PERU, 1553

ATAHUALPA c.1502–33

Atahualpa, the last Sapa Inca, was a younger son of Sapa Inca Huayna Cápac, whose death in c.1525 triggered a civil war. From his stronghold in Quito, Atahualpa imprisoned his half-brother Huáscar, based in Cusco, to become the 13th Sapa Inca in 1532. As such, he was revered – the representative of the sun god Inti, with a divine right to rule. In November 1532, Pizarro captured Atahualpa, demanding an unprecedented ransom of gold and silver, which the Sapa Inca paid. Reneging on his promise to release Atahualpa, Pizarro executed him on dubious charges in 1533.

An 18th-century portrait of Atahualpa

"There were many roads here that entered into the interior of the land, very fine highways."

GASPAR DE CARVAJAL, DESCRIBING THE SOCIETY OF THE ONIGUAYAL PEOPLE, 1542

△ **Portolan chart for coastal navigation**
The Atlas Portulano de América del Sur is a nautical chart of South America created by Juan (or Joan) Martínez in 1587. It incorporates information that was gathered on Orellana's expedition along the Amazon in the early 1540s.

ORELLANA'S AMAZON JOURNEY

Spanish conquistador Francisco de Orellana led the first European expedition to sail the length of the Amazon River. His voyage provided some of the earliest European accounts of Indigenous Amazonian peoples and his arrival began the colonization of the region.

Francisco de Orellana took part in conquistador Francisco Pizarro's toppling of the Inca empire and the early Spanish colonization of South America, before re-establishing the port of Guayaquil in present-day Ecuador. In 1541–42, he undertook an arduous expedition east of the city of Quito, initially in tandem with Pizarro's half-brother, Gonzalo. By the time it was completed, Orellana's party had sailed the full length of the Amazon River.

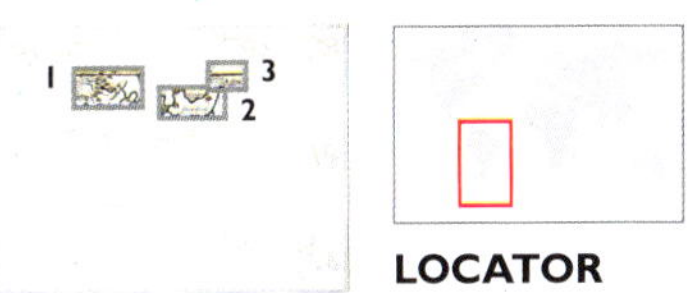

LOCATOR

KEY

1 At the junction of the Napo and Marañón rivers, defying orders, Orellana pressed on.

2 Orellana named the river and region "Amazon" after fighting the Piratapuyo.

3 On reaching the Atlantic, the party sailed to Cubagua island, then returned to Spain.

According to an account by the missionary Gaspar de Carvajal, the Spanish saw large towns and cities, as well as extensive road and canal networks built by Indigenous peoples. Their urban developments may have housed hundreds of thousands of people. These societies flourished in a challenging environment, but were destroyed by the violence and disease introduced by European colonialism and catalysed by Orellana's voyage.

The Spanish often sparked conflict, not least by raiding villages for supplies. On one occasion, this led to a fight with a Piratapuyo community whose women fought with the men. This gave rise to Orellana's epithet the "river of the Amazons", a reference to the warrior-women of classical Greek myth and the origin of the region's modern name.

The explorer's second Amazonian expedition ended in disaster: shipwrecks, conflicts, starvation and, ultimately, Orellana's death by drowning when his vessel capsized in 1546.

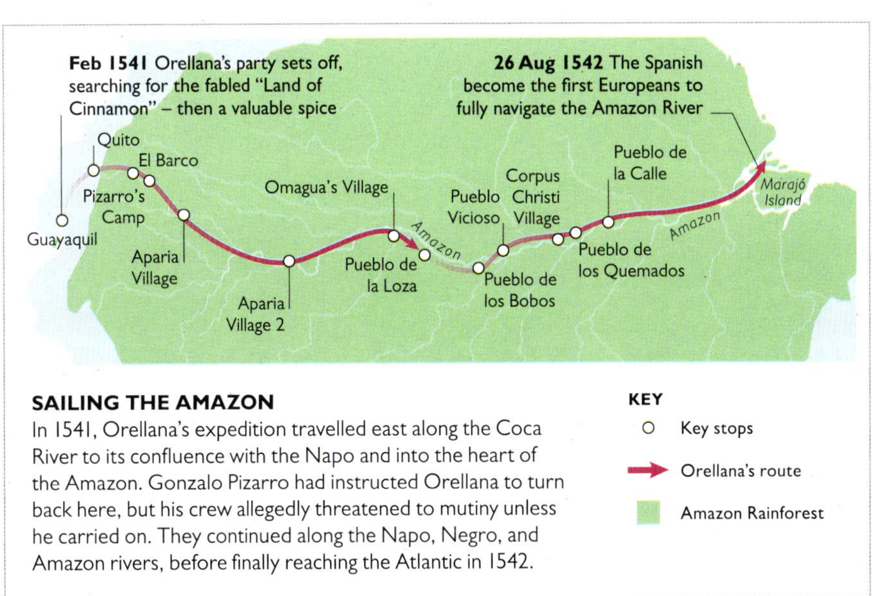

SAILING THE AMAZON

In 1541, Orellana's expedition travelled east along the Coca River to its confluence with the Napo and into the heart of the Amazon. Gonzalo Pizarro had instructed Orellana to turn back here, but his crew allegedly threatened to mutiny unless he carried on. They continued along the Napo, Negro, and Amazon rivers, before finally reaching the Atlantic in 1542.

KEY

- Key stops
- Orellana's route
- Amazon Rainforest

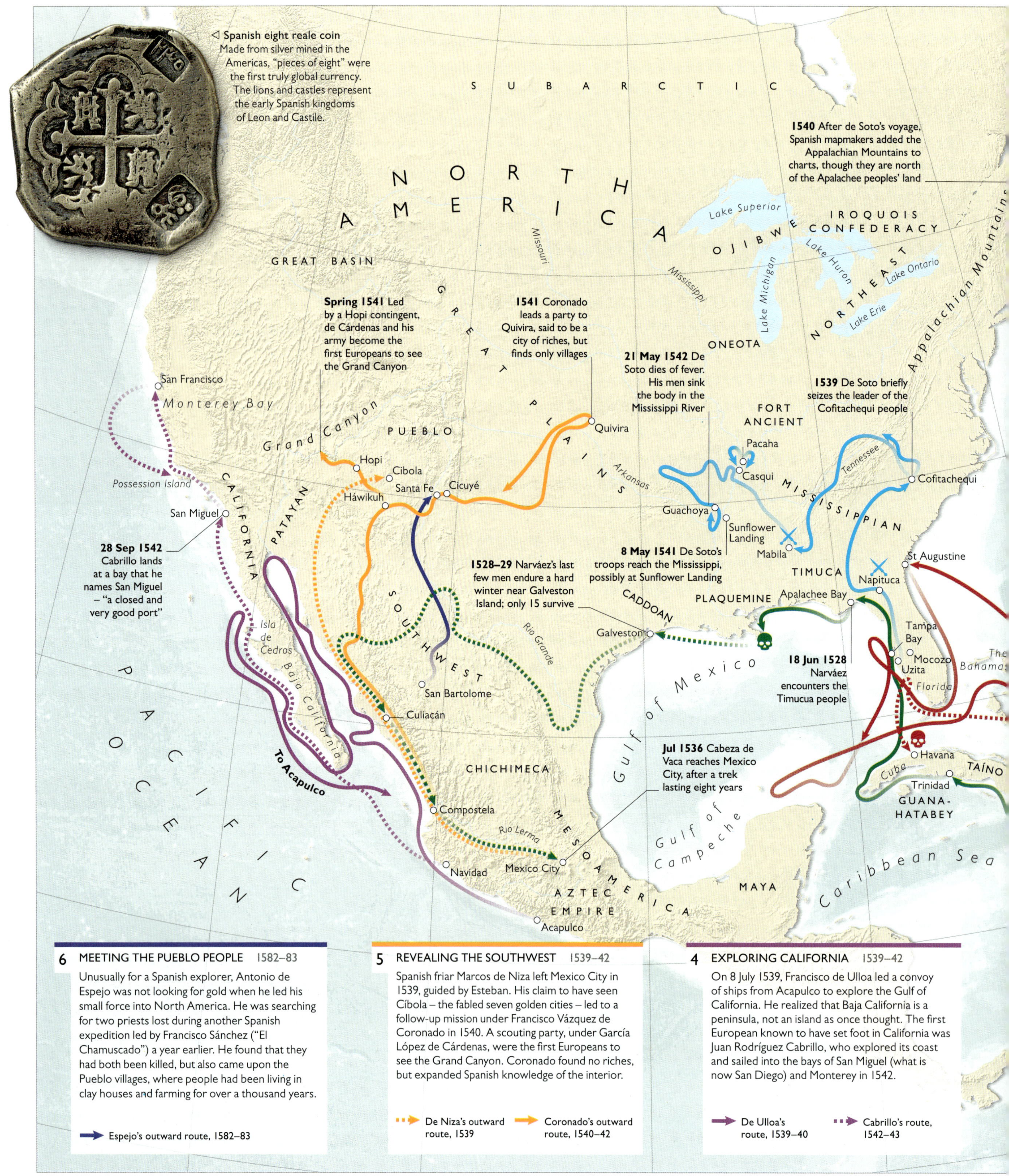
◁ Spanish eight reale coin
Made from silver mined in the Americas, "pieces of eight" were the first truly global currency. The lions and castles represent the early Spanish kingdoms of Leon and Castile.
SUBARCTIC
NORTH AMERICA
GREAT BASIN
GREAT PLAINS
1540 After de Soto's voyage, Spanish mapmakers added the Appalachian Mountains to charts, though they are north of the Apalachee peoples' land
Lake Superior
OJIBWE
IROQUOIS CONFEDERACY
Lake Huron
Lake Ontario
Lake Michigan
Lake Erie
NORTHEAST
Appalachian Mountains
Missouri
Mississippi
Spring 1541 Led by a Hopi contingent, de Cárdenas and his army become the first Europeans to see the Grand Canyon
1541 Coronado leads a party to Quivira, said to be a city of riches, but finds only villages
ONEOTA
21 May 1542 De Soto dies of fever. His men sink the body in the Mississippi River
1539 De Soto briefly seizes the leader of the Cofitachequi people
FORT ANCIENT
San Francisco
Monterey Bay
Grand Canyon
PUEBLO
Quivira
Pacaha
Casqui
Tennessee
Cofitachequi
Hopi
Cibola
Santa Fe
Cicuyé
Háwikuh
Arkansas
MISSISSIPPIAN
Possession Island
CALIFORNIA
San Miguel
PATAYAN
Guachoya
Sunflower Landing
Mabila
28 Sep 1542 Cabrillo lands at a bay that he names San Miguel – "a closed and very good port"
1528–29 Narváez's last few men endure a hard winter near Galveston Island; only 15 survive
8 May 1541 De Soto's troops reach the Mississippi, possibly at Sunflower Landing
TIMUCA
Napituca
St Augustine
CADDOAN
PLAQUEMINE
Apalachee Bay
SOUTHWEST
Rio Grande
Galveston
Isla de Cedros
Baja California
Tampa Bay
Mocozo
Uzita
The Bahamas
Florida
18 Jun 1528 Narváez encounters the Timucua people
San Bartolome
Gulf of Mexico
Culiacán
PACIFIC OCEAN
To Acapulco
CHICHIMECA
Jul 1536 Cabeza de Vaca reaches Mexico City, after a trek lasting eight years
Havana
Cuba
Trinidad
TAÍNO
GUANAHATABEY
Compostela
Rio Lerma
MESOAMERICA
Gulf of Campeche
Navidad
Mexico City
Caribbean Sea
AZTEC EMPIRE
MAYA
Acapulco

6 MEETING THE PUEBLO PEOPLE 1582–83

Unusually for a Spanish explorer, Antonio de Espejo was not looking for gold when he led his small force into North America. He was searching for two priests lost during another Spanish expedition led by Francisco Sánchez ("El Chamuscado") a year earlier. He found that they had both been killed, but also came upon the Pueblo villages, where people had been living in clay houses and farming for over a thousand years.

➔ Espejo's outward route, 1582–83

5 REVEALING THE SOUTHWEST 1539–42

Spanish friar Marcos de Niza left Mexico City in 1539, guided by Esteban. His claim to have seen Cíbola – the fabled seven golden cities – led to a follow-up mission under Francisco Vázquez de Coronado in 1540. A scouting party, under García López de Cárdenas, were the first Europeans to see the Grand Canyon. Coronado found no riches, but expanded Spanish knowledge of the interior.

➔ De Niza's outward route, 1539
➔ Coronado's outward route, 1540–42

4 EXPLORING CALIFORNIA 1539–42

On 8 July 1539, Francisco de Ulloa led a convoy of ships from Acapulco to explore the Gulf of California. He realized that Baja California is a peninsula, not an island as once thought. The first European known to have set foot in California was Juan Rodríguez Cabrillo, who explored its coast and sailed into the bays of San Miguel (what is now San Diego) and Monterey in 1542.

➔ De Ulloa's route, 1539–40
➔ Cabrillo's route, 1542–43

SEARCHING FOR GOLD
In the early 16th century, Spanish expeditions scoured North America on a futile quest for imagined riches. They devastated the lives of Indigenous groups through violence and by spreading disease.
KEY
ABC Indigenous groups
TIMELINE
1
2
3
4
5
6
1510
1530
1550
1570
1590
1 FIRST EUROPEAN LANDING AT FLORIDA 1513–21
Despite the myth, there is little evidence to support claims that Ponce de León was seeking a fountain of youth when he sailed from Puerto Rico to Florida in 1513. In 1521, he returned to Florida to set up a colony at what is now Safety Harbor on Old Tampa Bay on Florida's west coast, but the local Calusa people fought back fiercely. De León was badly wounded and later died in Cuba.
De León's route, 1513
De León's route, 1521
De León's death, 1521
2 CROSSING OF NORTH AMERICA 1527–36
Pánfilo de Narváez lost a third of his 600 crew before even landing in Florida in April 1528. After a fruitless foray north, Narváez and most of his men later drowned off the Mississippi mouth. The last 15 landed near present-day Tampa Bay, Florida. A harsh inland trek to the west left four survivors, including Cabeza de Vaca and Esteban (see box).
De Narváez's route, 1527–28
De Narváez's death, 1528
De Vaca's route, 1528–36
ATLANTIC OCEAN
Santiago de Cuba
Hispaniola
Santo Domingo
Puerto Rico
Jamaica
SOUTH AMERICA
3 HUNT FOR "CITIES OF GOLD" 1538–43
De Vaca told tales of the fabled "seven cities of gold". In search of them, Hernando de Soto led a military unit from Florida northwest, torturing Indigenous people to extract information. They reached the Mississippi River (then called *Misi-ziibi* by the Ojibwe people) and Arkansas, but fought local peoples such as the Caddo. After five years, they retreated; de Soto died of fever en route.
De Soto's route, 1539–43
Conflict with Indigenous people

THE SPANISH IN NORTH AMERICA

Starting with settlements in the Caribbean in 1493, Spanish colonization spread to the Central and South American mainland. Spain went on to become the first European nation to make its mark on North America. But armed Spanish expeditions had a disastrous impact on Indigenous peoples they encountered.

Spain's presence in North America began in 1513, when Juan Ponce de Léon sailed from Puerto Rico to Florida, landing on Easter Day and naming it after the Easter festival Pascua Florida ("Feast of Flowers"). The Spanish hoped that the region would boast similar riches to those of Aztec Mexico (see pp.90–91). The next major expedition was in 1528 when Panfilo de Narváez sailed to Florida from Cuba, but this soon unravelled. Its survivors, including Álvar Núñez Cabeza de Vaca, spent eight years trekking across Texas and the southwest. In 1538, Marcos de Niza explored the Sierra Madre's western edge with an African scout named Esteban, and claimed to have seen the fabled golden Cíbola cities. The period also saw expeditions by Francisco de Ulloa (Gulf of California, 1539–40) and Francisco Vázquez de Coronado (Arizona, 1540–42).

Spanish encounters with Indigenous peoples were often brutal. In 1540, Hernando de Soto came to a town in what is now South Carolina. Its Chief, "The Lady of Cofitachequi", welcomed the Spaniards and offered them gifts and hospitality. In return, de Soto raided her town, looted burial sites, and held her captive – although she escaped.

After three decades, disappointed in their search for gold, the Spanish state's interest in exploring North America waned. However, its settlers and missionaries would later colonize Florida, the southwest, and California.

> *"There are seven great Cities in this Province… the houses whereof are made of Lyme & Stone."*
>
> THE JOURNEY OF ÁLVAR NÚÑEZ CABEZA DE VACA, 1542

ESTEBAN ("ESTEVANICO") c.1500–1539

Born Mustafa Azemmouri, likely in Morocco, Esteban was one of North America's first non-Indigenous explorers. In 1522, he was enslaved by Andrés Dorantes, who served on Panfilo de Narváez's ill-fated expedition. Dorantes, de Vaca, and Esteban were among its few survivors. They endured years of hard labour under various Indigenous peoples, but Esteban also learned regional languages, becoming a useful interpreter. He and his companions heard tales of the wealthy cities of Cíbola, which they shared on reaching Mexico City in 1536. Three years later, Esteban set off with Marcos de Niza to find them, but was killed in the pueblo of Hawikuh.

A 19th-century representation of Esteban

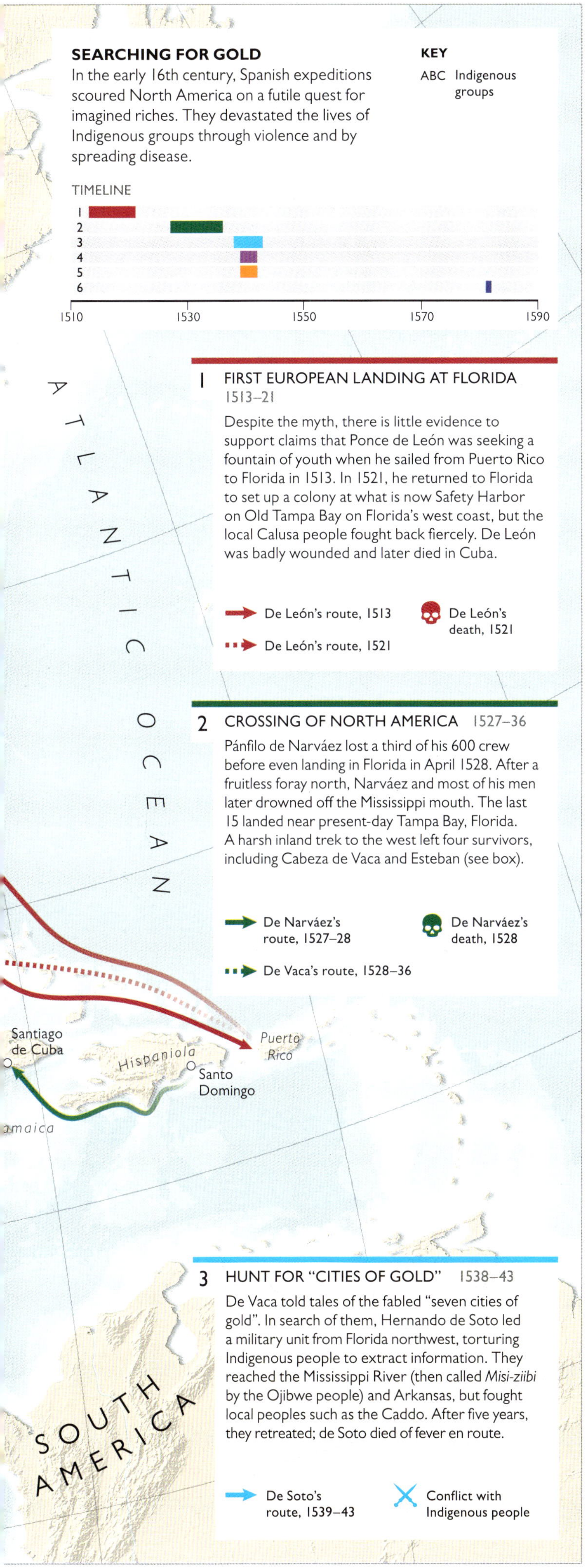

THE FRENCH IN NORTH AMERICA

In 1524, Italy's Giovanni da Verrazzano explored the east coast of North America for King Francis I of France. Thereafter, the French became a strong presence on the continent, pushing into the interior, initiating a valuable fur trade, and negotiating or battling with Indigenous people.

In 1534, the French explorer Jacques Cartier erected a large wooden cross, bearing the coat of arms of Francis I, on the shores of what is now eastern Canada, thus initiating French claims to this part of America. From here, the French ventured north into Labrador, south, and west, until they reached the mouth of the Mississippi and the Gulf of Mexico.

Samuel de Champlain made several expeditions to North America before founding Quebec in 1608. Forging alliances with Indigenous communities, he reached two of the Great Lakes – Huron and Ontario – in 1615. Such relations also involved him in local conflicts, however (see right).

In 1673, the Jesuit missionary Father Jacques Marquette and fur trader Louis Jolliet canoed far down the Mississippi with guidance from local Indigenous people, including Quapaw villagers. The detailed knowledge they gained was crucial in enabling René-Robert Cavelier, Sieur de La Salle to set up a vast network of trading posts to exploit the interior's riches – starting from Fort Frontenac on Lake Ontario.

In 1682, de La Salle travelled to the Mississippi Delta and claimed the territory. French colonists founded New Orleans in 1718, by which time France had gained a strong foothold on the continent.

"The fairest land that may possibly be seen, full of goodly meadows and trees."

JACQUES CARTIER, ON FIRST SEEING NEW BRUNSWICK IN 1534

A RIFT IN RELATIONS

Jacques Cartier sailed up the Montreal River in 1535, seeking gold, silver, copper, and spices. He was led by the sons of the Haudenosaunee (Iroquois) Chief Donnacona, but, possibly to corner trade with France or to deter French imperialism in the area, they refused to take him on to the village of Hochelaga. He reached it without them, on 2 October.

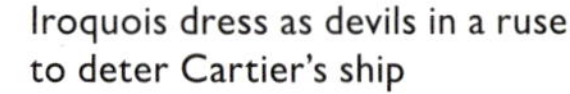

Iroquois dress as devils in a ruse to deter Cartier's ship

Battle at Lake Champlain
In 1609, a party of Montagnais, Algonquin, and Huron peoples recruited Champlain and a few of his men to attack the Haudenosaunee (Iroquois), turning them against the French for years to come. Champlain drew this interpretation of the fray.

THE MEXICA

The Mexica (whom Europeans would later call the Aztecs) migrated from their ancestral homeland and created one of the most influential empires in Mesoamerica, before it was destroyed by the Spanish conquest of the region.

△ **Fearsome giver of life**
A 15th-century ceramic vessel depicting the rain and fertility god, Tláloc, from the main temple in Tenochtitlán. The god's defining features are fangs and ringed eyes.

In the 12th or 13th century, the Mexica left their homeland of Aztlán, which was probably in what is now northern Mexico or the southwestern United States. The Mexica people belonged to a larger Indigenous group, the Nahuas. Pushed south by conflict, drought, or perhaps even divine inspiration, they migrated to central Mexico but did not find a permanent place to settle until 1325, when they saw a sign: an eagle perched on a cactus and eating a snake. In oral traditions, this was a message from the god Huitzilopochtli telling them they had reached their final destination.

The Mexica founded the city-state of Tenochtitlán on two islands in Lake Texcoco. Over the centuries, it grew into the capital of one of the most powerful empires in Mesoamerica. An awe-inspiring city of magnificent temples, palaces, and plazas and an ingenious network of canals, causeways, and artificial islands, it was home to more than 200,000 people at its height. However, in 1519, Spanish conquistadors led by Hernán Cortés came with their more numerous Indigenous allies and overthrew the Mexica in a period of violence and epidemics (see p.90). By 1522, the Spanish had razed most of Tenochtitlán, and used the rubble to build a new colonial capital on the site: Mexico City.

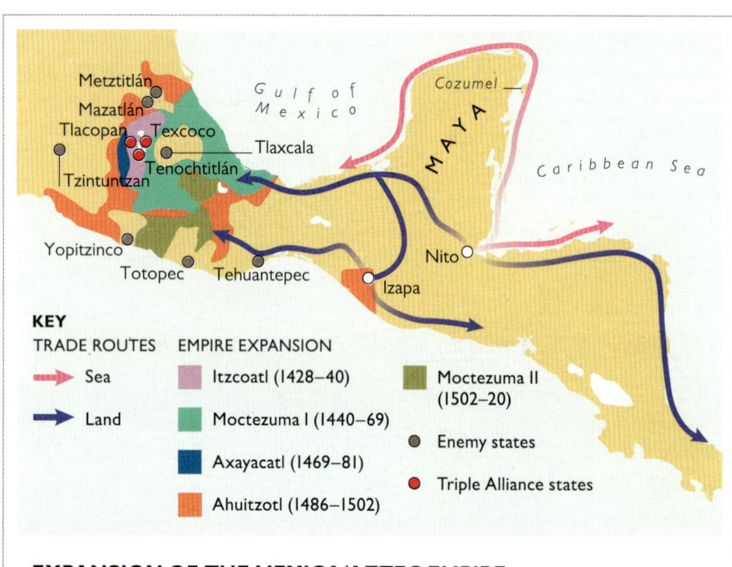

EXPANSION OF THE MEXICA/AZTEC EMPIRE
Tenochtitlán steadily grew and, in 1428 under Itzcoatl's rule, the Mexica formed the Triple Alliance with the Nahua city-states of Texcoco and Tlacopan. Together they dominated the region, bringing around 500 states under their cultural, political, and military sway and collecting huge tributes.

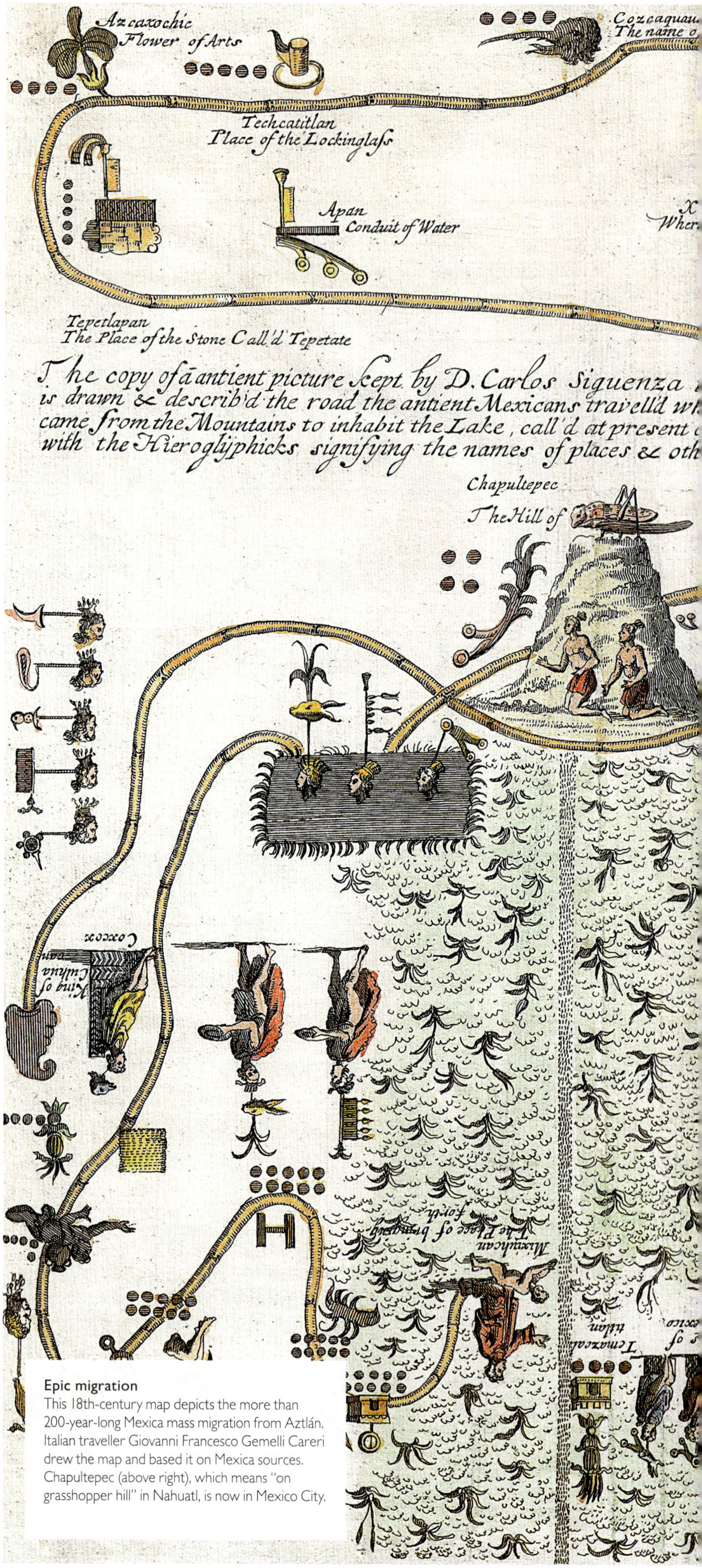

Epic migration
This 18th-century map depicts the more than 200-year-long Mexica mass migration from Aztlán. Italian traveller Giovanni Francesco Gemelli Careri drew the map and based it on Mexica sources. Chapultepec (above right), which means "on grasshopper hill" in Nahuatl, is now in Mexico City.

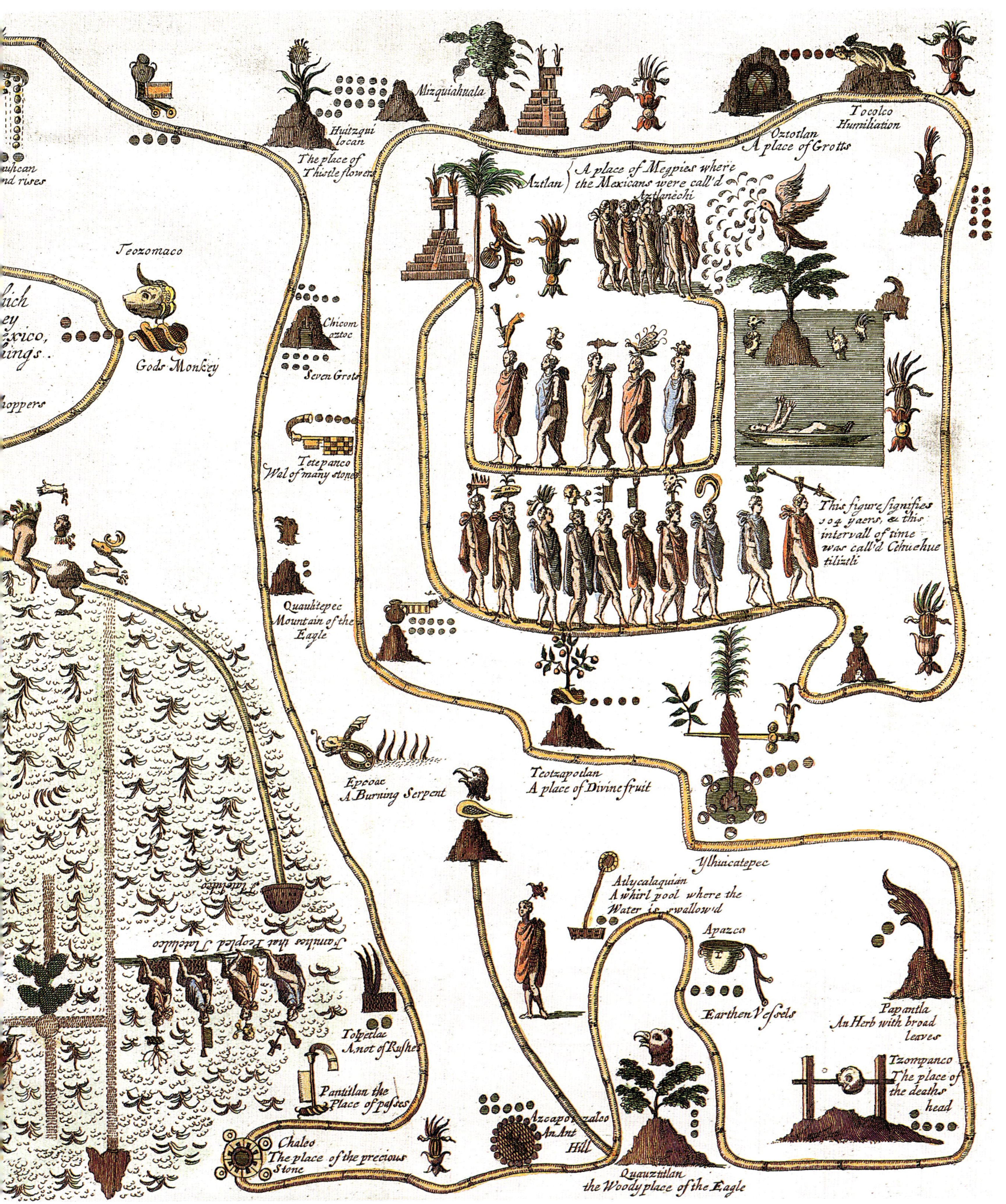

Mizquiahuala
Huitzqui locan
The place of Thistle flowers
Tocolco Humiliation
Oztotlan A place of Grotts
Aztlan
A place of Megpies where the Mexicans were call'd Aztlanechi
Teozomaco
Gods Monkey
Chicom oztoc
Seven Grots
Tetepanco Wal of many stones
This figure signifies 104 yaers, & this intervall of time was call'd Cehuehue tiliztli
Quauhtepec Mountain of the Eagle
Epeoac A Burning Serpent
Teotzapotlan A place of Divine fruit
Ylhuicatepec
Atlycalaquian A whirl pool where the Water is swallow'd
Apazco
Earthen Vessels
Papantla An Herb with broad leaves
Tzompanco The place of the deaths head
Families that Peopled Tlatelulco
Tolpetlac A not of Rushes
Pantitlan the Place of passes
Chalco The place of the precious Stone
Azcapotzalco An Ant Hill
Quauztitlan the Woody place of the Eagle

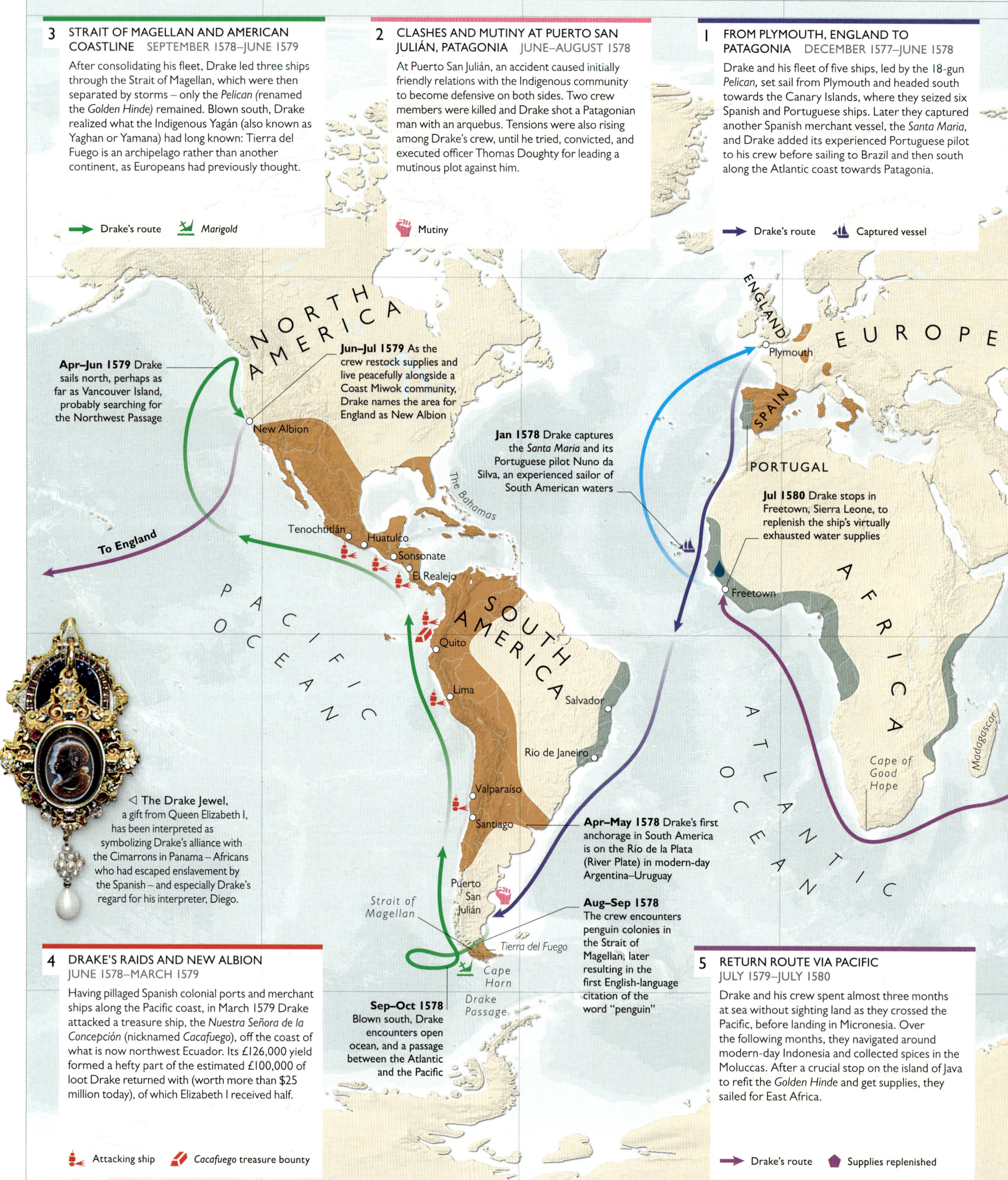

1 FROM PLYMOUTH, ENGLAND TO PATAGONIA DECEMBER 1577–JUNE 1578

Drake and his fleet of five ships, led by the 18-gun *Pelican*, set sail from Plymouth and headed south towards the Canary Islands, where they seized six Spanish and Portuguese ships. Later they captured another Spanish merchant vessel, the *Santa Maria*, and Drake added its experienced Portuguese pilot to his crew before sailing to Brazil and then south along the Atlantic coast towards Patagonia.

2 CLASHES AND MUTINY AT PUERTO SAN JULIÁN, PATAGONIA JUNE–AUGUST 1578

At Puerto San Julián, an accident caused initially friendly relations with the Indigenous community to become defensive on both sides. Two crew members were killed and Drake shot a Patagonian man with an arquebus. Tensions were also rising among Drake's crew, until he tried, convicted, and executed officer Thomas Doughty for leading a mutinous plot against him.

3 STRAIT OF MAGELLAN AND AMERICAN COASTLINE SEPTEMBER 1578–JUNE 1579

After consolidating his fleet, Drake led three ships through the Strait of Magellan, which were then separated by storms – only the *Pelican* (renamed the *Golden Hinde*) remained. Blown south, Drake realized what the Indigenous Yagán (also known as Yaghan or Yamana) had long known: Tierra del Fuego is an archipelago rather than another continent, as Europeans had previously thought.

4 DRAKE'S RAIDS AND NEW ALBION JUNE 1578–MARCH 1579

Having pillaged Spanish colonial ports and merchant ships along the Pacific coast, in March 1579 Drake attacked a treasure ship, the *Nuestra Señora de la Concepción* (nicknamed *Cacafuego*), off the coast of what is now northwest Ecuador. Its £126,000 yield formed a hefty part of the estimated £100,000 of loot Drake returned with (worth more than $25 million today), of which Elizabeth I received half.

5 RETURN ROUTE VIA PACIFIC JULY 1579–JULY 1580

Drake and his crew spent almost three months at sea without sighting land as they crossed the Pacific, before landing in Micronesia. Over the following months, they navigated around modern-day Indonesia and collected spices in the Moluccas. After a crucial stop on the island of Java to refit the *Golden Hinde* and get supplies, they sailed for East Africa.

◁ **The Drake Jewel,** a gift from Queen Elizabeth I, has been interpreted as symbolizing Drake's alliance with the Cimarrons in Panama – Africans who had escaped enslavement by the Spanish – and especially Drake's regard for his interpreter, Diego.

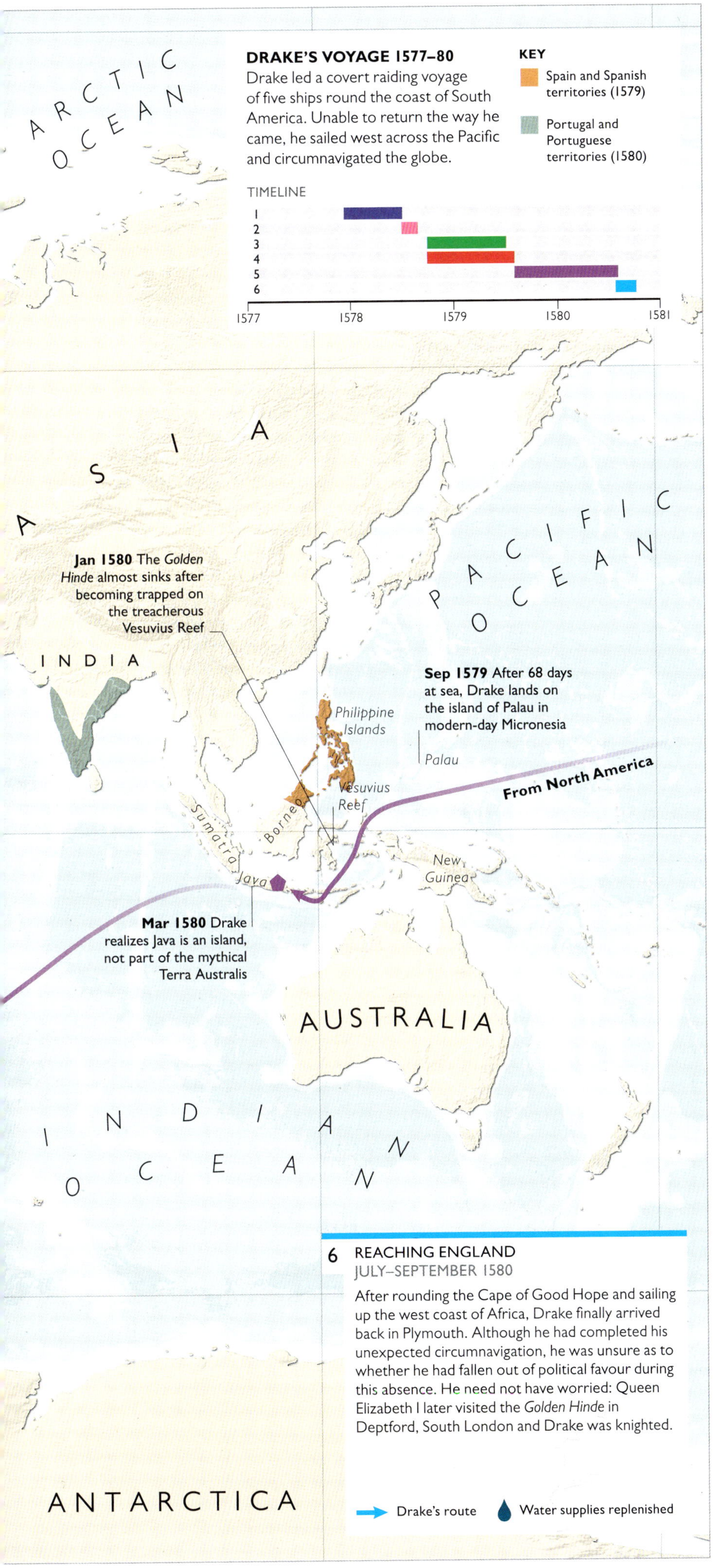

TRAVERSING THE WORLD

A renowned navigator of the Elizabethan Age (1558–1603), Sir Francis Drake was the first Englishman to circumnavigate the globe. He was also one of Britain's earliest traders of enslaved peoples, as well as a notorious privateer and pirate.

Born in Tavistock, Devon, England in about 1540, Francis Drake was brought up by relatives, the Hawkins – a wealthy maritime family of Plymouth. His cousin John Hawkins was the first English sea captain to sell enslaved Africans to Spanish colonies and, early in his career, Drake took part in slave-trading voyages to West Africa. He also raided merchant ships in order to seize their cargo, and was hated by the Spanish, who called him "El Draque" (The Dragon) because of the ferocity of his privateering against their colonial settlements and ships in the Caribbean, Central America, and beyond.

However, Drake's plundering endeared him to his English queen, who was keen to break the Spanish- and Portuguese-held monopoly on trade with the East. In 1577, Drake set off on an expedition unofficially backed by Queen Elizabeth I to raid Spanish ports and the treasure ships along the Pacific coast of South America. He later continued north, searching in vain for a sea route that would take him back east to the Atlantic. After failing to find such a route – now known as the Northwest Passage, which cuts through the Arctic Ocean – he instead headed west across the vast Pacific, ultimately becoming the first Englishman to circumnavigate the globe.

> *"Is the Queen alive and well?"*
>
> AFTER AN ABSENCE OF 1,046 DAYS, DRAKE CAUTIOUSLY RE-ENTERS ENGLISH WATERS IN 1580

THE SELK'NAM

The Selk'nam, also known as the Onawo or Ona people, are one of several Indigenous peoples who lived on the archipelago at the tip of South America. Magellan first encountered their bonfires, lighting up the coastline, in 1520 and thus named the place *Tierra del Fuego* (Land of Fire). They had inhabited the region for around 10,000 years before the arrival of Europeans and are likely to have encountered Drake during his passage through the Strait of Magellan.

A Selk'nam man in guanaco fur, c.1902

DRAKE'S VOYAGE 1577–80

Drake led a covert raiding voyage of five ships round the coast of South America. Unable to return the way he came, he sailed west across the Pacific and circumnavigated the globe.

KEY

- Spain and Spanish territories (1579)
- Portugal and Portuguese territories (1580)

TIMELINE: 1, 2, 3, 4, 5, 6 — 1577, 1578, 1579, 1580, 1581

ARCTIC OCEAN · ASIA · INDIA · PACIFIC OCEAN · Philippine Islands · Palau · Vesuvius Reef · From North America · Sumatra · Borneo · Java · New Guinea · AUSTRALIA · INDIAN OCEAN · ANTARCTICA

Jan 1580 The *Golden Hinde* almost sinks after becoming trapped on the treacherous Vesuvius Reef

Sep 1579 After 68 days at sea, Drake lands on the island of Palau in modern-day Micronesia

Mar 1580 Drake realizes Java is an island, not part of the mythical Terra Australis

6 REACHING ENGLAND
JULY–SEPTEMBER 1580

After rounding the Cape of Good Hope and sailing up the west coast of Africa, Drake finally arrived back in Plymouth. Although he had completed his unexpected circumnavigation, he was unsure as to whether he had fallen out of political favour during this absence. He need not have worried: Queen Elizabeth I later visited the *Golden Hinde* in Deptford, South London and Drake was knighted.

Drake's route · Water supplies replenished

RALEIGH'S EXPEDITIONS

As Spain grew rich from exploiting the resources of its colonies in the Americas, England was keen to follow suit. With the Spanish and Portuguese entrenched in South and Central America, the English looked to the North. Under a royal charter, Walter Raleigh sent three missions to establish colonies there.

On 27 April 1584, Raleigh sent the first expedition to North America. The colonists landed in July, in what is now North Carolina, naming the land "Virginia", in honour of Elizabeth I, the "Virgin Queen". Relations between the English and the Indigenous Secotan and Croatan peoples were good at first and, when a contingent returned home, two men – Manteo, a Croatan Chief, and Wanchese, a Roanoke – also travelled to England.

In 1585, Raleigh sent a larger force: seven ships under Sir Richard Greville. Leaving 100 men at Roanoke with Ralph Lane as governor, Greville returned to England to stock up. Lane's men built shelters before winter, but scarce provisions and deteriorating relations with the Secotan people meant that, when Sir Francis Drake arrived in the spring, the entire colony boarded his ship to return to England.

Raleigh tried again. He recruited 118 men, women, and children to create the "City of Raleigh" at Roanoke, led by John White, and in August 1587, Virginia Dare, White's daughter, became its first child born in the Americas. In late 1587, White sailed to England to replenish supplies, but when he returned in 1590, the colony was deserted. Its fate remains a mystery.

Yet, Raleigh refused to give up on his colonial ambitions. In 1595, he went to South America in search of El Dorado only to return empty-handed.

LOCATOR

KEY

1 In 1595, Raleigh set off up the Orinoco River in northeast South America, hoping to find El Dorado.

2 Raleigh wrote a popular book that set El Dorado by "Lake Parime" (based on a report by one of his men). It soon appeared on maps.

3 This drawing of a headless man typifies the exoticism of the Americas as seen from a European viewpoint.

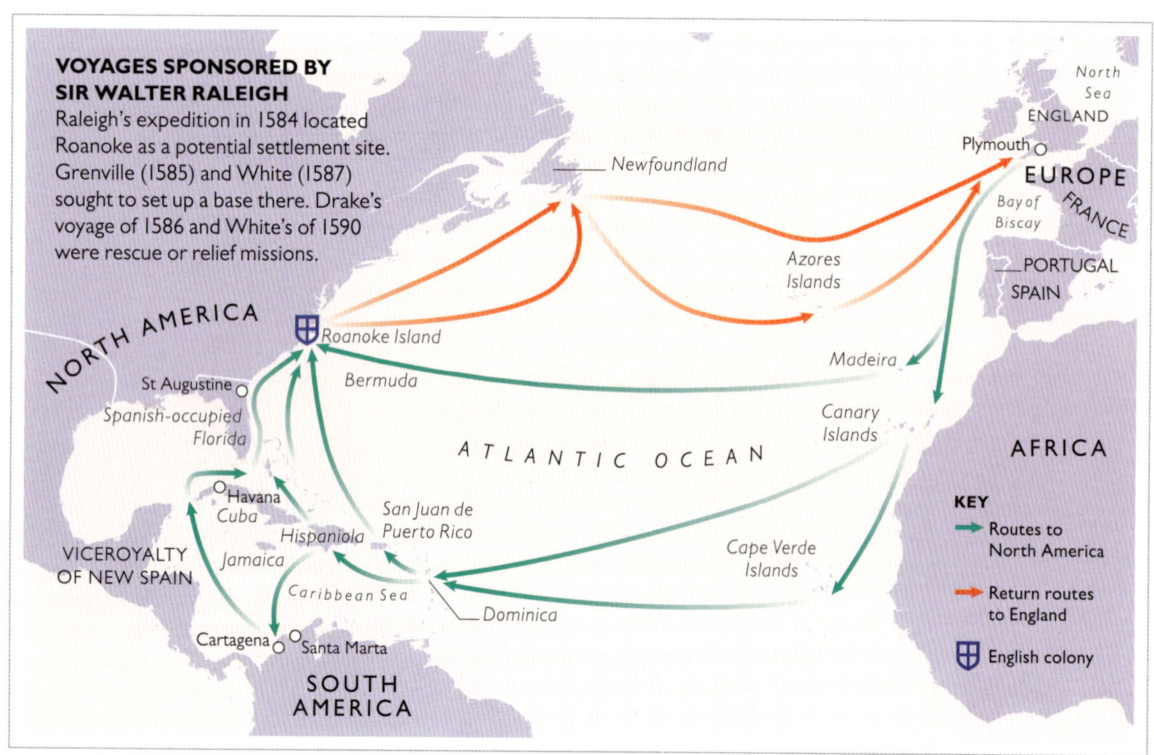

△ **City of gold**
Raleigh hoped to boost his standing at the English court by finding El Dorado, a mythical city of gold. Some stories placed it in Guyana, as in this map of 1598, a mix of fact and fantasy by Jodocus Hondius. Raleigh's trek along the Orinoco River in 1595 uncovered rich jungle life but no golden realm.

"Whoever commands the sea, commands the trade; whosoever commands the trade of the world commands the riches of the world, and consequently the world itself."

SIR WALTER RALEIGH, "A DISCOURSE OF THE INVENTION OF SHIPS, ANCHORS, COMPASS, &C.", 1650

Inuit as seen by Elizabethans
In 1577, seeking the Northwest Passage (see pp.78–79), explorer Martin Frobisher came across an Inuit man, woman, and their baby. An unknown artist based this image on anecdotal descriptions of their appearance and attire.

ARCTIC MIGRATIONS

For much of the last thousand years, Europeans, from the Vikings onward, moved westwards and colonized North America. But in the far north, it was a different story, with Arctic populations migrating east.

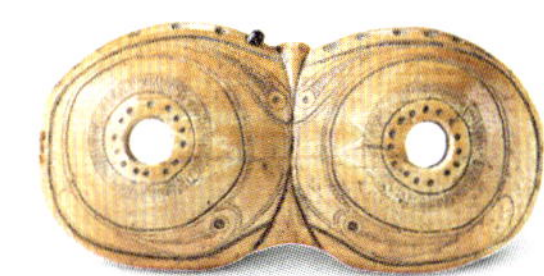

△ **Snow goggles**
Carved from walrus ivory, between 200 and 400 CE, these snow goggles may have been worn for funeral or hunting rituals.

It takes a hardy culture to live in the icy far north in the Arctic. With few plants available, its inhabitants rely on hunting. Between 2,500 and 1,500 years ago, ancestors of the Dorset and Inuit peoples came to North America and spread to the Arctic. The Dorset people chiefly hunted sea mammals through ice holes. As the climate warmed, *c.*1,000 years ago, they moved north and east to follow the ice, reaching Greenland. The Thule people were better able to survive the warming as they hunted on open water in skin-covered boats, using harpoons to catch sea mammals.

In 986 CE, Erik the Red led the Vikings to Greenland. Dwindling sea ice had reduced the Dorset population, but for centuries the Vikings shared the island with the Thule, who spread down the west coast. The Vikings also vanished, perhaps because the climate cooled and they failed to adapt, leaving just the Thule – Greenland's Inuit ancestors.

> *"The Thule people displayed a perfect mastery of their environment, both on land and sea."*
>
> JACQUES PRIVAT, *MYSTERIES OF THE FAR NORTH*, 2023

THE GREENLAND UMIAQ

The umiaq was key to the Thule's success. Made of sealskin stretched over a driftwood frame, the boat was light enough to carry over the ice, yet could accommodate up to 15 people and their possessions. This enabled the Thule to travel quickly and easily to their seasonal hunting grounds. With an umiaq and dogsleds at their disposal, the Thule could move readily as conditions changed.

▷ **Fragment of the world**
Drawn on gazelle skin, only a third of the Piri Reis map remains. Despite claims that it was based on unique, up-to-date knowledge, much of its geography draws on centuries-old maps that Piri Reis would have found in the Topkapi Palace library.

◁ Drawing on Columbus
It is thought that the positioning of this Caribbean island comes from one of Columbus's lost maps, which depicted Cuba as part of the Asian mainland and Hispaniola in line with Marco Polo's description of Japan.

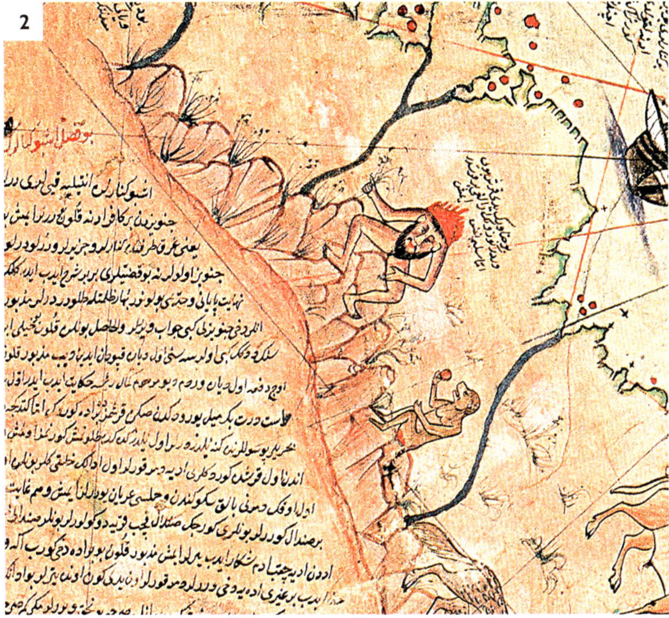

△ Strange monsters
Along the map's western edge, beyond an accurately drawn coast of South America, Piri Reis introduced various weird mythical monsters, including a headless man or "Blemmye" sitting alongside a monkey.

△ Mythical southern continent
Despite wild theories that Piri Reis somehow knew of Antarctica, the southern land mass is most likely showing Terra Australis – a hypothetical continent based on a theory that the world's hemispheres must balance.

OTTOMAN TRAVELS

The Ottomans emerged from Turkish tribes in the 14th century to control a vast empire that reached its peak in the 16th century, when military conquests of Egypt and Mesopotamia led to Ottoman interest in the outside world expanding, as the Piri Reis map shows.

Piri Reis, or Ahmed Muhiddin Piri, was first a privateer and later an admiral in the Ottoman Navy, with extensive experience sailing the Mediterranean. He was also a cartographer and drew a portolan-style (see pp.68–69) world map in 1513 that is so detailed for its time that scholars are baffled as to how its creator acquired the knowledge to show the Americas with such accuracy – only 21 years after Columbus's voyage (see pp.80–81) and barely a decade after Cabral arrived in Brazil (see pp.86–87).

The Ottoman conquests, which enabled access to the Red Sea, Persian Gulf, and Indian Ocean, also heightened rivalry with the Portuguese; acquiring knowledge about this part of the world became essential. In 1554, an admiral of the Ottoman Egyptian fleet, Seydi Ali Reis, led two battles against the Portuguese. However, caught in a typhoon at sea, he became marooned on the Gujarat coast of India and was forced to return home overland with 50 men. His account of this journey, *Mir'ât ül Memâlik* ("The Mirror of Countries") describes the peoples, cultures, and customs he encountered as he travelled through Turkestan, Kipchak Steppe, Khorasan, and Iran.

During 1579–89, an Ottoman corsair, Mir Ali Beg, also led naval expeditions to East Africa and was instrumental in making contact with Muslim lands and traders in the Indian Ocean.

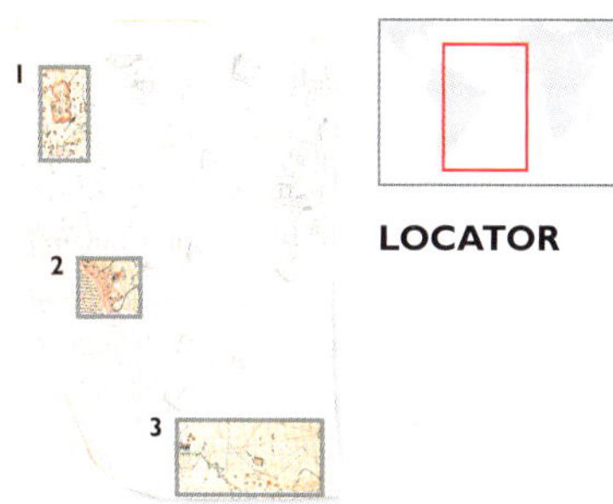

LOCATOR

EVLIYA ÇELEBI 1611–1682

In the 17th century, Evliya Çelebi journeyed throughout the Ottoman Empire and Europe, from Crimea to Cairo, for over 40 years. His epic, 10-volume travelogue *Seyahatnâme* ("Book of Travels") colourfully recounts "tall tales", in the style of a traveller returning from strange lands, but also includes detailed observations of daily life. He censors the treatment of Indigenous Americans in the Netherlands and Tatar slaves in Crimea.

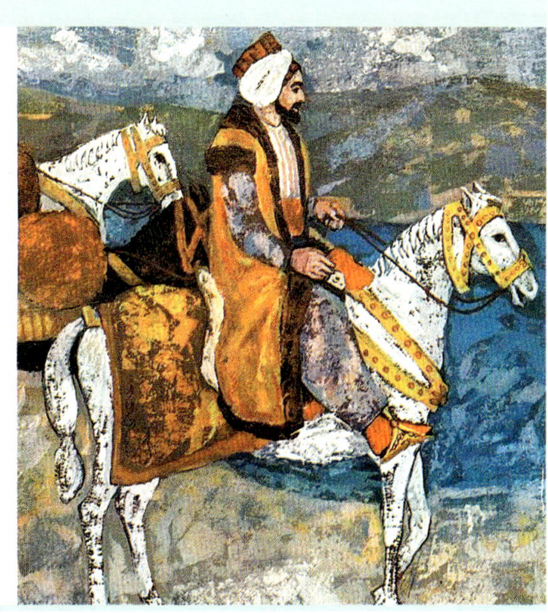

A contemporary portrayal of Çelebi in traditional dress

"Whoever is lord of Malacca has his hand on the throat of Venice."

16TH-CENTURY PORTUGUESE WRITER TOMÉ PIRES HIGHLIGHTING THE IMPORTANCE OF MELAKA (MALACCA) TO THE SPICE TRADE

△ **The "Venice of the East"**
At the centre of Asian trade, Melaka in present-day Malaysia became a Portuguese colony in 1511, when the Portuguese seized the bridge at the port's entry (above left), effectively dividing the city into two. Pedro Barreto de Resende drew this map over a century later in 1641, after the Dutch took over.

THE CRAZE FOR SPICES

Spices grown in Asia, such as nutmeg and cloves, were in huge demand in Europe in the 1400s. Finding a direct sea route to Asia was therefore a key driver behind European exploration, and efforts to control the spice trade involved coercion, force, and violence.

The Maluku Islands or Moluccas in present-day Indonesia were a pivotal destination as the Indian Ocean was home to Asian and African traders, who brought spices by ancient land and sea routes to the eastern Mediterranean. From 1500, the Portuguese sailed around the Cape of Good Hope with the intention of claiming such lucrative commerce for themselves. However, having no goods that the spice merchants wanted to trade, they resorted to warfare and built forts, starting with Kochi in India in 1503.

The capture of the key trading port of Melaka in 1511 helped the Portuguese wrest control of the spice trade. Still, Portuguese controls on trading ships crossing the Indian Ocean failed, and there were battles with Asian rulers and merchants, who also collaborated with the Europeans when it was in their own interests to do so.

The Dutch arrived in 1599 and allied with Maluku sultans, leading to the Dutch–Portuguese wars. When the Banda people resisted a Dutch monopoly on their nutmeg-producing islands in 1621, it led to a genocide. The British had ambitions there too, having taken control of Run in 1616. Yet in 1677, the British agreed to Dutch claims on the island in exchange for Manhattan – then thought comparatively worthless.

LOCATOR

KEY

1 The old quarter, with its mosques, lay over the river, west of the Portuguese settlement.

2 The central Portuguese citadel features St Paul's Catholic Church, perched on a hilltop.

3 The newer Dutch fortifications are located to the east of the old Portuguese city.

NUTMEG

Common nutmeg is the seed of a fruit that grows on the *Myristica fragrans* tree. Originally it grew only on a few tiny islands in the Malukus, but it is now also cultivated in the Caribbean. The pulp of the fruit is eaten locally, yet it is the dried nutmeg stones that are valued and ground into a nutty, slightly sweet powder. Mace comes from nutmeg's reddish seed covering. The Romans used nutmeg as incense, but in the late Middle Ages, Western Europeans used it as both a flavouring and as a hallucinogenic drug. It was also sought after for folk remedies.

A botanical drawing of nutmeg and mace

◁ **Celestial surveyor** Tsar Peter the Great (r.1682–1725) encouraged the import of sophisticated English astrolabes in the early 1700s to make accurate maps.

6 THE EASTERN CAPE 1648

In 1643, Cossack leaders Semyon Dezhnev and Mikhail Stadukhin reached the Kolyma River and built an ostrog at Srednekolymsk. In 1648, Dezhnev led kochs into the Arctic Ocean and Bering Strait. Most of the expedition perished in storms or battles with locals; Dezhnev and a few men reached shore and built a boat to sail home up the Anadyr River.

5 FINDING BAIKAL 1643

One of the great discoveries for Russia was Lake Baikal, the world's deepest lake. Guided by the Tungus prince Mozheul, Cossack leader Kurbat Ivanov and his men set off from Verkholensk to climb the Primorsky Ridge, sighting the lake during their descent on the other side. Ivanov was the first Russian to map the lake.

BUILDING AN EMPIRE

After 1585, Russian expansion paused until the late 1620s, when the empire grew steadily. Rivers were key to the conquest of Siberia. The Russians moved between the Ob, Yenisei, and Lena rivers, creating routes that became crucial to the Russian fur trade.

KEY

RUSSIAN EMPIRE BY

1505 | 1599 | 1672 | 1689

TIMELINE

1, 2, 3, 4, 5, 6

1580 1595 1610 1625 1640 1655

1 KHANATE CONQUERED 1582–85

Russian expansion gathered pace when rich merchants, the Stroganovs, hired the Cossack ataman or leader Yermak Timofeyevich to take control of the nearest part of Siberia – the Khanate of Sibir. This northernmost Muslim state was under the control of Khan Kuchum. Yermak captured Sibir Khanate and its capital, Qashliq. Yermak was later defeated and killed, but the venture meant Russians had explored the river routes of the West Siberian Plain and could push east.

2 BURYATS AND SAKHA 1628–31

In 1628, Cossack adventurer Pyotr Beketov encountered Buryat people (a Mongolic ethnic group native to southeastern Siberia) at the future Bratsk. He forced a tribute from them. Beketov created the first of many "ostrogs", small wooden forts built by Russian adventurers across Siberia. A few years later, in 1631, Beketov moved further east, into the land of the Sakha people, and built an ostrog at Yakutsk, which later became one of the key Siberian cities.

THE EXPANSION OF RUSSIA

In the 1400s, Russia comprised Muscovy, a small dukedom centred on Moscow under Tatar domination. Towards the end of the 1400s, it began to expand, first westwards, then east, through the territory of the Tatars and beyond. Eventually, Russia took over all of Siberia, building a vast empire made up of hundreds of ethnicities.

The drive east began under Tsar Ivan the Terrible (r.1547–84), who was determined to expel the Muslim khanates and obtain rumoured riches beyond the Urals. His armies took Kazan in 1551, Astrakhan in 1557, and in 1582, a Cossack leader, Yermak, was sent to crush the largest territory, Sibir.

Russian expeditions pushed on into Siberia, using the region's huge rivers to navigate its forests. Leaders such as Pyotr Beketov and Semyon Dezhnev penetrated ever further east, extending Russia's boundaries as far as the Pacific.

Russian history often portrays these individuals as explorers. However, the Russian tsars employed what were in effect mercenaries so they could expand their domains and dominate trade. The impact of the invasions on Indigenous peoples was catastrophic as huge numbers were slaughtered and millions more died from smallpox introduced by the Russians. So-called Russian settlers were often serfs forced to replace local people. In 1742, Russian Empress Elizabeth (r.1741–62) ordered that the Chukchi people of the far east be "totally extirpated". This devastated Siberia's native population, three-quarters of whom were killed, wiping out at least 12 native groups. The Russian advance decimated wildlife, too, as animals were slaughtered for their coveted fur.

> *"The history of Russia is the history of a country being colonized."*
>
> VASILY KLYUCHEVSKY, *A HISTORY OF RUSSIA*, 1904–10

FUR FEVER

Russian expansion east was driven partly by its demand for furs, which by the mid 1600s made up more than a tenth of the country's entire revenue. The most sought-after fur was sable from the northeast. A single sable pelt cost as much as a peasant would earn in ten years. Few Russians had the skill to catch sable without damaging the animal's fur, so they exacted "yasak" – a "tribute", or tax, in furs – from Indigenous hunters.

A Russian demands fur in this 19th-century artwork by an unknown artist

"The fourteenth day, we landed all our men, which were set to work about the fortification, and others some to watch and ward as it was convenient."

GEORGE PERCY, AN ENGLISH COLONIST, ON BUILDING JAMESTOWN'S FORT. DIARY, MAY 1607

△ **Colonies in North America**
Produced by Dutch cartographer Johannes Vingboons in 1639, this map shows the early European colonies on the northeastern seaboard. The English settlements are in green, the Dutch in mustard, and the French in maroon. Today, the region includes the states of New York, New Jersey, Delaware, and Connecticut.

EUROPEANS IN NORTH AMERICA

From the early 17th century, European settlers began to colonize eastern North America, led by the English, Dutch, and French. The land was inhabited by Indigenous people so, as was their practice, the colonists acquired territory by trade-offs or force.

In 1607, the English established their first colony, "Jamestown", on Chesapeake Bay, Virginia. The French set up Quebec on the Saint Lawrence River in 1608, while the Dutch West India Company founded New Amsterdam (now New York) in 1624 as the centre for its New Netherland colony. Many French and Dutch colonists set up trading posts to acquire furs – a valuable commodity – from Indigenous people. But soon companies and investors brought in many more settlers, including families. They built farms that generated their own wealth and supported the outposts.

To the newcomers, the terrain appeared to be "unused", as it was not intensively farmed. But to Indigenous populations it was a rich resource that had sustained them for millennia in their hunting, foraging, and light agriculture. When the Dutch paid or traded for this land, they presumed it was theirs; but private ownership of land was an alien concept to Indigenous nations, who assumed they were only granting the rights to use it. Ultimately, the Europeans took the land by force.

Enslaved African people provided the hard labour of colony building. It was their hands that built a defensive barrier for New Amsterdam in 1653, remembered in the name "Wall Street".

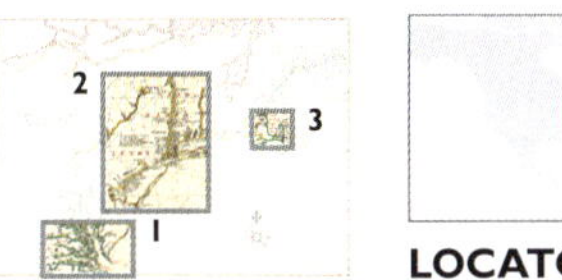

KEY

1 Chesapeake Bay became a major focus for English settlement in North America.

2 Dutch colonists built New Netherland around the Hudson River.

3 Plymouth was the northern focus of English colonization. In 1620, the pilgrims on the *Mayflower* landed at nearby Cape Cod.

MATOAKA (POCAHONTAS) c.1596–1617

Much is unknown about the Powhatan woman called Matoaka, best known by her childhood nickname "Pocahontas". But she played a key role in enabling the Jamestown settlers to survive and taught them to grow tobacco, a valuable cash crop. Although she was a willing visitor to the colony as a young girl, and led cordial exchanges, she was later captured and held for ransom. Matoaka then married – possibly unwillingly – Englishman John Rolfe, who wrote that he did so partly for "the benefit of this Plantation". They had a child and Rolfe took his family to London, where Matoaka became a celebrity. But after a brief illness, she died there, aged only 21.

Matoaka portrayed as a European

Portrait of a fakir
British merchant Peter Mundy spent three years with the Levant Company in Constantinople. His *A Briefe Relation of the Turckes* (1618) comprised illustrated vignettes of the city's diverse ethnic groups, including Persian, Armenian, and Jewish people.

TRAVELLERS IN ISLAMIC EMPIRES

In the 17th century, Europeans visited the Safavid and Mughal empires, which dominated Persia and India. Both realms had a refinement and wealth unrivalled in Europe, especially Mughal India under Shah Jahan. Travellers from these territories also visited Europe.

European conceptions of India in the late 17th century were widely formed by the book *Travels in the Mughal Empire* (1670–71). It was written by François Bernier, a Frenchman who worked as a physician in the subcontinent, first to Prince Dara Shikoh, and then to Aurangzeb, the last great Mughal emperor.

In 1653, English soldier Henry Bard was sent to Persia, to recover an alleged debt from Shah Abbas for the exiled Charles II. Bard died three years later, but his teenage secretary, Venetian Niccolao Manucci, spent fifty years in Mughal India and documented his experiences in the travelogue *Storia do Mogor* ("Stories of the Mughal").

△ **Dagger handle**
This equine Mughal knife hilt is made of jade that has been inset with gold and jewels. It was probably produced during the later 17th century or early 18th century.

From *c.*1600, the activities of companies such as the powerful British East India Company (EIC), which fought wars using its own army and navy, inserted themselves into the subcontinent, directly linking it with the West. Over the next 250 years, some 20,000–40,000 Indian men and women travelled to Europe as visitors, tutors, or in the employ of such companies (see pp.194–95). Bengali-born diplomat Mirza Sheikh I'tesamuddin worked variously for the EIC and Emperor Shah Alam II. In 1765–66, he visited Britain, France, and southern Africa on behalf of the shah. He recorded his sharp-eyed impressions of Britain in *Shigurf Namah-I-Vilaet* ("Wonderful Tales about Europe"), published in 1785 – the first travelogue about the West written by an Indian author.

MIRZA ABU TALEB KHAN 1752–1806

In 1799, Persian-born Indian Mirza Abu Taleb Khan visited Britain. He was struck by Ireland's rural poverty but dazzled by Dublin's streetlights. In London, he met Queen Charlotte and was awed by the new industrial factories. Abu Taleb later visited France, Italy, the Ottoman Empire, Kurdistan and Persia. In 1803, he finally returned to India, where he wrote an account of his travels.

Abu Taleb in a 19th-century portrait

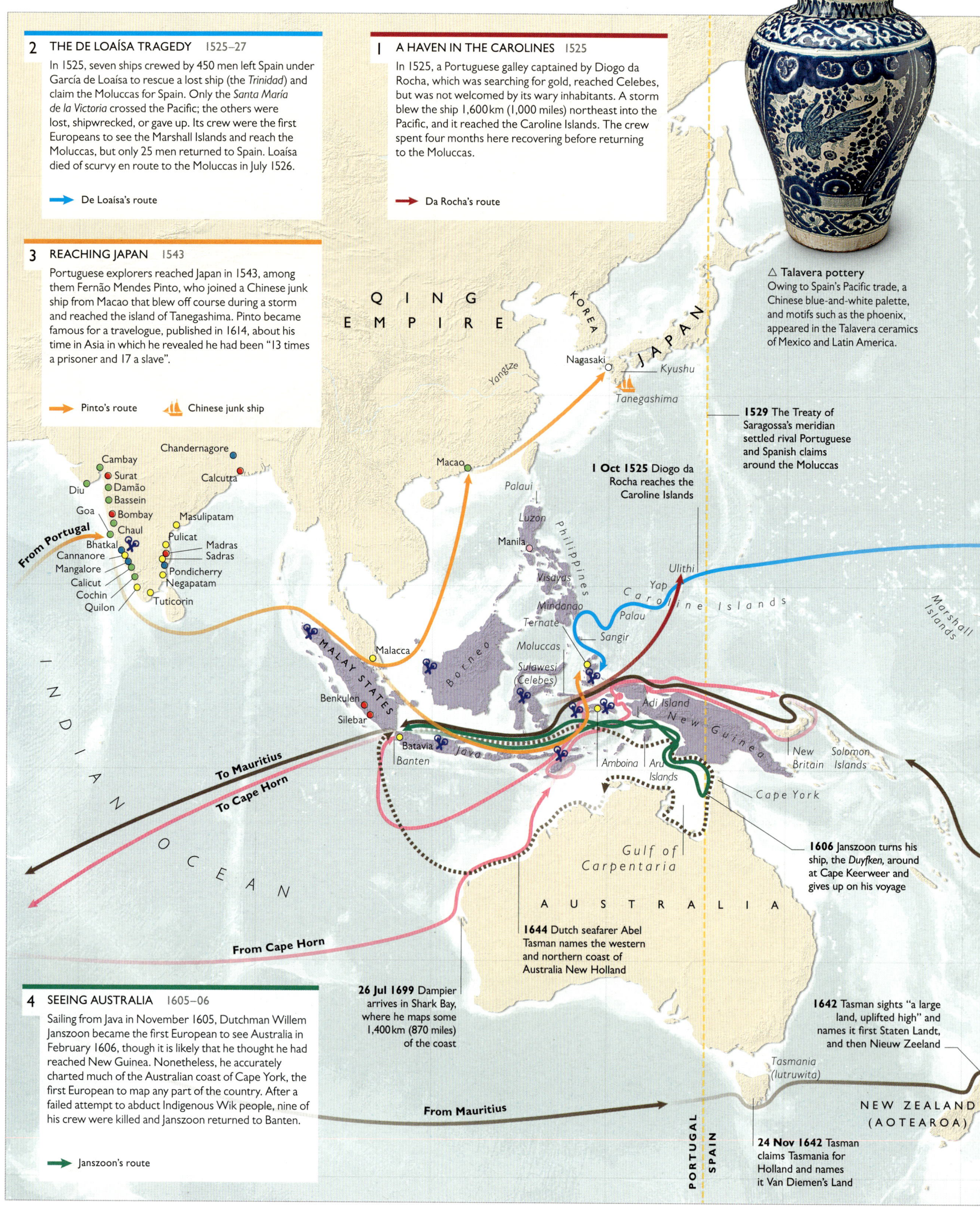

2 THE DE LOAÍSA TRAGEDY 1525–27

In 1525, seven ships crewed by 450 men left Spain under García de Loaísa to rescue a lost ship (the *Trinidad*) and claim the Moluccas for Spain. Only the *Santa María de la Victoria* crossed the Pacific; the others were lost, shipwrecked, or gave up. Its crew were the first Europeans to see the Marshall Islands and reach the Moluccas, but only 25 men returned to Spain. Loaísa died of scurvy en route to the Moluccas in July 1526.

De Loaísa's route

1 A HAVEN IN THE CAROLINES 1525

In 1525, a Portuguese galley captained by Diogo da Rocha, which was searching for gold, reached Celebes, but was not welcomed by its wary inhabitants. A storm blew the ship 1,600 km (1,000 miles) northeast into the Pacific, and it reached the Caroline Islands. The crew spent four months here recovering before returning to the Moluccas.

Da Rocha's route

3 REACHING JAPAN 1543

Portuguese explorers reached Japan in 1543, among them Fernão Mendes Pinto, who joined a Chinese junk ship from Macao that blew off course during a storm and reached the island of Tanegashima. Pinto became famous for a travelogue, published in 1614, about his time in Asia in which he revealed he had been "13 times a prisoner and 17 a slave".

Pinto's route Chinese junk ship

△ **Talavera pottery**
Owing to Spain's Pacific trade, a Chinese blue-and-white palette, and motifs such as the phoenix, appeared in the Talavera ceramics of Mexico and Latin America.

4 SEEING AUSTRALIA 1605–06

Sailing from Java in November 1605, Dutchman Willem Janszoon became the first European to see Australia in February 1606, though it is likely that he thought he had reached New Guinea. Nonetheless, he accurately charted much of the Australian coast of Cape York, the first European to map any part of the country. After a failed attempt to abduct Indigenous Wik people, nine of his crew were killed and Janszoon returned to Banten.

Janszoon's route

EUROPEANS IN THE PACIFIC

During the 16th century, Portuguese vessels from the east and Spanish ships from the west explored the South Pacific for spices and set up colonies and trading posts. In the following century, the British and Dutch joined them.

KEY

Source of spices | East Indies | Treaty of Saragossa meridian, 1529

COLONIES AND TRADING POSTS

British | Portuguese | French | Dutch | Spanish

TIMELINE

1, 2, 3, 4, 5, 6

1500 | 1550 | 1600 | 1650 | 1700

6 AUSTRALIA BOUND 1699

From 1678 to 1691, Englishman William Dampier got to know the Pacific well as a pirate, reaching Australia in 1688. In 1699, he was sent there again by the British and made the first scientific study of its terrain, seas, and biology. The voyage was curtailed when Dampier's ship, the *Roebuck*, rotted, but he went on to become the first person to circle the globe three times.

Dampier's route

PACIFIC OCEAN

From Spain

Fiji Islands

21 Jan 1643 Sailing northeast, Tasman sees Tonga and then, on 6 February, the Fiji Islands

Tonga

Anaa

Amanu

Tuamotu

5 SOUTH OF AUSTRALIA 1642–44

In 1642, Dutch seaman Abel Tasman left Java to explore a landmass far south, sailing west to Mauritius, then south to catch the eastward-blowing Roaring Forties winds. Meeting snowstorms, he headed northeast, reaching Tasmania, but a second storm blew them to New Zealand. Tasman went on to Tonga and the Fiji Islands. In 1644, he explored the north Australian coast.

Tasman's 1642–43 route

Tasman's 1644 route

SAILING THE PACIFIC

After Ferdinand Magellan's round-the-world expedition of 1521, Portuguese and Spanish vessels voyaged widely throughout the Pacific Ocean during the 16th century, in search of spices and other valuable trade goods. As they did so, they came across numerous Pacific islands.

Just four years after Magellan's voyage (see pp.88–89), Diogo da Rocha and Gomes de Sequeira sailed east from the Moluccas (Malukus) to reach the Caroline Islands. Also in 1525, a Spanish expedition sailed west from South America under García Jofre de Loaísa. After his death, command passed to Juan Sebastián Elcano (see pp.88–89), then Alonso de Salazar, who sighted Taongi in the Marshall Islands in 1526.

Trade was established with Asia when Portugal began trading with China *c.*1513, and with Japan in 1543. Chinese and Japanese ships regularly traded in the Philippines and traffic along the Luzon–Kyushu route grew, especially in wheat and silver.

From 1565, huge Spanish ships known as "Manila Galleons" traded across the Pacific, bringing goods such as Chinese porcelain to Mexico, via the Philippines. Drake's circumnavigation (see pp.102–03) and the Spanish Armada's 1588 defeat broke Spain's dominance, and Dutch and British ships began to explore the Pacific from the west.

In 1606, Dutch captain Willem Janszoon saw Australia, but probably mistook it for New Guinea; in 1616, Dutchman Dirk Hartog explored an island off Australia. In 1688, Englishman William Dampier also mistook Australia for New Guinea, but he reached the west coast in 1699. Captain Cook reached its east coast 70 years later (see pp.134–35).

"RED SEAL" SHIPS

After 1550, the Portuguese ran the "nanban trade" that notably exchanged Chinese silk for Japanese silver in giant galleons ("nanban" is a Sino-Japanese word denoting European visitors). By 1600, piracy by Japanese junks led the Japanese ruler Tokugawa Ieyasu to issue "red seal" permits to favoured Japanese, Chinese, and European junks, enabling them to trade around Southeast Asia protected by Portuguese, Spanish, Dutch, English, and Japanese vessels.

Conflict on the high seas
Pirates Attack an English Warship was painted in the style of William van de Velde, a leading 17th-century seascape painter. In reality, most pirate ships avoided confrontations with naval vessels, which heavily outgunned them.

TERROR FROM RHODE ISLAND

Some of the longest voyages in the 17th century were made not by explorers but by pirates from the Rhode Island colony, such as Thomas Tew and Edward Low, who sailed halfway around the world in search of plunder.

From 1650 until the end of the century, piracy was perhaps the main industry in Rhode Island, on the northeastern coast of North America. It was a haven for English pirates who equipped their ships there. Many took commissions (or "letters of marque") from England's High Court of Admiralty, making them "privateers". This licensed them to capture enemy vessels, bring the spoils back to Rhode Island, and take legal ownership of them – minus a cut for the government of the English colony.

△ **Flintlock pistol**
Handguns such as this one, dating to c.1700, were used in close combat by pirates (including the English pirate Blackbeard), but were not accurate at long range.

Initially, the pirates attacked only foreign shipping, not local vessels. They made extraordinary voyages known as "pirate cruises" right across the Atlantic, around the Cape of Good Hope, and into the Indian Ocean, to prey on ships there. The crews would be away for a year or more; Madagascar became infamous as one of their bases. They targeted ships laden with exotic goods voyaging from India to Southwest Asia (the Middle East), as well as the port city of Mocha on the Red Sea, where rich Muslims landed on their Hajj pilgrimage (see pp.122–23).

The infamous pirate Thomas Tew embarked on two epic pirate voyages, in 1692 and 1694. He sailed from Bermuda, round Africa and up to the Red Sea, then raided shipping across the Indian Ocean. The heyday of the Rhode Island pirates ended in 1723, when brutal English pirate Edward Low and sidekick Charles Harris were bested by HMS *Greyhound*. Low escaped, but 26 others, including Harris, were taken to Newport, a town on Rhode Island, and hanged.

ANNE BONNY c.1698–1782

Bonny became a pirate in 1720 after running away with "Calico Jack" Rackham on his ship *William*. According to legend, she dressed as a man in battle, and was known for her bravado and brutality. Bonny was later joined by the equally tough Mary Read. The *William* was finally seized by a privateer and both women were sent to jail, where Read died. Bonny's fate is less clear, though some reports suggest that she was released and lived into old age.

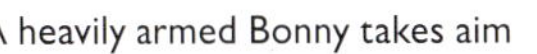

A heavily armed Bonny takes aim

▷ **The Great Mosque**
A French 19th-century lithograph taken from an Arab miniature of the Masjid-al-Haram at Mecca. This section depicts the Great Mosque and Kaaba. Such images featured on scrolls given to pilgrims to certify visits to shrines and the completion of the Hajj.

△ **The Kaaba**
Veiled in black cloth, Islam's holiest shrine is the focal point. The crescent represents a metal wall supporting lamps. Outside is the *mataf*, where pilgrims circulate around the Kaaba.

◁ **Coloured porticoes**
Domed *riwaq* (arcades) fringe the image. These arched structures were built around the mosque's central courtyard during the 16th-century Ottoman period.

▷ **Minarets**
Each tower has a balcony from which a *muezzin* calls the faithful to prayer. Other minarets have been added over time; the Great Mosque has more than any other.

▽ ***Maqam***
Four *maqamat* (chapels) once stood near the Kaaba, each one devoted to a separate Sunni school of thought. Imams from each of the four schools led daily prayers from them.

HAJJ ACCOUNTS

Performing the Hajj – the annual pilgrimage to Mecca – is a sacred duty in Islam that stretches back some 1,400 years. All believers are expected to complete the trek at least once. In the 17th and 18th centuries, the Muslim realm was vast, stretching from Morocco to India, and many pilgrims wrote narratives recording their long journeys.

Sometimes Hajj journeys were taken alone, but more often pilgrims joined huge caravans many thousands strong. When scholar Muhammad Ibn al-Tayyib began his journey in Fez, Morocco, in 1726, he was alone – but by the time he reached Mecca, in what is now Saudi Arabia, he had joined a vast convoy. He documented this journey in the book *al-Rihla al-Hijaziyya*, commenting on the landscapes, politics, and cultures he came across, the qualities a caravan's leader should possess, and hazards along the way.

Yusuf Rumi's Hajj journey in the early 18th century took him from his hometown of Sarajevo to Mecca via Plovdiv, Istanbul, and Damascus. It, too, reads like a modern traveller's guide, with reviews of places to stay and the quality of their water. Rumi also cautions against making 15-hour treks across the arid Arabian desert.

◁ **Ivory qibla mechanism**
This late 18th-century device is a hybrid sundial, compass, and "qibla" indicator pointing to Mecca. In the centre is a depiction of Kaaba, the sacred edifice at the heart of Masjid al-Haram, Mecca's "Great Mosque".

One of the earliest known Hajj narratives by a woman was written in the late 17th-century by the well-educated widow of the Safavid bureaucrat Mirza Khalil. Her name is unknown but she is sometimes described as "the lady of Isfahan". Initially, she used the Hajj journey to visit relatives in the wake of her husband's death and ventured north to the Southern Caucasus near the frontier fortress of Kars. After crossing into the Ottoman Empire and reaching Aleppo and Damascus, she joined the vast official caravan and memorably describes it as like "a vast sea", with its coloured banners filling the barren plain like "a wild garden of tulips".

> *"Who has not seen with his own eyes this Iraqi caravan has not experienced one of the genuine marvels of the world."*
>
> THE TRAVELS OF IBN JUBAYR, 12TH CENTURY

RAIDING PARTIES

The Hajj was not a safe journey. Assaults on caravans by Bedouin raiders occurred regularly and could be ferocious. The worst case took place in 1757, when Banu Sakhr robbers attacked and looted a cavalcade en route from Damascus in the Jordanian desert near Dhat al-Hajj, a convoy way station. An estimated 20,000 pilgrims are thought to have been killed in the raid, or to have died of hunger and thirst afterwards.

A Hajj stops off en route to Mecca

UPHEAVAL AND INDUSTRY

FROM 1700, ADVANCES IN KNOWLEDGE AND TECHNOLOGY SPURRED EXPLORATION AND TRAVEL, AS DID THE RACE TO CHART THE PACIFIC OCEAN AND TO SETTLE THE QUESTION OF A HYPOTHETICAL LARGE SOUTHERN CONTINENT. INDUSTRIALIZATION IN THE WEST ALSO DROVE THE SEARCH FOR RESOURCES OVERSEAS.

EXPANDING HORIZONS

In the 1700s and 1800s, Western explorers built on the experience of the previous two centuries to enlarge their understanding of the world. Lands and seas previously uncharted by Europeans were explored and new flora and fauna studied and scientifically recorded.

◁ **Lady Mary Wortley Montagu**
English aristocrat, writer, and medical pioneer, Lady Mary Wortley Montagu (1689–1762) is best remembered for writing about her experiences as a woman in Ottoman Constantinople (present-day Istanbul).

The voyages of the Portuguese navigators and Christopher Columbus in the late 1400s and the circumnavigators in the following two centuries had opened up the world to more people. However, a great deal was still unknown. The Romanov tsars, who ruled Russia from 1613, did not know whether their vast Siberian territory ended in water or crossed over a land bridge to North America. Likewise, citizens of the newly created republic of the United States (the land east of the Mississippi River), who had travelled across the ocean from Europe, had little idea what lay west across the river. The interior of Australia was known to its First Nations inhabitants, whose "songlines" stored ancestral knowledge of their terrain, yet it was "terra incognita" to its European colonizers; and a vast area of the African continent remained uncharted by Europeans.

△ **Galapagos findings**
Charles Darwin's 1889 drawing shows variations between finches on different Galapagos islands. His doubts about the stability of the species led to his theory of evolution by natural selection, one of the greatest 19th-century scientific discoveries.

New knowledge

To close the gaps in their knowledge of the globe, European explorers set out to map in detail the coasts and rivers of the parts of the continents they knew. The Pacific Ocean and its islands were charted by French explorer Lapérouse in the 1780s, and the idea that there was a large southern continent – to balance the landmass of the northern hemisphere – was finally dismissed when, in 1803, British navy officer Matthew Flinders proved Australia to be an island. Voyages from the eastern extremes of Siberia revealed that seas separated it from the American continent, while expeditions across America laid out its full extent. In the 19th century, the four great rivers of Africa were mapped to their sources, and the interior of that continent was slowly charted.

The success of these advances often relied on the knowledge and experience of Indigenous peoples. The first US expedition into the American West followed in the footsteps of a Yazoo elder, Moncacht-Apé, who had undertaken the same journey to the Pacific in the late 1600s. The Corps of Discovery, a unit in the United States army, was guided by Sacagawea, who initiated the group in survival techniques and diplomacy.

A scientific age

Alongside geographical exploration came an increased interest in the natural world's varied flora and fauna, notably in Asia and the Amazon. Naturalist explorers, including many women, made significant botanical discoveries, while geologists probed the natural forces shaping the globe. These findings gave rise to such revolutionary theories as evolution by natural selection, while an emerging understanding of fossils began to reveal the age of life on Earth.

AN INTERCONNECTED WORLD

The arrival of new technologies such as the steamship meant that ever-greater numbers of people were able to travel long distances. Towards the end of the 19th century, the globe was increasingly interconnected. In addition, exploration into new lands across the globe brought not only discoveries and increased knowledge, but also the mass colonization of countries and the displacement of native communities.

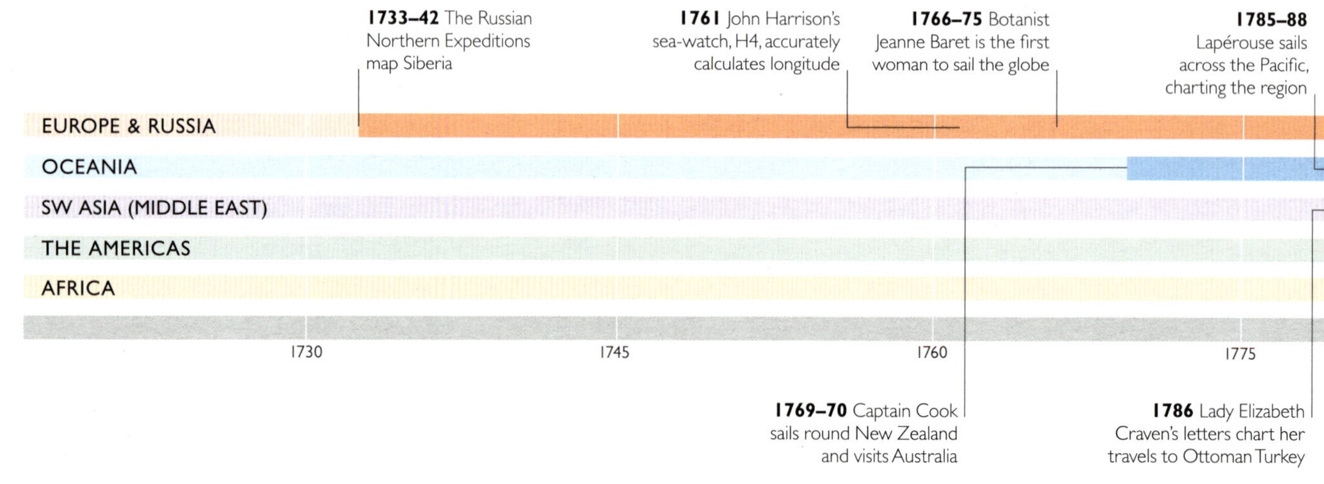

1733–42 The Russian Northern Expeditions map Siberia
1761 John Harrison's sea-watch, H4, accurately calculates longitude
1766–75 Botanist Jeanne Baret is the first woman to sail the globe
1785–88 Lapérouse sails across the Pacific, charting the region
EUROPE & RUSSIA
OCEANIA
SW ASIA (MIDDLE EAST)
THE AMERICAS
AFRICA
1730
1745
1760
1775
1769–70 Captain Cook sails round New Zealand and visits Australia
1786 Lady Elizabeth Craven's letters chart her travels to Ottoman Turkey

◁ **Marine timekeeper**
In 1761, John Harrison developed a "sea watch" – a chronometer, now known as H4 – that made ocean-going navigation far safer by accurately plotting longitude while at sea.

The atmosphere of the Enlightenment in Europe (1688–1789) also created a large readership for travel writing, which became a popular genre during this period. The travelogues of explorers, missionaries, and researchers provided first-hand accounts of newly encountered lands and peoples, as well as scientific and natural discoveries. Women in a position of privilege at the time were increasingly able to travel and their writing provided a new perspective.

New technologies

In the 19th century, industrial revolutions in Europe and the US accelerated Europe's engagement with the rest of the world. British inventor Richard Trevithick built the world's first steam locomotive in 1801 and the first passenger train service began in 1825, connecting coal mines across northeast England. By 1838, the first steamship purpose-built for crossing the Atlantic could travel from Bristol to New York in just over a fortnight, making possible the mass migrations that occurred throughout the 1800s.

However, the technological ascendance of the West sharpened its appetite for global dominance. European colonizers followed not far behind the explorers, seizing lands from their original inhabitants and suppressing resistance in their determination to build vast empires.

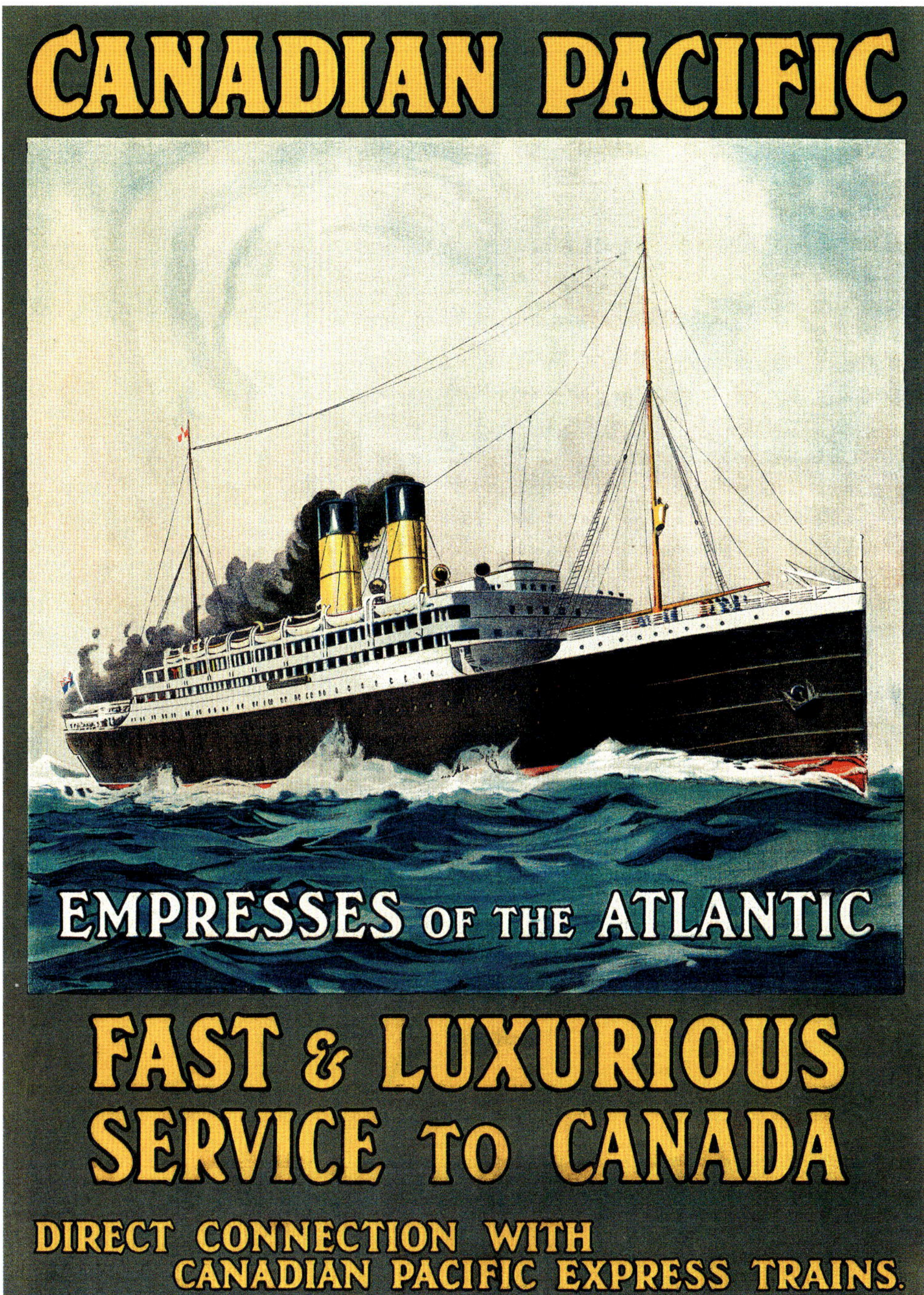

▷ **Spanning the continent**
Transatlantic shipping lines – such as the Canadian Pacific in this 1915 poster – allowed passengers to cross the Atlantic Ocean and then connect directly with the new railways. Those arriving in North America could easily traverse the continent by trains straight to the Pacific coast in the far west.

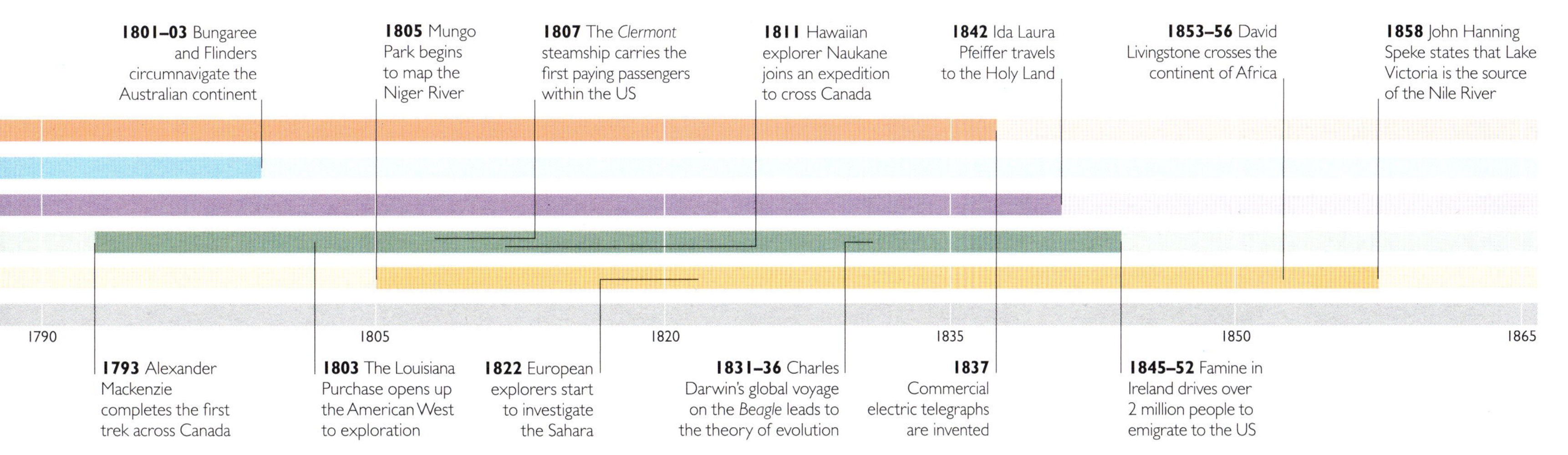

1801–03 Bungaree and Flinders circumnavigate the Australian continent
1805 Mungo Park begins to map the Niger River
1807 The *Clermont* steamship carries the first paying passengers within the US
1811 Hawaiian explorer Naukane joins an expedition to cross Canada
1842 Ida Laura Pfeiffer travels to the Holy Land
1853–56 David Livingstone crosses the continent of Africa
1858 John Hanning Speke states that Lake Victoria is the source of the Nile River
1790
1805
1820
1835
1850
1865
1793 Alexander Mackenzie completes the first trek across Canada
1803 The Louisiana Purchase opens up the American West to exploration
1822 European explorers start to investigate the Sahara
1831–36 Charles Darwin's global voyage on the *Beagle* leads to the theory of evolution
1837 Commercial electric telegraphs are invented
1845–52 Famine in Ireland drives over 2 million people to emigrate to the US

RUSSIAN NORTHERN EXPEDITIONS

In 1703, Peter the Great (r.1682–1725) founded St Petersburg in the far European west of the Russian empire. What lay to the east, in Siberia and beyond, was largely uncharted – it was unknown whether Russia was attached to the Americas or separated by sea. From 1725, a series of expeditions expanded Russian geographical knowledge.

In 1724, Peter the Great asked the Danish explorer and Russian naval officer Vitus Bering to explore the eastern lands and to find out whether they were attached to the Americas. However, the first expedition to Kamchatka and the seas beyond (1725–30), led by Bering and his deputy, Aleksei Chirikov, was inconclusive.

Empress Anna (r.1730–40) and her successor Elizabeth (r.1741–62) continued Peter's work and commissioned Bering to lead a second Kamchatka expedition. This expanded into what was later called the Great Northern Expedition of 1733–42, one of the largest exploration enterprises in history. Its aims included surveying the Arctic coastline of Siberia and finding a sea route to Japan and North America. A wider aim was to study the nature, history, and peoples of Siberia.

The expedition was organized by Bering into three groups: an academic group (on a separate itinerary) led by three professors, to research the plants, animals, and minerals of the region; a northern group to chart the northern Siberian coast and its rivers; and a Pacific naval group, led by Bering with Chirikov, to map the eastern coast of Siberia and, hopefully, the northern shores of

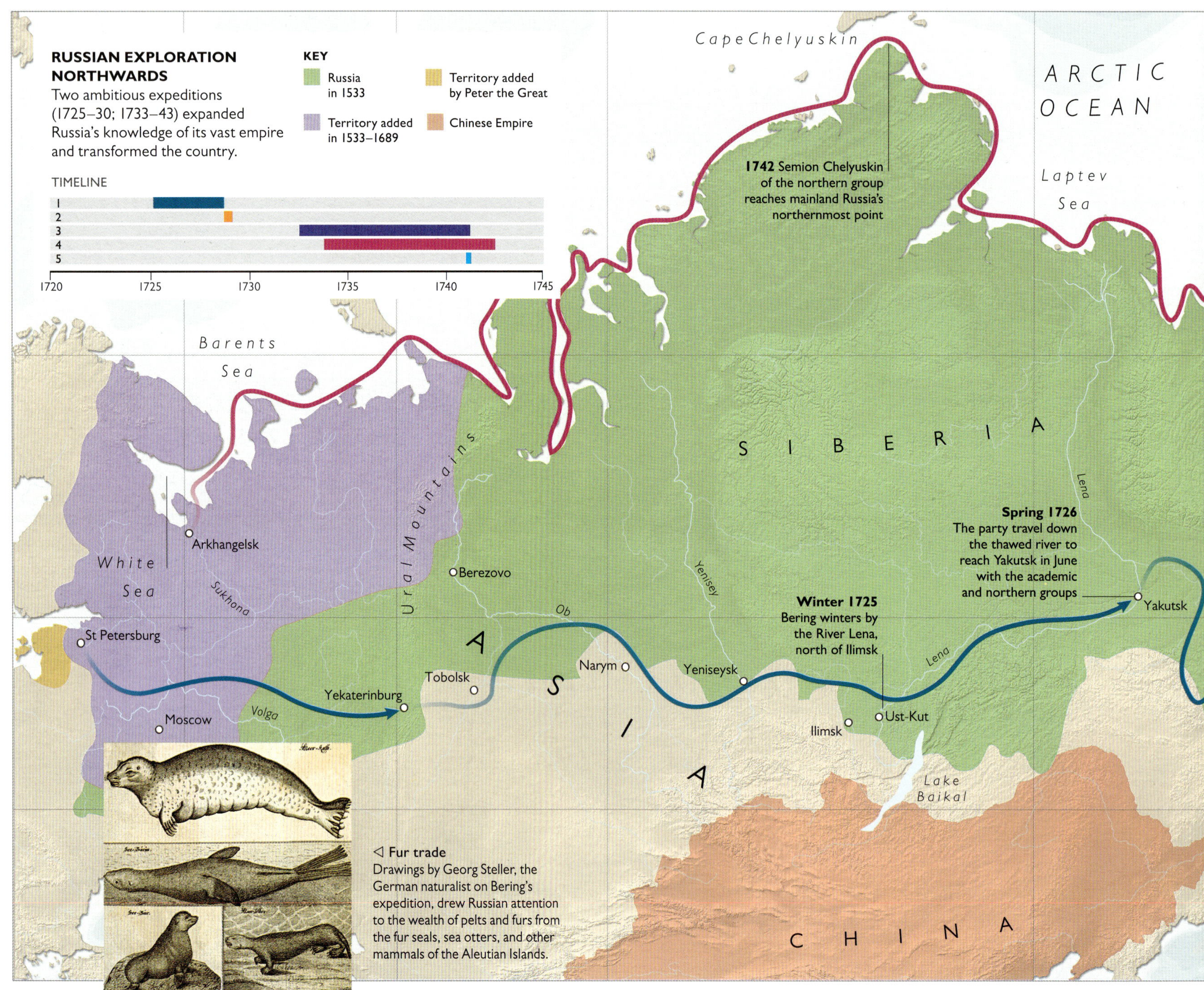

◁ **Fur trade**
Drawings by Georg Steller, the German naturalist on Bering's expedition, drew Russian attention to the wealth of pelts and furs from the fur seals, sea otters, and other mammals of the Aleutian Islands.

North America. More than 3,000 people were involved; the total cost, funded entirely by the Russian government, was around 1.5 million roubles, or around one-sixth of the entire income of Russia in 1724.

The expedition was considered to be a great success: vast amounts of scientific data were recorded, thousands of kilometres of the north and east Siberian coasts were mapped, and inland rivers explored. The North American coastline was discovered for the first time by Europeans, and Russian expansion into Siberia accelerated. However, this was followed by exploitation of the Aleutians via the fur trade. It also came at the cost of Bering's life, and that of many of his crew.

> *"At two o'clock in the afternoon we saw land ahead of us, with high mountains on it."*
>
> WRITTEN IN THE DECK LOG ON THE *ST PAUL* BY CAPTAIN ALEKSEI CHIRIKOV, 1741

MEETING "THE AMERICANS"

In September 1741, Bering and his Russian crew first met people from the Aleutian Islands (in present-day Alaska), whom they termed "the Americans". They shared no common language. Within a few years, the relationship centred around acquiring furs. Aleutian men worked hard to catch sea mammals, while the women turned pelts into clothing and household goods. Islanders soon became addicted to the tobacco and alcohol that the Europeans traded.

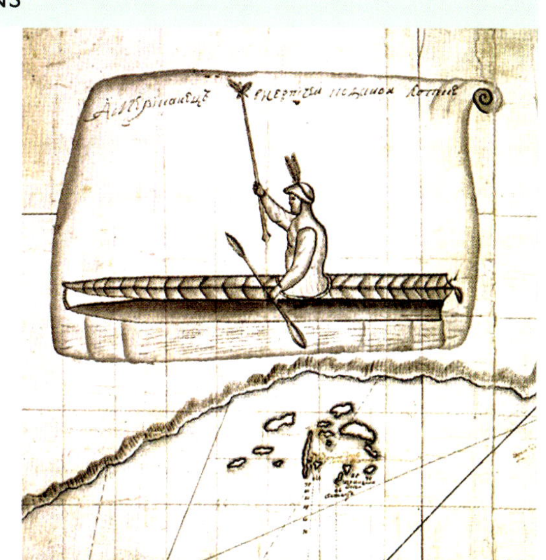

Drawing of an Aleutian islander, 1744

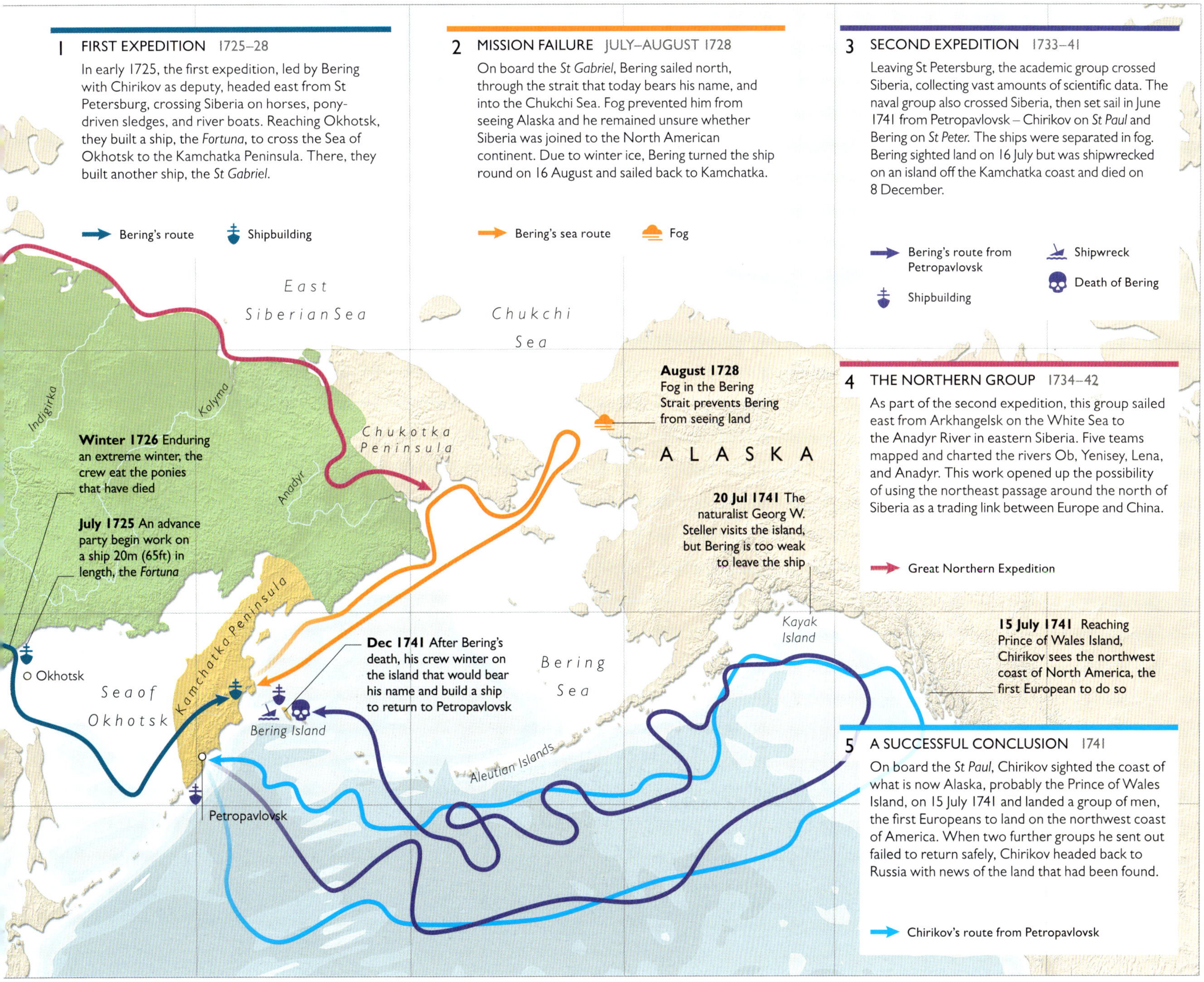

1 FIRST EXPEDITION 1725–28

In early 1725, the first expedition, led by Bering with Chirikov as deputy, headed east from St Petersburg, crossing Siberia on horses, pony-driven sledges, and river boats. Reaching Okhotsk, they built a ship, the *Fortuna*, to cross the Sea of Okhotsk to the Kamchatka Peninsula. There, they built another ship, the *St Gabriel*.

Bering's route
Shipbuilding

2 MISSION FAILURE JULY–AUGUST 1728

On board the *St Gabriel*, Bering sailed north, through the strait that today bears his name, and into the Chukchi Sea. Fog prevented him from seeing Alaska and he remained unsure whether Siberia was joined to the North American continent. Due to winter ice, Bering turned the ship round on 16 August and sailed back to Kamchatka.

Bering's sea route
Fog

3 SECOND EXPEDITION 1733–41

Leaving St Petersburg, the academic group crossed Siberia, collecting vast amounts of scientific data. The naval group also crossed Siberia, then set sail in June 1741 from Petropavlovsk – Chirikov on *St Paul* and Bering on *St Peter*. The ships were separated in fog. Bering sighted land on 16 July but was shipwrecked on an island off the Kamchatka coast and died on 8 December.

Bering's route from Petropavlovsk
Shipbuilding
Shipwreck
Death of Bering

East Siberian Sea
Chukchi Sea
Chukotka Peninsula
Indigirka
Kolyma
Anadyr
ALASKA
Kamchatka Peninsula
Sea of Okhotsk
Okhotsk
Petropavlovsk
Bering Island
Bering Sea
Aleutian Islands
Kayak Island

Winter 1726 Enduring an extreme winter, the crew eat the ponies that have died

July 1725 An advance party begin work on a ship 20m (65ft) in length, the *Fortuna*

August 1728 Fog in the Bering Strait prevents Bering from seeing land

20 Jul 1741 The naturalist Georg W. Steller visits the island, but Bering is too weak to leave the ship

Dec 1741 After Bering's death, his crew winter on the island that would bear his name and build a ship to return to Petropavlovsk

15 July 1741 Reaching Prince of Wales Island, Chirikov sees the northwest coast of North America, the first European to do so

4 THE NORTHERN GROUP 1734–42

As part of the second expedition, this group sailed east from Arkhangelsk on the White Sea to the Anadyr River in eastern Siberia. Five teams mapped and charted the rivers Ob, Yenisey, Lena, and Anadyr. This work opened up the possibility of using the northeast passage around the north of Siberia as a trading link between Europe and China.

Great Northern Expedition

5 A SUCCESSFUL CONCLUSION 1741

On board the *St Paul*, Chirikov sighted the coast of what is now Alaska, probably the Prince of Wales Island, on 15 July 1741 and landed a group of men, the first Europeans to land on the northwest coast of America. When two further groups he sent out failed to return safely, Chirikov headed back to Russia with news of the land that had been found.

Chirikov's route from Petropavlovsk

LATER PACIFIC EXPLORATION

European navigators scoured the Pacific during the 18th century in search of new lands, trade, bullion, and commercial and naval advantage over their rivals. In doing so, they came across islands that were new to them, or that previously had been forgotten.

In 1721–22, Jacob Roggeveen arrived at Rapa Nui (Easter Island) and Samoa during the last great Dutch expedition to the Pacific. Two decades later, British commodore George Anson attacked Spanish possessions in the Pacific, seizing a treasure-laden Manila galleon and challenging Spain's hegemony.

In 1764, John Byron, who had been in Anson's crew, returned to the Pacific to seek out an island for a British naval base. After encountering many low-lying islands, he reached the island of Nikunau in July 1765 – the first recorded European encounter with the island's Indigenous population.

The British Royal Navy continued its interest in the Pacific with voyages by Samuel Wallis and Philip Carteret. In 1767, Wallis became the first European navigator to encounter Tahiti. He proposed the island as a good site for studying the upcoming transit of Venus, a task given to James Cook (see pp.134–35). Carteret, a lieutenant on Byron's voyage, returned to the Pacific in 1766. He became the first European to reach Pitcairn and rediscovered the Solomon Islands, originally sighted in 1574 by Spain's Álvaro de Mendaña.

Louis-Antoine de Bougainville also explored the South Seas while making the first French circumnavigation in 1766–69 (see pp.132–33).

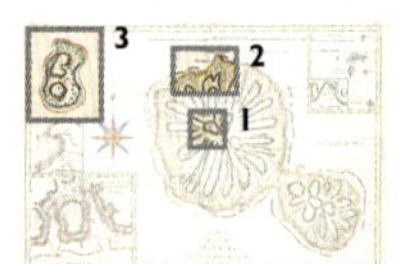

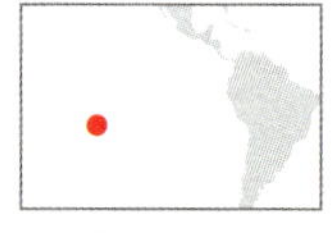

LOCATOR

KEY

1 In June 1767, Wallis's crew sighted Mou'a Orohena, the highest peak in Tahiti.

2 Matavai Bay, where Wallis's ship HMS *Dolphin* anchored on 24 June.

3 Leaving Tahiti, Wallis passed the isle Eimeo, and named it "Duke of York Island".

TRAVERSING THE PACIFIC

Two centuries after Europeans first explored the Pacific (see pp.118–19), Dutch, English, and French navigators were criss-crossing the ocean. Spanish galleons also sailed the Pacific between colonies in South America and the Philippines, taking luxury Asian goods east from Manila to Acapulco in Mexico in return for "New World" silver.

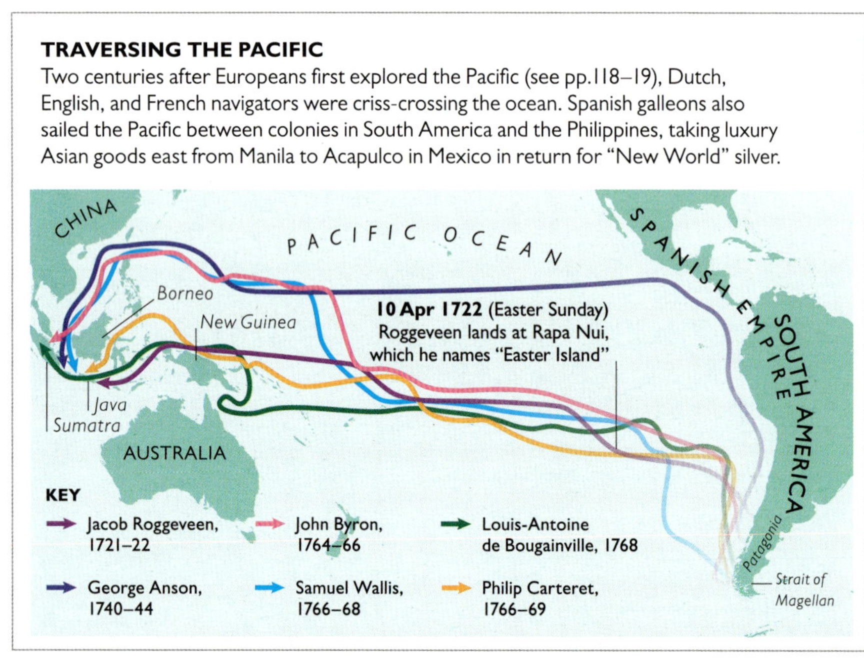

△ **Tahiti**
The island was first colonized by Polynesians c.200 BCE, but the first European contact was only in 1767 with the arrival of Samuel Wallis. This map was created in 1780 by the prominent French cartographer Rigobert Bonne.

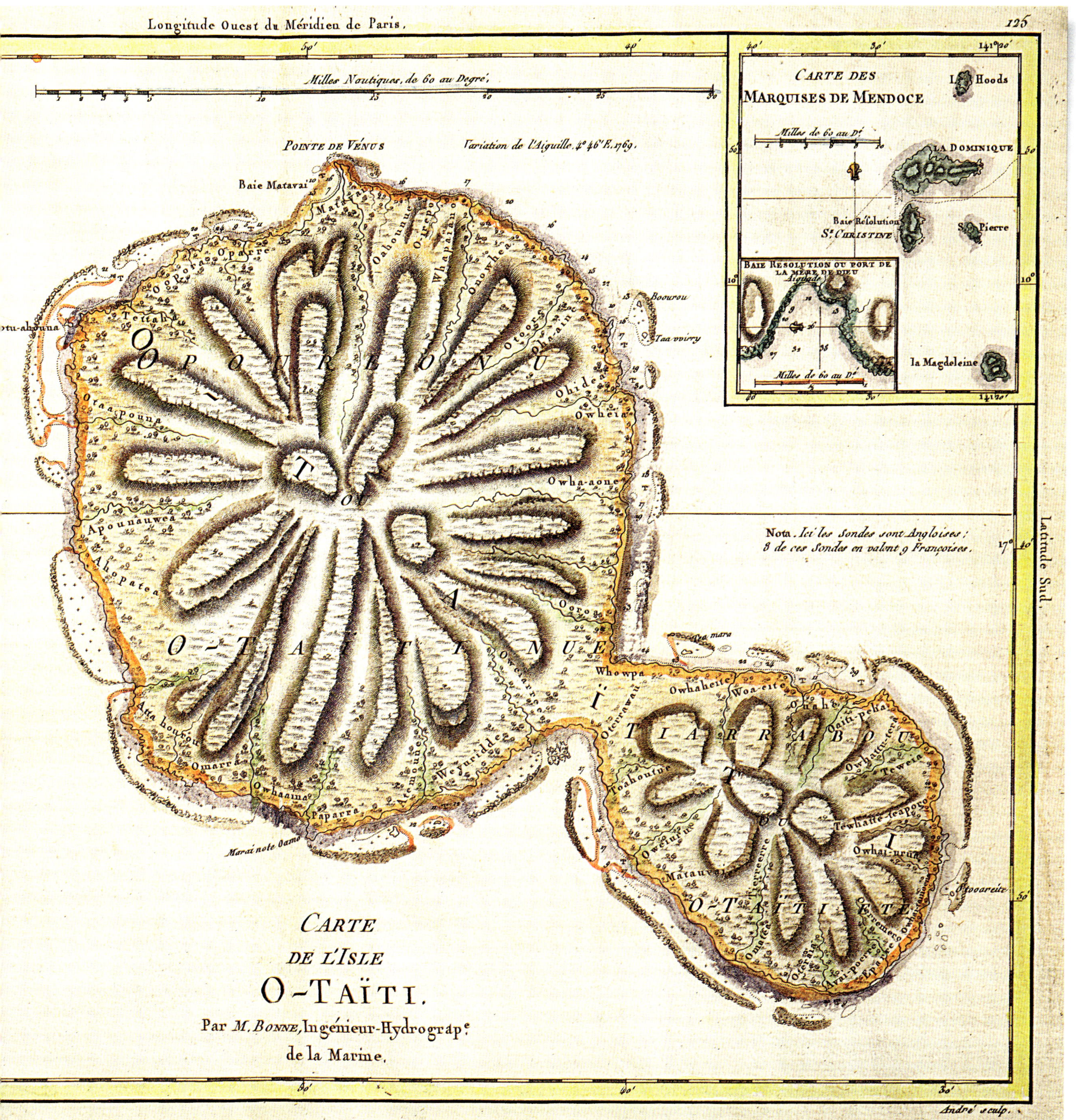

"When [the fog] cleared away, we were much surprised to find ourselves surrounded by some hundreds of canoes."

SAMUEL WALLIS, FROM THE LOG OF HMS *DOLPHIN*, JUNE 1767

AN EXPLORER IN DISGUISE

Botanist Jeanne Baret never intended to go to sea – yet disguised as a man, she set off in 1766, eventually returning to her native France in 1775. Inadvertently, she had become the first woman to circumnavigate the world.

△ **A daring disguise**
This allegorical portrait of Jeanne Baret from 1861 shows her dressed as a sailor and carrying plant specimens.

Little is known about Jeanne Baret's early life. She was born, likely in poverty, in rural Burgundy, France, on 27 July 1740. Around 1760 Baret found work as a servant – later promoted to housekeeper – with Philibert Commerson, a naturalist. When his wife died in 1762, Baret became his companion. Informed about plants from childhood, she began to help the naturalist.

In 1766, Commerson was asked to join Louis de Bougainville in the first scientific voyage to circumnavigate the world. Commerson was in poor health and relied on Baret as his nurse as well as his assistant. Since women were forbidden on French naval ships, Baret dressed as a man and kept her identity hidden for 18 months. The pair collected over 6,000 plant species, among them the vine *Bougainvillea* (named after Bougainville). Baret did fieldwork when Commerson was ill and found many plants. When the expedition docked in Isle de France (now Mauritius), Baret and Commerson remained in the country. He died in 1773; Baret married and returned to France in 1775, where she died in 1807. In 2012 a species of vine, *Solanum baretiae*, was named after her.

BOUGAINVILLE'S CIRCUMNAVIGATION

Louis de Bougainville left France with two ships in November 1766 and sailed down the Atlantic, via Rio de Janeiro and Montevideo, then around Cape Horn into the Pacific. Sailing west, he entered the Indian Ocean, stopping in November 1768 at Mauritius, where Baret and the unwell Commerson left the fleet. Bougainville returned to France in spring 1769.

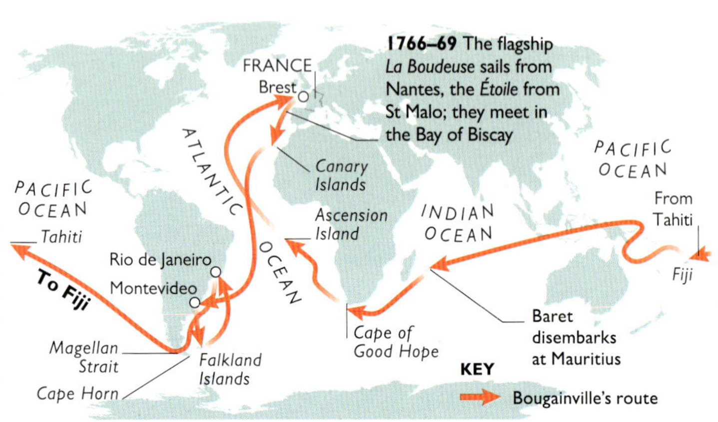

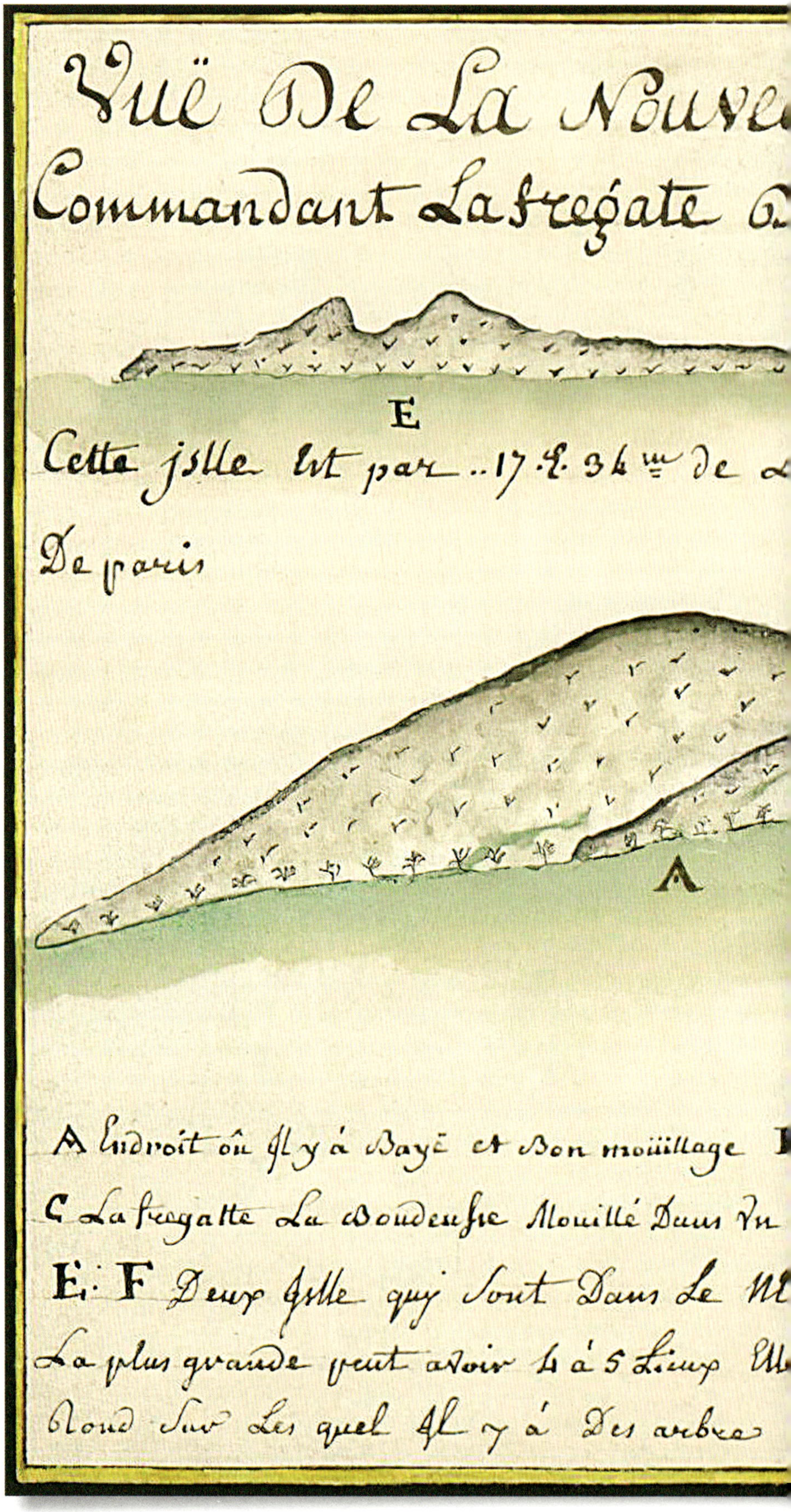

△ **Unmasked in Tahiti**
An extract from Louis de Bougainville's navigation diary records the expedition's arrival on 7 April 1768 in Tahiti, which he renamed La Nouvelle Cythera. Rumours that "Jean" was a woman in disguise had circulated on board the *Étoile* for some time, but many accounts suggest that Baret's true identity was revealed on Tahiti by the curious islanders.

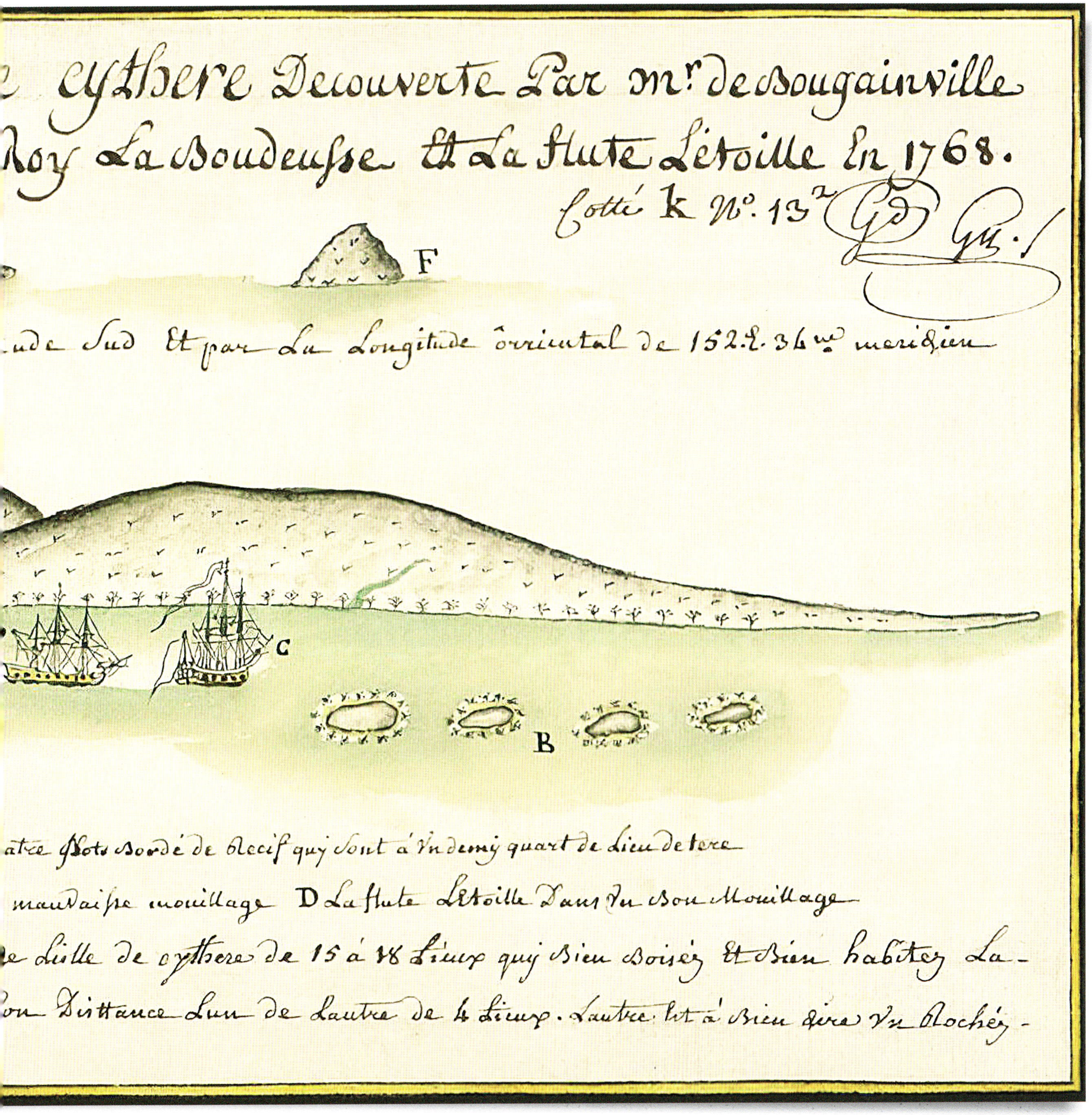

"I thought I had been transported into the garden of Eden… beautiful landscapes, covered with the richest productions of nature."

LOUIS DE BOUGAINVILLE'S NAVIGATION DIARY, 7 APRIL 1768

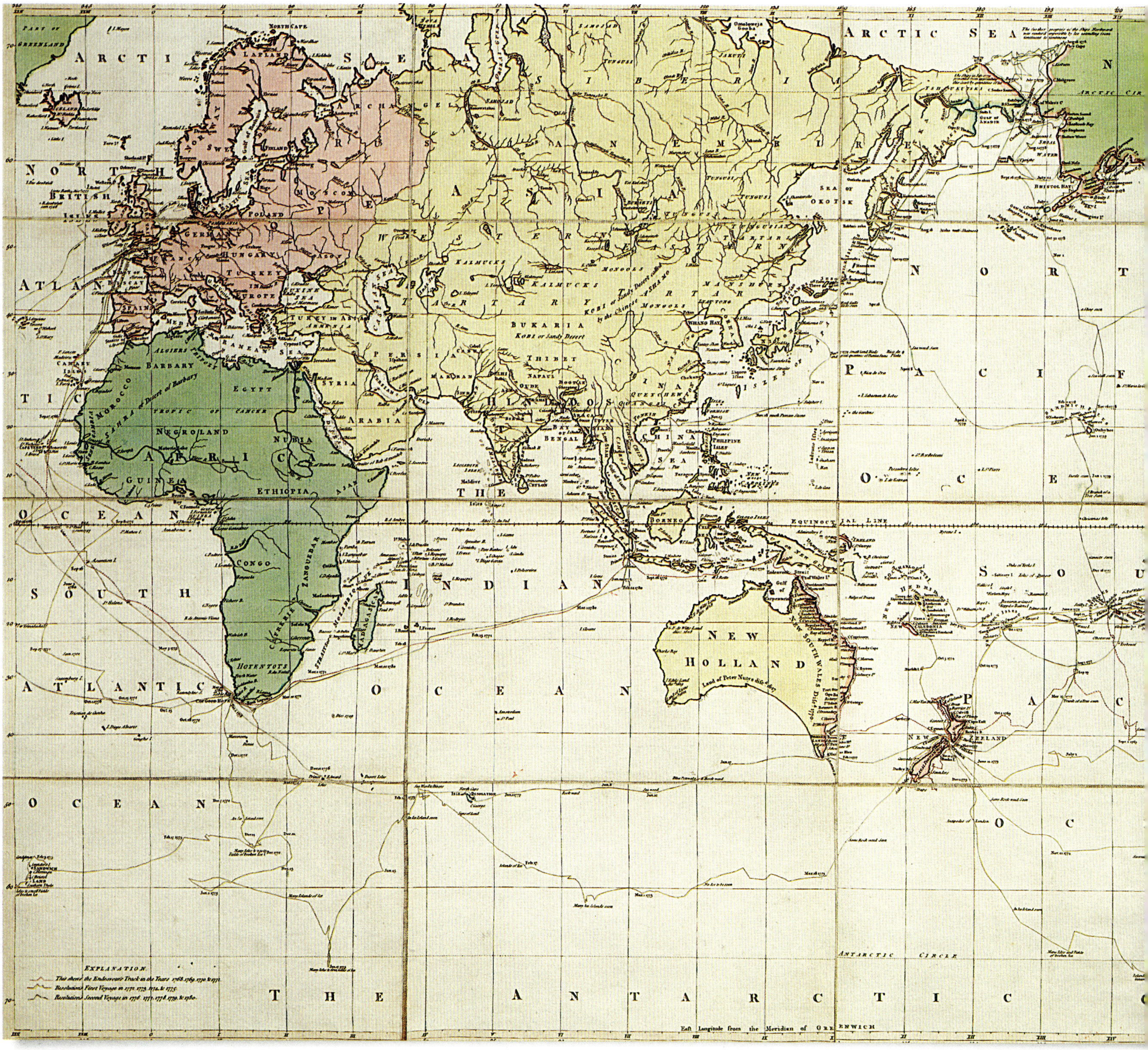

▷ **First voyage 1768–71**
Cook spent six months mapping the islands that, since 1645, had been named Nova Zeelandia by Dutch cartographers, after the Dutch coastal province of Zeeland, before heading to New Holland (Australia).

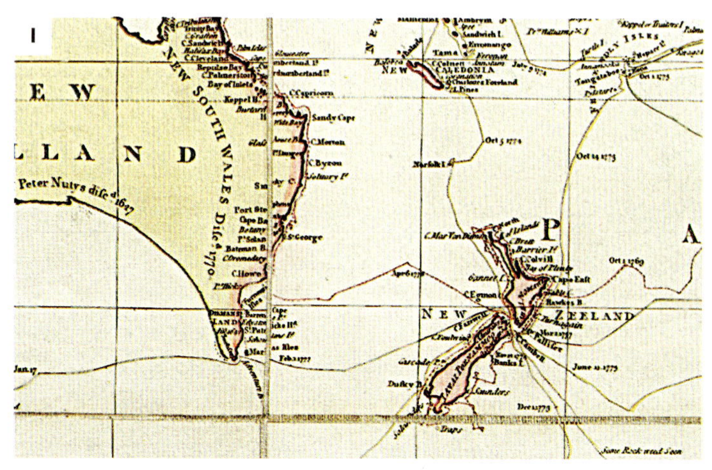

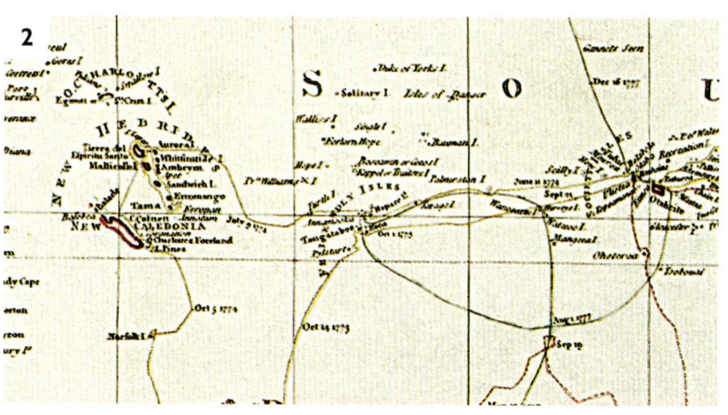

△ **Second voyage 1772–75**
After sailing the Antarctic coastline, Cook sailed north then west from Tahiti to Tonga and the New Hebrides, becoming the first European to sight New Caledonia in September 1774.

THE VOYAGES OF COOK

Over three long voyages, navigator James Cook sailed across oceans and surveyed lands previously unknown to Europeans. The scientific and geographical discoveries were groundbreaking, yet the expedition's brutal treatment of Indigenous peoples overshadows Cook's legacy.

An officer in the British Royal Navy, James Cook had surveying skills that attracted attention during the Seven Years' War (1756–63), when he mapped the Newfoundland coast. In 1768, when the Admiralty sponsored an expedition to the Pacific, Cook was their choice to command HMS *Endeavour*, with a crew of 94 men and a group of Royal Society scientists on board, including the botanist Joseph Banks.

The ostensible purpose of the first voyage was to observe in Tahiti the transit of Venus across the Sun. Yet after completing this task, Cook continued on his more clandestine mission – to search the South Pacific for Terra Australis, the southern continent that ancient Greeks believed must exist to counterbalance the northern hemisphere. In this, the new recruits to Cook's party – the Polynesian High Priest Tupaia, and his apprentice, Taiata – proved invaluable. Using their knowledge of the seas and navigating by the stars, they guided the *Endeavour* to circumnavigate both islands of "Nova Zeelandia" (Aotearoa) and, on 29 April 1770, to land in eastern Australia.

In 1772–75 Cook journeyed again in the southern oceans, still in search of Terra Australis. In 1773 he was the first navigator to cross the Antarctic Circle.

His third voyage of 1776 aimed to locate a northwest passage around North America. In this he failed, but in 1778 he was the first European to land in the Hawaiian Islands. A year later, his punitive, overbearing treatment of the islanders led to his fatal stabbing.

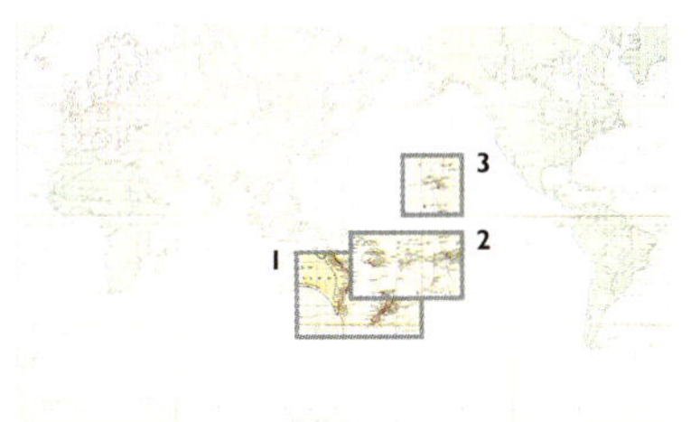

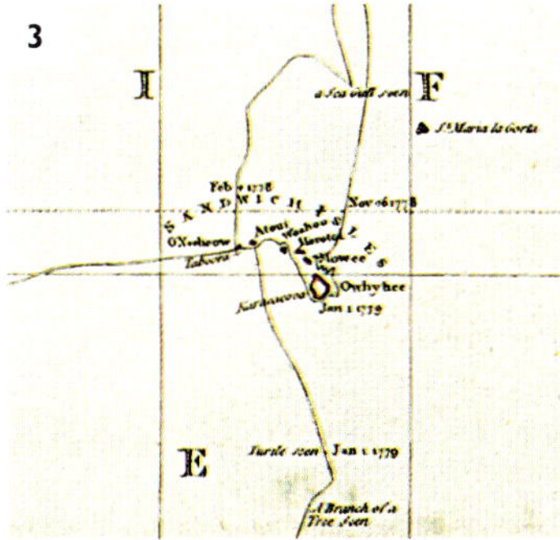

△ Henry Roberts's map (1784)
In his three Pacific voyages (Roberts served on the latter two), Cook sailed further south and north than earlier European navigators, coming close to Antarctica and sailing out of the Pacific into the Arctic Ocean.

△ Third voyage 1776–79
Cook became the first European to visit the Hawaiian Islands (which he named Sandwich Islands), where he died on 14 February 1779.

THE *ENDEAVOUR*

Launched in 1764 as the *Earl of Pembroke*, the ship was renamed when the Royal Navy refitted it for Cook's first historic voyage to Australia in 1768. Built to carry coal, the *Endeavour* was a slow but safe ship, its long 29.77-m (97 ft 8-in) box-like, oak hull driven forward by the three masts' sails. The flat bottom made it well suited to shallow waters. In 1770 the ship had a near miss when it ran aground within the Great Barrier Reef. After the voyage, the *Endeavour* was decommissioned and, as the transport ship *Lord Sandwich II*, sank off Rhode Island in 1778 during the American War of Independence.

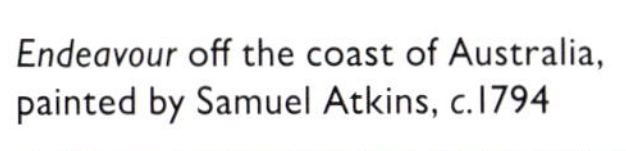

Endeavour off the coast of Australia, painted by Samuel Atkins, c.1794

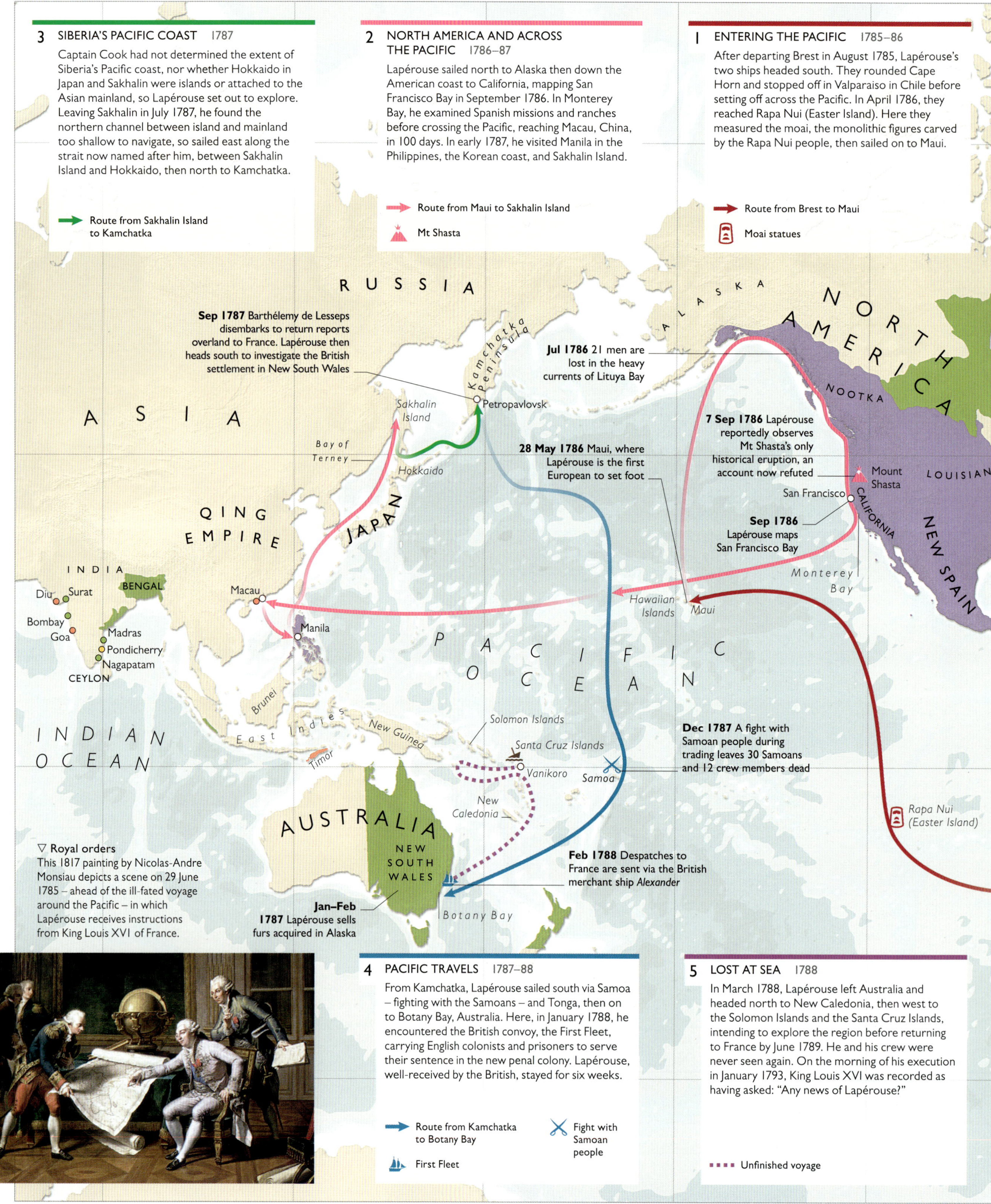

1 ENTERING THE PACIFIC 1785–86

After departing Brest in August 1785, Lapérouse's two ships headed south. They rounded Cape Horn and stopped off in Valparaiso in Chile before setting off across the Pacific. In April 1786, they reached Rapa Nui (Easter Island). Here they measured the moai, the monolithic figures carved by the Rapa Nui people, then sailed on to Maui.

2 NORTH AMERICA AND ACROSS THE PACIFIC 1786–87

Lapérouse sailed north to Alaska then down the American coast to California, mapping San Francisco Bay in September 1786. In Monterey Bay, he examined Spanish missions and ranches before crossing the Pacific, reaching Macau, China, in 100 days. In early 1787, he visited Manila in the Philippines, the Korean coast, and Sakhalin Island.

3 SIBERIA'S PACIFIC COAST 1787

Captain Cook had not determined the extent of Siberia's Pacific coast, nor whether Hokkaido in Japan and Sakhalin were islands or attached to the Asian mainland, so Lapérouse set out to explore. Leaving Sakhalin in July 1787, he found the northern channel between island and mainland too shallow to navigate, so sailed east along the strait now named after him, between Sakhalin Island and Hokkaido, then north to Kamchatka.

4 PACIFIC TRAVELS 1787–88

From Kamchatka, Lapérouse sailed south via Samoa – fighting with the Samoans – and Tonga, then on to Botany Bay, Australia. Here, in January 1788, he encountered the British convoy, the First Fleet, carrying English colonists and prisoners to serve their sentence in the new penal colony. Lapérouse, well-received by the British, stayed for six weeks.

5 LOST AT SEA 1788

In March 1788, Lapérouse left Australia and headed north to New Caledonia, then west to the Solomon Islands and the Santa Cruz Islands, intending to explore the region before returning to France by June 1789. He and his crew were never seen again. On the morning of his execution in January 1793, King Louis XVI was recorded as having asked: "Any news of Lapérouse?"

▽ **Royal orders**
This 1817 painting by Nicolas-Andre Monsiau depicts a scene on 29 June 1785 – ahead of the ill-fated voyage around the Pacific – in which Lapérouse receives instructions from King Louis XVI of France.

LAPÉROUSE IN THE PACIFIC

Anxious to restore his navy to glory after its defeat in the Seven Years' War of 1756–63, Louis XVI of France (r.1774–92) commissioned an ambitious – and expensive – expedition to the Pacific Ocean, to enrich knowledge and trade. France's prestige was at stake.

In 1785, Louis XVI appointed Jean-François de Galaup, Comte de Lapérouse, to lead a global expedition. A titled naval officer, Lapérouse had joined the French navy in 1756 as a teenager, gaining experience in the Seven Years' War and going on to serve with distinction in various campaigns. The expedition's aim was to further the findings of the British explorer James Cook; correct and complete the oceanic maps; open new maritime routes and establish trading links; and enrich French scientific knowledge and collections. Lapérouse was to command two ships, *L'Astrolabe* and *La Boussole*, each able to carry over 500 tonnes. On board were around 225 officers and crew, including a physicist, an astronomer and mathematician, a geologist, a botanist, three naturalists, and three scientific illustrators. A young Napoleon Bonaparte applied to join the voyage but was not chosen.

The expedition left Brest, France, in August 1785, heading south, then rounding South America to reach the Pacific. Following Cook's example, they compiled accurate calculations of longitude using precision timekeeping devices – chronometers – and drew up precise maps and charts. Lapérouse sent regular progress reports to Paris before disaster struck in 1788, when the fleet disappeared in the Santa Cruz Islands, bringing the expedition to an end.

> *"I shall sail... from Botany Bay... until December, by when I hope to arrive at the Isle de France."*
>
> LAPÉROUSE, WRITING TO EXPLORER FLEURIEU, 1788

TRAVELS IN THE PACIFIC

Lapérouse's 1785–88 voyage sailed from southern Europe, round Cape Horn, and across the Pacific. Its final stop was in Australia before the journey ended in the Santa Cruz Islands. The fate of his final voyage was discovered some 40 years later.

KEY

TERRITORIES AND COLONIES, 1788

- British
- French
- Portuguese
- Spanish

TIMELINE

1, 2, 3, 4, 5, 6

1780 1790 1800 1810 1820 1830

BRITISH NORTH AMERICA
UNITED STATES
Saint Pierre and Miquelon
GREAT BRITAIN
IRELAND
EUROPE
Brest
FRANCE
PORTUGAL
SPAIN
Sardinia
Gibraltar
Ceuta
Azores
Madeira
ATLANTIC OCEAN
Bermuda
Canary Islands
Bahamas
Puerto Rico
St Kitts
Montserrat
Barbuda
Antigua
Guadeloupe
Dominica
Martinique
Barbados
St Lucia
Tobago
Trinidad
Cuba
Santo Domingo
Jamaica
Saint Domingue
AFRICA
Cape Verde
St Louis
Gorée
CACHEU
FREEDOM PROVINCE
Príncipe
Fernando Po
São Tomé
Annobón
NEW GRANADA
FRENCH GUIANA
SOUTH AMERICA
BRAZIL
PERU
ANGOLA
MATUMBA
St Helena
Trindade Island
RIO DE LA PLATA
Santa Catarina Island
Valparaíso
CHILE
Eastern Falklands
Cape Horn

6 DISCOVERY 1825–27

The French Wars of 1789–1815 hindered any search for Lapérouse. However, in 1825 an English whaler found that the inhabitants of Tikopia, an island between New Caledonia and New Guinea, appeared to have items from the expedition. In 1826, Peter Dillon, an Irish sea captain, traced more items to the neighbouring island of Vanikoro, where in 1827 he found the wrecks of two ships in the coral reefs. In 2005, one of the ships was formally identified as Lapérouse's ship, *Boussole*.

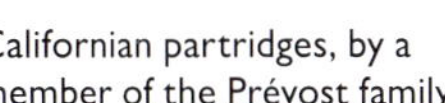
Lapérouse's shipwreck

A RECORD OF DISCOVERIES

The three illustrators on Lapérouse's 1785 expedition sent drawings back to France of the flora and fauna they encountered on the voyage. They also drew local canoes and sailboats, the moai statues on Rapa Nui (Easter Island), and some of the local people they encountered.

Californian partridges, by a member of the Prévost family

CHARTING AUSTRALIA

Two decades after Captain Cook first made landfall in Australia in 1770, and the first settlers and convicts had arrived in 1788, British explorers began systematically to chart the coast, circumnavigating the country by 1803, and confirming Van Diemen's Land (Tasmania) as an island.

In the late 18th century, the true extent of Australia was still unknown to colonizing Europeans. Dutch explorers in the 17th century had named the western region of Australia "New Holland" and James Cook had mapped the eastern coastline of what he called "New South Wales", while the south coast of "Van Diemen's Land" (Tasmania), or "lutruwita" to its Indigenous population, had been charted by Abel Tasman (see pp.118–19) in 1642. However, the way these areas related to each other remained unclear to Europeans. Were they separated by a strait, or part of the fabled "Southern Continent"?

Within a very few years, the extent of Australia was confirmed by the joint and solo expeditions of British naval officer Matthew Flinders and naval surgeon George Bass. By 1803, Flinders had completed the first inshore circumnavigation of the continent and could confirm that it was an island – a single landmass and not part of a large southern continent. When labelling his map, Flinders chose the name Australia or *Terra Australis* (Southern Land), and the British officially adopted the term in 1817.

> *"I call the whole island Australia, or Terra Australis."*
>
> MATTHEW FLINDERS, IN A LETTER TO HIS BROTHER, 1804

BUNGAREE c.1775–1830

Bungaree, an Indigenous Australian man, originally from Ku-ring-gai country, north of Sydney, moved to Sydney in the 1790s and interacted with the colonists, quickly picking up their language. He was hired as an interpreter on Flinders' survey of the north coast, advising him on Aboriginal protocol. Recruited again as an essential mediator on Flinders' major expedition, Bungaree was the first Australian to circumnavigate the continent where he was born.

Bungaree in European attire

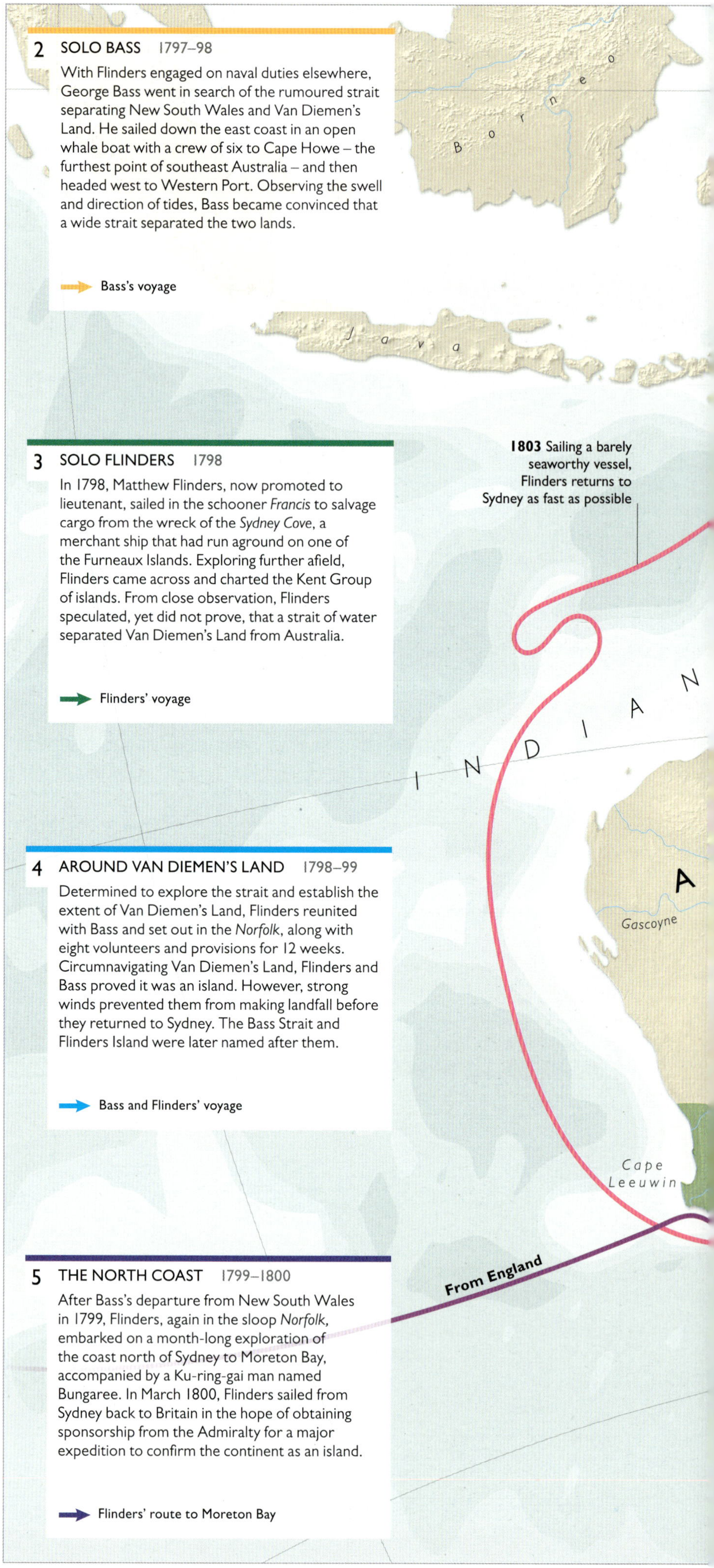

2 SOLO BASS 1797–98

With Flinders engaged on naval duties elsewhere, George Bass went in search of the rumoured strait separating New South Wales and Van Diemen's Land. He sailed down the east coast in an open whale boat with a crew of six to Cape Howe – the furthest point of southeast Australia – and then headed west to Western Port. Observing the swell and direction of tides, Bass became convinced that a wide strait separated the two lands.

Bass's voyage

3 SOLO FLINDERS 1798

In 1798, Matthew Flinders, now promoted to lieutenant, sailed in the schooner *Francis* to salvage cargo from the wreck of the *Sydney Cove*, a merchant ship that had run aground on one of the Furneaux Islands. Exploring further afield, Flinders came across and charted the Kent Group of islands. From close observation, Flinders speculated, yet did not prove, that a strait of water separated Van Diemen's Land from Australia.

Flinders' voyage

4 AROUND VAN DIEMEN'S LAND 1798–99

Determined to explore the strait and establish the extent of Van Diemen's Land, Flinders reunited with Bass and set out in the *Norfolk*, along with eight volunteers and provisions for 12 weeks. Circumnavigating Van Diemen's Land, Flinders and Bass proved it was an island. However, strong winds prevented them from making landfall before they returned to Sydney. The Bass Strait and Flinders Island were later named after them.

Bass and Flinders' voyage

5 THE NORTH COAST 1799–1800

After Bass's departure from New South Wales in 1799, Flinders, again in the sloop *Norfolk*, embarked on a month-long exploration of the coast north of Sydney to Moreton Bay, accompanied by a Ku-ring-gai man named Bungaree. In March 1800, Flinders sailed from Sydney back to Britain in the hope of obtaining sponsorship from the Admiralty for a major expedition to confirm the continent as an island.

Flinders' route to Moreton Bay

1803 Sailing a barely seaworthy vessel, Flinders returns to Sydney as fast as possible

Borneo

Java

INDIAN

A

Gascoyne

Cape Leeuwin

From England

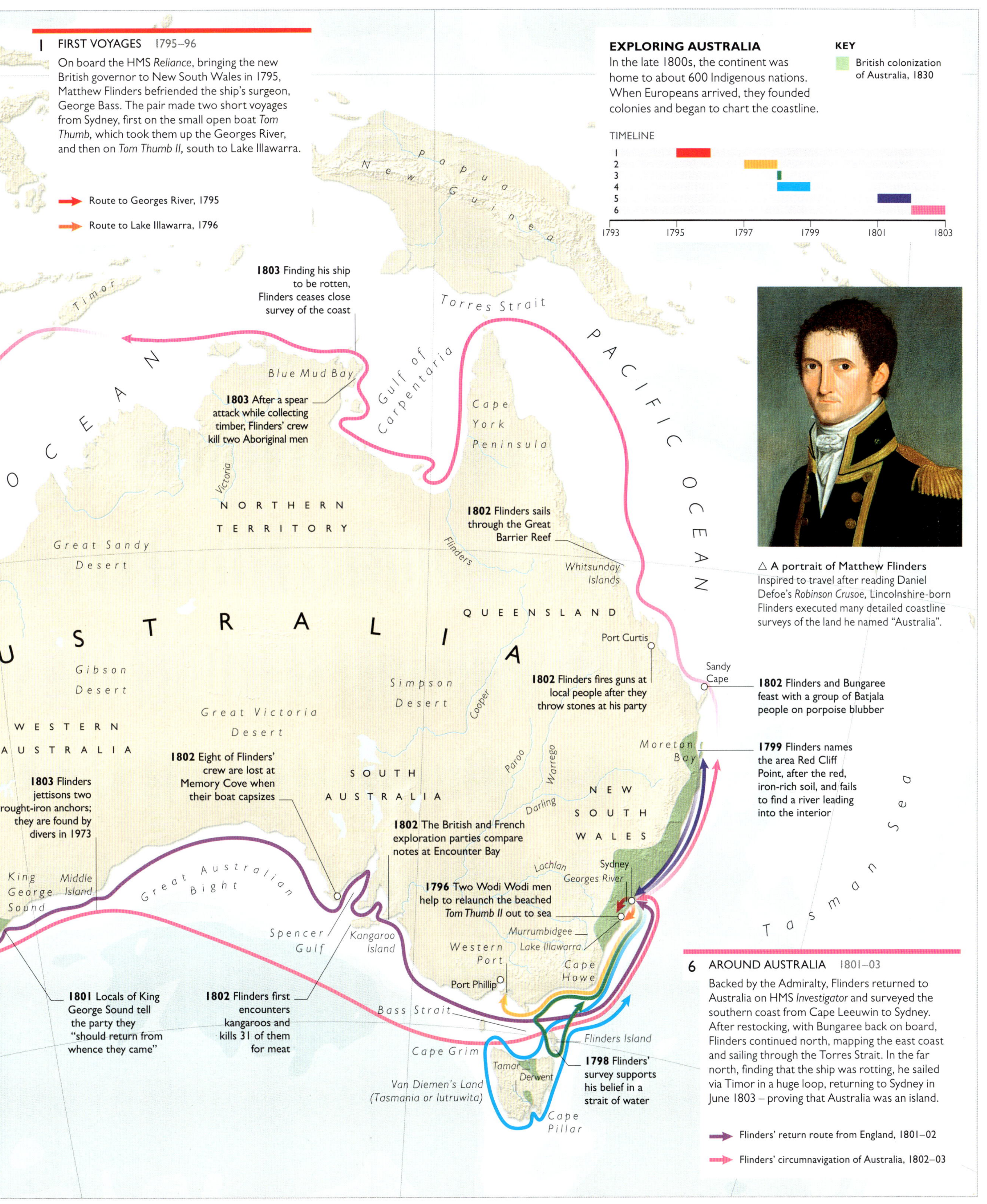

1 FIRST VOYAGES 1795–96

On board the HMS *Reliance*, bringing the new British governor to New South Wales in 1795, Matthew Flinders befriended the ship's surgeon, George Bass. The pair made two short voyages from Sydney, first on the small open boat *Tom Thumb*, which took them up the Georges River, and then on *Tom Thumb II*, south to Lake Illawarra.

EXPLORING AUSTRALIA

In the late 1800s, the continent was home to about 600 Indigenous nations. When Europeans arrived, they founded colonies and began to chart the coastline.

△ **A portrait of Matthew Flinders**
Inspired to travel after reading Daniel Defoe's *Robinson Crusoe*, Lincolnshire-born Flinders executed many detailed coastline surveys of the land he named "Australia".

6 AROUND AUSTRALIA 1801–03

Backed by the Admiralty, Flinders returned to Australia on HMS *Investigator* and surveyed the southern coast from Cape Leeuwin to Sydney. After restocking, with Bungaree back on board, Flinders continued north, mapping the east coast and sailing through the Torres Strait. In the far north, finding that the ship was rotting, he sailed via Timor in a huge loop, returning to Sydney in June 1803 – proving that Australia was an island.

Turkish ships
Lady Elizabeth Craven travelled alone across Europe to St Petersburg and Moscow, then south to Crimea. From Sebastopol, she sailed on an Ottoman vessel to Constantinople. This image is from her account of the journey, published in 1789.

EARLY WOMEN TRAVEL WRITERS

Some notable European women who travelled the world in the 18th century kept journals, which captured fascinating details of the people and places they visited. These travel journals found a wide readership when published.

In the early 1700s, English Eliza Justice travelled to St Petersburg as governess to an English merchant family. Observing the "laws, manners, and customs" of the Russian empire, she later wrote about her experiences in *A Voyage to Russia* (1739).

Lady Mary Wortley Montagu lived overseas as the wife of the British ambassador to the Ottoman Empire, and she and her husband travelled extensively during his posting. Her reflections in the posthumously published *Turkish Embassy Letters* (1763) led many people to view her as the first true travel writer. She observed the Eastern practice of smallpox inoculation and, having tested it successfully on her own daughter in 1721, introduced the idea to the medical establishment in England. Later in the century, the aristocrat Lady Elizabeth Craven, in self-exile from the strict mores of English society, travelled extensively in the same region, a period that she recorded in *A Journey through the Crimea to Constantinople* (1789).

△ **Portrait of Odonais**
Odonais was a multilinguist who understood Incan *quipu* – nonverbal communication that used knots and strings.

In 1750, Isabel Godin des Odonais, who lived in the Viceroyalty of Peru, was separated from her French husband by the divisive politics of the time. After a wait of 20 years, she set off in a 20-person party across the perilous Andes Mountains and Amazon basin to find him. Everyone died, except for Isabel who, after receiving help from Indigenous Amazonian people, was finally reunited with her husband. The couple recounted her extraordinary story in a letter to the French explorer Charles Marie de La Condamine, and this was published in an anthology of expedition literature, *Perils and Captivity*, in 1827.

ART, SCIENCE, AND TRAVEL

In 1699, aged 52, Dutch-born Maria Sibylla Merian set out for Dutch Guiana (now known as Suriname), where she made repeated expeditions into the interior to scientifically observe and record the wildlife in her sketchbooks. The fine illustrations in her self-published *Metamorphosis Insectorum Surinamensium* detailed the interaction of organisms and contributed greatly to the modern science of ecology.

A harlequin beetle and a moth's life cycle

HUMBOLDT IN THE AMERICAS

Charles Darwin described Alexander von Humboldt as "the greatest scientific traveller who ever lived". The German scholar was a true polymath, interested in all aspects of natural history, and is rightly seen as the father of ecology and environmentalism.

Born in Berlin in 1769, Alexander von Humboldt took a youthful delight in collecting and labelling plants, insects, and shells, for which he was nicknamed "the little apothecary". He studied law at university but later taught himself chemistry and palaeontology. His mother's death in 1797 left him enough money to pursue his two great loves: natural history and exploration.

In 1798, Humboldt teamed up with Aimé Bonpland, a French botanist, first visiting Venezuela in 1799–1800 and then Cuba, the Andes, and Mexico in 1800–04. Humboldt's main interest was the geography of plants – what grew where, and why. The pair collected many new plants, which they recorded, dried, and stored.

Humboldt was one of the first to suggest that Africa and South America were once joined together and that climatic conditions in different countries could be recorded by isotherms (lines on maps linking places of equal temperature). His most influential discovery, however, was that everything on the planet is interconnected in ecosystems.

LOCATOR

KEY

1 The mountainous zone is inhabited by flora and fauna suited to drier, cooler climes.

2 The Amazonian zone contains flora and fauna able to cope with the hot, wet climate.

3 The flora and fauna of the coastal zone are swept by sea winds and bathed in sun.

Humboldt's legacy is vast. Among other things, a Peruvian penguin is named after him, as is the Humboldt current off the west coast of South America. This current is responsible for Chile's Atacama Desert, and its disruption by the El Niño weather effect has fatal social and economic impacts.

▷ **Humboldt's *Naturgemälde* ("picture of nature")** In 1805 Humboldt and Bonpland published this striking cross-section of Chimborazo volcano in Ecuador. They meticulously mapped the animals and plants found at each elevation, and recorded details of temperature, altitude, humidity, and atmospheric pressure.

AMERICAN DISCOVERIES

In 1802 Humboldt and Bonpland climbed Chimborazo, a volcano just below the equator in Ecuador, and recorded its diverse plant life. During their five-year stay in South and Central America the pair collected over 6,000 plant species, among them cinchona, the bark of which cures malaria. They reported 14 species with medicinal uses, including treatments for snake bites and digestive diseases.

Humboldt at the foot of Chimborazo

ABORIGINAL EXPLORATION

Without the use of a common language or the written word, around 60,000 years ago the First Nations Australian people created an extraordinary oral network that helped them explore the lands where they lived, using pathways known as "songlines".

According to Indigenous Australian tradition, these verbal songlines or oral maps were based on the creator beings and their formation of the world during an ancient creation time known as "the Dreaming" or "Dreamtime". They carried crucial knowledge about the geography, mythology, and culture of the land, as well as the direction of the sunrise and the location of the stars. Passed down the generations through dancing, stories, and cave art, songlines – some of which were up to 3,500 km (2,200 miles) long – created a network linking around 250 different language-based groups, and helped the Indigenous people to navigate, explore, and trade across their vast continent.

Knowledge of the terrain was essential for finding seasonal foods and sources of water. It was also key to maintaining the intricate network of exchange that criss-crossed the country. Over time, distinct areas had become known for the skilled production of certain goods, such as stone axes, pearl shells, and red ochre, and these were transported along trade routes that covered immense distances.

From at least 1700 until 1907, the Yolngu people also traded with Makassan fishermen, who sailed on monsoon winds from Sulawesi to the Arnhem Land coast to harvest trepang (sea cucumber), which was highly prized in China. Reports suggest that many Aboriginal people travelled to Sulawesi, Manila, and Singapore on Makassan boats.

> *"Those who lose dreaming are lost."*
>
> FIRST NATIONS AUSTRALIAN PROVERB

EXPEDITION GUIDES

Indigenous guides were key to European expeditions in Australia. In 1791, Balloderry and Colbee explored the hinterland with Watkin Tench, and Bungaree (see pp.138–39) sailed with Matthew Flinders in 1802–03. There are few accounts of female guides, but Turandurey, a Wiradjuri woman, proved crucial to surveyor Thomas Mitchell in New South Wales in 1836. She guided Mitchell's party to food and water sources and to camp spots.

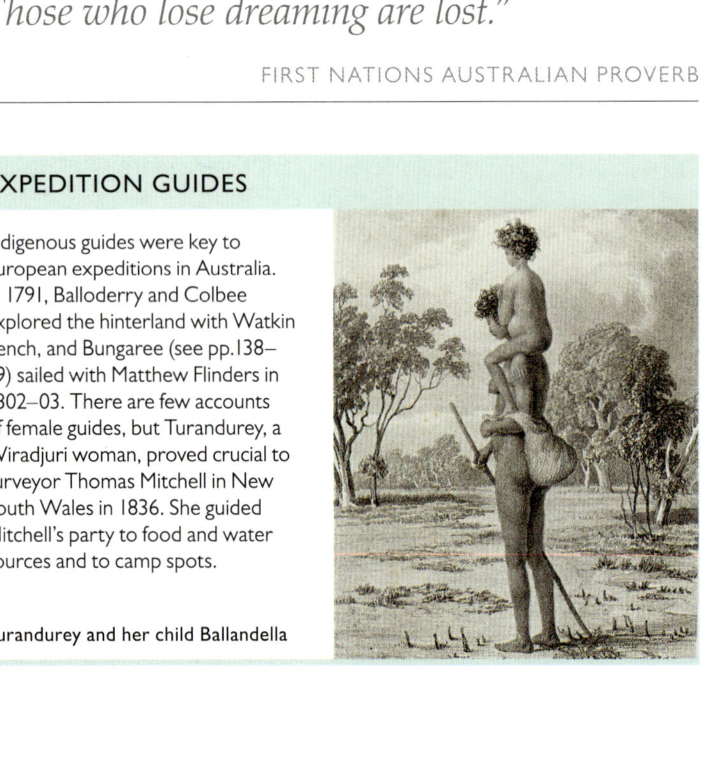

Turandurey and her child Ballandella

Navigating the Australian outback
For Aboriginal people, the vast Australian landscape becomes navigable through songlines. These hold information on features and landmarks, and lead them to water sources, food supplies, and safe places to take shelter.

MAPPING CANADA

Europeans had been settling in Canada since the early 1500s, but it was not until 1793 that fur trader Alexander Mackenzie completed the first trek across the country to the Pacific Ocean. In the early 1800s, David Thompson spent many years charting the vast interior.

In the late 18th century, Europeans in Canada confined themselves to a few isolated towns and settlements, mostly in Quebec and Ontario in the east. The fur trade was a key driver in exploration of the Canadian west. Europeans made contact with Indigenous peoples and brought them into the trade as hunters, guides, and interpreters.

Between 1769 and 1772, over the course of three trips, the Hudson's Bay Company sponsored an English-born fur trader, Samuel Hearne, to search for copper. Leaving Fort Churchill and led by a Chipewyan Chief, Matonabbee, Hearne became the first European to reach the Arctic Ocean overland in 1771.

The North West Company tasked the Scottish-born fur trader Alexander Mackenzie to find a trade route to the Pacific Ocean. His first attempt in 1789 failed when, by mistake, he canoed down a river to the Arctic Ocean. But on the next expedition, which set out in 1792, Mackenzie reached Bella Coola, situated on an inlet of the Pacific Ocean, on 22 July 1793 – completing the first recorded transcontinental crossing of North America north of Mexico.

Another fur trader, the Anglo-Canadian David Thompson later crossed Canada from east to west (see below), accompanied by the Hawaiian explorer Naukane (see pp.150–51). Thompson became the first European to sail the full length of the Columbia River in 1811.

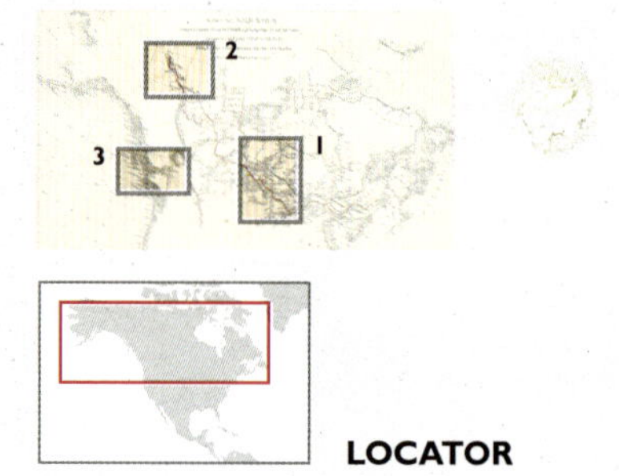

LOCATOR

△ **Mackenzie's routes**
This map from 1801 shows Alexander Mackenzie's accidental route to the Arctic Ocean in 1789, in red, and his successful route to the Pacific in 1793, in yellow.

▷ **Heading north**
Sent by the North West Company to handle fur-trading affairs in the far west, Mackenzie travelled upriver through miles of great rapids.

DAVID THOMPSON'S EXPLORATIONS
The fur trader and cartographer David Thompson mapped more of North America than anyone else. Working first for the Hudson's Bay Company in 1784–97, and then the North West Company in 1797–1812, he travelled west on foot or by horseback, canoe, and dog sled over some 90,000 km (55,000 miles), equivalent to circling the world twice.

KEY
Expeditions under the Hudson's Bay Company, 1784–97
Expeditions under the North West Company, 1797–1812

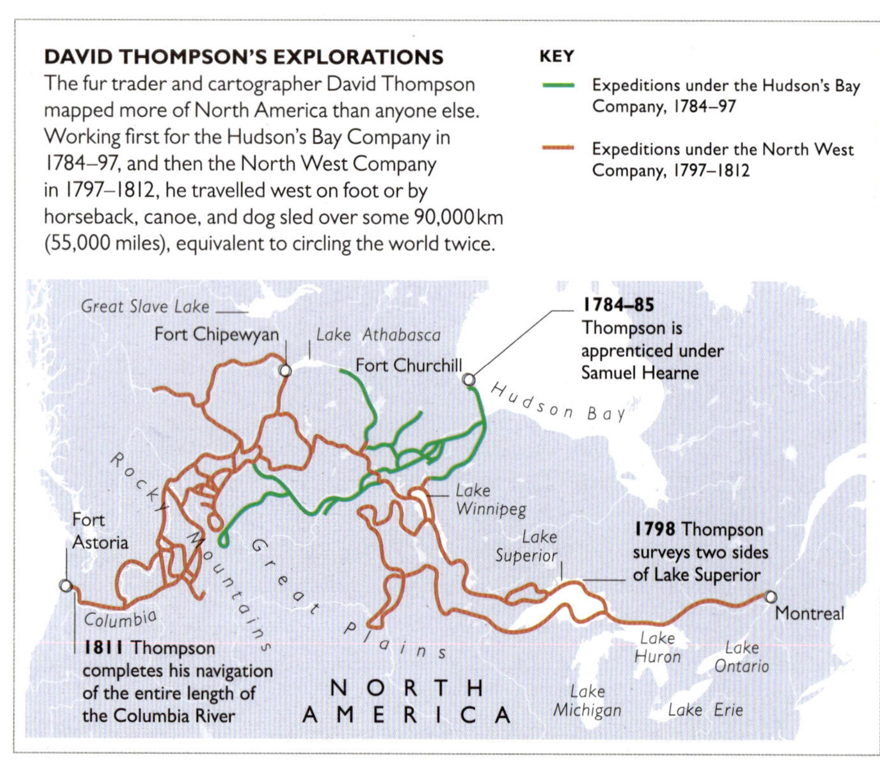

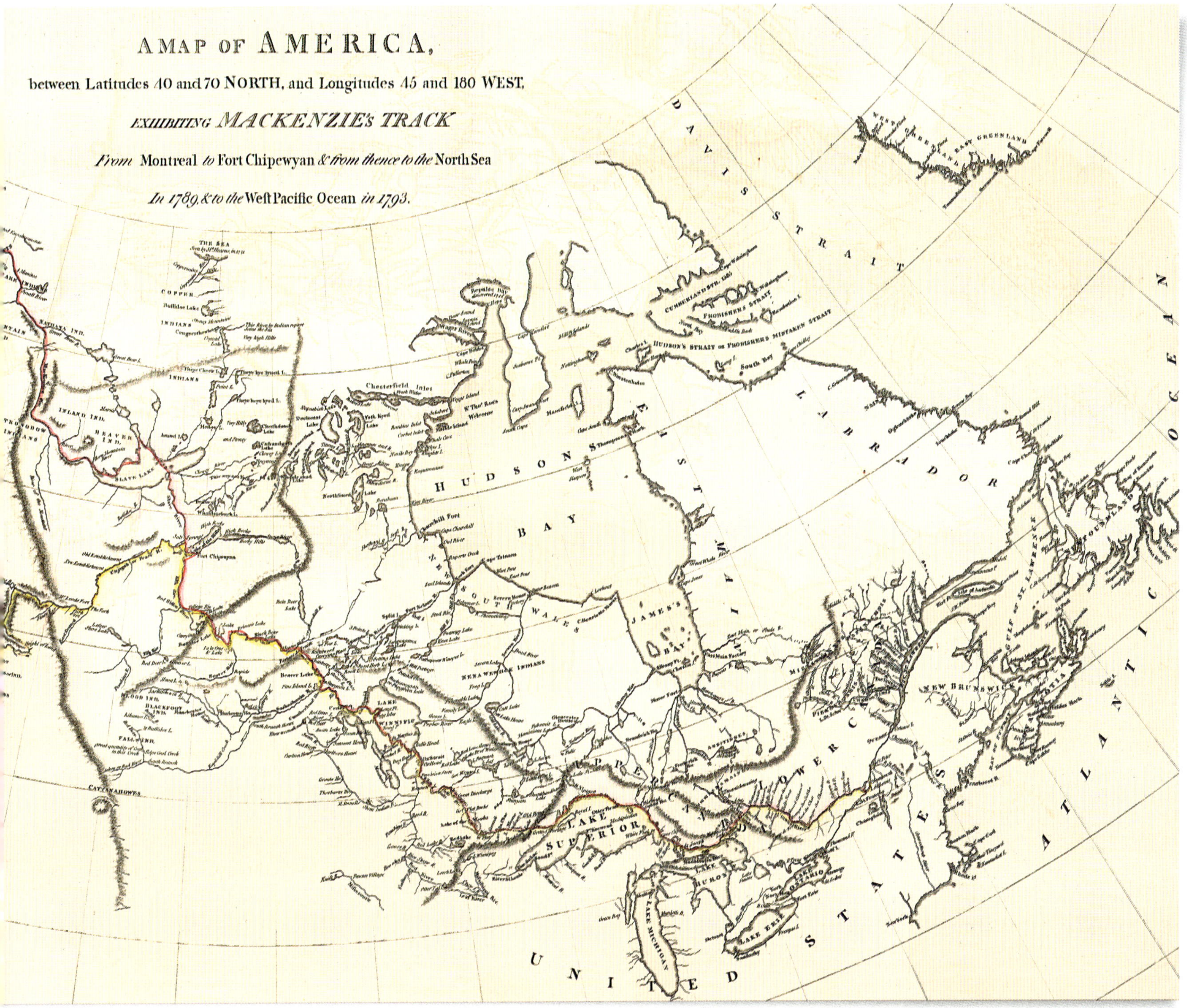

A MAP OF AMERICA,
between Latitudes 40 and 70 NORTH, and Longitudes 45 and 180 WEST.
EXHIBITING MACKENZIE's TRACK
From Montreal to Fort Chipewyan & from thence to the North Sea
In 1789, & to the West Pacific Ocean in 1793.

2

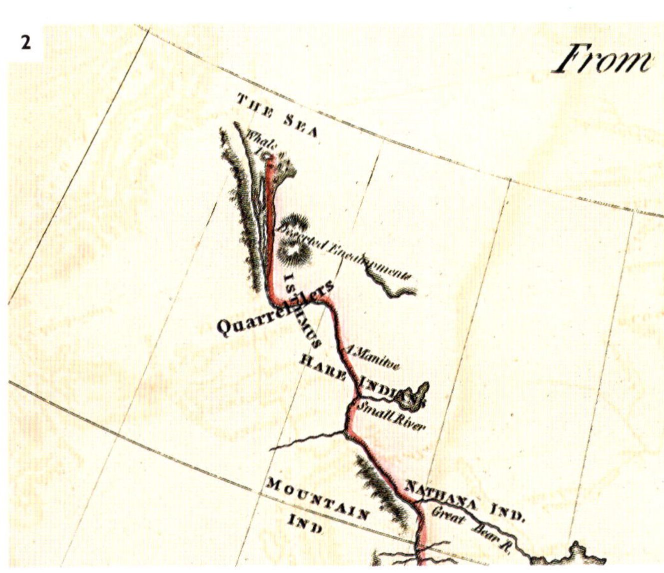

3

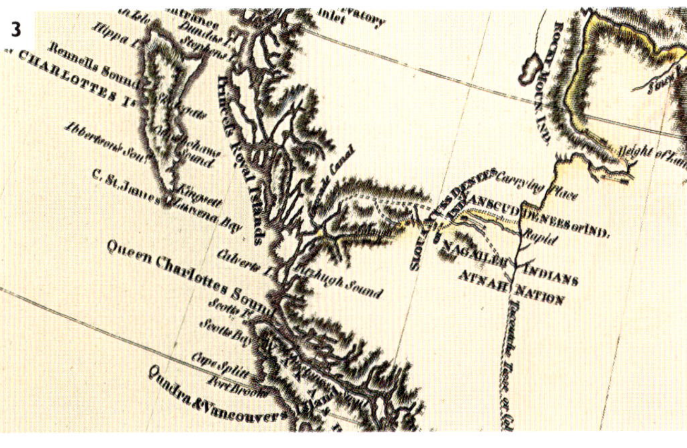

◁ **The wrong ocean**
Testing the theory of his North West Company predecessor in the Athabasca area, Peter Pond, Mackenzie followed what he called the "Grand River", believing it flowed west to the Pacific, but found it went north instead, to the Arctic Ocean.

◁ **Reaching the Pacific**
Borrowing canoes from Nuxalk (Bella Coola) people, Mackenzie followed Bella Coola River to the Pacific. On a rock, using grease and vermilion pigment, he wrote: "Alexander Mackenzie, from Canada by land, 22 July, 1793".

THE CORPS OF DISCOVERY

In 1803, US president Thomas Jefferson bought a vast tract of land from France for $15 million. The Louisiana Purchase almost doubled the size of the USA. Jefferson appointed an expedition to survey the Missouri River and find a transcontinental trade route to the Pacific Ocean.

Having read Scottish explorer Alexander Mackenzie's book about his 1792–93 traverse of Canada, Jefferson assembled a US expeditionary force to find a practical commercial route across the continent and to establish an American presence there before European nations intervened. The group was also tasked with learning about the region's geography, fauna and flora, mineral wealth, and inhabitants.

In 1803, Jefferson commissioned the Corps of Discovery led by Meriwether Lewis, a US Army captain. Lewis chose a soldier, William Clark, as his partner, along with five officers, thirty enlisted men, a number of civilians, and York, an enslaved African-American man. Sacagawea, a young Lemhi Shoshone woman, joined the party as a guide and interpreter during their first winter at Fort Mandan.

The expedition sailed from the American bank of the Mississippi on 14 May 1804. Their route took them to the Missouri headwaters, across the Rockies and down the Clearwater, Snake, and Columbia rivers to the Pacific coast, which they first sighted on 7 November 1805. They then faced a second tough winter before returning home to St Louis on 23 September 1806. The party had succeeded in their mission. The American route to the Pacific Ocean was now open.

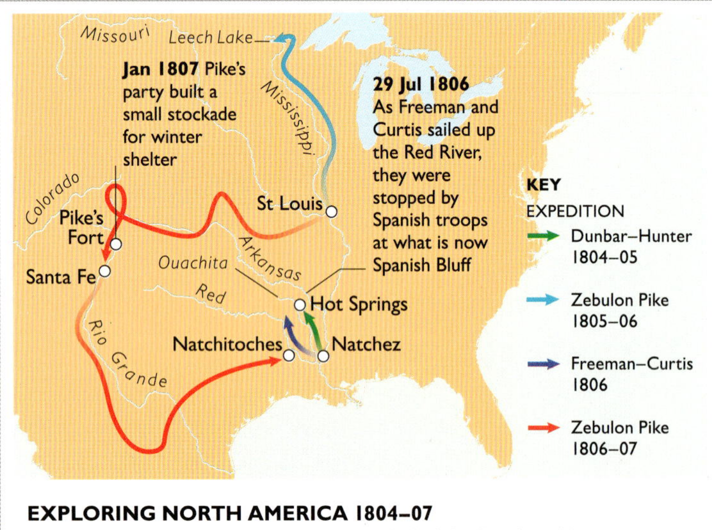

EXPLORING NORTH AMERICA 1804–07
Other surveys undertaken at the time included the Dunbar–Hunter Expedition of 1804–05, which was limited in scope. Zebulon Pike explored the upper Mississippi during 1805–06, but a second trek in 1806–07 ended when the Spanish authorities escorted him back across the border. The Freeman–Curtis or Red River Expedition of 1806 also had to retire after confronting Spanish troops. Each gained valuable information about the far west, however, enabling the US to expand further across the continent.

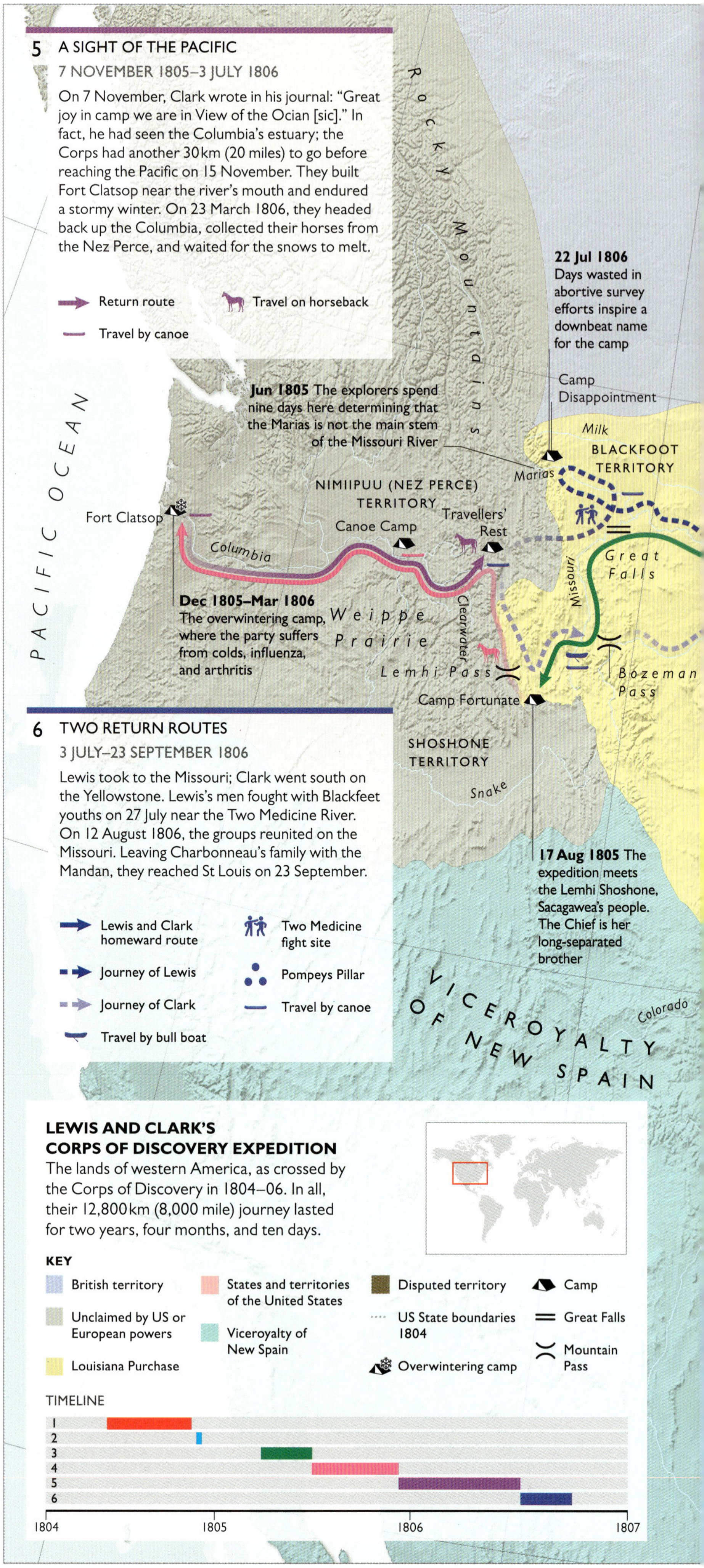

5 A SIGHT OF THE PACIFIC

7 NOVEMBER 1805–3 JULY 1806

On 7 November, Clark wrote in his journal: "Great joy in camp we are in View of the Ocian [sic]." In fact, he had seen the Columbia's estuary; the Corps had another 30km (20 miles) to go before reaching the Pacific on 15 November. They built Fort Clatsop near the river's mouth and endured a stormy winter. On 23 March 1806, they headed back up the Columbia, collected their horses from the Nez Perce, and waited for the snows to melt.

Return route
Travel on horseback
Travel by canoe

Rocky Mountains
PACIFIC OCEAN

22 Jul 1806 Days wasted in abortive survey efforts inspire a downbeat name for the camp

Camp Disappointment
Milk
BLACKFOOT TERRITORY
Marias

Jun 1805 The explorers spend nine days here determining that the Marias is not the main stem of the Missouri River

NIMIIPUU (NEZ PERCE) TERRITORY
Fort Clatsop
Travellers' Rest
Canoe Camp
Columbia
Great Falls
Missouri

Dec 1805–Mar 1806 The overwintering camp, where the party suffers from colds, influenza, and arthritis

Weippe Prairie
Clearwater
Lemhi Pass
Bozeman Pass
Camp Fortunate

6 TWO RETURN ROUTES

3 JULY–23 SEPTEMBER 1806

Lewis took to the Missouri; Clark went south on the Yellowstone. Lewis's men fought with Blackfeet youths on 27 July near the Two Medicine River. On 12 August 1806, the groups reunited on the Missouri. Leaving Charbonneau's family with the Mandan, they reached St Louis on 23 September.

Lewis and Clark homeward route
Two Medicine fight site
Journey of Lewis
Pompeys Pillar
Journey of Clark
Travel by canoe
Travel by bull boat

SHOSHONE TERRITORY
Snake

17 Aug 1805 The expedition meets the Lemhi Shoshone, Sacagawea's people. The Chief is her long-separated brother

VICEROYALTY OF NEW SPAIN
Colorado

LEWIS AND CLARK'S CORPS OF DISCOVERY EXPEDITION
The lands of western America, as crossed by the Corps of Discovery in 1804–06. In all, their 12,800km (8,000 mile) journey lasted for two years, four months, and ten days.

KEY
British territory
States and territories of the United States
Disputed territory
Camp
Unclaimed by US or European powers
US State boundaries 1804
Great Falls
Viceroyalty of New Spain
Mountain Pass
Louisiana Purchase
Overwintering camp

TIMELINE
1
2
3
4
5
6
1804 1805 1806 1807

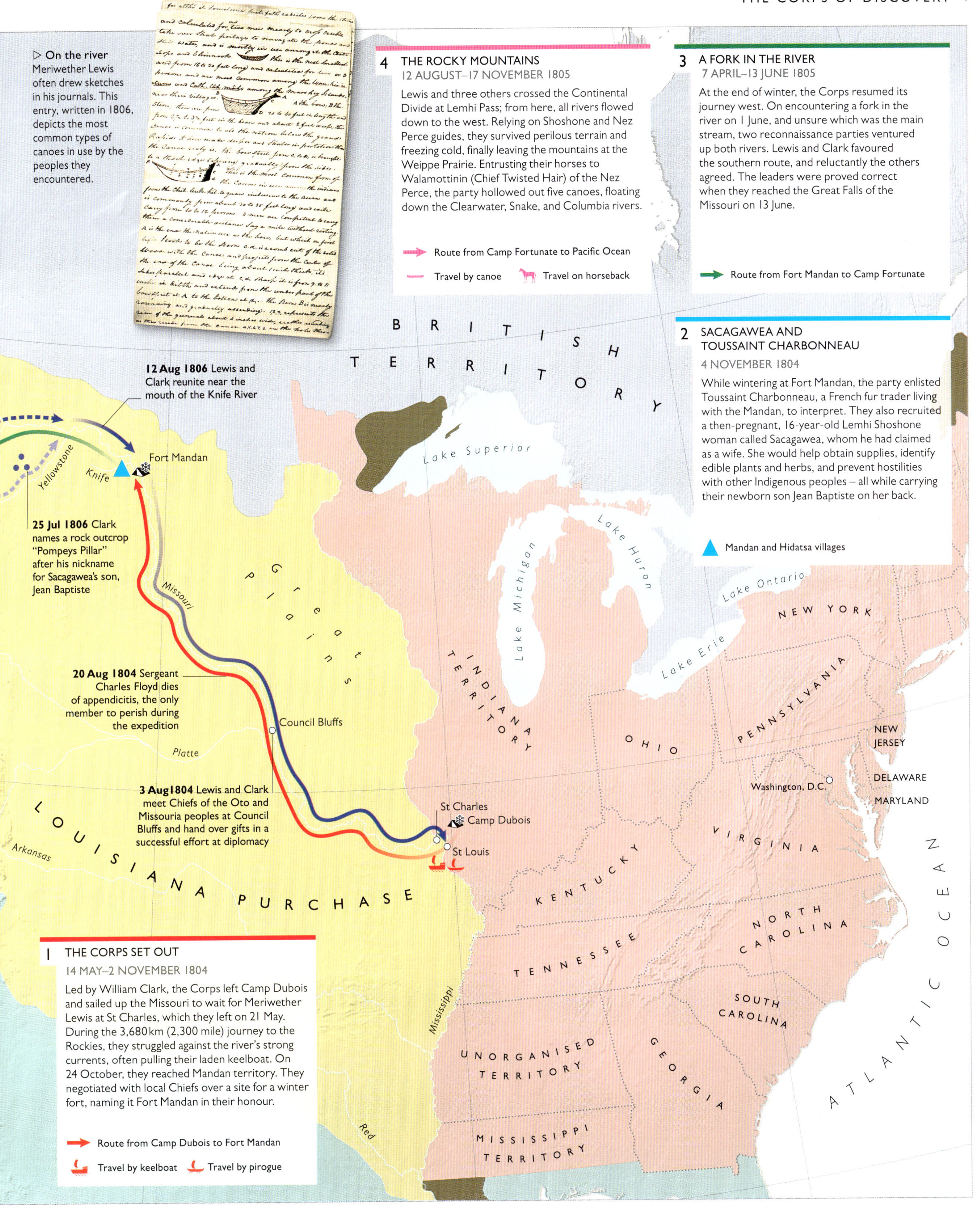

▷ **On the river**
Meriwether Lewis often drew sketches in his journals. This entry, written in 1806, depicts the most common types of canoes in use by the peoples they encountered.

4 THE ROCKY MOUNTAINS

12 AUGUST–17 NOVEMBER 1805

Lewis and three others crossed the Continental Divide at Lemhi Pass; from here, all rivers flowed down to the west. Relying on Shoshone and Nez Perce guides, they survived perilous terrain and freezing cold, finally leaving the mountains at the Weippe Prairie. Entrusting their horses to Walamottinin (Chief Twisted Hair) of the Nez Perce, the party hollowed out five canoes, floating down the Clearwater, Snake, and Columbia rivers.

Route from Camp Fortunate to Pacific Ocean
Travel by canoe
Travel on horseback

3 A FORK IN THE RIVER

7 APRIL–13 JUNE 1805

At the end of winter, the Corps resumed its journey west. On encountering a fork in the river on 1 June, and unsure which was the main stream, two reconnaissance parties ventured up both rivers. Lewis and Clark favoured the southern route, and reluctantly the others agreed. The leaders were proved correct when they reached the Great Falls of the Missouri on 13 June.

Route from Fort Mandan to Camp Fortunate

2 SACAGAWEA AND TOUSSAINT CHARBONNEAU

4 NOVEMBER 1804

While wintering at Fort Mandan, the party enlisted Toussaint Charbonneau, a French fur trader living with the Mandan, to interpret. They also recruited a then-pregnant, 16-year-old Lemhi Shoshone woman called Sacagawea, whom he had claimed as a wife. She would help obtain supplies, identify edible plants and herbs, and prevent hostilities with other Indigenous peoples – all while carrying their newborn son Jean Baptiste on her back.

Mandan and Hidatsa villages

1 THE CORPS SET OUT

14 MAY–2 NOVEMBER 1804

Led by William Clark, the Corps left Camp Dubois and sailed up the Missouri to wait for Meriwether Lewis at St Charles, which they left on 21 May. During the 3,680 km (2,300 mile) journey to the Rockies, they struggled against the river's strong currents, often pulling their laden keelboat. On 24 October, they reached Mandan territory. They negotiated with local Chiefs over a site for a winter fort, naming it Fort Mandan in their honour.

Route from Camp Dubois to Fort Mandan
Travel by keelboat
Travel by pirogue

INDIGENOUS EXPLORERS

For many explorers, the most significant item in their kitbag was their journal, to record events and observations for posterity. In contrast, first-hand accounts by Indigenous explorers are mostly missing from the historical record, yet we can still perceive glimpses of their fascinating journeys.

△ **Food on the go**
Sacagawea identified edible plants, such as the "buffaloberries" of *Shepherdia argentea*.

In the early 18th century, a French ethnographer, Antoine-Simon Le Page du Pratz, spent time researching Indigenous American history in Louisiana. In his memoirs, he writes of meeting an old man, Moncacht-Apé of the local Yazoo tribe, who told him of a journey he had undertaken in the late 1600s, crossing the continent alone from the Mississippi to the Pacific to learn about the origins of his people. Le Page's account, published in 1758, though unverified officially, became well-known.

Captain Meriwether Lewis and Lieutenant William Clark carried this narrative of Moncacht-Apé's travels with them on their Corps of Discovery expedition (1804–06), and also relied heavily on the knowledge of their Lemhi-Shoshone interpreter and guide, Sacagawea. Her presence of mind is noted in the expedition journals – she saved essential supplies when a boat capsized on the Missouri River, and her successful bartering acquired the party the horses that were so essential for crossing the Rockies and reaching the Pacific. On 13 July 1806, William Clark wrote in his journal that Sacagawea had been their "pilot through this country".

Innumerable journeys were made possible thanks to the expertise of the Indigenous guides, and their knowledge of terrain and mapping skills were acknowledged in many instances. In other cases, they received little credit. Explorers in the Arctic, for example, relied on the Inuit to survive, yet when Seegloo, Egingwah, Ooqueah, and Ootah travelled with Robert Peary and Matthew Henson on their 1908–09 expedition, they were merely recorded as "four Polar Eskimos".

ACROSS CANADA

In 1811, Naukane, a Hawaiian leader, was working for the Pacific Fur Company at Fort Astoria on the west coast of Canada when the cartographer David Thompson hired him to join his continent-crossing expedition (see pp.146–47). Naukane travelled with Thompson as far as Fort William on Lake Superior, and then independently journeyed east as far as Quebec, and then onwards to England.

Portrait of Naukane by Paul Kane, 1847

Smoothing the waters
This 1905 painting shows the moment, in 1805, when the Corps of Discovery met a Chinook group on the Lower Columbia River. Without Sacagawea, the only female in their group, the male explorers would have been seen as a war party and treated with suspicion.

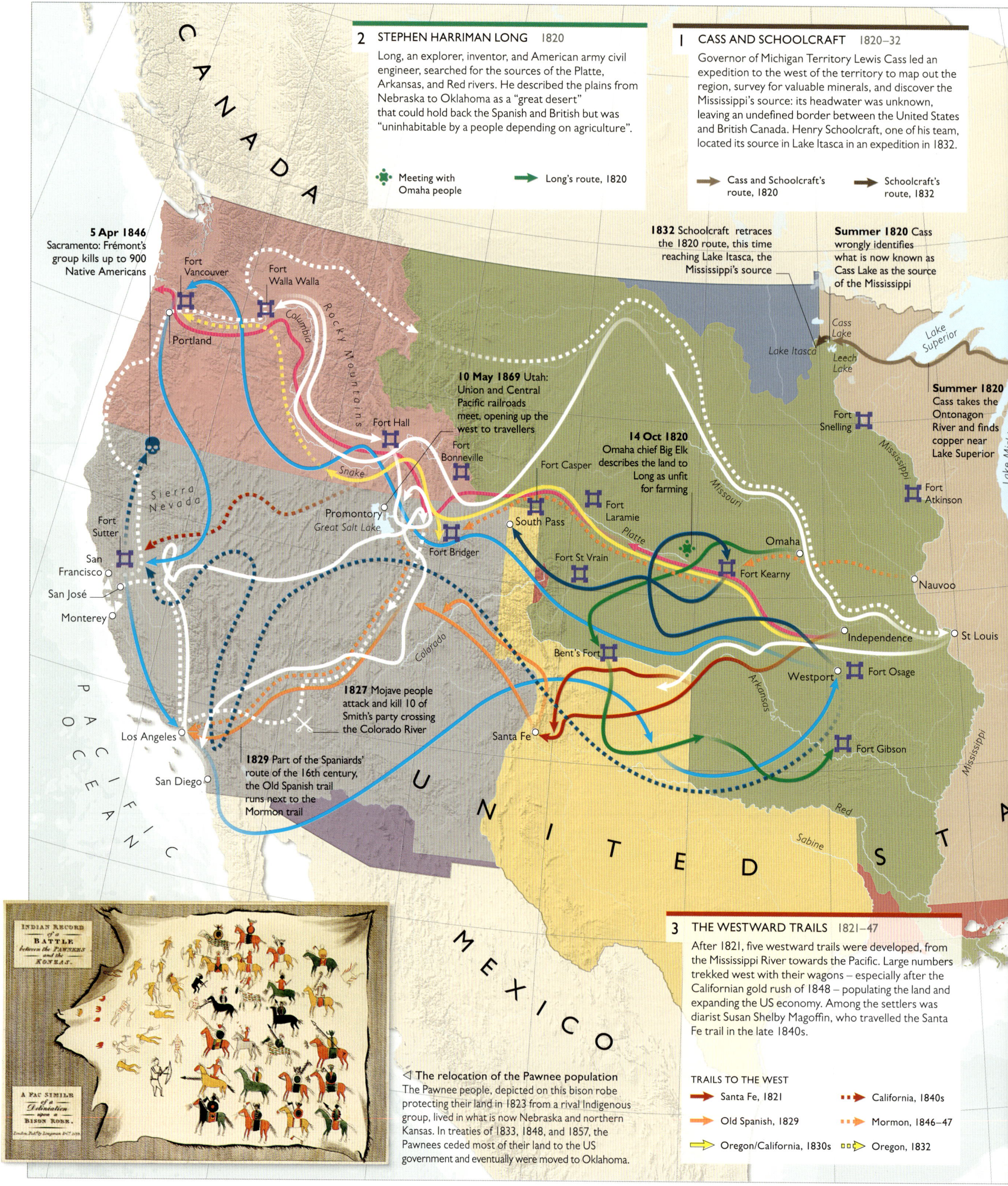

2 STEPHEN HARRIMAN LONG 1820

Long, an explorer, inventor, and American army civil engineer, searched for the sources of the Platte, Arkansas, and Red rivers. He described the plains from Nebraska to Oklahoma as a "great desert" that could hold back the Spanish and British but was "uninhabitable by a people depending on agriculture".

Meeting with Omaha people

Long's route, 1820

1 CASS AND SCHOOLCRAFT 1820–32

Governor of Michigan Territory Lewis Cass led an expedition to the west of the territory to map out the region, survey for valuable minerals, and discover the Mississippi's source: its headwater was unknown, leaving an undefined border between the United States and British Canada. Henry Schoolcraft, one of his team, located its source in Lake Itasca in an expedition in 1832.

Cass and Schoolcraft's route, 1820

Schoolcraft's route, 1832

3 THE WESTWARD TRAILS 1821–47

After 1821, five westward trails were developed, from the Mississippi River towards the Pacific. Large numbers trekked west with their wagons – especially after the Californian gold rush of 1848 – populating the land and expanding the US economy. Among the settlers was diarist Susan Shelby Magoffin, who travelled the Santa Fe trail in the late 1840s.

TRAILS TO THE WEST

- Santa Fe, 1821
- California, 1840s
- Old Spanish, 1829
- Mormon, 1846–47
- Oregon/California, 1830s
- Oregon, 1832

◁ **The relocation of the Pawnee population**
The Pawnee people, depicted on this bison robe protecting their land in 1823 from a rival Indigenous group, lived in what is now Nebraska and northern Kansas. In treaties of 1833, 1848, and 1857, the Pawnees ceded most of their land to the US government and eventually were moved to Oklahoma.

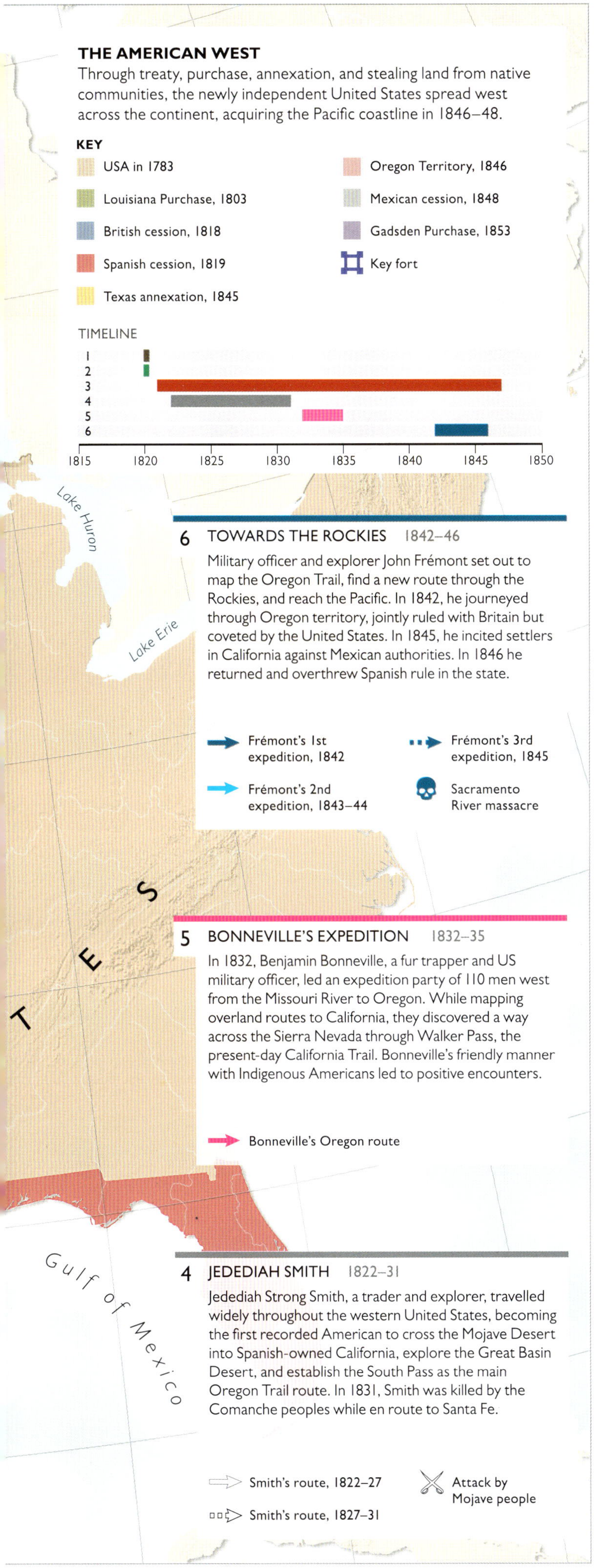

THE AMERICAN WEST

Through treaty, purchase, annexation, and stealing land from native communities, the newly independent United States spread west across the continent, acquiring the Pacific coastline in 1846–48.

KEY

- USA in 1783
- Louisiana Purchase, 1803
- British cession, 1818
- Spanish cession, 1819
- Texas annexation, 1845
- Oregon Territory, 1846
- Mexican cession, 1848
- Gadsden Purchase, 1853
- Key fort

TIMELINE

1 2 3 4 5 6

1815 1820 1825 1830 1835 1840 1845 1850

Lake Huron

Lake Erie

Gulf of Mexico

T E S

6 TOWARDS THE ROCKIES 1842–46

Military officer and explorer John Frémont set out to map the Oregon Trail, find a new route through the Rockies, and reach the Pacific. In 1842, he journeyed through Oregon territory, jointly ruled with Britain but coveted by the United States. In 1845, he incited settlers in California against Mexican authorities. In 1846 he returned and overthrew Spanish rule in the state.

- Frémont's 1st expedition, 1842
- Frémont's 3rd expedition, 1845
- Frémont's 2nd expedition, 1843–44
- Sacramento River massacre

5 BONNEVILLE'S EXPEDITION 1832–35

In 1832, Benjamin Bonneville, a fur trapper and US military officer, led an expedition party of 110 men west from the Missouri River to Oregon. While mapping overland routes to California, they discovered a way across the Sierra Nevada through Walker Pass, the present-day California Trail. Bonneville's friendly manner with Indigenous Americans led to positive encounters.

- Bonneville's Oregon route

4 JEDEDIAH SMITH 1822–31

Jedediah Strong Smith, a trader and explorer, travelled widely throughout the western United States, becoming the first recorded American to cross the Mojave Desert into Spanish-owned California, explore the Great Basin Desert, and establish the South Pass as the main Oregon Trail route. In 1831, Smith was killed by the Comanche peoples while en route to Santa Fe.

- Smith's route, 1822–27
- Smith's route, 1827–31
- Attack by Mojave people

ACROSS THE AMERICAN WEST

At the end of the 18th century, the west of the United States, across the Mississippi River, was uncharted and unknown territory to most European settlers. In 1803, the US purchased around 827,000 square miles to the west of the river from the French. Over the next 40 years, the newcomers mapped out this land and sought a route to the Pacific.

The US expeditions that set out to explore the west of the country mixed scientific curiosity with commercial gain and territorial greed. Lewis Cass, the US governor of Michigan Territory, embarked on his expedition in 1820. One of his aims was to find the source of the Mississippi River, and from here, delineate a national border with British Canada. He also intended to prospect for copper and other ores and minerals.

During John Frémont's expedition of 1845, he incited Californians to rebel against Spanish rule and join the US.

All of the many expeditions west encountered Indigenous Americans, leading to battles, skirmishes, and massacres. They also introduced new diseases into populations who had no immunity, leading to the deaths of thousands of Indigenous Americans.

As the explorers headed west, pioneer settlers followed. Five major westward trails were established by 1847, along which thousands of settlers – together with their families, belongings, and livestock – trekked in their wagons, in search of new lives. These Americans believed that, in the words of a newspaper editor in 1845, they had "a manifest destiny" to possess and control the whole of the continent. This belief shaped the country, allowing the new citizens to claim the land from coast to coast, displacing the Indigenous peoples who had lived there for thousands of years.

> *"And that claim is by the right of our manifest destiny to overspread and to possess the whole of the continent..."*
>
> JOHN O'SULLIVAN, *NEW YORK MORNING NEWS*, 27 DECEMBER 1845

EXPLORING THE FAR NORTH

The United States bought Alaska from Russia in 1867 for $7.2 million (around $160 million today), but had little idea of the land it was purchasing. To remedy this, a number of expeditions set out. The Northern Alaska Exploring Expedition of 1884–86 was led by Lieutenant George Stoney of the US Navy. He explored the Kobuk River and northern Alaska, sending mineral samples to the United States National Museum.

Fort Cosmos, built by Lt Stoney

NATURALIST DISCOVERIES

In the footsteps of the explorers of this era, who charted and exploited the lands they visited, came specialists in the fields of science, botany, and biology. They investigated the wild fauna and flora of the regions into which they ventured, notably southeast Asia and the Amazon rainforest.

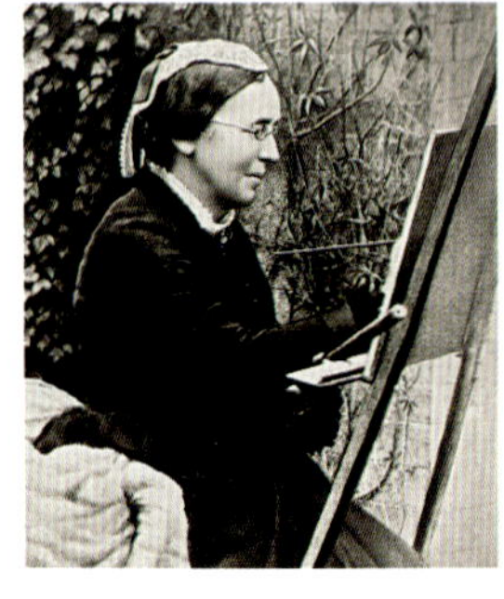

△ **North at the easel**
Marianne North's paintings of flora are valuable botanic studies. London's Kew Gardens has her work on permanent display.

Many of these naturalist explorers produced a considerable body of work, cataloguing and sending home numerous samples for study. Englishman Alfred Russel Wallace lived in the Amazon basin from 1848 to 1852, although he lost his entire collection of insect and animal specimens, and most of his notes, on the return voyage when his ship caught fire. From 1854 to 1862, he worked in the East Indies.

Henry Bates sailed from England in 1848 to join Wallace, staying in the Amazon until 1859. He collected 14,712 specimens (mostly insects) and wrote the first scientific account of mimicry in animals. From 1849, English botanist Richard Spruce spent 15 years studying Amazonian plants.

English botanical artist Marianne North spent most of her life after 1871 painting the flora of distant countries. She travelled widely (and mostly alone) – notably in Europe, North America, Australasia, and India. Her last trip, in 1884–85, took her to the Seychelles and Chile.

The British plant collector E.H. Wilson dedicated his life to Asian plant species, travelling to China and Japan on seven occasions between 1899 and 1918. He introduced approximately 2,000 Asian plant species to Europe, of which around 60 bear his name.

THE WALLACE LINE

In his travels around the Malay Archipelago, Alfred Wallace noticed profound differences between the mammals and birds he encountered, often within short distances of each other. In 1859, he sketched out a rough boundary line between Asia and Australia, which delineated two ecozones of fauna: on the northwest side are Asian species and on the southeast side, Australasian and Asian species.

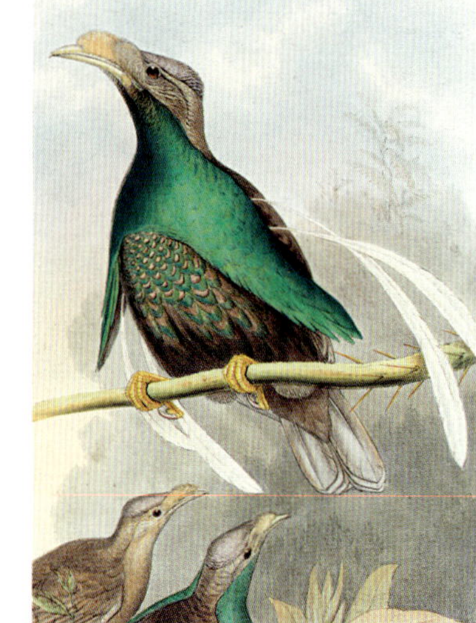

A *Semioptera wallacii* bird of paradise, identified by Wallace in 1858

538 Oxysternon – feeble ♂? – probably an undesc.
found in dung, Aveyros, riv. Tapajos
539 Oxyst. festivum (doubtless) 1 fully developed ♂
have found this sp. only at Sant.
day time
540 Oxyst. very fine, cobalt blue sp. – a ♀ –
541 –„– very fine, brilliant green sp. one ♂ 2♀
recollect to have found it near
542 „ – very pretty little sp. in dung at
1 feeble ♂ 1♀
543 – 1 feeble ♂ – also in dung at Ega
544 – „ – 1 feeble ♂ – very brilliant little sp.
altogether 7 species of Oxysternon – Ega 6 May 18
545 Myrmicidae – peculiar genus – bispinose
running over decayed branches of trees,
tail elevated like them – Ega May 11
546-7 8-9 Isonychus – 4 specs. each one different
of one species, as found in same
then by beating the lower trees.
They are evidently Isonychus – from canalicul
from their scaly clothing I judged them
cannot be as they want the projecting piec
550 –„– another, scaleless sp. on foliage
but in a few days not one was to be
551 – largest sp. I have seen – foliage of
552 –„ 2nd largest sp. common at Ega
553 – „ blk. sp. found Ega
554 – prettily variegated sp. quite distinct
Ega
555 –„ a small Villa Nova sp.
556 –„ Santarem has not any peculiarity
nearly allied sp.
there are 4 or 5 more species in my box which
557 / 558 New genus of Rutelidae. Its facies is different from
tarsi are different – the terminal jt. being slender
it cannot be anything but a Rutelida, & from its unbordered
in one of the two groups Anticheirida or Rutel

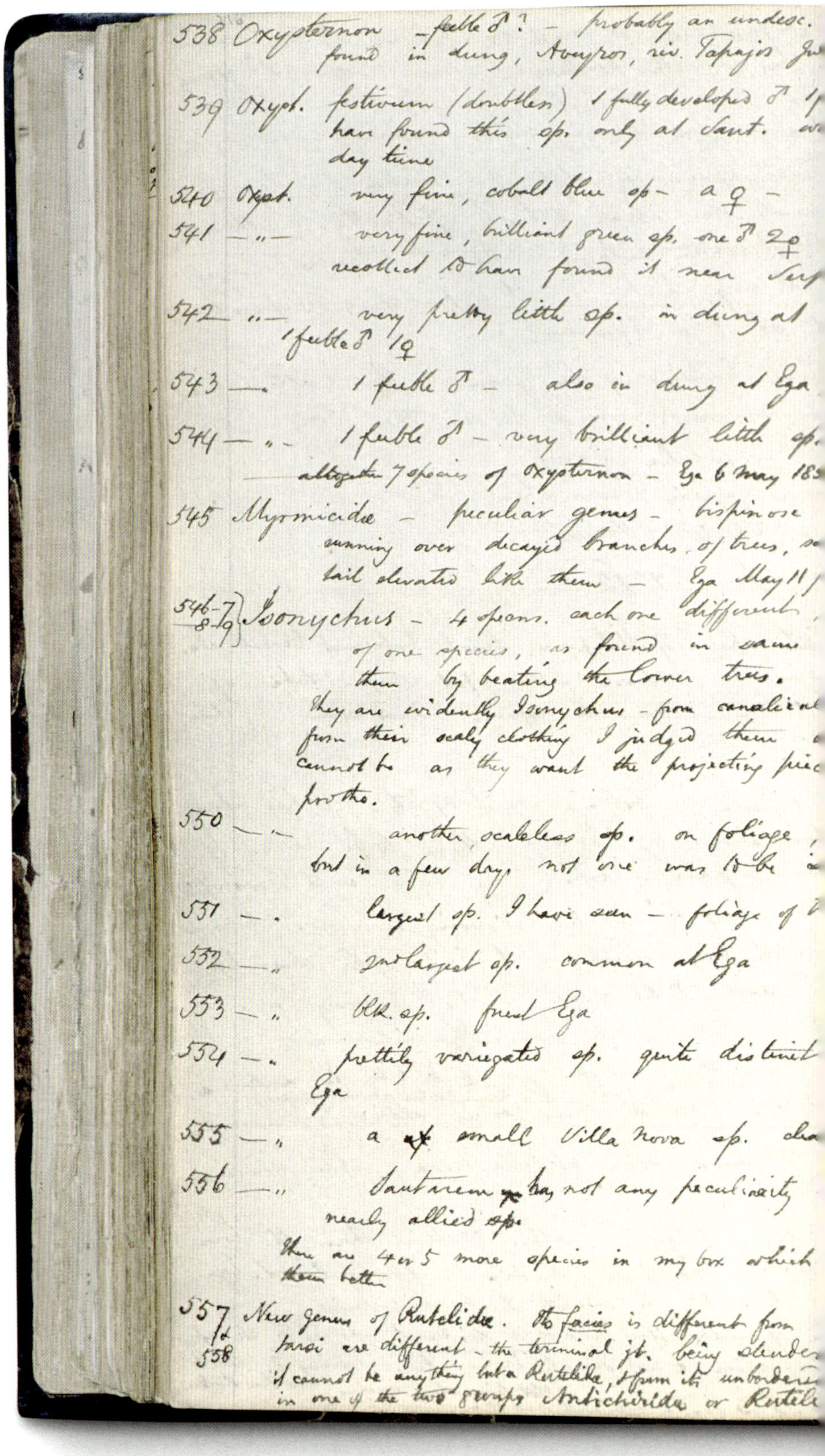

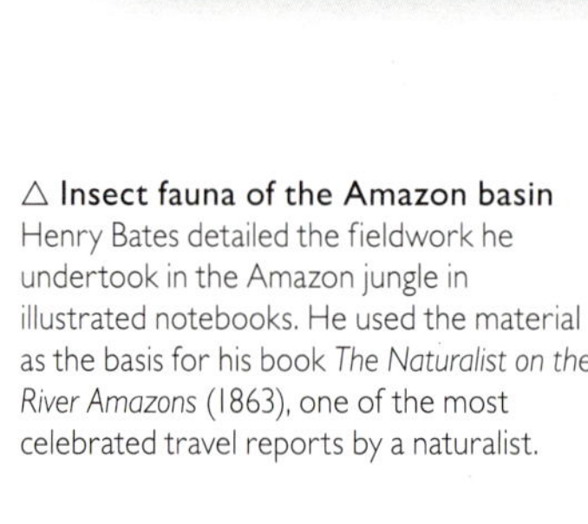

△ **Insect fauna of the Amazon basin**
Henry Bates detailed the fieldwork he undertook in the Amazon jungle in illustrated notebooks. He used the material as the basis for his book *The Naturalist on the River Amazons* (1863), one of the most celebrated travel reports by a naturalist.

"There is something in a tropical forest akin to the ocean… Man feels so completely his insignificance there and the vastness of nature."

HENRY BATES, QUOTED IN *THE NATURALIST ON THE RIVER AMAZONS*, 1863

SOUTH AMERICA.
English Miles
0 100 200 300 400 500 1000
Caribbean Sea
Barbadoes
Grenada
Tobago
Trinidad
Margarita
Curacoa
La Guayra
Barcelona
Cumana
G. of Paria
Pt. Barima
CARACAS
Valencia
Maracaybo
Merida
Carthagena
C. Gallinas
G. of Venezuela
G. of Darien
S. Juan de Nicaragua or Greytown
Porto Bello
Aspinwall
Panama
G. of Panama
Buenaventura
B. of Choco
UNITED STATES OF COLOMBIA
Granadian Confederation
Sta. Fé de Bogota
VENEZUELA
Orinoco
Ciudad Bolivar
Sa. Parima
Parime
Negro
Branco
BRITISH GUIANA
Dutch
French
Georgetown & R. Demerara
New Amsterdam & R. Berbice
Nassau & R. Saramacca
Paramaribo & R. Surinam
Maroni
Sinnamary
Cayenne
C. Orange
Oyapok
Mouth of the Amazons, or Maranon
Macapa
Caviana I.
Rio Para
Marajo or Joannes I.
Para
Braganza
Maranhao
S. Luis de Paranahyba
Ceara or Fortaleza
Equator
C. S. Francisco
QUITO
Chimborazo
Riobamba
Guayaquil
G. of Guayaquil
ECUADOR
Cuenca
Napo
Iça
Yapura
Amazons
Manaos
Santarem
Obydos
Cameta
Olivença
Teffé
Nauta
Yavari
Ucayali
Huallaga
Juruá
Purus
Coary
Madeira
Crato
Tapajos
Xingu
Tocantins
Araguay
S. Thomé
Theresina
Pt. Allegre
Natal
C. S. Roque
C. Branco
Olinda
Recife
Pernambuco
Maceio
S. Francisco
Sergipe del Rey
Jacobina
Cachoeira
Bahia
B. de todos os Santos
Olivença
Belmonte
Porto Seguro
S. Joze do Pto. Allegre
C. Parina
Sechura
Caxamarca
Truxillo
Huanuco
Huacho
Callao
LIMA
Huancavelica
Cuzco
Chincha I.
Pisco
P. S. Nicolas
PERU
Puno
Arequipa
Islay
Arica
Tacna
L. Titicaca
Sorata
Illimani
La Paz
ORURO
Cochabamba
Beni
Mamore
BOLIVIA
(SUCRE)
Chuquisaca
Potosi
Balsamo
BRAZIL
Diamantina
Matto Grosso or Villa Bella
Cuyaba
Goyaz or Villa Boa
Villa do Pto. Imperiale
Januaria
Diamantina
Porteira
Doce
Victoria
S. Joao da Barra
C. S. Thome
Rio Janeiro
Taquary
Miranda
Nova Coimbra
Iquique
Cobija
Tropic of Capricorn
Antofagasta
Desert of Atacama
Chimbo
Chaneral
Copiapo
Coquimbo
Felix
Ambrose
Atacama
Oran
Tarija
Pilcomayo
Vermejo
Gran Chaco
Salta
Tucuman
Salado
Santiago del Estero
Catamarca
Rioja
PARAGUAY
Villa Real
Asuncion
Corrientes
Parana
Uruguay
S. Paulo
Santos
Curitiba
Paranagua
S. Catherina I. & Pt. of Desterro
Porto Allegre
L. dos Patos
S. Pedro
ARGENTINE REPUBLIC
S. Juan
Cordova
Sta. Fé
Parana
Mendoza
Aconcagua
Valparaiso
SANTIAGO
S. Luis
Rosario
URUGUAY or BANDA ORIENTAL
Monte Video
Maldonado
BUENOS AYRES
Rio de la Plata
C. S. Antonio
Masafuera
Juan Fernandez
Concepcion
Biobio
Pampas
Colorado
Bahia Blanca
C. Corrientes
Mocha I.
Valdivia
Negro
Carmen
Mt. Osorno
Port Montt
Port S. Antonio
S. Carlos
Ancud
Chiloe
G. of S. Matias
Pena. of S. Josef
Chonos Archipelago
Camerones B.
Pena. of Tres Montes
G. of St. George
G. of Penas
Port Desire
Wellington I.
Port S. Julian
PATAGONIA
Madre I.
Hanover I.
Port Sta. Cruz
Falkland Is.
Port Gallegos
Stanley
Qn. Adelaide I.
Str. of Magellan
Desolation I.
Tierra del Fuego
Sta. Inez
Staten I.
Cockburn Ch.
Mt. Sarmiento
Hoste I.
Navarin I.
Diego Ramirez Is.
C. Horn
South Georgia I.
PACIFIC OCEAN
ATLANTIC OCEAN
Longitude 50 West from 40 Greenwich
90
80
70
60
50
40
30
20
10
0
10
20
30
40
50

DARWIN'S VOYAGE ON THE *BEAGLE*

In 1831, the British Royal Navy sloop HMS *Beagle* set off on a five-year mission to survey and chart "the southern coasts of South America". On board was a young scientist who was to rethink the origin of species.

Charles Darwin (1809–82) had graduated from the University of Cambridge in 1831 and wished to visit the tropics before becoming a Church of England minister. He was a keen natural scientist, and, aged 22 and at a loose end, he accepted an invitation from Robert Fitzroy, commander of HMS *Beagle*, to join his surveying expedition as a naturalist and geologist. The role allowed Darwin to leave the ship for extended periods to pursue his own interests. While HMS *Beagle* mapped the South American coast, Darwin ventured inland, discovering and mapping the region's complex geology and uncovering many prehistoric fossils. Here, he began to understand that the mountainous landscape was being uplifted and was still in the process of formation and transformation.

Darwin's greatest revelation came when HMS *Beagle* stopped at the Galapagos Islands in the Pacific Ocean. Darwin soon noticed that the mockingbirds, giant tortoises, and Galapagos finches varied from island to island. When he studied his findings later, he came to wonder just how stable individual species were and whether they could slowly evolve by themselves. From this insight, Darwin began to develop his theory of natural selection, whereby traits that best facilitate the reproduction and survival of individual species are passed on. The publication of his book *On the Origin of Species* in 1859, and the later development of the science of genetics, transformed how we view the evolution of species on our planet.

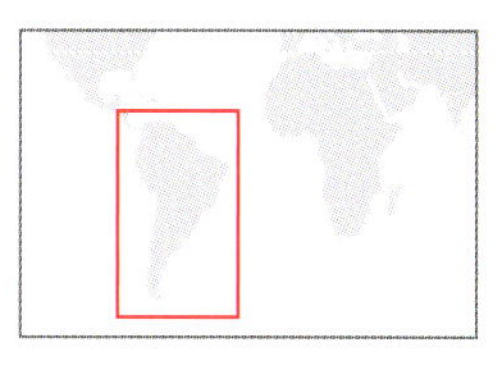

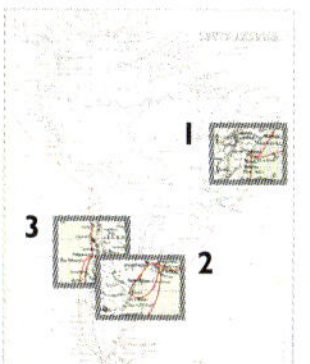

LOCATOR

◁ Darwin in South America
Darwin's intended two-year voyage around the world on HMS *Beagle* in fact lasted for five years, and three of those were spent in South America (1832–35). While the ship systematically surveyed the coastline, Darwin went ashore to explore and collect specimens.

▽ Brazilian rainforests
Soon after the *Beagle* reached Brazil in 1832, at Salvador de Bahia (modern-day Salvador), Darwin was excited by his first encounter with the ecosystem of a tropical forest, recording in his journal: "The mind is a chaos of delight."

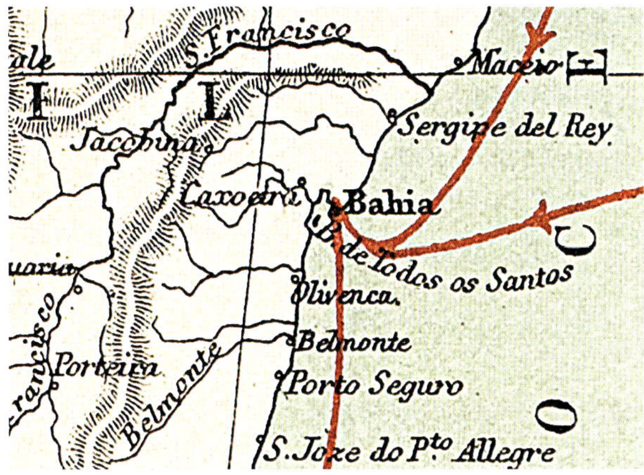

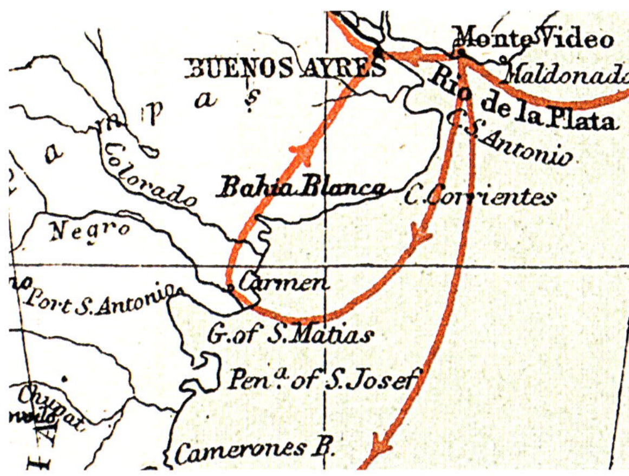

△ Prehistoric *Megatherium* fossil
In September 1832, Darwin found a giant fossilized skull, embedded in coastal cliffs south of Bahia Blanca. He chipped it out of the soft rock and later identified it as a giant ground sloth, a species that went extinct 11,000 years ago.

▽ A petrified fir forest
Climbing the Andes from Valparaiso in March 1835, Darwin discovered fossilized tree trunks standing in volcanic sandstone, and theorized that mountains were formed by a gradual series of earthquakes.

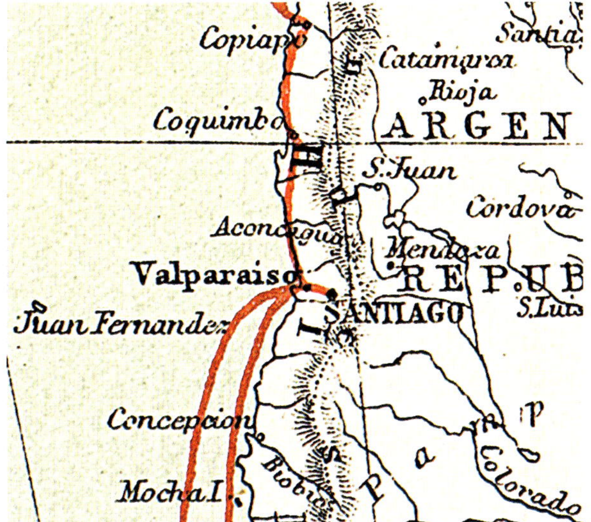

AROUND THE WORLD ON HMS *BEAGLE*

On 27 December 1831, Charles Darwin left Plymouth in England on board HMS *Beagle*. The ship slowly sailed south and then west around the world, returning to Falmouth, west of Plymouth, almost five years later, on 2 October 1836. Darwin himself spent only 18 months at sea, devoting the rest of his time to making his own explorations on land, particularly in South America and the Galapagos Islands.

TRAVERSING INLAND AFRICA

African travellers had journeyed across the continent since prehistoric times; in the 8th century, regular trans-Saharan trade routes became established. Europeans had navigated the coast since they sailed around the Cape of Good Hope in the 15th century. In the mid 19th century, they embarked on a series of expeditions inland.

The African continent posed a challenge to explorers, with its predominantly desert or jungle terrain, its waterfalls and challenging rapids, its hot and humid climate, and unfamiliar wildlife and diseases. Many of the first European explorers of the interior were motivated by religious zeal, often disregarding local religions and customs. Scotsman David Livingstone travelled to Africa in 1841 as a missionary doctor. He was the first European to journey from coast to coast and went on to explore the Kalahari basin, the Zambezi River, and many East African Lakes.

In 1847, German missionaries Johannes Rebmann and Johann Krapf were the first Europeans both to enter Africa from the Indian Ocean and to see the snow-capped mountains of East Africa.

After Livingstone's death in 1873, American Welsh-born Henry Morton Stanley determined to search for the Nile's source, inadvertently becoming the first foreigner to sail the Congo's length in 1877. He was accused of cruelty towards Africans and was an agent for King Leopold II of Belgium, helping to set up his brutal colony that used forced labour. This contributed to the 1880s Scramble for Africa, when European nations divided Africa between their respective empires, suppressing resistance from African kingdoms.

THE PERILS OF EXPLORATION

Danger was ever-present on the African continent. This European illustration shows Livingstone's boat being attacked by a hippopotamus on the Orange River in Southern Africa on one of his early expeditions. The African guides and porters who accompanied Europeans were put at direct risk. When Stanley set off down the Congo in 1874, his expedition party was 228-strong, yet many died en route from disease, hunger, drowning, or fighting.

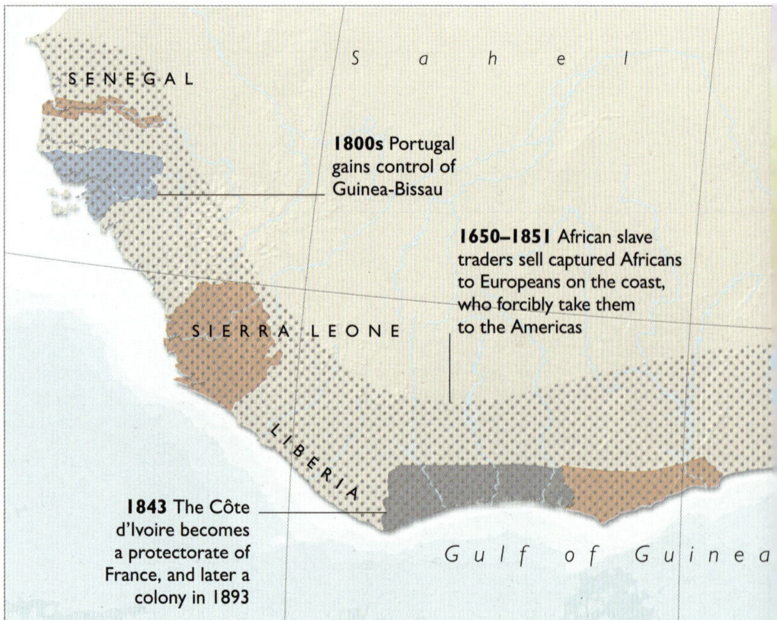

SENEGAL

Sahel

1800s Portugal gains control of Guinea-Bissau

1650–1851 African slave traders sell captured Africans to Europeans on the coast, who forcibly take them to the Americas

SIERRA LEONE

LIBERIA

1843 The Côte d'Ivoire becomes a protectorate of France, and later a colony in 1893

Gulf of Guinea

ATLANTIC

2 TO THE MOUNTAINS 1846–49

From 1846, Johannes Rebmann journeyed from Mombasa into the East African interior, alongside his fellow missionary Dr Johann Krapf. In 1848, Rebmann was led by the renowned caravan leader Bwana Kheri to the great mountain Kilimanjaro. And in 1849, Krapf became the first European to see Mount Kenya. Their reports of snowy mountains so close to the equator led to a surge of European interest in the African continent.

Rebmann and Krapf's route — Mountain

1 CHRISTIAN TRAVELS 1841–52

David Livingstone arrived in Southern Africa in 1841, intent on spreading Christianity through commerce and on introducing European practices in irrigation and medicine. Over the next decade, he made a series of journeys north of the Cape frontier, up to the Zambezi River. Countless times, he witnessed the trade of enslaved people, which strengthened his abolitionist beliefs.

Livingstone's early journeys

INSIDE AFRICA

After 1840, missionaries and explorers from Europe and America journeyed extensively into the vast interior of the African continent.

KEY

Slave trading areas

EUROPEAN POSSESSIONS, 1850s

British — Portuguese — Spanish — French

TIMELINE

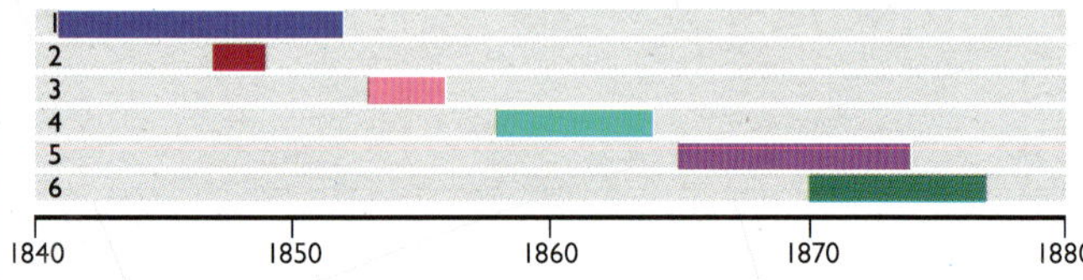

1 2 3 4 5 6

1840 1850 1860 1870 1880

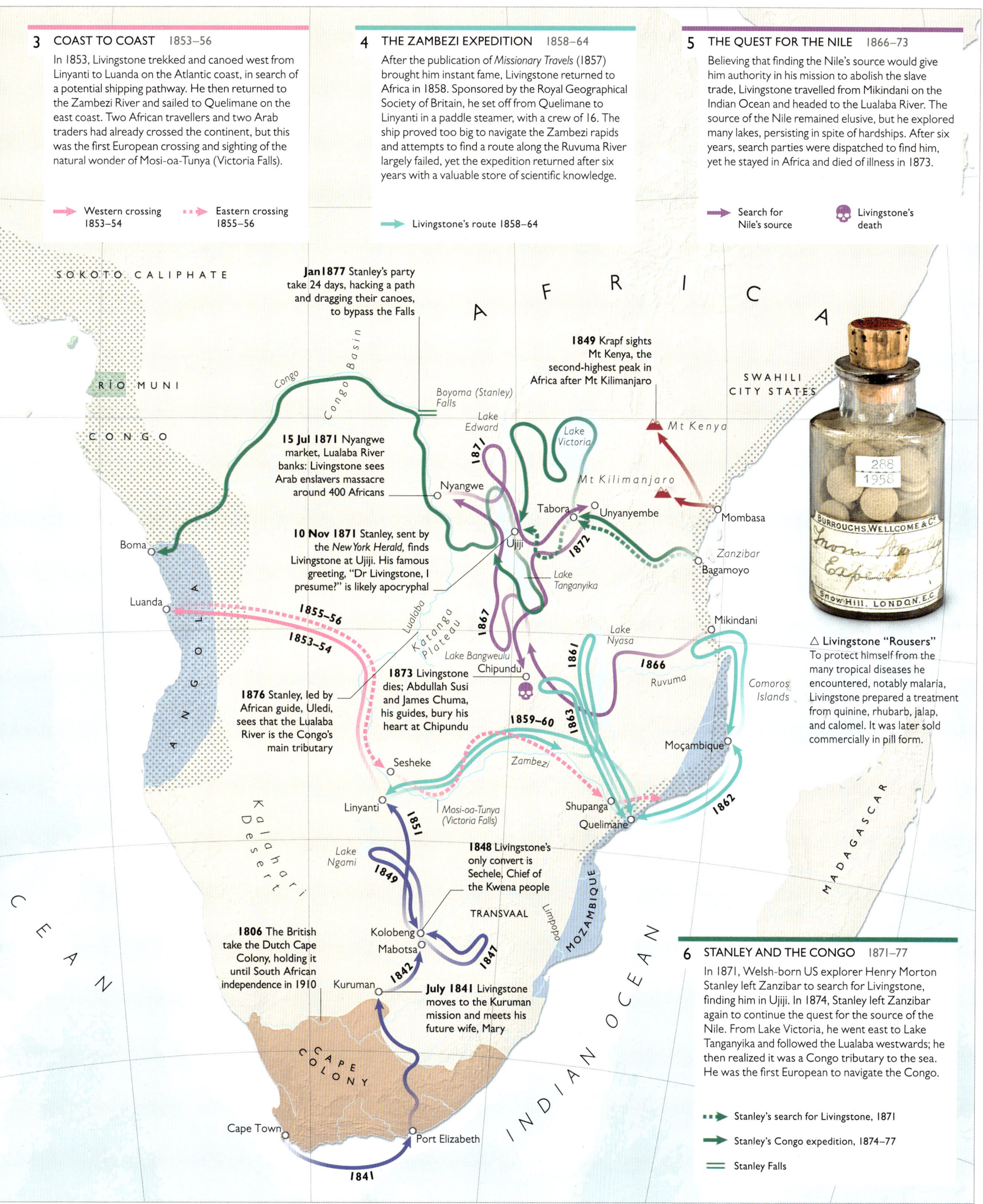

3 COAST TO COAST 1853–56

In 1853, Livingstone trekked and canoed west from Linyanti to Luanda on the Atlantic coast, in search of a potential shipping pathway. He then returned to the Zambezi River and sailed to Quelimane on the east coast. Two African travellers and two Arab traders had already crossed the continent, but this was the first European crossing and sighting of the natural wonder of Mosi-oa-Tunya (Victoria Falls).

Western crossing 1853–54

Eastern crossing 1855–56

4 THE ZAMBEZI EXPEDITION 1858–64

After the publication of *Missionary Travels* (1857) brought him instant fame, Livingstone returned to Africa in 1858. Sponsored by the Royal Geographical Society of Britain, he set off from Quelimane to Linyanti in a paddle steamer, with a crew of 16. The ship proved too big to navigate the Zambezi rapids and attempts to find a route along the Ruvuma River largely failed, yet the expedition returned after six years with a valuable store of scientific knowledge.

Livingstone's route 1858–64

5 THE QUEST FOR THE NILE 1866–73

Believing that finding the Nile's source would give him authority in his mission to abolish the slave trade, Livingstone travelled from Mikindani on the Indian Ocean and headed to the Lualaba River. The source of the Nile remained elusive, but he explored many lakes, persisting in spite of hardships. After six years, search parties were dispatched to find him, yet he stayed in Africa and died of illness in 1873.

Search for Nile's source

Livingstone's death

△ **Livingstone "Rousers"** To protect himself from the many tropical diseases he encountered, notably malaria, Livingstone prepared a treatment from quinine, rhubarb, jalap, and calomel. It was later sold commercially in pill form.

6 STANLEY AND THE CONGO 1871–77

In 1871, Welsh-born US explorer Henry Morton Stanley left Zanzibar to search for Livingstone, finding him in Ujiji. In 1874, Stanley left Zanzibar again to continue the quest for the source of the Nile. From Lake Victoria, he went east to Lake Tanganyika and followed the Lualaba westwards; he then realized it was a Congo tributary to the sea. He was the first European to navigate the Congo.

Stanley's search for Livingstone, 1871

Stanley's Congo expedition, 1874–77

Stanley Falls

MAPPING THE NIGER

Along with the Nile, Zambezi, and Congo, the Niger is one of Africa's four great rivers. Towards the end of the 18th century, European efforts to chart the river's course and find its source and estuary were stepped up. After 1795, several determined attempts were made to plot the river in its entirety.

European interest in the Niger was sparked by the Association for Promoting the Discovery of the Interior Parts of Africa. Founded in London in 1788 and commonly known as the African Association, the society aimed to explore the Niger and locate Timbuktu – a centre of Islamic scholarship not yet visited by Europeans.

In 1796, the Scotsman Mungo Park was the first European to see the Niger, observing that it flowed east "towards the rising sun". This caused some confusion as, while he established that it was not a Nile tributary, he still thought it travelled south to the Congo in central Africa. Before his death in 1806, he navigated two-thirds of it and found that it took a sharp turn south, beyond Timbuktu.

After Park, explorations continued. Alexander Laing, a British army officer serving in Sierra Leone, came close to the Niger's source in the Guinea Highlands in 1822. Eight years later, the English Lander brothers observed that the Niger emptied into the Atlantic Ocean via a vast delta in the Gulf of Guinea. Two German explorers added to the picture: Heinrich Barth established that Lake Chad to its east, and its many rivers, was unconnected to the Niger, while Eduard Flegel traced the Benue River, the Niger's main tributary, to its source in 1882.

"I saw... the great object of my mission: the long sought for, majestic Niger, glittering to the morning sun."

MUNGO PARK, *TRAVELS IN THE INTERIOR DISTRICTS OF AFRICA*, 1799

MUNGO PARK IN WEST AFRICA

Europeans maintained a trading presence on the coast of West Africa from the mid-15th century, where they were required to deal with intermediaries and remain at their trading posts. As an explorer, Park was free to travel but had to pay transit duties to the rulers of the territories he passed through. Islam was spreading through the region in the 1790s. When Park found himself in Benowm, he was held hostage for three months by Ali, Emir of Ludamar (depicted below in his tent). Park eventually escaped and continued his journey.

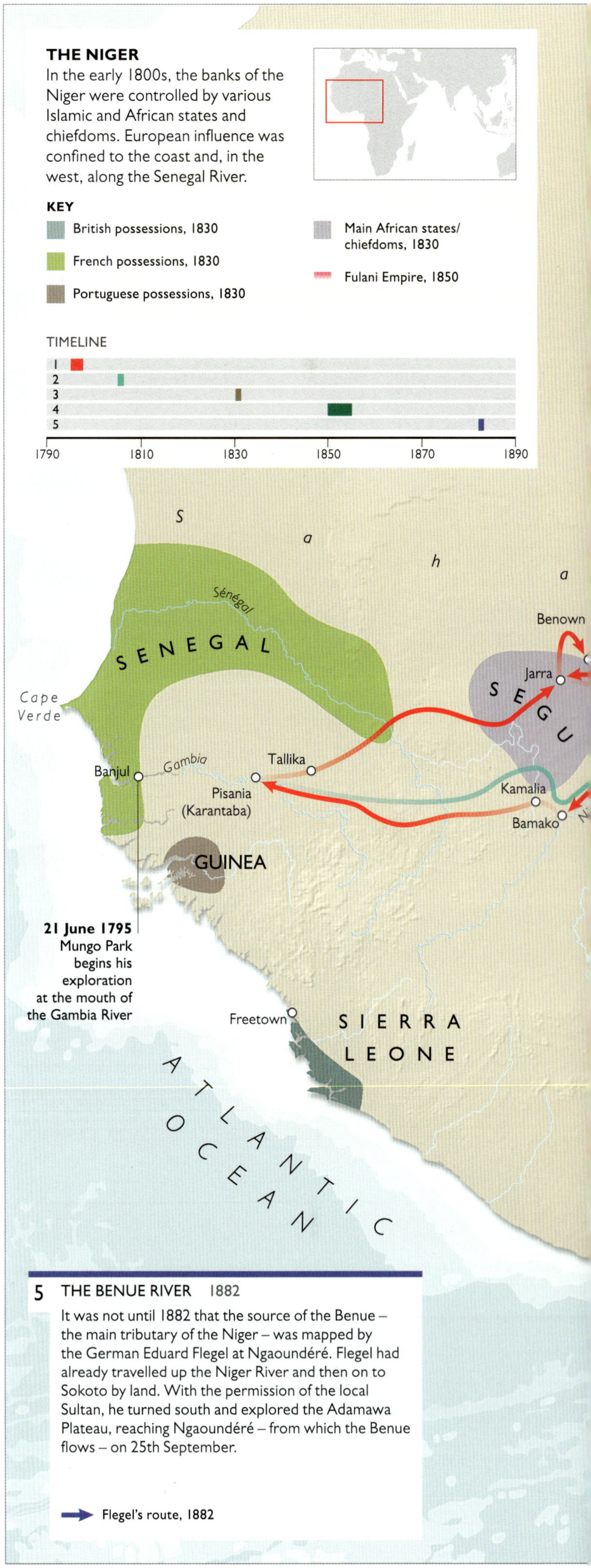

THE NIGER

In the early 1800s, the banks of the Niger were controlled by various Islamic and African states and chiefdoms. European influence was confined to the coast and, in the west, along the Senegal River.

KEY

- British possessions, 1830
- French possessions, 1830
- Portuguese possessions, 1830
- Main African states/ chiefdoms, 1830
- Fulani Empire, 1850

TIMELINE

1 2 3 4 5

1790 1810 1830 1850 1870 1890

Sahara

SENEGAL

Sénégal

Cape Verde

Banjul

Gambia

Tallika

Pisania (Karantaba)

GUINEA

Benown

Jarra

SEGU

Kamalia

Bamako

Freetown

SIERRA LEONE

ATLANTIC OCEAN

21 June 1795 Mungo Park begins his exploration at the mouth of the Gambia River

5 THE BENUE RIVER 1882

It was not until 1882 that the source of the Benue – the main tributary of the Niger – was mapped by the German Eduard Flegel at Ngaoundéré. Flegel had already travelled up the Niger River and then on to Sokoto by land. With the permission of the local Sultan, he turned south and explored the Adamawa Plateau, reaching Ngaoundéré – from which the Benue flows – on 25th September.

Flegel's route, 1882

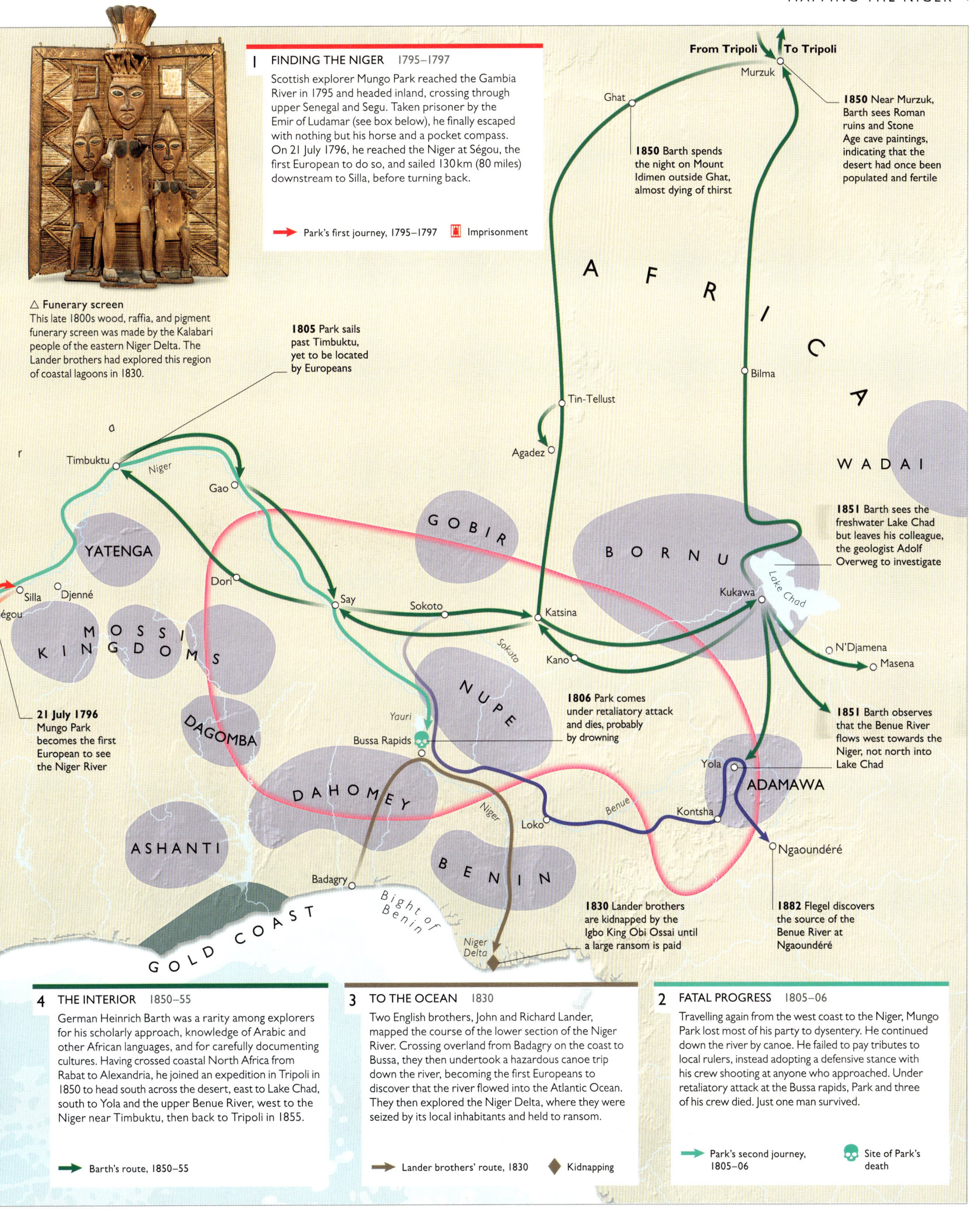
1 FINDING THE NIGER 1795–1797
Scottish explorer Mungo Park reached the Gambia River in 1795 and headed inland, crossing through upper Senegal and Segu. Taken prisoner by the Emir of Ludamar (see box below), he finally escaped with nothing but his horse and a pocket compass. On 21 July 1796, he reached the Niger at Ségou, the first European to do so, and sailed 130km (80 miles) downstream to Silla, before turning back.
Park's first journey, 1795–1797
Imprisonment
From Tripoli
To Tripoli
Murzuk
Ghat
1850 Near Murzuk, Barth sees Roman ruins and Stone Age cave paintings, indicating that the desert had once been populated and fertile
1850 Barth spends the night on Mount Idimen outside Ghat, almost dying of thirst
A F R I C A
Bilma
Tin-Tellust
Agadez
WADAI
1805 Park sails past Timbuktu, yet to be located by Europeans
Timbuktu
Niger
Gao
GOBIR
BORNU
1851 Barth sees the freshwater Lake Chad but leaves his colleague, the geologist Adolf Overweg to investigate
YATENGA
Dori
Say
Sokoto
Katsina
Kukawa
Lake Chad
Silla
Djenné
Ségou
MOSSI KINGDOMS
Sokoto
Kano
N'Djamena
Masena
NUPE
1806 Park comes under retaliatory attack and dies, probably by drowning
1851 Barth observes that the Benue River flows west towards the Niger, not north into Lake Chad
21 July 1796 Mungo Park becomes the first European to see the Niger River
DAGOMBA
Yauri
Bussa Rapids
Yola
ADAMAWA
DAHOMEY
Niger
Benue
Kontsha
Loko
Ngaoundéré
ASHANTI
BENIN
Badagry
Bight of Benin
GOLD COAST
Niger Delta
1830 Lander brothers are kidnapped by the Igbo King Obi Ossai until a large ransom is paid
1882 Flegel discovers the source of the Benue River at Ngaoundéré
4 THE INTERIOR 1850–55
German Heinrich Barth was a rarity among explorers for his scholarly approach, knowledge of Arabic and other African languages, and for carefully documenting cultures. Having crossed coastal North Africa from Rabat to Alexandria, he joined an expedition in Tripoli in 1850 to head south across the desert, east to Lake Chad, south to Yola and the upper Benue River, west to the Niger near Timbuktu, then back to Tripoli in 1855.
Barth's route, 1850–55
3 TO THE OCEAN 1830
Two English brothers, John and Richard Lander, mapped the course of the lower section of the Niger River. Crossing overland from Badagry on the coast to Bussa, they then undertook a hazardous canoe trip down the river, becoming the first Europeans to discover that the river flowed into the Atlantic Ocean. They then explored the Niger Delta, where they were seized by its local inhabitants and held to ransom.
Lander brothers' route, 1830
Kidnapping
2 FATAL PROGRESS 1805–06
Travelling again from the west coast to the Niger, Mungo Park lost most of his party to dysentery. He continued down the river by canoe. He failed to pay tributes to local rulers, instead adopting a defensive stance with his crew shooting at anyone who approached. Under retaliatory attack at the Bussa rapids, Park and three of his crew died. Just one man survived.
Park's second journey, 1805–06
Site of Park's death

△ **Funerary screen**
This late 1800s wood, raffia, and pigment funerary screen was made by the Kalabari people of the eastern Niger Delta. The Lander brothers had explored this region of coastal lagoons in 1830.

PFEIFFER'S SOLO TRAVELS

In 1845, Austrian-born Ida Laura Pfeiffer wrote: "When I was but a little child, I had already a strong desire to see the world. Whenever I met a travelling-carriage, I would stop involuntarily and gaze after it until it had disappeared." She never lost the urge to travel.

△ **Portrait of Pfeiffer**
Pfeiffer often financed expeditions by collecting plant, animal, and mineral specimens, which she sold to museums in Vienna and Berlin.

Born in Vienna in 1797, Pfeiffer made her first long journey, to Palestine and Egypt with her family, aged just five. She then waited 40 years, when her sons were grown, before a small inheritance enabled her to fund her first solo trip. Between 1842 and 1845, she spent nine months touring first Southwest Asia (the Middle East) and then Iceland. Possibly to counter disapproval from family about travelling alone, she claimed her initial journey was a pilgrimage to the Holy Land, an acceptable undertaking for a woman. She also journeyed through Egypt, Syria, Jordan, and Turkey. In Iceland, she travelled by horseback in the company of an Icelandic guide, setting out from Reykjavik to the volcanic region of Kleifarvatn Lake.

Pfeiffer's travel journals were a great success; they were translated into seven languages and helped to fund further exploration. In 1846–48, she circled the globe, and again in 1851–55. Mostly travelling without fellow Europeans, she covered some 32,000km (20,000 miles) by land and 240,000km (150,000 miles) by sea. An intrepid traveller, she endured perilous conditions, in Sumatra trekking through swamps and jungles. And through her encounters with Indigenous peoples, she saw how their way of life was under threat from colonialism.

PIONEERING MOUNTAINEERS

The French explorer Henriette d'Angeville was an avid walker who had long aspired to scale Mont Blanc, the highest peak in western Europe. She achieved her ambition on 4 September 1838, becoming the first woman to climb it unaided. Her success inspired many others, including the British alpinist Lucy Walker, who undertook 98 expeditions over the course of her 21-year career and claimed the first female ascent of the Matterhorn in 1871.

Henriette d'Angeville, 1838

Ida Pfeiffer and Queen Ranavalona
The sovereign of Madagascar wished to protect the island from European interference, yet Pfeiffer gained permission to travel around with a degree of freedom. She was presented before the Queen at the royal palace, shown in this 1857 illustration.

Trading relations
The Māori engaged in trade with early European arrivals, as seen in this watercolour drawing by Tupaia, a Tahitian navigator on Cook's voyage in 1769. It shows a Māori man trading a crayfish for a piece of cloth with Joseph Banks, one of Cook's naval officers.

EXPLORING NEW ZEALAND

When Europeans came across New Zealand (Aotearoa) in the 17th century, they knew little about the Māori, the descendants of Polynesians whose last wave of canoes had arrived by the early 1300s (see pp.24–25 and pp.58–59).

Dutchman Abel Tasman was the first European to sight New Zealand (Aotearoa) in 1642 (see pp.118–19) before British captain James Cook came in 1769, 1773, and 1777 (see pp.134–35). Others followed, and with them came whalers, traders, and European diseases. Sealers landed in Tamatea on the South Island in 1792, and built the first English house.

In the early 1800s, Māori-European contact was rare, but isolated violence fuelled the idea that conflict was more widespread. In 1807, in retaliation for the mistreatment of a local chief, the Ngati Uru attacked the British ship *Boyd*, killing 70 crew. Whalers killed 60 Māori in 1810.

However, many Māori welcomed the Europeans and some intermarried. They participated in trade and were employed in whaling crews, yet often suffered ill-treatment. From the 1810s, the Europeans traded muskets with the Māori that were used in wars between rival groups, killing up to 20,000. In the 1840s, British and colonial forces waged war to open up land for settlement and many Māori died. By 1870, most Māori land on the South Island had been bought or confiscated and sold to settlers. The Treaty of Waitangi, which ostensibly created a partnership between the Māori and the British, abjectly failed to protect their rights.

> *"I will not agree to the mana [authority] of a strange people being placed over this land."*
>
> MĀORI CHIEF TE HEUHEU TUKINO II ON THE WAITANGI TREATY, 1840

THE FIRST KNOWN MAORI MAP

In 1793 the British abducted northern Māori chief Tuki te Terenui Whare Pirau and took him to Norfolk Island, a penal colony in the Pacific. To communicate with the governor, Tuki drew the first known Māori map of New Zealand. The South Island, to which he had never been, was drawn overly small. The double dotted line marks the path thought to have been taken by dead spirits to the underworld.

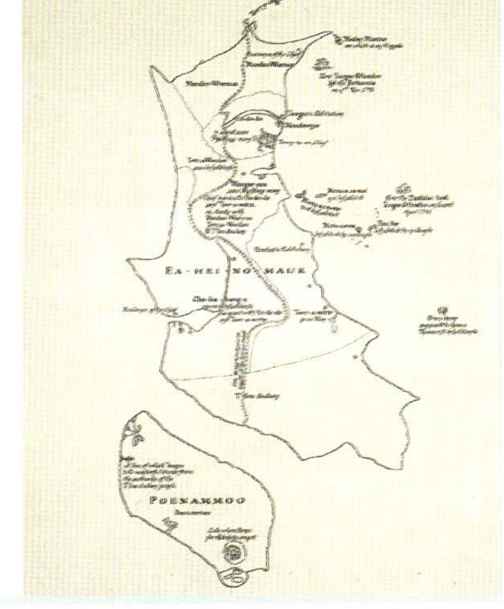

Reproduction of Tuki's map, 1940

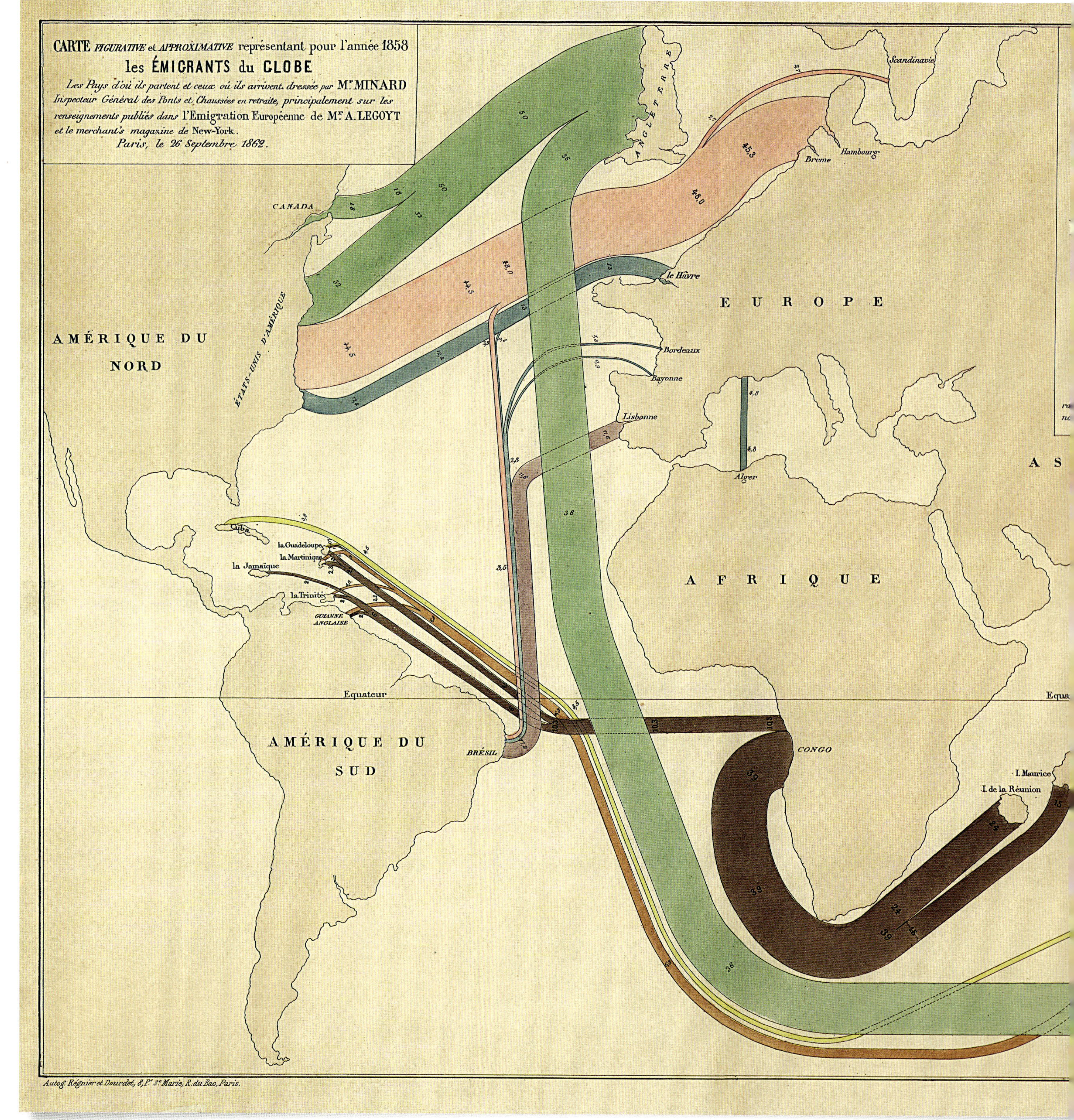

"A great emigration... implies unhappiness... in the country that is deserted."

THOMAS MALTHUS, ENGLISH ECONOMIST AND DEMOGRAPHER,
AN ESSAY ON THE PRINCIPLE OF POPULATION, 1798

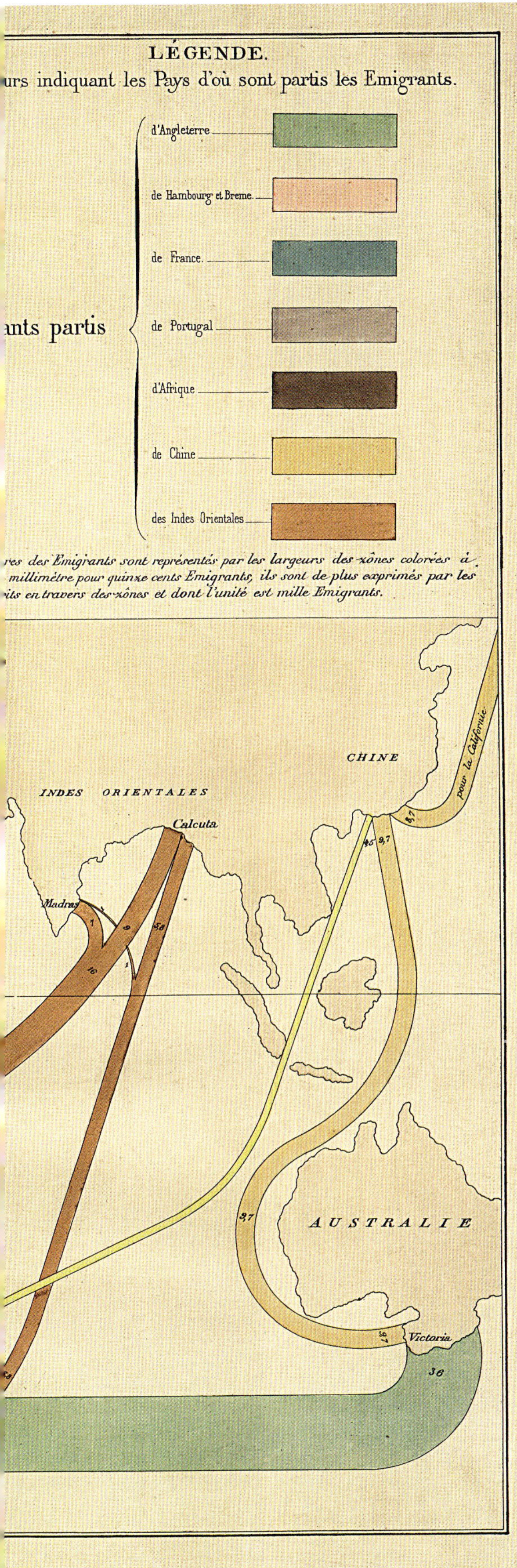

△ **Mass migration**
Produced by French engineer Charles Joseph Minard in 1862, this map, with its crude colour-coding, shows the flow of European migrants, enslaved Africans, and Asian labourers. The numbers represent the movement of people in their thousands.

MIGRATION IN THE 19TH CENTURY

During the 1800s, increasing numbers of people left their homelands. Many Europeans went voluntarily, emigrating across the oceans to begin a new life. Others had little or no choice – whether fleeing persecution in Europe, enslaved and transported from Africa, or signed on as indentured labourers from Asia.

After 1813, commercially viable steamships enabled people to emigrate more cheaply and easily than ever before. Millions left Europe for Canada, the United States, Latin America, New Zealand, and Australia, many fleeing poverty, repression, and religious persecution, including anti-Jewish pogroms. Thirty million people went to the United States – 10 million of whom were from Ireland and the UK – incentivized by cheap passages and free land. The discovery of gold in 1848 in California (see pp.176–77) and after 1851 in Australia was also a draw.

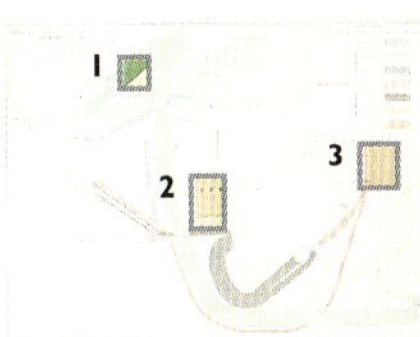

KEY

1 In 1815–60, c.5 million Europeans migrated.

2 In 1821–50, c.2 million enslaved Africans were forcibly transported.

3 More than 1 million Indians migrated "freely" in 1834–1917.

The largest documented European migration was the 13 million who left Italy between 1880 and 1914 for a new life in the Americas, leaving a life of poverty in the rural south and post-war hardship in the urban north. These economic migrants were joined by millions of enslaved people who were forcibly transported from Africa before the transatlantic slave trade ended in 1873. Slavery was finally abolished in the Americas in 1888, with Brazil the last to outlaw the practice.

A final group uprooted was the 1.6 million indentured labourers from India. After the British abolition of the slave trade in 1807, and slavery itself in 1833, these labourers were imported to work for a fixed term and for low wages in sugar plantations and sites across the islands of the Pacific and Indian oceans, Southeast Asia, South Africa, and the Caribbean. In addition, around 250,000 Chinese indentured labourers were sent to the Australian gold mines, Caribbean plantations, and the silver mines and guano fields of Peru.

IMMIGRANTS ON AN ATLANTIC LINER

Steamships sailing regular timetables carried migrants across the Atlantic to the Americas. These passengers travelled steerage – the lowest and cheapest accommodation on board – sleeping in long rows of large, shared bunk beds with straw mattresses and no bed linen. Up to 900 people could be packed into a hold, making walking on deck almost impossible. Some died on the voyage. Once ashore, migrants were processed by immigration control, checked for diseases, and granted access to their new land. Ellis Island in New York harbour alone processed nearly 12 million immigrants between 1892 and 1954.

Liner carrying migrants c.1900

THE SOURCE OF THE NILE

One of the world's great rivers, the Nile was a lifeline for Ancient Egyptian and Nubian civilizations. Yet in the 18th century, the origins of its two tributaries, the Blue Nile and the White Nile, were unknown to geographers. The source of the Blue Nile was eventually identified, but that of the White Nile remains in doubt.

The search for the source of the Nile began in at least the 3rd century BCE when Egyptian Pharaoh Ptolemy II Philadelphus (r. 284–246 BCE) sent an expedition to the Ethiopian mountains. In 1770, Scottish explorer James Bruce claimed to be the first European to see the source of the Blue Nile, although a Jesuit priest, Pedro Paez, had already visited it in 1618. Bruce argued that, while the White Nile was the longer river, the Blue Nile was the Nile of the ancient Nubians and Egyptians, and thus its true source.

The origin of the White Nile was far more difficult to ascertain, and is still in dispute. The English explorer John Hanning Speke confirmed that it flows out of the Ripon Falls into a lake he named Victoria after the British monarch (the lake had multiple names in local dialects). A few years later, the explorers Samuel Baker and Florence von Szász traced its flow through another large lake to the north, and named it Albert, after Queen Victoria's Prince Consort. However, many rivers flow into Lake Victoria, any of which could be called the true source.

As recently as 2006, a British-New Zealander expedition determined that the Rukarara River in the Nyungwe Forest in southwest Rwanda could be the Nile's most distant headwater.

"The vast expanse of the pale-blue waters of the N'yanza burst suddenly upon my gaze."

JOHN HANNING SPEKE, *THE DISCOVERY OF THE SOURCE OF THE NILE*, 1863

FLORENCE BAKER (NÉE VON SZÁSZ) 1841–1914

Orphaned in the Hungarian Revolution (1848–9), 14-year-old Florence von Szász became part of Samuel Baker's expedition after he bribed the guards at a white slave auction where she was being sold. Together they searched for the Nile's source, later marrying and returning to Africa in 1869 to campaign against slavery. Baker later said it was to Florence that he "owed success and life".

Florence von Szász and Samuel Baker in Africa, c.1863

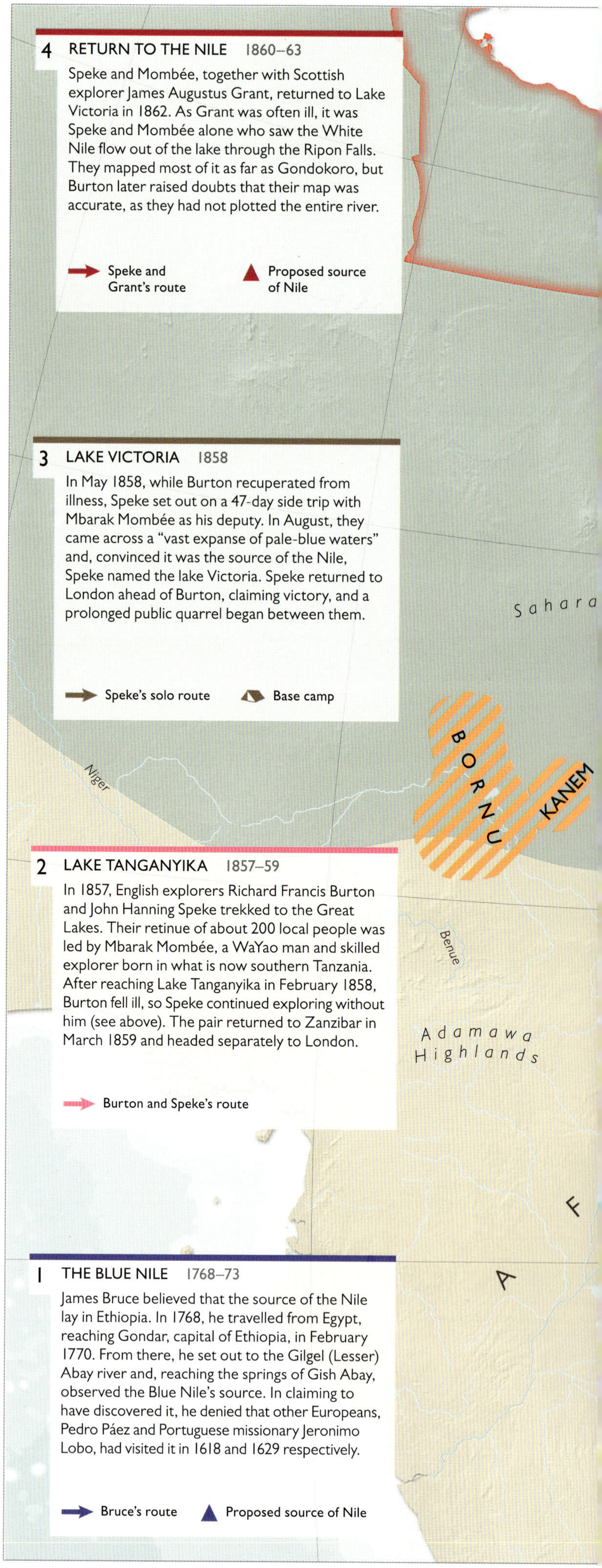

4 RETURN TO THE NILE 1860–63

Speke and Mombée, together with Scottish explorer James Augustus Grant, returned to Lake Victoria in 1862. As Grant was often ill, it was Speke and Mombée alone who saw the White Nile flow out of the lake through the Ripon Falls. They mapped most of it as far as Gondokoro, but Burton later raised doubts that their map was accurate, as they had not plotted the entire river.

Speke and Grant's route
Proposed source of Nile

3 LAKE VICTORIA 1858

In May 1858, while Burton recuperated from illness, Speke set out on a 47-day side trip with Mbarak Mombée as his deputy. In August, they came across a "vast expanse of pale-blue waters" and, convinced it was the source of the Nile, Speke named the lake Victoria. Speke returned to London ahead of Burton, claiming victory, and a prolonged public quarrel began between them.

Speke's solo route
Base camp

2 LAKE TANGANYIKA 1857–59

In 1857, English explorers Richard Francis Burton and John Hanning Speke trekked to the Great Lakes. Their retinue of about 200 local people was led by Mbarak Mombée, a WaYao man and skilled explorer born in what is now southern Tanzania. After reaching Lake Tanganyika in February 1858, Burton fell ill, so Speke continued exploring without him (see above). The pair returned to Zanzibar in March 1859 and headed separately to London.

Burton and Speke's route

1 THE BLUE NILE 1768–73

James Bruce believed that the source of the Nile lay in Ethiopia. In 1768, he travelled from Egypt, reaching Gondar, capital of Ethiopia, in February 1770. From there, he set out to the Gilgel (Lesser) Abay river and, reaching the springs of Gish Abay, observed the Blue Nile's source. In claiming to have discovered it, he denied that other Europeans, Pedro Páez and Portuguese missionary Jeronimo Lobo, had visited it in 1618 and 1629 respectively.

Bruce's route
Proposed source of Nile

Sahara
Niger
BORNU
KANEM
Benue
Adamawa Highlands

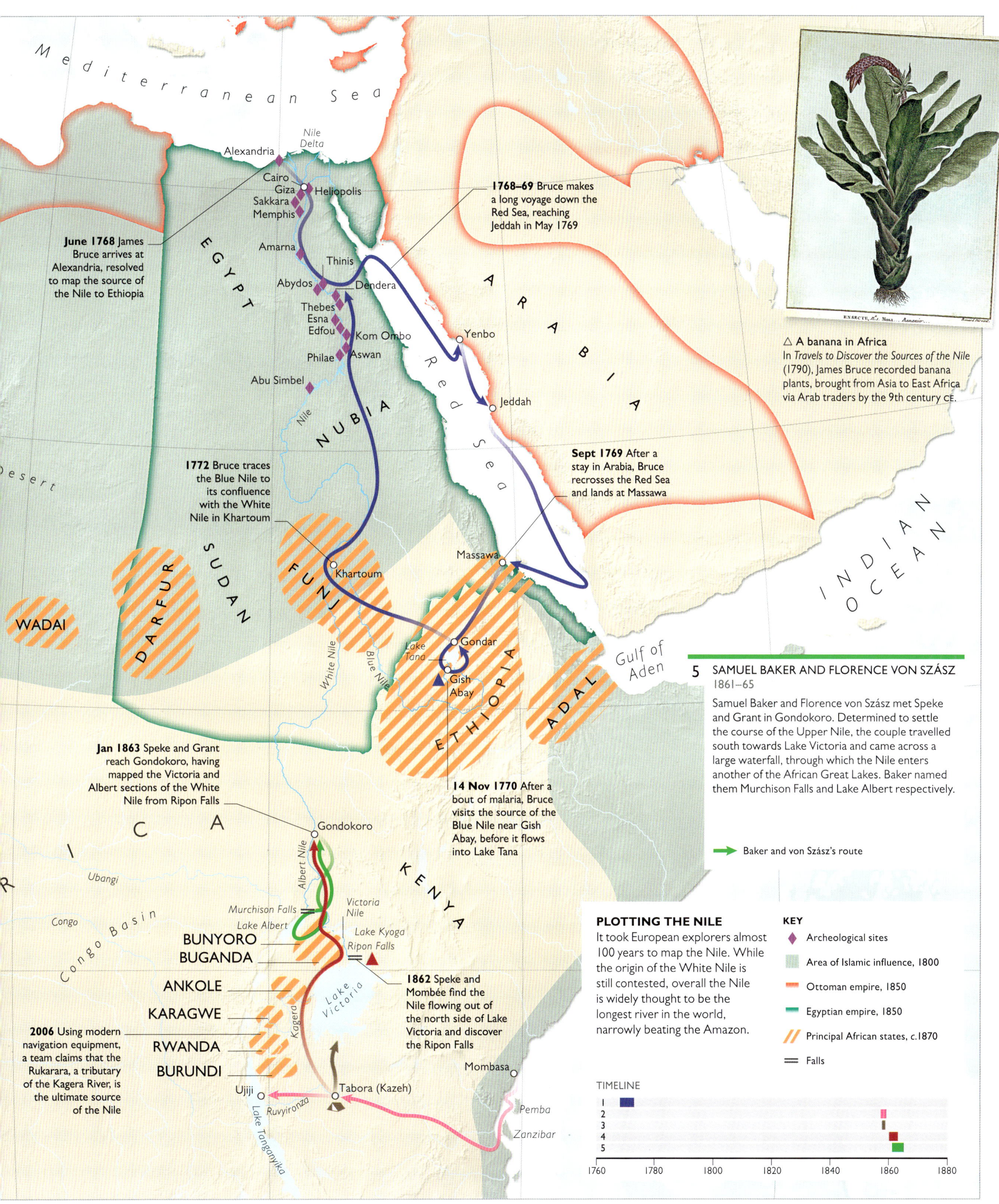

△ **A banana in Africa**
In *Travels to Discover the Sources of the Nile* (1790), James Bruce recorded banana plants, brought from Asia to East Africa via Arab traders by the 9th century CE.

5 SAMUEL BAKER AND FLORENCE VON SZÁSZ
1861–65

Samuel Baker and Florence von Szász met Speke and Grant in Gondokoro. Determined to settle the course of the Upper Nile, the couple travelled south towards Lake Victoria and came across a large waterfall, through which the Nile enters another of the African Great Lakes. Baker named them Murchison Falls and Lake Albert respectively.

PLOTTING THE NILE
It took European explorers almost 100 years to map the Nile. While the origin of the White Nile is still contested, overall the Nile is widely thought to be the longest river in the world, narrowly beating the Amazon.

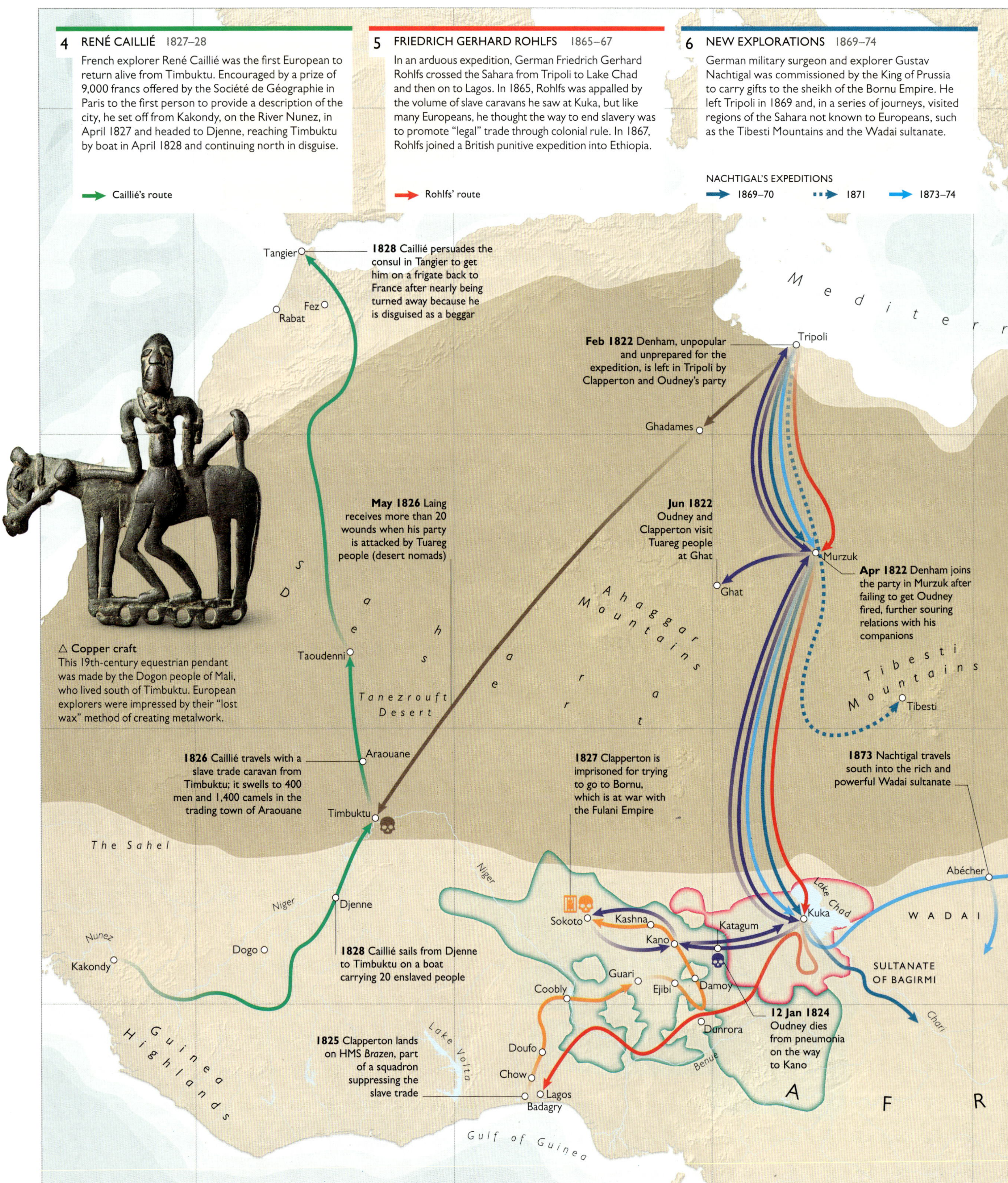
Caillié's route
Rohlfs' route
NACHTIGAL'S EXPEDITIONS
1869–70
1871
1873–74
Tangier
Fez
Rabat
1828 Caillié persuades the consul in Tangier to get him on a frigate back to France after nearly being turned away because he is disguised as a beggar
Mediterr
Tripoli
Feb 1822 Denham, unpopular and unprepared for the expedition, is left in Tripoli by Clapperton and Oudney's party
Ghadames
May 1826 Laing receives more than 20 wounds when his party is attacked by Tuareg people (desert nomads)
Jun 1822 Oudney and Clapperton visit Tuareg people at Ghat
Murzuk
Ghat
Apr 1822 Denham joins the party in Murzuk after failing to get Oudney fired, further souring relations with his companions
Sahara Desert
Ahaggar Mountains
Tibesti Mountains
Tibesti
Taoudenni
Tanezrouft Desert
Araouane
1826 Caillié travels with a slave trade caravan from Timbuktu; it swells to 400 men and 1,400 camels in the trading town of Araouane
Timbuktu
1827 Clapperton is imprisoned for trying to go to Bornu, which is at war with the Fulani Empire
1873 Nachtigal travels south into the rich and powerful Wadai sultanate
The Sahel
Niger
Niger
Djenne
Nunez
Kakondy
Dogo
1828 Caillié sails from Djenne to Timbuktu on a boat carrying 20 enslaved people
Sokoto
Kashna
Kano
Katagum
Kuka
Lake Chad
Abécher
WADAI
SULTANATE OF BAGIRMI
Chari
Guari
Ejibi
Damoy
Coobly
Dunrora
12 Jan 1824 Oudney dies from pneumonia on the way to Kano
Benue
Guinea Highlands
Lake Volta
Doufo
Chow
Lagos
Badagry
1825 Clapperton lands on HMS *Brazen*, part of a squadron suppressing the slave trade
Gulf of Guinea
A F R

4 RENÉ CAILLIÉ 1827–28

French explorer René Caillié was the first European to return alive from Timbuktu. Encouraged by a prize of 9,000 francs offered by the Société de Géographie in Paris to the first person to provide a description of the city, he set off from Kakondy, on the River Nunez, in April 1827 and headed to Djenne, reaching Timbuktu by boat in April 1828 and continuing north in disguise.

5 FRIEDRICH GERHARD ROHLFS 1865–67

In an arduous expedition, German Friedrich Gerhard Rohlfs crossed the Sahara from Tripoli to Lake Chad and then on to Lagos. In 1865, Rohlfs was appalled by the volume of slave caravans he saw at Kuka, but like many Europeans, he thought the way to end slavery was to promote "legal" trade through colonial rule. In 1867, Rohlfs joined a British punitive expedition into Ethiopia.

6 NEW EXPLORATIONS 1869–74

German military surgeon and explorer Gustav Nachtigal was commissioned by the King of Prussia to carry gifts to the sheikh of the Bornu Empire. He left Tripoli in 1869 and, in a series of journeys, visited regions of the Sahara not known to Europeans, such as the Tibesti Mountains and the Wadai sultanate.

△ **Copper craft**
This 19th-century equestrian pendant was made by the Dogon people of Mali, who lived south of Timbuktu. European explorers were impressed by their "lost wax" method of creating metalwork.

CROSSING THE SAHARA

Until the early 19th century, North Africa's Sahara Desert had a thriving trade network independent of outsiders. The region was a mystery to Europeans – a vast and sandy wasteland with inhabitants adapted to its unforgiving terrain and harsh climate.

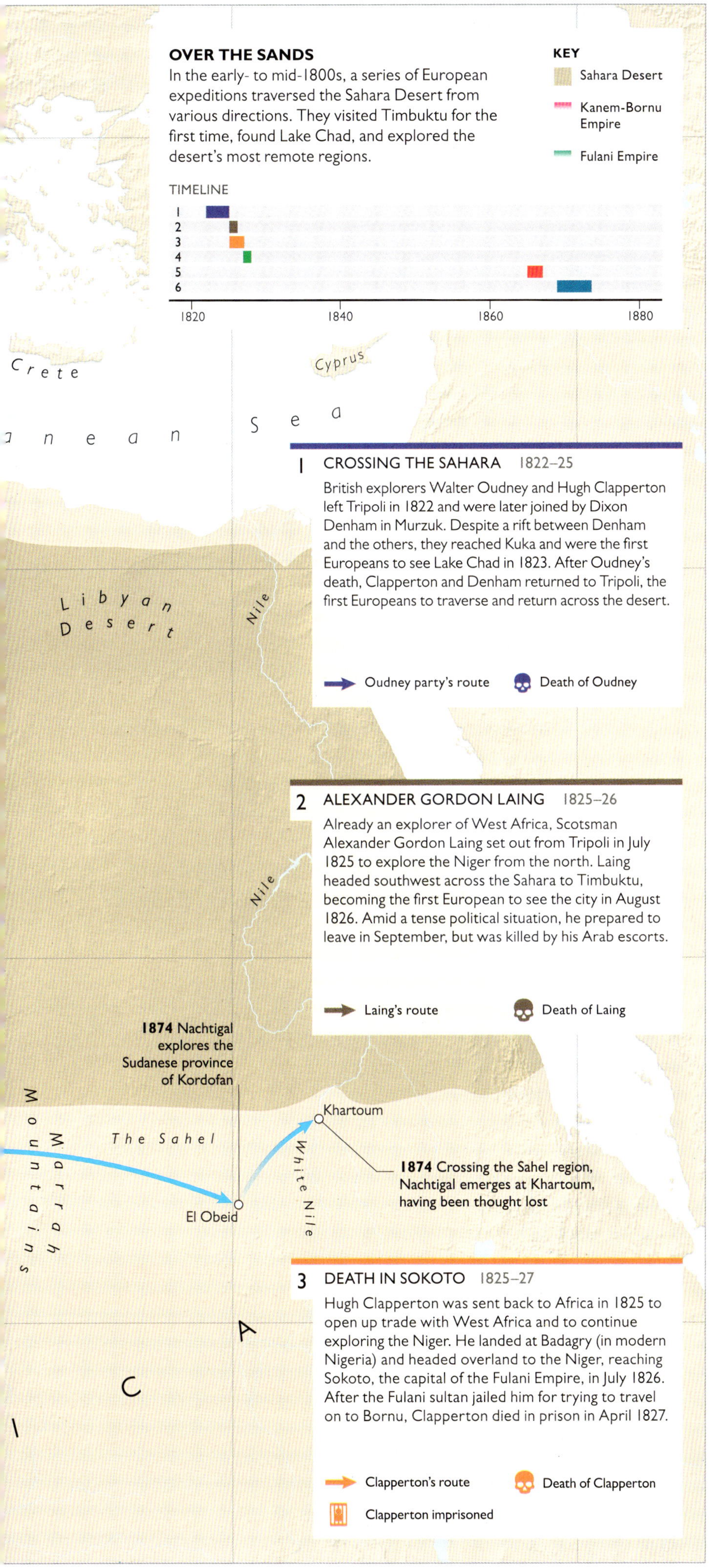

The formidable size of the Sahara Desert attracted European explorers. So too did the city of Timbuktu (see box below) – its location unknown to Europeans until Scottish explorer Alexander Gordon Laing reached it in 1826. He lost his life trying to leave, allowing Frenchman René Caillié to claim the title of the first European to return alive from the city in 1828. The first European crossing and recrossing of the Sahara itself was completed by a fractious British expedition in 1822–25, led by doctor Walter Oudney, naval captain Hugh Clapperton, and a soldier, Major Dixon Denham. Becoming the first Europeans to see Lake Chad in west-central Africa in 1823, they opened the door to a number of other European explorers over the next 30 years. The highly experienced German explorer Friedrich Gerhard Rohlfs crossed the desert from north to south in 1865–67; in a series of expeditions from 1869 to 1874, another German, Gustav Nachtigal, explored some of the lesser-known regions in the eastern Sahara, although he was also on a mission to establish German colonial regimes. All of these expeditions made use of local guides who knew the locations of wells and oases, with some explorers, such as Caillié, joining the huge camel caravans in which merchants and other travellers crossed the desert.

OVER THE SANDS

In the early- to mid-1800s, a series of European expeditions traversed the Sahara Desert from various directions. They visited Timbuktu for the first time, found Lake Chad, and explored the desert's most remote regions.

KEY

- Sahara Desert
- Kanem-Bornu Empire
- Fulani Empire

TIMELINE

1
2
3
4
5
6

1820 1840 1860 1880

Crete
Cyprus
Mediterranean Sea
Libyan Desert
Nile
Nile
The Sahel
Marrah Mountains
Khartoum
El Obeid
White Nile
AFRICA

1874 Nachtigal explores the Sudanese province of Kordofan

1874 Crossing the Sahel region, Nachtigal emerges at Khartoum, having been thought lost

1 CROSSING THE SAHARA 1822–25

British explorers Walter Oudney and Hugh Clapperton left Tripoli in 1822 and were later joined by Dixon Denham in Murzuk. Despite a rift between Denham and the others, they reached Kuka and were the first Europeans to see Lake Chad in 1823. After Oudney's death, Clapperton and Denham returned to Tripoli, the first Europeans to traverse and return across the desert.

Oudney party's route
Death of Oudney

2 ALEXANDER GORDON LAING 1825–26

Already an explorer of West Africa, Scotsman Alexander Gordon Laing set out from Tripoli in July 1825 to explore the Niger from the north. Laing headed southwest across the Sahara to Timbuktu, becoming the first European to see the city in August 1826. Amid a tense political situation, he prepared to leave in September, but was killed by his Arab escorts.

Laing's route
Death of Laing

3 DEATH IN SOKOTO 1825–27

Hugh Clapperton was sent back to Africa in 1825 to open up trade with West Africa and to continue exploring the Niger. He landed at Badagry (in modern Nigeria) and headed overland to the Niger, reaching Sokoto, the capital of the Fulani Empire, in July 1826. After the Fulani sultan jailed him for trying to travel on to Bornu, Clapperton died in prison in April 1827.

Clapperton's route
Death of Clapperton
Clapperton imprisoned

TIMBUKTU

From the 14th century, Timbuktu in Mali was a centre of Islamic learning with its university, madrasahs, and three great mosques. The city's proximity to the River Niger ensured its importance on the caravan trade route and Timbuktu became rich via the gold and salt trade. But from the 16th century, this trade moved to the coast and, when René Caillié arrived in the city in 1828 (and made this drawing in 1830), he was unimpressed by its lack of grandeur.

THE AGE OF GLOBAL EMPIRES

FROM 1800, WITH STEAM POWER EXPANDING THE SCOPE OF TRAVEL, EXPLORATION FOCUSED ON THE REMOTE REGIONS OF THE EARTH. IN THE EARLY 20TH CENTURY, EXPEDITIONS FORGED A WAY THROUGH THE NORTHERN PASSAGE, REACHED THE NORTH AND SOUTH POLES, BEGAN TO CHART THE OCEAN FLOOR, AND INAUGURATED THE AGE OF FLIGHT.

△ **Louis Daguerre**
In 1837, Frenchman Louis Daguerre invented the technology of daguerreotype photography. It was taken up by many explorers and archaeologists, who recorded the first photographic images of their travels.

IMPERIAL EXPANSION

In the 19th century, Western colonization turned to imperialism as the West's growing industrial dominance led to territorial expansion and the creation of vast global empires under the rule of a handful of nations. Exploration and conquest were often two sides of the same coin, and the whole world was drawn into the imperial project.

For centuries, European activities overseas had been dominated by trade and the establishment of a chain of staging posts, via which the riches of the world could be transported to Europe. However, two industrial revolutions (1760–1830 and 1870–1914) started in Britain and spread to the rest of the world, creating a colossal demand for raw materials. The West sought control of territories to exploit their resources and gain land for their growing populations at home. Many imperial nations sponsored expeditions into the interiors of countries that had previously been settled only along the coastline, with a view to future settlement.

Social reform movements spread across the West and, gradually, anti-slavery legislation was signed into law by almost every imperial government, although Europeans also used abolition as an argument for extending their colonial rule. Other forms of exploitation dramatically increased. As European empires seized more territories, they brutally swept aside the rights of Indigenous peoples. Many died from foreign diseases or were removed from their lands by military conquest and genocide. A vast gap emerged between the world's rich in the industrial nations, and the world's poor, both in the West and in the colonies, and many people travelled overseas in search of a better life, resulting in repeated mass migrations.

▷ **The opening of the Suez Canal**
After ten years of construction, the Suez Canal in Egypt opened on 17 November 1869, creating a direct shipping route between Europe and the Indian and Western Pacific oceans.

Industrial innovation

Hand in hand with industrial expansion came dramatic advances in technology that connected the world in unprecedented ways. By 1850, there were over 40,000km (25,000 miles) of railtrack around the world, and by 1950 this had increased to over 1,200,000km (750,000 miles). From the mid-19th century, ocean liners became the primary mode of intercontinental travel and the opening of the Suez Canal in 1869 dramatically reduced the travel time between Europe and Asia. Hundreds of millions of people were on the move every day, and, in 1890, American journalist Nellie Bly proved that it was possible to travel around the whole world in just 72 days, largely on steam ships and steam trains.

The coming of flight further increased the speed of travel when, in America in 1903, the Wright brothers made the world's first sustained controlled and powered flight. And in 1919, British aviators John Alcock and Arthur Brown flew across the Atlantic, and the world's first international

THE IMPERIAL WORLD

From the 19th century on, much of the world was transformed by the industrialization of the West, which led to the subjugation of many nations under foreign rule. The expansion of trade spawned the growth of cities on every continent, and a mass migration from rural areas to the new urban centres. Trains, steamships, planes, and cars opened up regional and global travel to huge numbers of people, and countless explorers set off to investigate the last uncharted corners of the world.

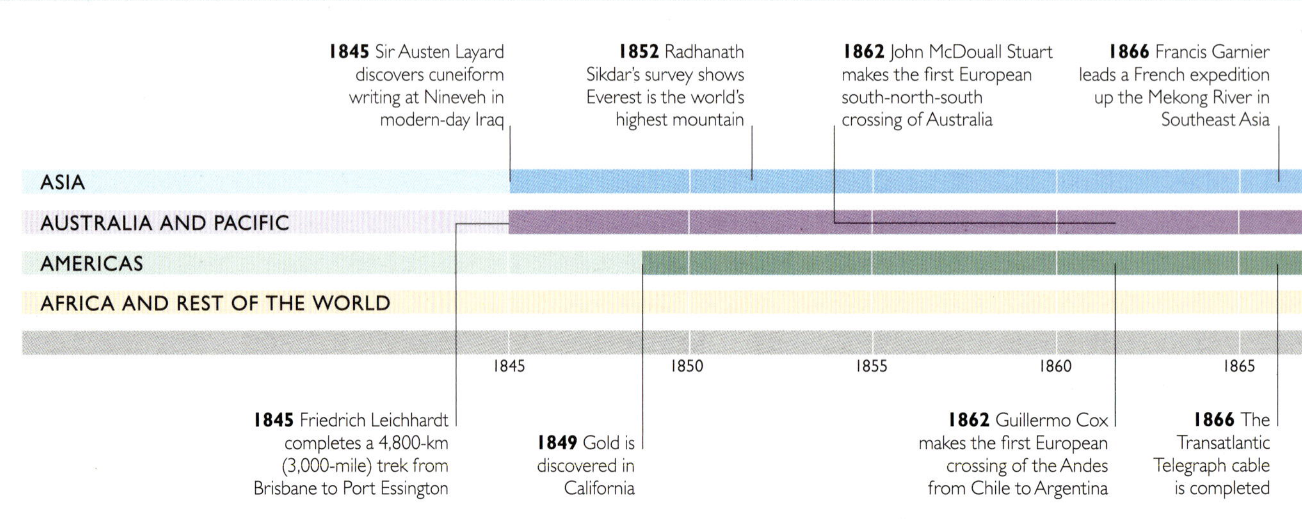

◁ **The Survey of India**
From 1767, the British East India Company systematically surveyed and mapped the Indian subcontinent as a means of acquiring knowledge to assist them in their administrative rule.

passenger flight flew from London to Paris. But if people began to move fast, messages moved faster. The first electric telegraphs were sent in the 1830s, and by 1866, messages could be sent across the Atlantic in 67 minutes. By 1900, there was a global telegraph network, and the first transatlantic phone call was made in 1927. Meanwhile, the invention in France of photography in 1839, and moving pictures in 1895, created a mass media that enabled people to see just what far-flung places really looked like.

All this new information and knowledge inspired many explorers to penetrate further into lesser-known areas – to reach the exact co-ordinates of the North and South Poles, trek across the vast Australian interior, or into the wilds of Patagonia, and refuse to admit defeat in the search for a Northern Passage. Often these missions were fuelled by commercial or territorial ambitions, but the pursuit of knowledge also played its part. HMS *Challenger* explored the ocean deeps, and archaeologists searched for the ruins of "lost civilizations", such as the Mexica and Maya.

In China and Japan, centuries of isolationist policies had kept the modernizing influences of industrialization at bay. But in the 19th century, both countries began tentatively to engage with the wider world.

▽ **A new era of aviation**
This 1937 poster advertises the "empire class flying boats" of Imperial Airways. Powered by four engines, the planes could travel at a speed of up to 320 km (200 miles) an hour.

> *"The desire to fly... is an ideal handed down to us by our ancestors who... looked enviously on the birds soaring freely through space."*
>
> WILBUR WRIGHT, IN A SPEECH AT THE AERO-CLUB DE FRANCE, 1908

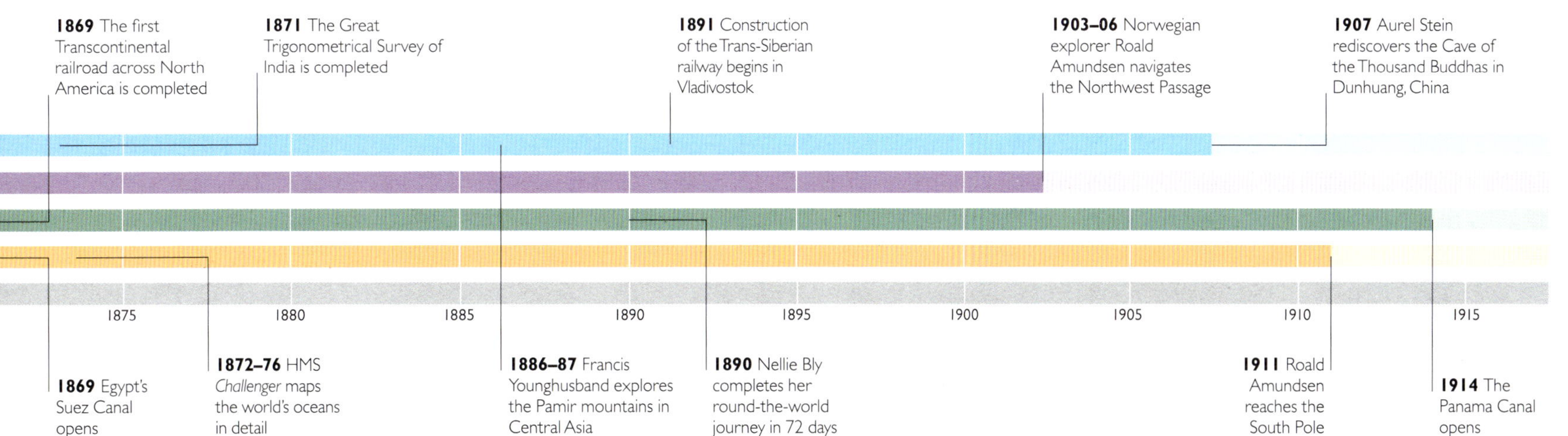

1869 The first Transcontinental railroad across North America is completed
1871 The Great Trigonometrical Survey of India is completed
1891 Construction of the Trans-Siberian railway begins in Vladivostok
1903–06 Norwegian explorer Roald Amundsen navigates the Northwest Passage
1907 Aurel Stein rediscovers the Cave of the Thousand Buddhas in Dunhuang, China
1875
1880
1885
1890
1895
1900
1905
1910
1915
1869 Egypt's Suez Canal opens
1872–76 HMS *Challenger* maps the world's oceans in detail
1886–87 Francis Younghusband explores the Pamir mountains in Central Asia
1890 Nellie Bly completes her round-the-world journey in 72 days
1911 Roald Amundsen reaches the South Pole
1914 The Panama Canal opens

"Forty-niners"
The discovery of gold in Coloma, California drew up to 100,000 newcomers in 1849, known as the "forty-niners". Chinese workers had to pay a Foreign Miners' Tax from 1850, but many stayed on to work on the railroads.

THE PRICE OF GOLD

When a farmer's son found a 7-kg (17-lb) yellow rock in North Carolina, USA in 1799, it sparked the first of a wave of "gold rushes" that, in 1848–55, led to one of the largest mass migrations of people in US history, and the systematic displacement and genocide of Indigenous peoples.

△ **Call of the gold**
A poster from 1850 promotes a new ship that cut the travel time from New York to the gold rush town of San Francisco from seven to four months.

At the turn of the 19th century, gold mining in North Carolina had transformed a previously sparsely populated area of backcountry. Fortunes were made not just in gold but also in the many support industries that created fast-growing boomtowns.

In 1829, when gold was discovered in the Appalachian Mountains, a similar influx of settlers joined the "Georgia Gold Rush" to seize on its potential. But fortunes were made at great human cost as tens of thousands of Indigenous people were violently evicted from their ancestral homelands. In 1848, nearly 100,000 people travelled across the country and from overseas to California, where much of the land in the west was still unsettled by Europeans. Again, the local population paid the price; by 1873, only 30,000 Indigenous people remained of around 150,000.

In 1896, the discovery of gold in the Yukon-Klondike region of northwest Canada prompted 100,000 people to brave the treacherous journey across icy and rocky terrain, while also transporting heavy mining and camping equipment. Less than a third succeeded, the remainder perished or turned back.

TRAIL OF TEARS

After gold was discovered in 1829 on land inhabited by the Cherokee Nation in Georgia, speculators lobbied Congress to reduce the land rights of Indigenous peoples across the US. This was one of the catalysts for the Indian Removal Act of 1830, which led to the forced relocation of Indigenous Americans to "Indian Territory". Their journeys across vast distances, during which some 15,000 died of starvation, exposure, and disease, became known as the "Trail of Tears."

The 1838 Cherokees' forced removal to the West, from a woodcut c.1850

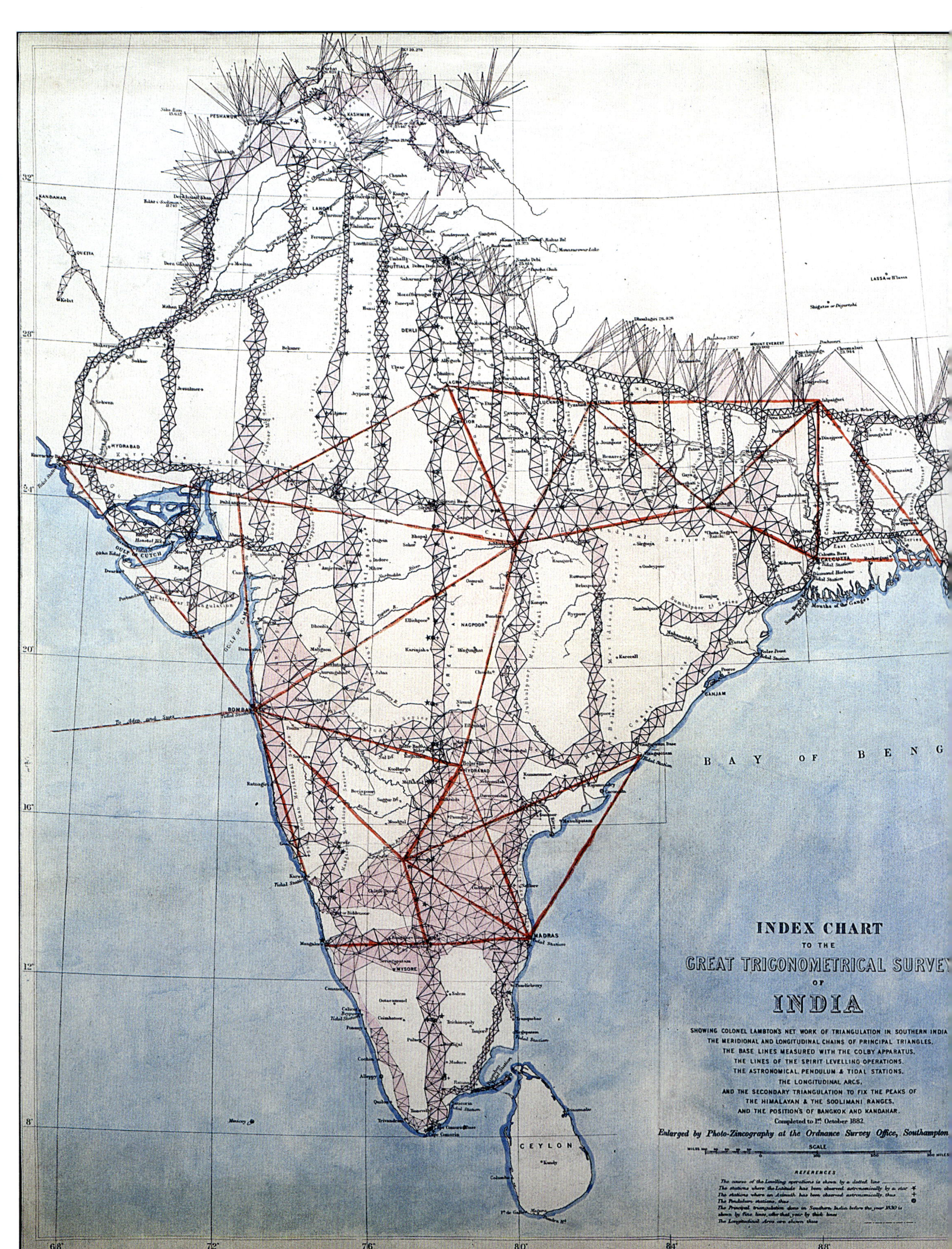

▷ **India by triangle**
This chart from 1882 shows the extraordinary grid of triangles created across India to map it accurately. Usually, triangulation points were simply high points on hills and buildings, but on the very flat plains of northern India, 15-m (50-ft) high masonry towers had to be built to get a clear view.

SURVEYING INDIA

In 1802 the British East India Company launched the Great Trigonometrical Survey (GTS) of India, the first effort to map the entire subcontinent with scientific precision. The British used this huge gathering of scientific information to administer and "legitimize" its colonial rule in the region.

The East India Company first began to map what it saw as its territory in 1767 when it set up the Survey of India. The company thought such surveys would enhance Britain's reputation by advancing scientific progress in the region, but measuring and naming the territory was also a declaration of ownership. Starting in 1802, when it became part of the larger Survey of India, the GTS took 70 years to complete and involved huge teams of surveyors and explorers, including thousands of local experts. The aim was to survey India using triangulation – measuring the angles in a triangle formed by three survey control points. Starting with a baseline of known length as one side of a triangle, the distance and direction of a third point can be calculated accurately using trigonometry.

British soldier and surveyor William Lambton began by measuring a baseline of 13km (7½ miles) across the flat plains from the Madras Observatory, with St Thomas Mount at the north end and Perumbakkam Hill at the southern end. Gradually, he created hundreds of triangles across India, using hills and high buildings to find sight lines and angles measured with a theodolite (a precision optical instrument). Entourages often had four elephants for the survey leaders to ride on, 30 horses for the military officers, and 42 camels for carrying supplies. The GTS was highly accurate, sometimes only inches out over many miles, and had a significant scientific impact, measuring, for example, an arc of longitude precisely for the first time. It also measured the height of the world's three highest mountain peaks: Everest, K2, and Kanchenjunga.

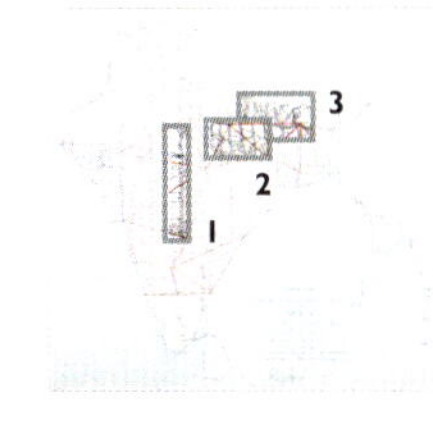

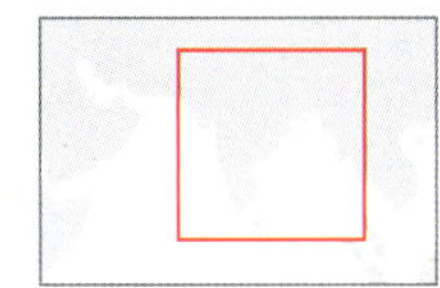

LOCATOR

KEY

1 The key reference line, called the Great Indian Arc of the Meridian.

2 Triangulation towers were crucial to measuring the Northern Plain.

3 Mt Everest's height was measured using triangulation from a distance.

"Lines must be sharp and clean, and the ink used must be perfectly black."

"NOTES FOR GUIDANCE" FROM *THE SURVEY OF INDIA*, 1904

PUNDITS AND EXPLORATION

British leaders William Lambton and Sir George Everest, who took over after Lambton died in 1823, got most credit for the survey, but countless Indian experts, known as "pundits", were also involved. Many were underpaid by the government and their work went unacknowledged. Some pundits also lost their lives when the British sent them to secretly explore areas north of British India, such as Tibet and Central Asia. Nain Singh, Abdul Hamid, and Radhanath Sikdar were the best-known of the pundits. Sikdar's calculations established Sagarmatha (Mount Everest, then known as Peak XV) as the highest mountain in the world.

Pundits predominate on a later survey in 1879

ACROSS AUSTRALIA

Less than 20 years after James Cook reached the east coast of Australia in 1770, the first British penal colony was set up in Sydney in 1788. Throughout the 19th century, Europeans embarked on a series of expeditions to explore and map the interior.

Inhabited by Indigenous Aboriginal peoples for tens of thousands of years, Australia is a huge landmass, much of it desert. In the 1800s, Europeans found the arid environment beyond their southeast coast settlements challenging. The land offered few rivers and resources, unless one knew how to find them – as Aboriginal people did – to ease travel. Journeys had to be made mostly on foot, occasionally on horseback, and, from the 1860s, with the help of imported camels.

Gradually, aided by Aboriginal guides, European (mainly British) explorers crossed the country. Hamilton Hume, William Hovell, and Charles Sturt, among others, mapped the interior's one major river system, the Murray–Darling. In 1840, Edward John Eyre journeyed from south to west Australia and, in 1845, German Friedrich Leichhardt reached the northeast. In 1860–61, William John Wills crossed from south to north with Robert O'Hara Burke, and John McDouall Stuart made a complete return trip in 1862. Ernest Giles reached Uluru in the 1870s, and John Forrest led an expedition across the west in 1874.

"… they cried out to inform me that the [Darling] river was so salt as to be unfit for drinking."

CHARLES STURT, *TWO EXPEDITIONS INTO THE INTERIOR OF SOUTHERN AUSTRALIA*, 1833

THE RABBIT-PROOF FENCE

In 1901–1907, the Australian government built a fence to keep rabbits out of Western Australia. It proved to be a helpful guide in one of the most remarkable of Australian treks. Molly Craig was one of over 100,000 mixed-race Aboriginal children taken forcefully from their families by the Australian government between the two world wars. She used survival skills learnt from her Aboriginal stepfather and, with two other children, followed the fence home for 1,600 km (990 miles), pursued by the authorities all the way.

A section of the rabbit-proof fence

AUSTRALIA REVEALED

Between about 1820 and 1880, European explorers trekked into Australia's harsh desert and bush outback, the home of Indigenous peoples for 50,000 years or more.

KEY

- British settlements, 1851
- Indigenous territory, 1851
- Mountain
- Overland Telegraph Station

TIMELINE

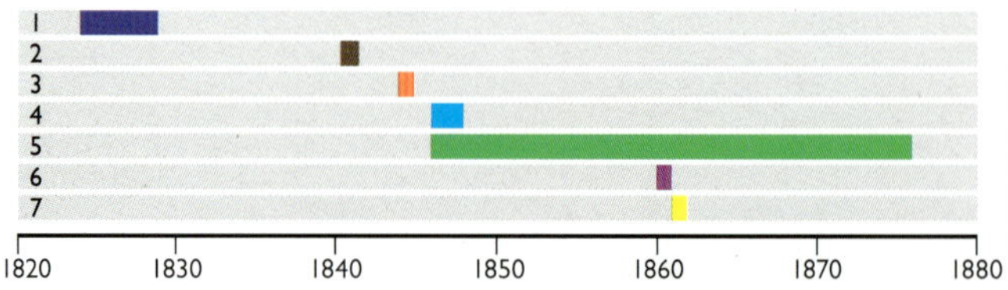

1, 2, 3, 4, 5, 6, 7

1820 1830 1840 1850 1860 1870 1880

1 FINDING THE DARLING 1824–29

In 1824, Hamilton Hume and William Hovell led an expedition to map the rivers west of the New South Wales colony. Aided by Indigenous people, the explorers crossed the Great Dividing Range mountains and met the Murrumbidgee River. In 1828–29, Charles Sturt went further; he came up on the Barka River (now known as the Darling) and mapped the Murray–Darling river system.

- Hume and Hovell's outward route, 1824
- Hume and Hovell's return route, 1824–25
- Sturt's route, 1828
- Sturt's route, 1829

2 ROUND THE BIGHT JUNE 1840–JULY 1841

Edward John Eyre was a tough "overlander" who drove sheep and cows to set distance records. He made several forays north. In 1840, guided by Wylie, an Indigenous Australian man, he led the first European expedition from Port Lincoln, round the Great Australian Bight and largely through desert, to reach Albany in July 1841. Wylie received a government pension, while Eyre was awarded a Royal Geographical Society gold medal.

- Eyre's route, 1840–41

3 FROM SEA TO SEA AUGUST 1844–DECEMBER 1845

Friedrich Leichhardt led the first European expedition inland across the Australian northeast, starting close to Meanjin (now Brisbane). The team of nine volunteers, including two Aboriginal guides, reached Port Essington on the north coast some 14 months later, after a 4,800-km (3,000-mile) trek, during which Leichhardt and his party had been given up for dead.

- Leichhardt's expedition, 1844–45

INDIAN OCEAN

Great Sandy Desert

AUS

Grea

Geraldton

Dongara

Perth (Boorloo)

WESTERN AUSTRALIA

Cape Leeuwin

Albany

King George Sound

7 July 1841 Eyre and Wylie arrive in Albany after the rest of the team have given up or died along the way

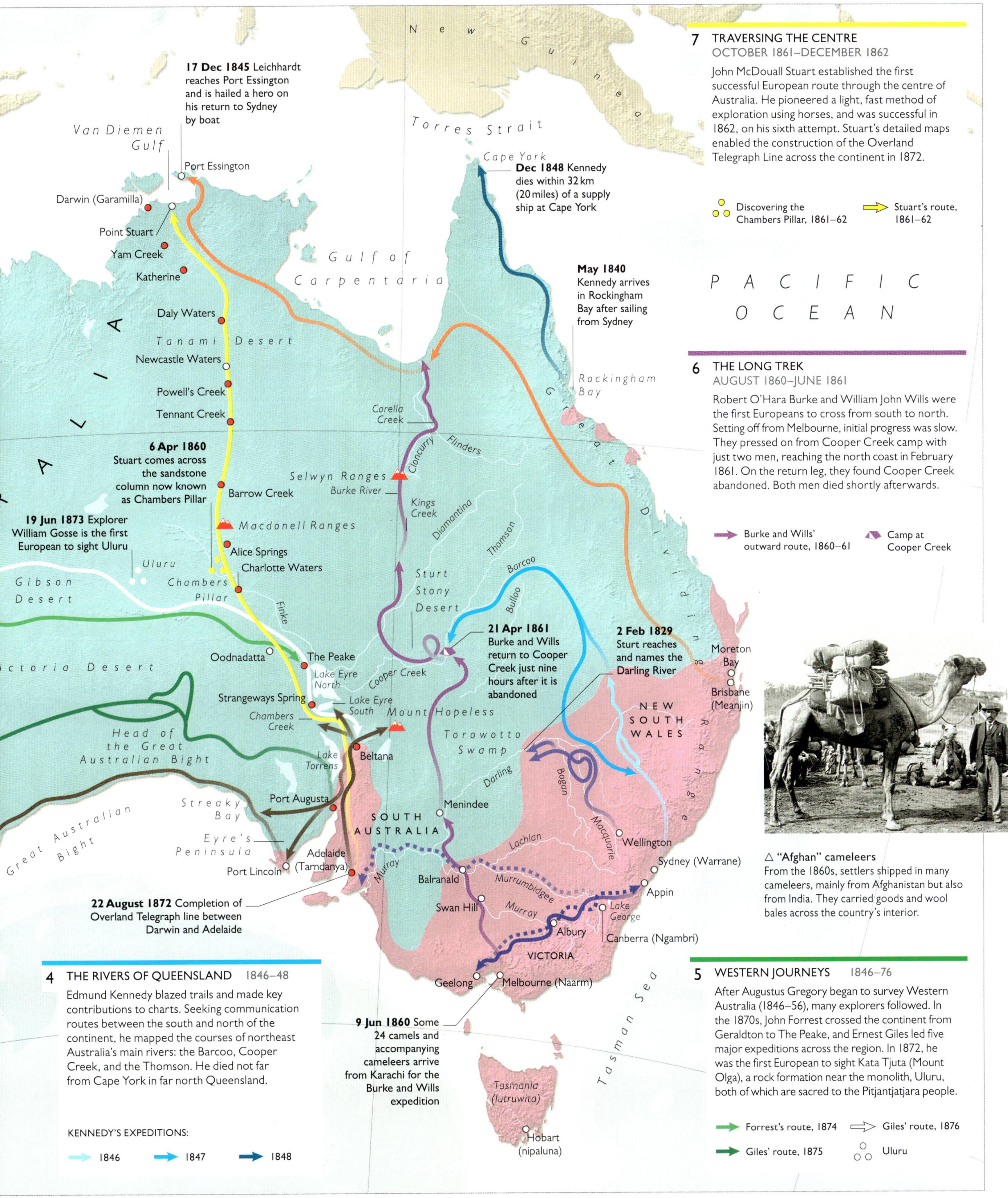

7 TRAVERSING THE CENTRE
OCTOBER 1861–DECEMBER 1862

John McDouall Stuart established the first successful European route through the centre of Australia. He pioneered a light, fast method of exploration using horses, and was successful in 1862, on his sixth attempt. Stuart's detailed maps enabled the construction of the Overland Telegraph Line across the continent in 1872.

Discovering the Chambers Pillar, 1861–62

Stuart's route, 1861–62

6 THE LONG TREK
AUGUST 1860–JUNE 1861

Robert O'Hara Burke and William John Wills were the first Europeans to cross from south to north. Setting off from Melbourne, initial progress was slow. They pressed on from Cooper Creek camp with just two men, reaching the north coast in February 1861. On the return leg, they found Cooper Creek abandoned. Both men died shortly afterwards.

Burke and Wills' outward route, 1860–61

Camp at Cooper Creek

△ **"Afghan" cameleers**
From the 1860s, settlers shipped in many cameleers, mainly from Afghanistan but also from India. They carried goods and wool bales across the country's interior.

4 THE RIVERS OF QUEENSLAND 1846–48

Edmund Kennedy blazed trails and made key contributions to charts. Seeking communication routes between the south and north of the continent, he mapped the courses of northeast Australia's main rivers: the Barcoo, Cooper Creek, and the Thomson. He died not far from Cape York in far north Queensland.

KENNEDY'S EXPEDITIONS:

1846 1847 1848

5 WESTERN JOURNEYS 1846–76

After Augustus Gregory began to survey Western Australia (1846–56), many explorers followed. In the 1870s, John Forrest crossed the continent from Geraldton to The Peake, and Ernest Giles led five major expeditions across the region. In 1872, he was the first European to sight Kata Tjuta (Mount Olga), a rock formation near the monolith, Uluru, both of which are sacred to the Pitjantjatjara people.

Forrest's route, 1874

Giles' route, 1876

Giles' route, 1875

Uluru

MEKONG JOURNEYS

Anglo–French colonial rivalry in Southeast Asia in the 1860s led the French to explore whether the Mekong River could provide a trade route from Saigon, Vietnam, into China, similar to the Yangtse River route into British Shanghai.

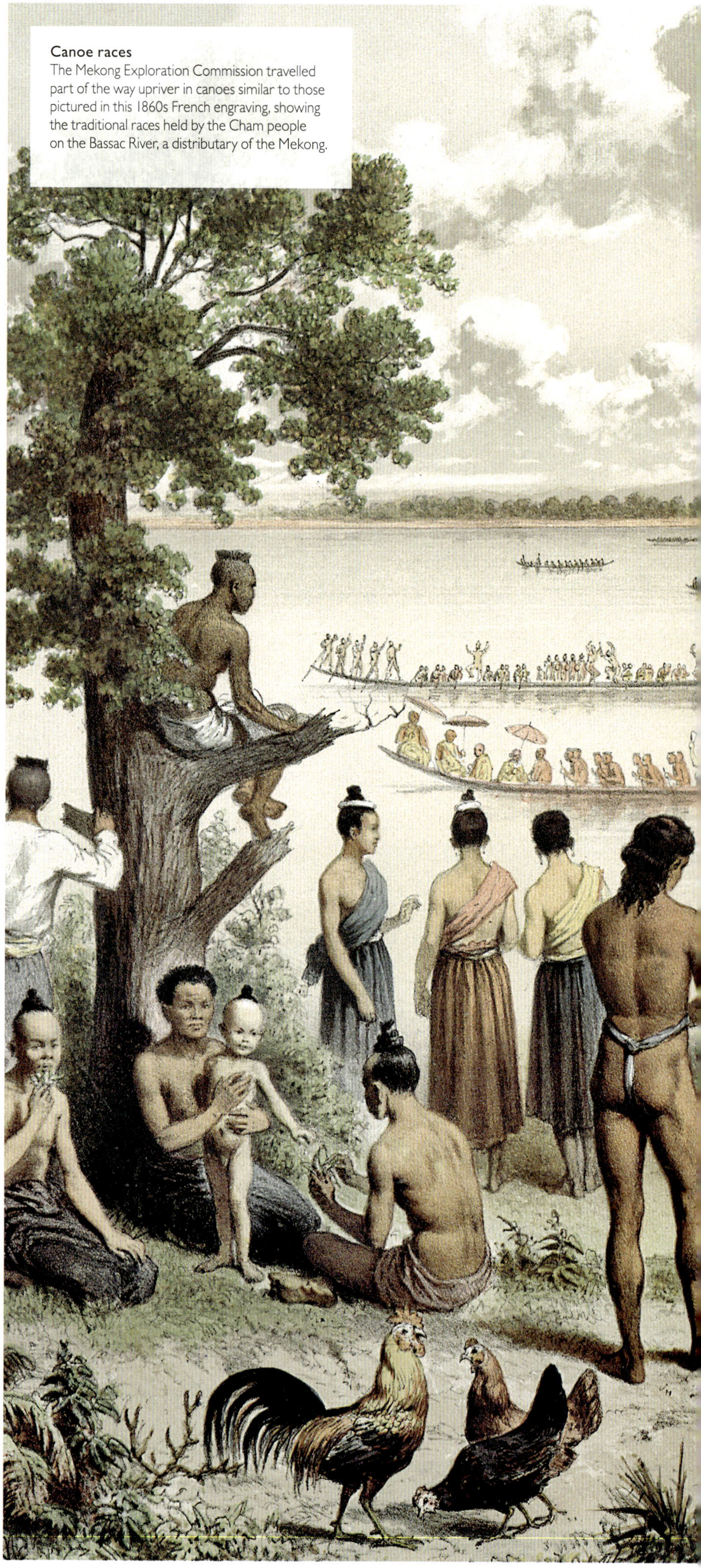

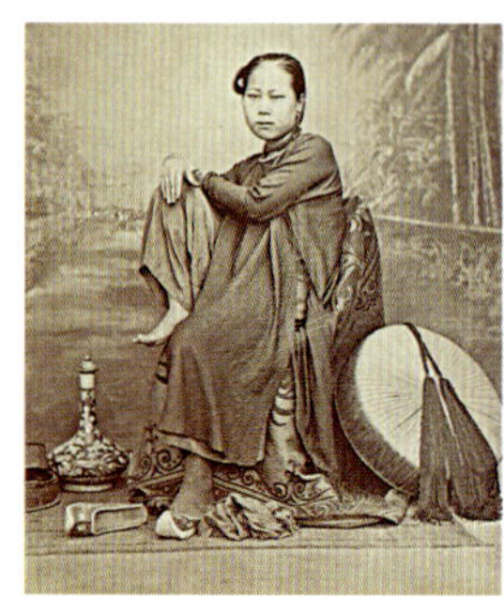

△ **Vietnamese elite**
A wealthy elite in Vietnam saw economic gains under French colonial rule, but for the majority of Vietnamese society, living standards and land ownership fell.

In 1866, a French expedition, the Mekong Exploration Commission, which included naval officer Francis Garnier, set off from Saigon up the Mekong River in two small gunboats. They soon met impassable rapids and had to switch to canoes. Over the following year, they canoed more than 1,000km (621 miles), until they reached the unnavigable Falls of Khon, which made it clear the Mekong could not be opened up for trade. Garnier returned to Saigon, walking 1,660km (994 miles) in 60 days, to seek government support to continue, with a new aim to map uncharted territory.

With his renewed mandate, Garnier led his crew through Cambodia, Vietnam, Laos, Siam (Thailand), Burma, and into China's Yunnan Province, braving tough conditions, including wading through leech-infested mud. Two years later, they arrived in Shanghai. Although the expedition failed in its goal of opening up the Mekong River, the mapping of more than 5,800km (3,604 miles) of territory previously unknown to Europeans was hailed as a scientific achievement.

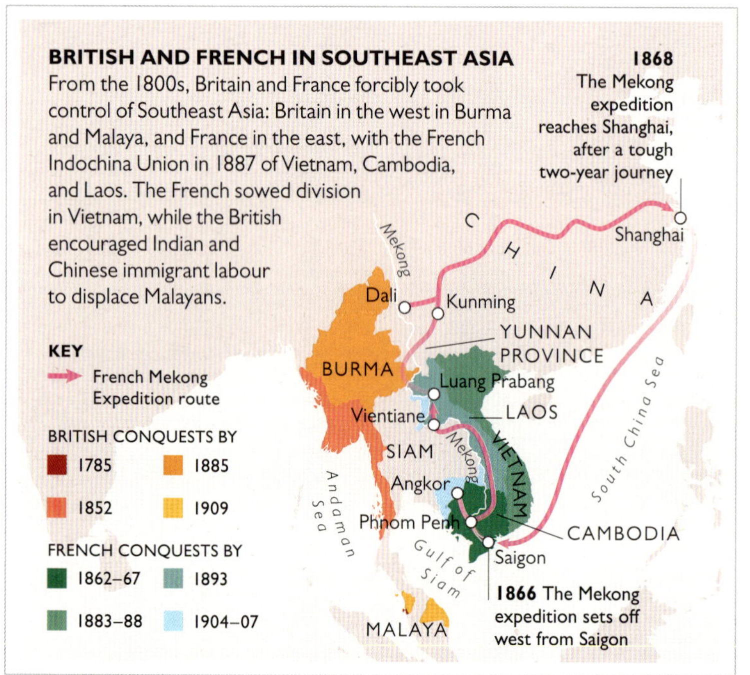

BRITISH AND FRENCH IN SOUTHEAST ASIA

From the 1800s, Britain and France forcibly took control of Southeast Asia: Britain in the west in Burma and Malaya, and France in the east, with the French Indochina Union in 1887 of Vietnam, Cambodia, and Laos. The French sowed division in Vietnam, while the British encouraged Indian and Chinese immigrant labour to displace Malayans.

KEY
French Mekong Expedition route

BRITISH CONQUESTS BY
1785
1885
1852
1909

FRENCH CONQUESTS BY
1862–67
1893
1883–88
1904–07

Canoe races
The Mekong Exploration Commission travelled part of the way upriver in canoes similar to those pictured in this 1860s French engraving, showing the traditional races held by the Cham people on the Bassac River, a distributary of the Mekong.

Overwhelmed by the jungle
Europeans were astonished to learn of the ruins of vast ancient buildings in Central America. This drawing by the English explorer Frederick Catherwood, published in 1844, depicts the temple at Tulum, a Mayan city on the Yucatán Peninsula.

SEEKING "LOST" CIVILIZATIONS

In the 19th and early 20th centuries, archaeologists from the West went in search of the sites of "lost" civilizations (although in most cases, Indigenous people already knew of them). Every continent yielded remarkable finds.

Archaeologists' "rediscoveries" during this time immeasurably increased knowledge of the ancient world. The wider world began to learn more about great cultures, including the Inca in South America; the Mexica (Aztec) and Maya in Mesoamerica; the Khmer (Southeast Asia) and Harappa (south Asia) societies; Africa's Carthaginians and ancient Egyptians; the ancient Greeks; and ancient Babylonian, Assyrian, and Persian civilizations in western Asia. European and American explorers took priceless artefacts from their countries of origin, placing them in their museums and private collections. These excited public imagination, but there are now demands that they be returned to their places of origin.

△ **The mask of Agamemnon**
In Mycenae in Greece, Schliemann found what he claimed was the gold funerary mask of Agamemnon, the legendary Greek leader in the Trojan War. In fact, it is centuries older.

One of the most famous archaeologists of the time was Hiram Bingham, an American professor at Yale University, USA. He publicized the high mountain refuge of the Incas at Macchu Picchu, Peru, which he climbed in 1911 after being guided there by local farmers. Another, John Lloyd Stephens, trekked through the Central American jungle to reach the Mayan ruins of Palenque on 11 May 1840, mapping this ancient site.

A few years later in Iraq, English archaeologist Sir Austen Henry Layard rediscovered the remains of the ancient Assyrian cities Nimrud and Nineveh. At Nimrud, he unearthed a huge winged bull, a valued treasure that now resides at the British Museum. At Nineveh, Layard uncovered tablets inscribed with cuneiform inscriptions, one of the earliest forms of writing.

THE SEARCH FOR TROY

In the 1860s and 1870s, the German businessman and amateur archaeologist Heinrich Schliemann undertook an obsessive hunt for the site of Troy, the city in Homer's *Iliad*. This quest took him to modern-day Hisarlik in Turkey, now thought to be the site of Troy. But his work damaged ancient sites and he stole many Trojan artefacts.

Schliemann at Troy's excavations

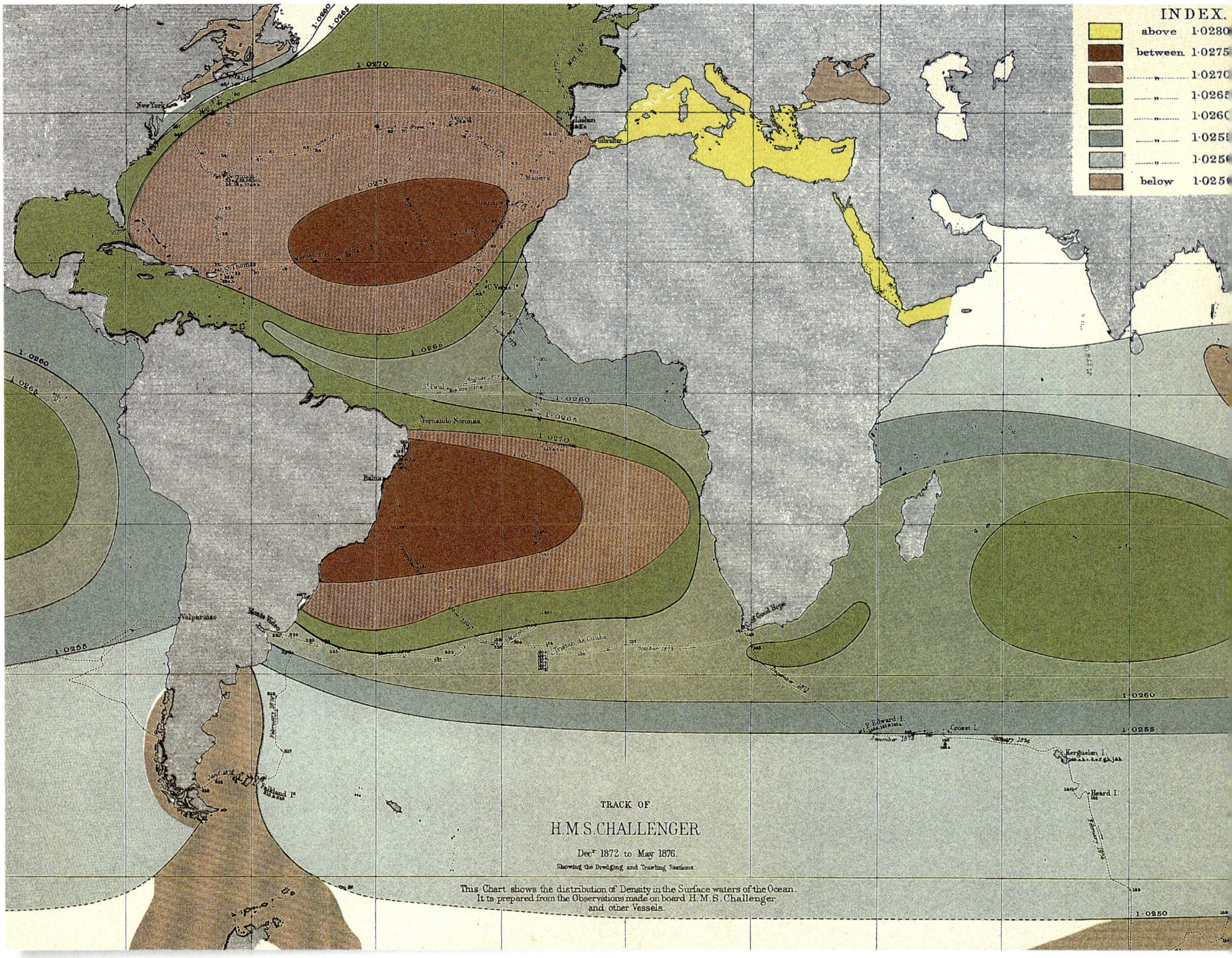

EXPLORING THE OCEANS

△ **Global picture**
The dotted line above indicates HMS *Challenger*'s course and the points at which it dredged and trawled the ocean bed, taking samples. The colours represent the variations in global ocean water surface density that HMS *Challenger* found, with the densest water, in brown, shown in the Atlantic. The crew also measured depth.

On 7 December 1872, HMS *Challenger* set off from Sheerness on England's Kent coast on a four-year mission to sail around the world, with the aim of exploring the physical, chemical, and biological characteristics of the deep sea.

HMS *Challenger* circled the world and mapped the sea floor using just ropes and wire, down to a depth of 8,184m (26,850ft). It also dredged up samples of known and unknown species, and observed currents, water temperatures, weather, and surface ocean conditions.

The ship had a crew of around 250 sailors, engineers, carpenters, marines, and officers, as well as a six-person science team led by Scottish naturalist Charles Wyville Thomson. Over four years, HMS *Challenger* sailed 127,580km (68,890 nautical miles) and made 374 deep-sea soundings, 255 observations of water temperature, 111 dredges, and 129 trawls. They collected water samples, marine plants and animals, and sea-floor deposits.

HMS *Challenger*'s soundings proved that the ocean floor had a highly varied landscape, with mountains, valleys, and vast plains. The team found warm and cold zones too, and discovered 5,000 unknown species of marine life. Formerly, scientific consensus had been that no life could survive the huge pressure, cold, and darkness of the ocean depths – but Thomson and his team proved otherwise, revolutionizing oceanography.

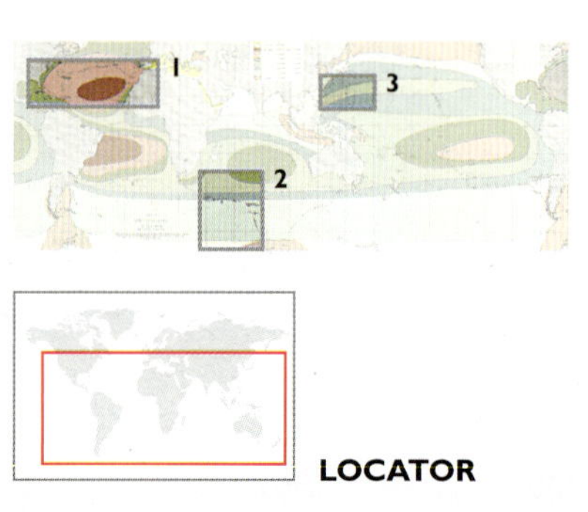

LOCATOR

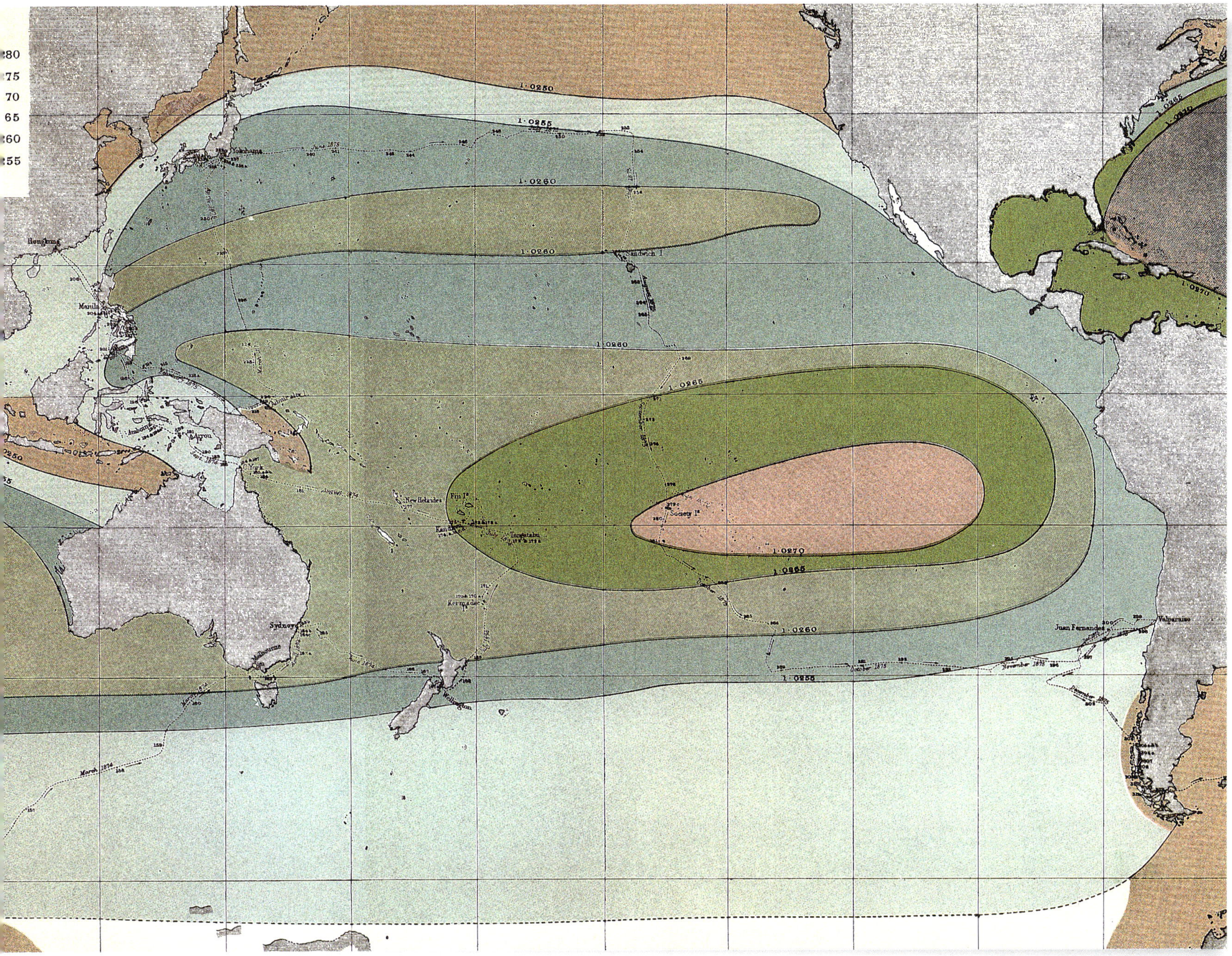

▷ Crossing the Gulf Stream
To build up a complete picture, HMS *Challenger* criss-crossed the Atlantic – as a result sailing over the Gulf Stream several times – taking soundings en route.

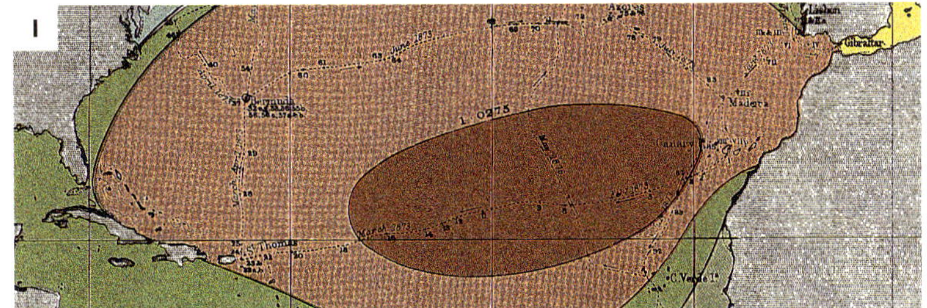

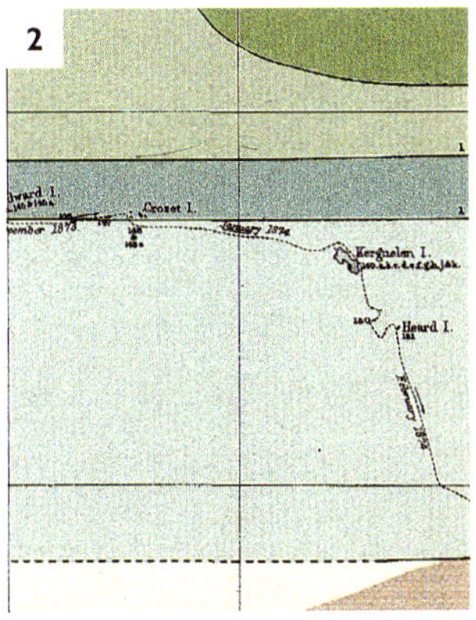

▷ Southern Ocean
In 1873–74, HMS *Challenger* travelled far into the Southern Ocean, sailing east, then turning sharply south at the Kerguelen Islands; the crew saw icebergs and whales. Having traversed the Antarctic Circle, the ship turned back north.

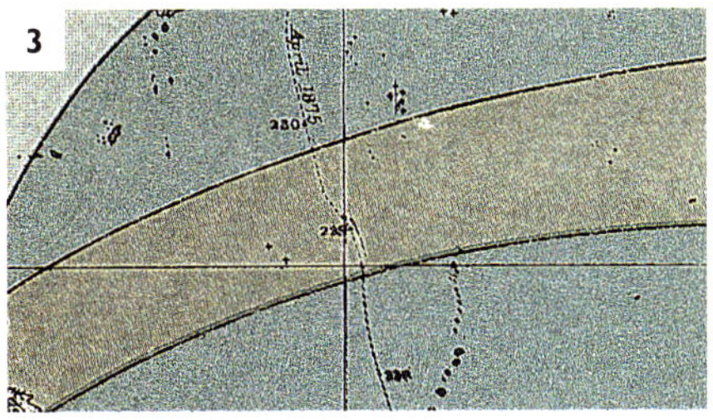

△ Mariana Trench
The crew also located Earth's deepest point, the Pacific's Mariana Trench, measuring it at 8,184 m (26,850 ft). Modern readings put its lowest point at c.11,000 m (36,090 ft).

FLOATING LABORATORIES

HMS *Challenger* was carefully adapted for its four-year scientific mission. In place of gun batteries, it had two laboratories on either side of the captain's cabin – one for chemistry and the other for natural history. The natural-history facility (below), where specimens were analysed, was as well equipped as any university laboratory on land. Equipment was firmly secured to cope with the rolling seas.

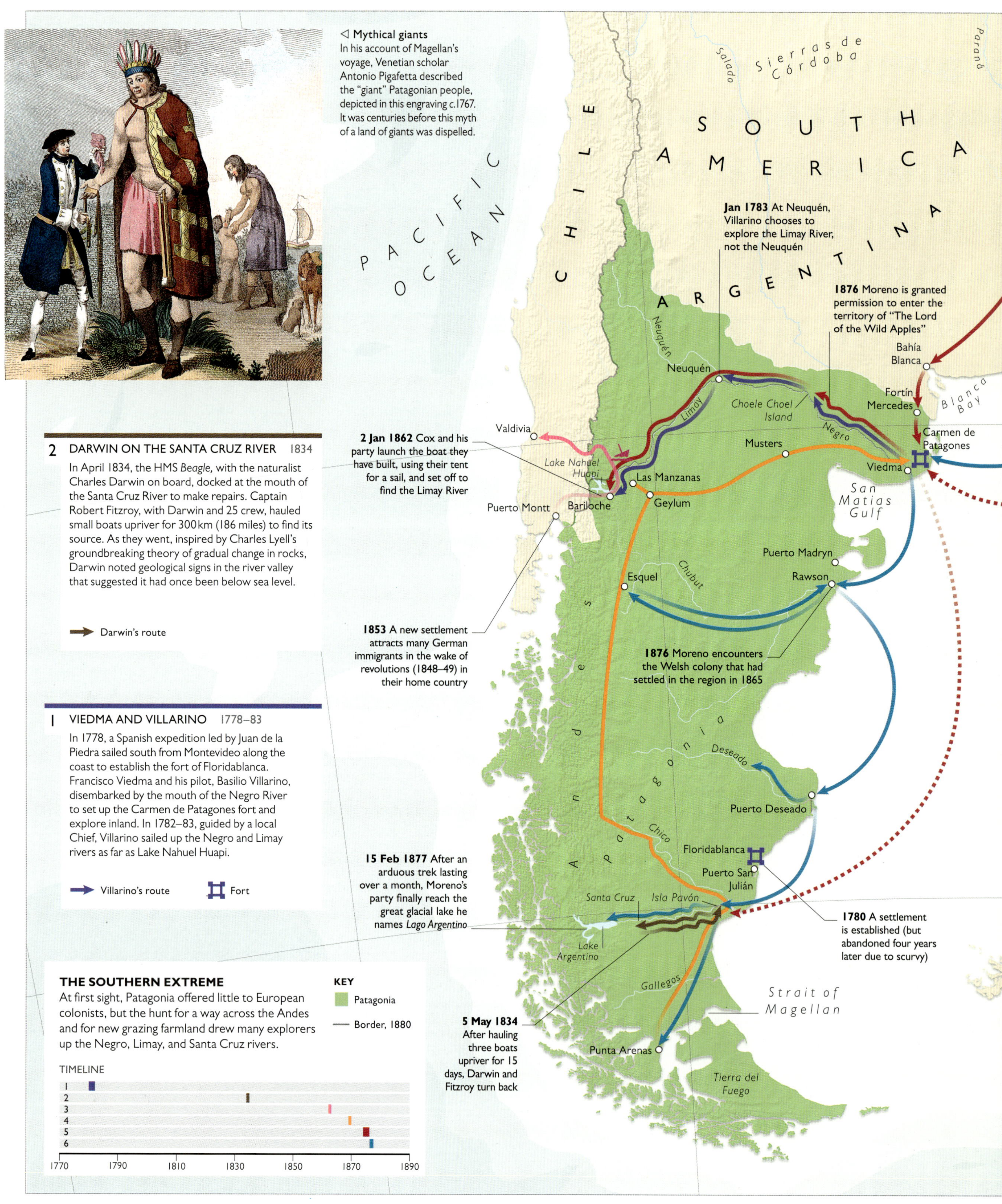
PACIFIC OCEAN
CHILE
SOUTH AMERICA
ARGENTINA
Sierras de Córdoba
Salado
Paraná
Jan 1783 At Neuquén, Villarino chooses to explore the Limay River, not the Neuquén
1876 Moreno is granted permission to enter the territory of "The Lord of the Wild Apples"
Bahía Blanca
Blanca Bay
Fortín Mercedes
Carmen de Patagones
Viedma
Neuquén
Neuquén
Choele Choel Island
Negro
Limay
Musters
Valdivia
Lake Nahuel Huapi
Bariloche
Las Manzanas
Geylum
Puerto Montt
2 Jan 1862 Cox and his party launch the boat they have built, using their tent for a sail, and set off to find the Limay River
San Matias Gulf
Puerto Madryn
Rawson
Esquel
Chubut
1853 A new settlement attracts many German immigrants in the wake of revolutions (1848–49) in their home country
1876 Moreno encounters the Welsh colony that had settled in the region in 1865
Andes
Patagonia
Deseado
Puerto Deseado
Chico
Floridablanca
Puerto San Julián
15 Feb 1877 After an arduous trek lasting over a month, Moreno's party finally reach the great glacial lake he names Lago Argentino
Santa Cruz
Isla Pavón
Lake Argentino
1780 A settlement is established (but abandoned four years later due to scurvy)
Gallegos
Strait of Magellan
5 May 1834 After hauling three boats upriver for 15 days, Darwin and Fitzroy turn back
Punta Arenas
Tierra del Fuego
TIMELINE
1
2
3
4
5
6
1770
1790
1810
1830
1850
1870
1890

◁ **Mythical giants**
In his account of Magellan's voyage, Venetian scholar Antonio Pigafetta described the "giant" Patagonian people, depicted in this engraving c.1767. It was centuries before this myth of a land of giants was dispelled.

2 DARWIN ON THE SANTA CRUZ RIVER 1834

In April 1834, the HMS *Beagle*, with the naturalist Charles Darwin on board, docked at the mouth of the Santa Cruz River to make repairs. Captain Robert Fitzroy, with Darwin and 25 crew, hauled small boats upriver for 300km (186 miles) to find its source. As they went, inspired by Charles Lyell's groundbreaking theory of gradual change in rocks, Darwin noted geological signs in the river valley that suggested it had once been below sea level.

→ Darwin's route

1 VIEDMA AND VILLARINO 1778–83

In 1778, a Spanish expedition led by Juan de la Piedra sailed south from Montevideo along the coast to establish the fort of Floridablanca. Francisco Viedma and his pilot, Basilio Villarino, disembarked by the mouth of the Negro River to set up the Carmen de Patagones fort and explore inland. In 1782–83, guided by a local Chief, Villarino sailed up the Negro and Limay rivers as far as Lake Nahuel Huapi.

→ Villarino's route
Fort

THE SOUTHERN EXTREME

At first sight, Patagonia offered little to European colonists, but the hunt for a way across the Andes and for new grazing farmland drew many explorers up the Negro, Limay, and Santa Cruz rivers.

KEY
- Patagonia
- Border, 1880

3 ACROSS THE ANDES 1862

In 1862, Guillermo Cox (Chilean-born but with an English father) self-funded an expedition to find a route between the new settlements on the Pacific coast and the Negro River. Setting off from Puerto Montt, his small party scaled the Andes, built a boat on the shores of Lake Nahuel Huapi, and began to sail the Limay River, until rapids wrecked the boat. An Indigenous chief offered help, but with nothing to offer in exchange, Cox was forced to turn back.

Cox's route
Shipwreck

4 GEORGE CHAWORTH MUSTERS 1869–70

A British naval officer serving on a ship off the coast of South America and inspired by reading accounts of Darwin's travels, Musters took leave to go ashore at Punta Arenas on the Magellan Strait and travel as far north as he could. *At Home with the Patagonians*, his account of his year-long journey that took him all the way to the Rio Negro northern province, is a tribute to the Tehuelche guides who accompanied his "wanderings".

Musters' expedition

5 MORENO'S EARLY EXPEDITIONS 1874–76

Concerned about Chile's interest in the region, the Argentine government commissioned fact-finding expeditions, with Moreno part of the team. The first excursion sailed from Buenos Aires to Carmen de Patagones and then onwards to the mouth of the Santa Cruz River. In 1875, Moreno sailed the Negro and Limay rivers in search of the trans-Andean route to Chile (which Argentina was keen to control), yet was refused permission to cross the mountains.

Moreno's 1874 expedition
Moreno's 1875–76 expedition

6 MORENO'S THIRD EXPEDITION 1876–77

In October 1876, Moreno set sail again from Buenos Aires to explore the Chubut, Deseado, and Santa Cruz rivers. Sailing the Santa Cruz in a small boat (rowing, or hauling with horses or by hand), he finally reached its source in 1877. In the 1880s, Moreno continued to explore and survey Patagonia, proving that it was a fertile region of vast pampas. However, by this time, Argentina's military expansion ("Conquest of the Desert") had destroyed relations with Indigenous peoples.

Moreno's third expedition

MAPPING PATAGONIA

In 1520, Portuguese explorer Ferdinand Magellan stopped on the Patagonian coast of Chile and Argentina on his global voyage. However, the vast interior, inhabited by the Tehuelche people, remained largely uncharted by Europeans until the 18th century, when Spanish and British expeditions explored the region.

After the discovery in 1616 of a sea route that went around Cape Horn rather than through the stormy Straits of Magellan, the Spanish lost interest in Patagonia and sailed past the southern tip of South America en route to their island colonies in the Pacific. But around 1740, Thomas Falkner, an English ship's surgeon who had become stranded in Buenos Aires due to ill health, began to explore the region as a Jesuit missionary.

Over nearly 40 years, Falkner recorded Patagonia's soil, flora, and fauna, and travelled to the southern interior and the now famous lakes. On his return to England in 1774, he published a memoir, *A Description of Patagonia*, which included one of the most detailed maps of the region at that time, and unnerved the Spanish as to Britain's colonial intentions in the country.

From 1778, the Spanish sent a series of expeditions into Patagonia's interior via its major rivers. But the most extensive explorations were conducted in the 19th century by the naturalist and geographer Francisco Moreno. Sponsored by the Argentine government, which considered Patagonia a desert, his fieldwork and maps proved that the land was far from this. He also provided information for Argentina's territorial wars with Chile and its genocidal campaign against the Indigenous people who lived on the land.

"The plains of Patagonia are boundless... and there appears no limit to their duration through future time..."

CHARLES DARWIN, *VOYAGE OF THE BEAGLE*, 1839

THE WELSH IN PATAGONIA

In the 1800s, as the Industrial Revolution encroached on the rural culture of Britain, groups of immigrants left Wales in search of places where they could preserve their identity. In May 1865, 150 Welsh men, women, and children sailed on the tea clipper *Mimosa* from Liverpool to Puerto Madryn, Patagonia. Helped by the Teheulche people, they trekked to the Chubut valley. There they created a settlement, Rawson, on the Chubut River (called *Afon Camwy* by the Welsh). Expecting fertile lands, they struggled on the barren pampas, but the colony exists to this day.

Early Welsh settlers of Patagonia, c.1880

EXPEDITIONS IN CENTRAL ASIA

After the decline of the Silk Road in the 1400s, the steppes, deserts, and mountains of Central Asia became largely desolate frontier territory. But the region became strategically important to the empire-building politics of European powers in the 19th century.

The British and Russian empires vied for influence in Central Asia. The British believed the Russians posed a threat to colonial India from the north, while the Russians feared British power encroaching on their territories. But rather than take part in open conflict, they engaged in an intelligence war (known as the Great Game), sending envoys to woo local rulers in the region or, crucially, embarked on exploratory expeditions that doubled as espionage.

In 1863, Hungarian Ármin Vámbéry ventured into the Emirate of Bukhara (in present-day Uzbekistan) and reached the old Silk Road (see pp.38–39) city of Samarkand. From 1879, Russian-Polish army officer Nikolai Przhevalsky mapped the eastern Tarim Basin. He was followed in the 1880s by another officer of the Imperial Russian Army, Bronislav Grombchevsky, who explored the mountains to the west and famously crossed paths with his British counterpart, Francis Younghusband, who was also collecting information for his country. These British and Russian adventurers were often heavily armed and local people suffered, most notoriously in 1904 when 600 Tibetans were massacred by Younghusband's troops at Chumik Shenko.

In the 1900s, the Great Game subsided, only to be replaced by another rivalry as European archaeologists, such as Sven Hedin and Marc Aurel Stein, competed to excavate ancient treasures buried in the sands of the Taklamakan Desert. In the process, a vast number of relics were removed from the region to be housed in the collections of European museums, where they remain today.

> *"I was content with nothing less than to tread paths where no European had ever set foot."*
>
> SVEN HEDIN, *MY LIFE AS AN EXPLORER*, 1926

THE THOUSAND BUDDHA CAVES

The Mogao Caves ("Thousand Buddha Caves") near Dunhuang, China, were an important site of worship on the ancient Silk Road, yet had fallen into disuse by 1899, when Chinese priest Wang Yuanlu became a custodian. In 1900, he discovered an ancient sealed library, hidden for almost a millennium. In 1907, Marc Aurel Stein arrived from Kashmir, India, and persuaded Wang to let him purchase around 10,000 scrolls from the library, including the Diamond Sutra, printed in 868 CE.

Stein's photograph of the "Library Cave"

STRATEGIC TERRITORY

Central Asia's Tarim Basin, an entirely landlocked plain, bounded by the world's highest mountains, drew the attention of imperial rivals Britain and Russia in the 19th century.

KEY

- Russian territory and protectorates, 1907
- British territory and dependent states, 1907
- Peak

TIMELINE

1
2
3
4
5
6

1855 1865 1875 1885 1895 1905 1915 1925 1935

Western Steppe
R U
From Sweden
Aral Sea
Turan Plain
Caspian Sea
Caucasus
Khiva
Samarkand
Karakum Desert
Elburz
Tehran
PERSIA
Paropamisus
Herat
AFGHANISTAN
Kirthar Range
Arabian Sea

1893 Sven Hedin travels through Russia on his first expedition to Central Asia

Mar 1864 Vámbéry returns to Europe via Tehran, and publishes *Travels in Central Asia* later that year

6 MARC AUREL STEIN 1900–07

Hungarian-born British archaeologist Marc Aurel Stein is credited with rediscovering the Silk Road. In a series of expeditions (1900–1907), he unearthed tens of thousands of ancient relics from old Silk Road cities, such as Khotan, Dandan-Uiliq, Niya, and Endere. Conditions were harsh and very dry in the Taklamakan, but by travelling only in winter he could bring a water supply in the form of blocks of ice, carried by camels.

Silk Road site

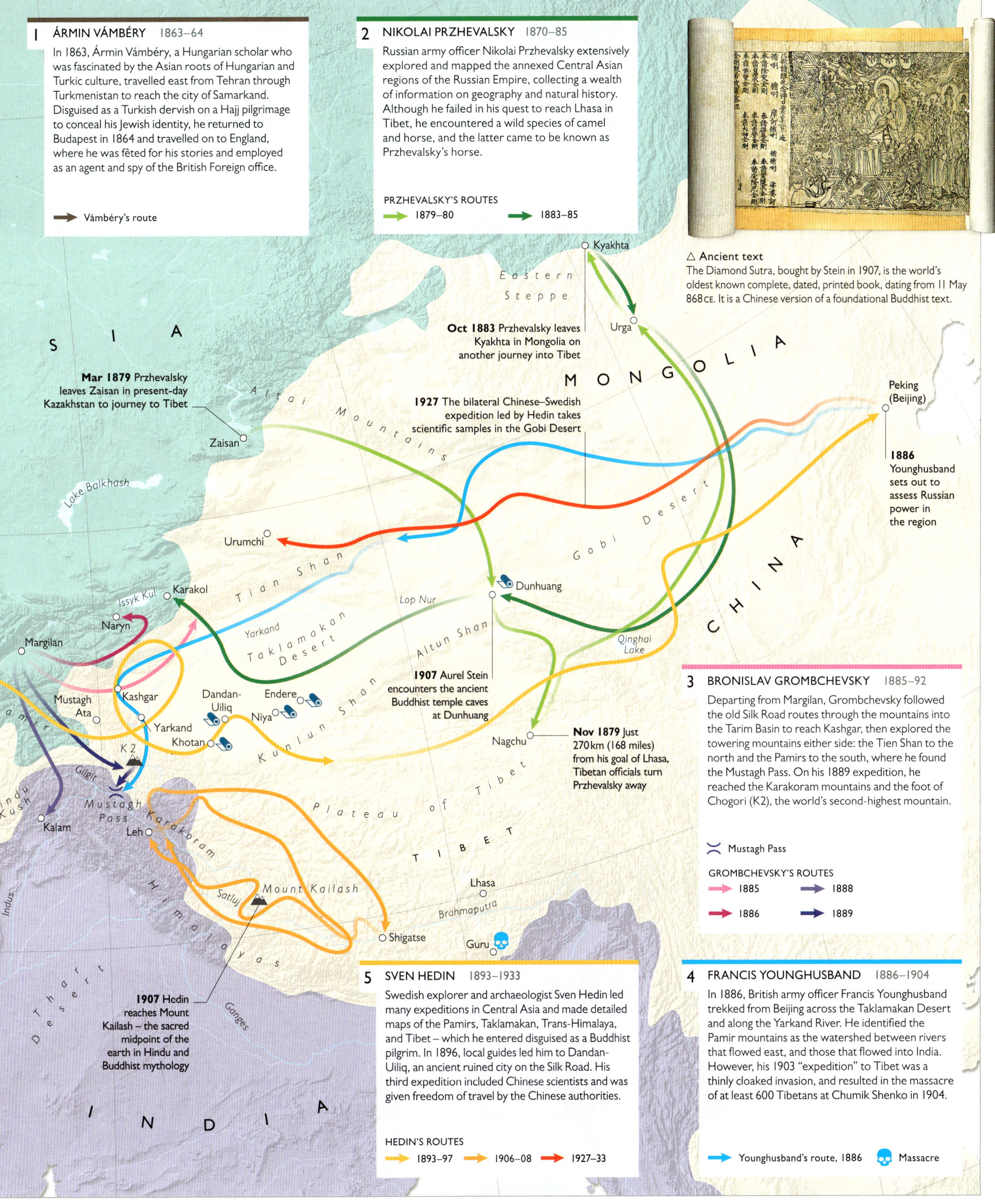

△ **Ancient text**
The Diamond Sutra, bought by Stein in 1907, is the world's oldest known complete, dated, printed book, dating from 11 May 868 CE. It is a Chinese version of a foundational Buddhist text.

1 ÁRMIN VÁMBÉRY 1863–64

In 1863, Ármin Vámbéry, a Hungarian scholar who was fascinated by the Asian roots of Hungarian and Turkic culture, travelled east from Tehran through Turkmenistan to reach the city of Samarkand. Disguised as a Turkish dervish on a Hajj pilgrimage to conceal his Jewish identity, he returned to Budapest in 1864 and travelled on to England, where he was fêted for his stories and employed as an agent and spy of the British Foreign office.

Vámbéry's route

2 NIKOLAI PRZHEVALSKY 1870–85

Russian army officer Nikolai Przhevalsky extensively explored and mapped the annexed Central Asian regions of the Russian Empire, collecting a wealth of information on geography and natural history. Although he failed in his quest to reach Lhasa in Tibet, he encountered a wild species of camel and horse, and the latter came to be known as Przhevalsky's horse.

PRZHEVALSKY'S ROUTES
1879–80
1883–85

3 BRONISLAV GROMBCHEVSKY 1885–92

Departing from Margilan, Grombchevsky followed the old Silk Road routes through the mountains into the Tarim Basin to reach Kashgar, then explored the towering mountains either side: the Tien Shan to the north and the Pamirs to the south, where he found the Mustagh Pass. On his 1889 expedition, he reached the Karakoram mountains and the foot of Chogori (K2), the world's second-highest mountain.

Mustagh Pass

GROMBCHEVSKY'S ROUTES
1885
1886
1888
1889

5 SVEN HEDIN 1893–1933

Swedish explorer and archaeologist Sven Hedin led many expeditions in Central Asia and made detailed maps of the Pamirs, Taklamakan, Trans-Himalaya, and Tibet – which he entered disguised as a Buddhist pilgrim. In 1896, local guides led him to Dandan-Uiliq, an ancient ruined city on the Silk Road. His third expedition included Chinese scientists and was given freedom of travel by the Chinese authorities.

HEDIN'S ROUTES
1893–97
1906–08
1927–33

4 FRANCIS YOUNGHUSBAND 1886–1904

In 1886, British army officer Francis Younghusband trekked from Beijing across the Taklamakan Desert and along the Yarkand River. He identified the Pamir mountains as the watershed between rivers that flowed east, and those that flowed into India. However, his 1903 "expedition" to Tibet was a thinly cloaked invasion, and resulted in the massacre of at least 600 Tibetans at Chumik Shenko in 1904.

Younghusband's route, 1886
Massacre

△ **A Nellie Bly board game**
Nellie Bly's round-the-world trip caught the public imagination. Not only were there competitions to guess her time of arrival at each port of call, but this board game enabled players to travel along with her.

> *"Start the man, and I'll start the same day... and beat him."*
>
> NELLIE BLY, IN A LETTER TO HER EDITOR, *AROUND THE WORLD IN SEVENTY-TWO DAYS*, 1890

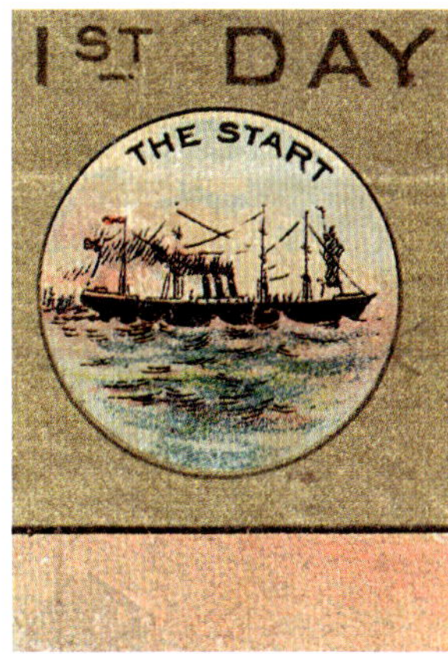

△ **The fastest liner in the Atlantic**
Heading east, Bly made a swift start by crossing the Atlantic on the newly launched luxury steamship *Auguste Victoria*, which had twin propellers for extra speed.

△ **A feat of modern engineering**
Completed just 20 years earlier, the Suez Canal provided a short cut from the Mediterranean to the Indian Ocean, saving 10 days or more on the journey.

A RACE AROUND THE WORLD

In 1889, American journalist Nellie Bly took on the challenge to circumnavigate the world solo, travelling by public transport. Inspired by her example, in 1894, Annie Londonderry, a Latvian immigrant to the US, travelled around the world by bicycle. Both won fame and victories for women's equality at a time when the world was deemed unsafe for a woman to travel alone.

Nellie Bly was the pseudonym of Elizabeth Jane Cochrane, a pioneering journalist who made her name in New York City, writing investigative reports on subjects such as the plight of female factory workers and the horrors of a mental health institution. But, aged 25, after reading Jules Verne's novel *Around the World in 80 Days*, Bly pitched to her editor at the *New York World* the idea of beating Phileas Fogg's record and travelling the world solo in just 75 days, sending reports back along the way.

Bly planned meticulously and travelled extremely light, taking a single small piece of luggage, no bigger than a modern airline carry-on – a riposte to her editor's belief that a woman's need for accoutrements prevented her from travelling both quickly and alone. She sailed from Hoboken, New Jersey just after 9:40am on 14 November 1889, crossing the Atlantic to Southampton. After briefly meeting the author Jules Verne, she travelled on through Italy and sailed to Egypt. More ships took her to Sri Lanka, Singapore, and Hong Kong, where she learned of a *Cosmopolitan* magazine-sponsored rival, Elizabeth Bisland, who was hot on her tail.

Bly set sail again from Yokohama in Japan for San Francisco. There, the *New York World* chartered a one-car train to race her across the country, and all along the way she was cheered by huge crowds. Bly beat Bisland, arriving in Jersey City at 3:51pm on 25 January 1890, only 72 days, 6 hours, 11 minutes, and 14 seconds after she had left it.

◁ **Nellie's worlwide celebrity**
The McLoughlin Brothers publishing firm distributed the game in 1890 during the height of Bly's fame. The cover image shows Nellie in her travelling dress and carrying her single piece of luggage.

△ **A competitor in the race**
In Hong Kong, Bly learned for the first time about the 28-year-old Elizabeth Bisland, who was circling the globe in the opposite direction (west–east) and had left the city three days previously.

△ **The last leg of the journey**
The *New York World* hired a special one-car train to transport Bly from San Francisco to Chicago. She then rode a standard train to Jersey City, where cannons were fired the moment that she landed on the platform.

CYCLING THE GLOBE

Inspired by Nellie Bly, in 1894 Annie Cohen Kopchovsky, a Jewish Latvian immigrant to the US, left her Boston home to cycle all the way around the world. Accepting sponsorship from the Londonderry bottled water company, she adopted the name Annie Londonderry and gave talks to pay her way. In fact, her journey was conducted mainly on ships and trains. However, it was a demonstration of independent travelling and the bicycle was a symbol of that. Although Londonderry had cycled very little before she started, by the time she returned home she was extremely adept. Indeed, she travelled the last leg across America with a broken wrist after crashing into some pigs, reaching Boston in September 1895.

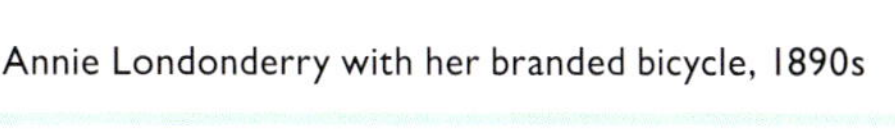
Annie Londonderry with her branded bicycle, 1890s

A Japanese view of London
In the tradition of *Namban byobu* (folding screens showing Europeans in Japan), during the 1860s Japanese illustrator Utagawa Yoshitora created popular panoramas of European cities, inspired by images from the *London Illustrated News*.

A CULTURAL EXCHANGE

Since the expansion of international trade and European colonies in the 1600s, the culture and influence of the West had spread throughout the world. However, by the 18th century, a more reciprocal cultural exchange was developing between the East and West.

△ **Japanese-style comb**
From the 1850s, when Japan ended its long isolation, Japonisme, a craze for Japanese art and design, swept through the West.

From 1603 until 1867, Japan was closed off to the outside world when its rulers, fearing that missionaries from the West would spread Christianity, closed its borders. Japanese were not allowed to leave and only a few foreigners were granted access. However, an interest in travel grew when Japanese writer Bashō chronicled his journey on foot in northern Japan in *The Narrow Road to the Deep North* in 1694. Later, in 1785–86, on behalf of the shogun rulers, Mogami Tokunai explored and reported on Hokkaido and parts of the Kuril Islands in northern Japan, where he met the Ainu people, the original inhabitants of the Kurils.

A breakthrough moment for relations between Japan and the West was when the boat of a 14-year-old fisherman, Manjiro, was swept out to sea in 1841. He was picked up by an American whaler and taken to the United States, where he adopted the name John Mung and became the first Japanese person to live there. A few decades later, Japanese diplomats began travelling regularly to the United States.

Increasingly, elite Indians also travelled to the West, returning with tales of their journeys. Travellers began to mix and swap habits – Europeans bought Chinese and Japanese art, and Asians adopted Western dress and reading habits.

Aside from wealthy travellers, cultures also merged when Chinese workers on ships travelled west. "Chinatowns" in cities such as New York and London grew, but districts were often run down and there was little prospect of returning home.

SAKE DEAN MAHOMED 1759–1851

Born to an elite Muslim family in India, Sake Dean Mahomed published *The Travels of Dean Mahomet* (1794). The first book written in English by an Indian, it described his sojourns with the East India Company and his later emigration to Ireland, where he married. In 1807 he and his family moved to London, where he opened the Hindoostane Coffee House, the city's first Indian restaurant, in 1810. Although this business failed, the enterprising Mahomed relocated to Brighton in 1814 and established a "shampoo" (steam and massage) baths that included the Prince of Wales among its clientèle.

A 19th-century portrait of Sake Dean Mahomed

A NORTHWEST ROUTE

For centuries, European explorers took huge risks amid the shifting ice as they navigated between the Arctic islands north of Canada in search of a Northwest Passage, linking the Atlantic and Pacific oceans. Many perished along the way, before Roald Amundsen succeeded in 1906.

The Inuit peoples were probably the first to explore the Northwest Passage, but no documentation survives. From the late 15th century onwards, European expeditions made many failed attempts to find a Northwest Passage (see pp.78–9) and open up a direct route to the East. They reached no further west than Baffin Bay and Hudson Bay.

Treacherous Arctic conditions were the greatest obstacle to this exploration. Ice strong enough to crush ships blocked the way between islands or locked vessels in, so that crews, stuck for months or even years, died of disease and starvation. The prize was to find a lucrative trading route between Europe and Asia. However, many explorers were drawn simply to the challenge of navigating extreme conditions. These voyages fascinated the public, too, inspiring folk songs and plays, as well as artists, who saw in the terror of the Arctic the existential confrontation between humanity and nature.

It took countless voyages, and cost many lives, before Norwegian explorer Roald Amundsen found a way through in 1906, aided by skills he had learned earlier from the Inuit. Although the Inuit peoples aided these early explorers, the presence of the outsiders and their large ships was disorientating to local communities.

Today, climate change keeps the Northwest Passage sufficiently ice-free for ships to pass through. Yet even now, aided by modern technology, it remains a hazardous route that few ships use, and recognition of Inuit and other First Nations peoples' long presence in the region brings the landscape some protection.

LEARNING FROM THE INUIT

Roald Amundsen spent 1903–05 on King William Island in the Arctic Archipelago, gathering scientific data and preparing to sail the Passage. His team of six were the first to prove that the magnetic North Pole moves regularly. A Netsilik Inuit camp set up next to Amundsen, the two groups forging close relations. The Inuit taught the explorers survival skills – how to hunt seals, build igloos, drive dogs, and dress in loose hide and furs. In return, the Norwegians gave the Inuit knives, needles, and matches.

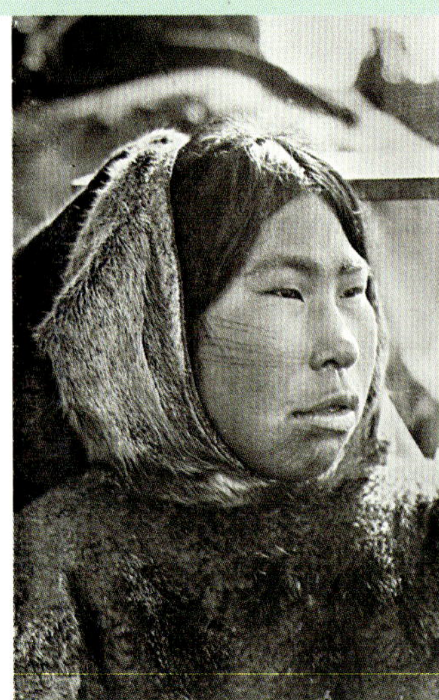

Kabloka, whose husband Ugpi was a friend of Amundsen, 1904

EXPLORING THE ARCTIC

In the 19th and early 20th centuries, European explorers persisted in their search for a northern passage between the Pacific and Atlantic oceans. Gathering scientific knowledge of the Arctic was an added incentive for these daring voyagers.

KEY
Arctic Circle

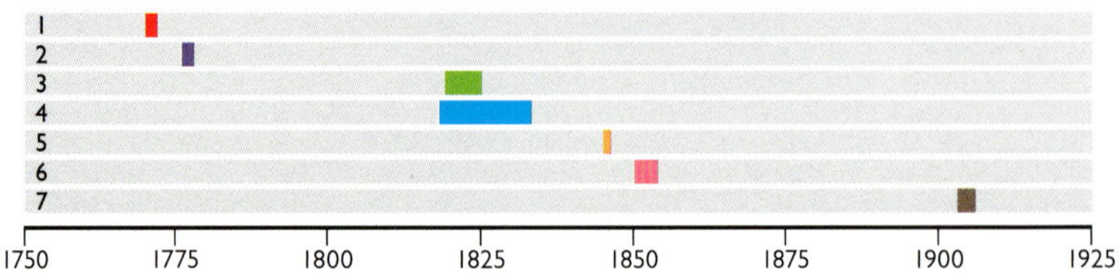

TIMELINE
1 2 3 4 5 6 7
1750 1775 1800 1825 1850 1875 1900 1925

◁ **Ice-bound**
Ship's artist Samuel Cresswell painted HMS *Investigator* trapped in ice off Bering Island in 1851 during a search for Franklin's missing expedition.

ARCTIC OCEAN
Beaufort Sea
Bering Strait
Icy Cape
Cape Shmidta
SIBERIA
Arctic Circle
Bering Sea
Alaska
To Norway
From England
PACIFIC OCEAN
To Hawaii

14 Feb 1779 After a long, welcoming sojourn on Hawaii, a broken mast forces Cook's fleet to return. The mood now changed, conflict over a boat leads to a battle and Cook dies

1 WALKING TO THE ARCTIC 1770–72

In 1770, inspired by a recent Indigenous expedition, the Canadian Hudson Bay Company commissioned British fur trader Samuel Hearne to search for copper mines and the fabled Northwest Passage. Travelling with the Chipewyan Chief Matonabbee and his people, he trekked 8,000km (5,000miles) from Hudson Bay to Coronation Gulf at the Arctic Ocean, yet did not find the passage – thus proving that, if a passage existed, it must lie further north.

Hearne's route, 1770
Hearne's route, 1770–72

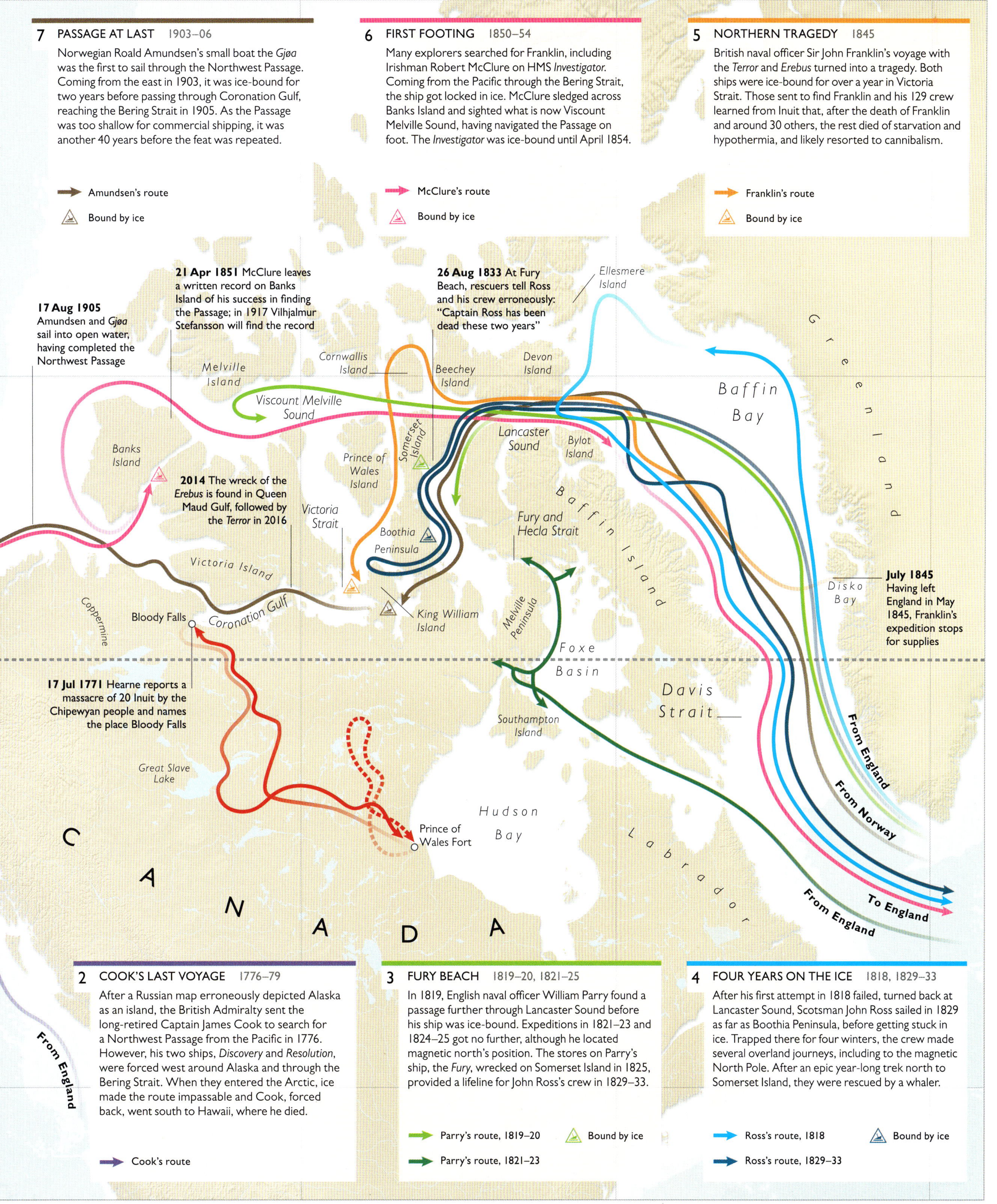

Amundsen's route
Bound by ice
McClure's route
Bound by ice
Franklin's route
Bound by ice
21 Apr 1851 McClure leaves a written record on Banks Island of his success in finding the Passage; in **1917** Vilhjalmur Stefansson will find the record
26 Aug 1833 At Fury Beach, rescuers tell Ross and his crew erroneously: "Captain Ross has been dead these two years"
Ellesmere Island
17 Aug 1905 Amundsen and *Gjøa* sail into open water, having completed the Northwest Passage
Greenland
Cornwallis Island
Beechey Island
Devon Island
Melville Island
Viscount Melville Sound
Baffin Bay
Lancaster Sound
Bylot Island
Somerset Island
Banks Island
Prince of Wales Island
2014 The wreck of the *Erebus* is found in Queen Maud Gulf, followed by the *Terror* in 2016
Victoria Strait
Baffin Island
Fury and Hecla Strait
Boothia Peninsula
Victoria Island
July 1845 Having left England in May 1845, Franklin's expedition stops for supplies
Disko Bay
Coppermine
Bloody Falls
Coronation Gulf
King William Island
Melville Peninsula
Foxe Basin
17 Jul 1771 Hearne reports a massacre of 20 Inuit by the Chipewyan people and names the place Bloody Falls
Davis Strait
Southampton Island
From England
Great Slave Lake
From Norway
Hudson Bay
Prince of Wales Fort
Labrador
To England
From England
CANADA
From England

2 COOK'S LAST VOYAGE 1776–79

After a Russian map erroneously depicted Alaska as an island, the British Admiralty sent the long-retired Captain James Cook to search for a Northwest Passage from the Pacific in 1776. However, his two ships, *Discovery* and *Resolution*, were forced west around Alaska and through the Bering Strait. When they entered the Arctic, ice made the route impassable and Cook, forced back, went south to Hawaii, where he died.

Cook's route

3 FURY BEACH 1819–20, 1821–25

In 1819, English naval officer William Parry found a passage further through Lancaster Sound before his ship was ice-bound. Expeditions in 1821–23 and 1824–25 got no further, although he located magnetic north's position. The stores on Parry's ship, the *Fury*, wrecked on Somerset Island in 1825, provided a lifeline for John Ross's crew in 1829–33.

Parry's route, 1819–20
Bound by ice
Parry's route, 1821–23

4 FOUR YEARS ON THE ICE 1818, 1829–33

After his first attempt in 1818 failed, turned back at Lancaster Sound, Scotsman John Ross sailed in 1829 as far as Boothia Peninsula, before getting stuck in ice. Trapped there for four winters, the crew made several overland journeys, including to the magnetic North Pole. After an epic year-long trek north to Somerset Island, they were rescued by a whaler.

Ross's route, 1818
Bound by ice
Ross's route, 1829–33

5 NORTHERN TRAGEDY 1845

British naval officer Sir John Franklin's voyage with the *Terror* and *Erebus* turned into a tragedy. Both ships were ice-bound for over a year in Victoria Strait. Those sent to find Franklin and his 129 crew learned from Inuit that, after the death of Franklin and around 30 others, the rest died of starvation and hypothermia, and likely resorted to cannibalism.

6 FIRST FOOTING 1850–54

Many explorers searched for Franklin, including Irishman Robert McClure on HMS *Investigator*. Coming from the Pacific through the Bering Strait, the ship got locked in ice. McClure sledged across Banks Island and sighted what is now Viscount Melville Sound, having navigated the Passage on foot. The *Investigator* was ice-bound until April 1854.

7 PASSAGE AT LAST 1903–06

Norwegian Roald Amundsen's small boat the *Gjøa* was the first to sail through the Northwest Passage. Coming from the east in 1903, it was ice-bound for two years before passing through Coronation Gulf, reaching the Bering Strait in 1905. As the Passage was too shallow for commercial shipping, it was another 40 years before the feat was repeated.

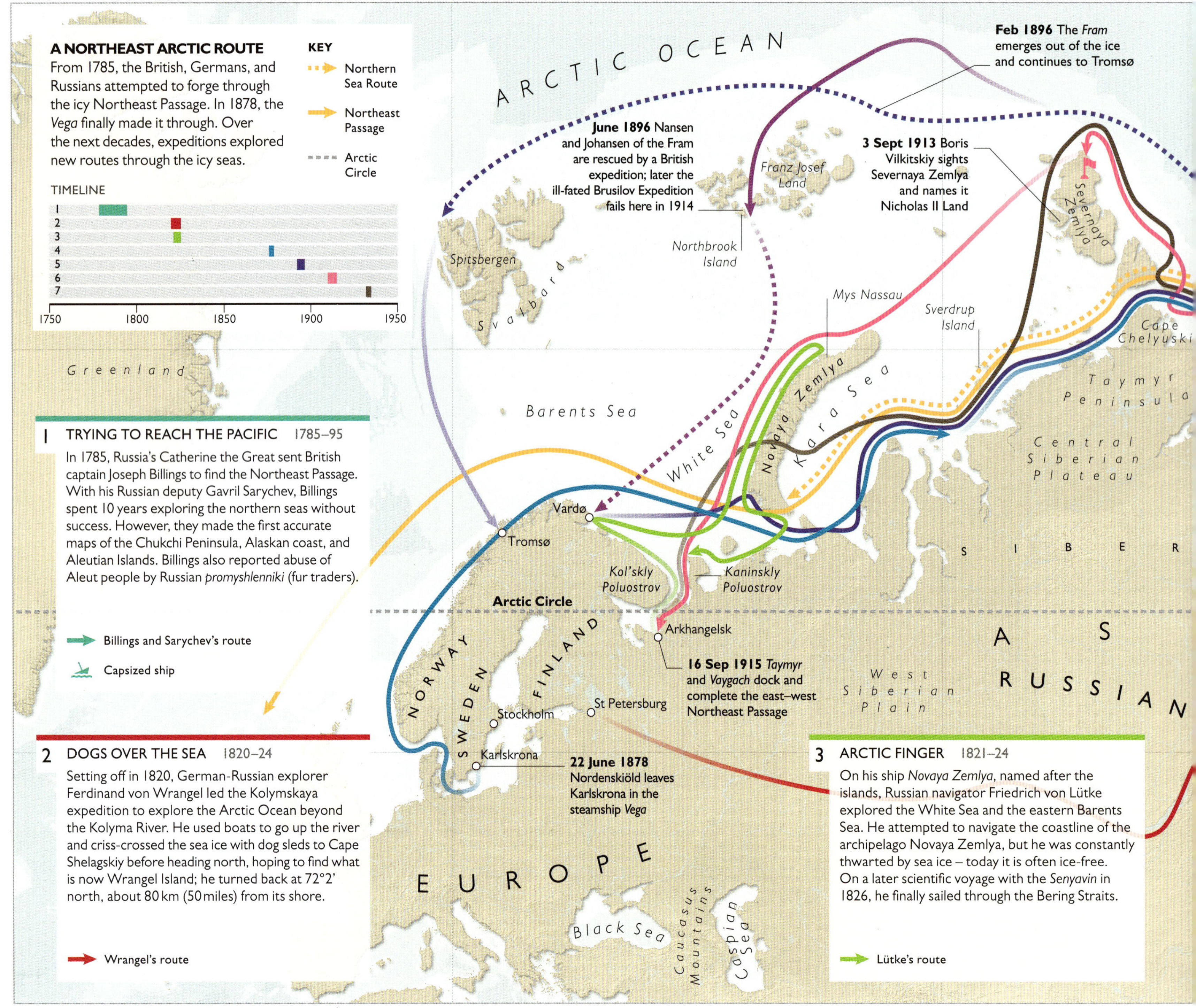

A NORTHEAST ARCTIC ROUTE
From 1785, the British, Germans, and Russians attempted to forge through the icy Northeast Passage. In 1878, the *Vega* finally made it through. Over the next decades, expeditions explored new routes through the icy seas.
KEY
Northern Sea Route
Northeast Passage
Arctic Circle
TIMELINE
1
2
3
4
5
6
7
1750
1800
1850
1900
1950
ARCTIC OCEAN
Feb 1896 The *Fram* emerges out of the ice and continues to Tromsø
June 1896 Nansen and Johansen of the Fram are rescued by a British expedition; later the ill-fated Brusilov Expedition fails here in 1914
3 Sept 1913 Boris Vilkitskiy sights Severnaya Zemlya and names it Nicholas II Land
Franz Josef Land
Severnaya Zemlya
Northbrook Island
Spitsbergen
Svalbard
Mys Nassau
Sverdrup Island
Cape Chelyuskin
Taymyr Peninsula
Greenland
Barents Sea
Novaya Zemlya
Kara Sea
White Sea
Central Siberian Plateau
1 TRYING TO REACH THE PACIFIC 1785–95
In 1785, Russia's Catherine the Great sent British captain Joseph Billings to find the Northeast Passage. With his Russian deputy Gavril Sarychev, Billings spent 10 years exploring the northern seas without success. However, they made the first accurate maps of the Chukchi Peninsula, Alaskan coast, and Aleutian Islands. Billings also reported abuse of Aleut people by Russian *promyshlenniki* (fur traders).
Billings and Sarychev's route
Capsized ship
Vardø
Tromsø
Kol'skiy Poluostrov
Kaninskiy Poluostrov
SIBERIA
Arctic Circle
Arkhangelsk
16 Sep 1915 *Taymyr* and *Vaygach* dock and complete the east–west Northeast Passage
ASIAN RUSSIA
West Siberian Plain
NORWAY
SWEDEN
FINLAND
Stockholm
St Petersburg
Karlskrona
22 June 1878 Nordenskiöld leaves Karlskrona in the steamship *Vega*
2 DOGS OVER THE SEA 1820–24
Setting off in 1820, German-Russian explorer Ferdinand von Wrangel led the Kolymskaya expedition to explore the Arctic Ocean beyond the Kolyma River. He used boats to go up the river and criss-crossed the sea ice with dog sleds to Cape Shelagskiy before heading north, hoping to find what is now Wrangel Island; he turned back at 72°2' north, about 80 km (50 miles) from its shore.
Wrangel's route
EUROPE
Black Sea
Caucasus Mountains
Caspian Sea
3 ARCTIC FINGER 1821–24
On his ship *Novaya Zemlya*, named after the islands, Russian navigator Friedrich von Lütke explored the White Sea and the eastern Barents Sea. He attempted to navigate the coastline of the archipelago Novaya Zemlya, but he was constantly thwarted by sea ice – today it is often ice-free. On a later scientific voyage with the *Senyavin* in 1826, he finally sailed through the Bering Straits.
Lütke's route

THE NORTHEAST PASSAGE

Since the mid-17th century, explorers had known that it was possible to sail east along northern Siberia, to reach the Pacific Ocean (see pp.112–13). However, it was not known whether it was possible to sail across open seas from the Atlantic Ocean, via the Arctic Ocean north of Siberia, to the Pacific. Like the fabled Northwest Passage north of Canada (see pp.78–79, 196–197), the Northeast Passage remained elusive.

The Northeast Passage, as it became known, stretches from the north of the Scandinavian Peninsula (Sweden, Norway, and most of northern Finland) through seas along the northern coast of Russia – the Barents, Kara, Laptev, East Siberian, and Chukchi seas – before turning southwards through the Bering Strait between northeastern Siberia and western Alaska. All these seas are essentially giant bays that form part of the Arctic Ocean.

While the Northwest Passage around North America threads its way through a complex maze of islands, the Northeast Passage is, in theory, an open sea route. The problem is that it is ice-bound for much of the year, so finding a way through potentially frozen waters was, for a long time, more challenging. Yet, the prize was considerable, for it would reduce the distance between Europe and Japan by a third, even compared to the Suez Canal route, which was opened in 1869.

In 1785, Empress Catherine the Great launched another of many Russian expeditions (see pp.128–29), this time to find a way through, led by the English navigator Joseph Billings. Although these voyages

▷ **Cold comfort**
A 19th-century illustration shows the cold-weather clothing worn on Adolf Nordenskiöld's voyage in 1878. Crew needed clothing suitable for walking over the ice should the ship get stuck, which, inevitably, it did.

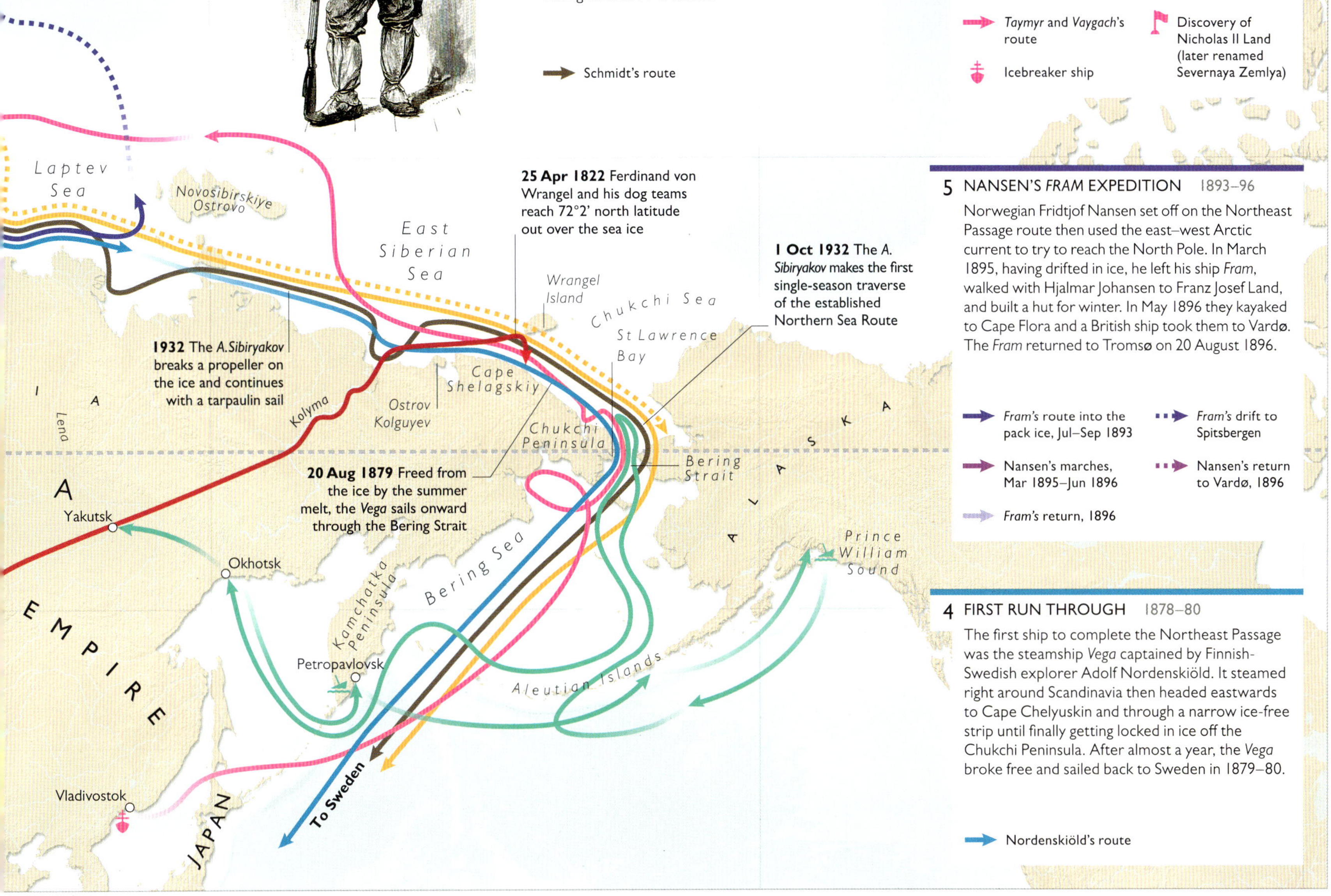

7 SINGLE-SEASON VOYAGE 1932

Soviet scientist Otto Schmidt pioneered the idea of the Northern Sea Route to keep an open pathway through the ice. He attempted the first traverse of the route in a single season on the icebreaker *Aleksandr Sibiryakov*, leaving Arkhangelsk on 28 July 1932. Close to the end, one of the ship's propellers broke on the ice. The crew tore tarpaulins to make a sail and the ship glided on safely through the Bering Strait on 1 October.

→ Schmidt's route

6 ICEBREAKERS EAST TO WEST 1910–15

Advances in technology meant early icebreakers *Taymyr* and *Vaygach* had strong hulls and powerful engines to push through ice. Led by Russian hydrographic surveyor Boris Vilkitskiy, over several years they set out from Vladivostok into the Arctic from the east, completing the first east–west northeast passage at Arkhangelsk in 1915.

→ *Taymyr* and *Vaygach*'s route
Icebreaker ship
Discovery of Nicholas II Land (later renamed Severnaya Zemlya)

25 Apr 1822 Ferdinand von Wrangel and his dog teams reach 72°2' north latitude out over the sea ice

1 Oct 1932 The *A. Sibiryakov* makes the first single-season traverse of the established Northern Sea Route

1932 The *A. Sibiryakov* breaks a propeller on the ice and continues with a tarpaulin sail

20 Aug 1879 Freed from the ice by the summer melt, the *Vega* sails onward through the Bering Strait

Laptev Sea
Novosibirskiye Ostrovo
East Siberian Sea
Wrangel Island
Chukchi Sea
St Lawrence Bay
Cape Shelagskiy
Ostrov Kolguyev
Kolyma
Lena
Chukchi Peninsula
Bering Strait
Yakutsk
Okhotsk
Kamchatka Peninsula
Bering Sea
Petropavlovsk
Aleutian Islands
Prince William Sound
ALASKA
EMPIRE
Vladivostok
JAPAN
To Sweden

5 NANSEN'S *FRAM* EXPEDITION 1893–96

Norwegian Fridtjof Nansen set off on the Northeast Passage route then used the east–west Arctic current to try to reach the North Pole. In March 1895, having drifted in ice, he left his ship *Fram*, walked with Hjalmar Johansen to Franz Josef Land, and built a hut for winter. In May 1896 they kayaked to Cape Flora and a British ship took them to Vardø. The *Fram* returned to Tromsø on 20 August 1896.

→ *Fram*'s route into the pack ice, Jul–Sep 1893
→ *Fram*'s drift to Spitsbergen
→ Nansen's marches, Mar 1895–Jun 1896
→ Nansen's return to Vardø, 1896
→ *Fram*'s return, 1896

4 FIRST RUN THROUGH 1878–80

The first ship to complete the Northeast Passage was the steamship *Vega* captained by Finnish-Swedish explorer Adolf Nordenskiöld. It steamed right around Scandinavia then headed eastwards to Cape Chelyuskin and through a narrow ice-free strip until finally getting locked in ice off the Chukchi Peninsula. After almost a year, the *Vega* broke free and sailed back to Sweden in 1879–80.

→ Nordenskiöld's route

revealed much about the geography of the remote and icy northern seas, none could make more than forays into the territory without their path being blocked by the ever-shifting ice floes.

It took more than a century before the Finnish-Swedish explorer Adolf Nordenskiöld completed the first successful navigation in 1878 through the Northeast Passage on the barque *Vega*. However, as the sea ice constantly shifts and blocks the route in a different configuration each season, it was not until the 1930s that icebreaker ships were able to regularly travel this way.

> *"One should not poke one's nose into places where Nature does not want the presence of man."*
>
> VALERIAN ALBANOV, *IN THE LAND OF WHITE DEATH*, 1917

DISASTER ON THE ICE

The Brusilov Expedition to find a northeast passage was a dark moment in Russian exploration. In 1912, the *Santa Anna*, captained by Georgy Brusilov, set sail too late and was ice-bound in the Kara Sea. Over 18 months, ice pushed it northwest of Franz Josef Land. After trying to cross land to safety, only navigator Valerian Albanov and one of the 22 crew members survived.

The *Santa Anna* crew in 1912

REACHING THE NORTH POLE

The challenge for explorers trying to reach the North Pole was that it is situated not on land, but far out across shifting sea ice. American explorers, using dedicated Inuit guides, claimed the prize in both 1908 and 1909, but the first fully documented exploration would come many decades later.

In September 1909, readers of the *New York Times* were thrilled by a front-page headline: "Peary Discovers the North Pole After Eight Trials in 23 Years." American explorer Robert Peary claimed to have achieved this feat in April 1909.

The problem was that a week earlier the *New York Herald* had run a headline "The North Pole is Discovered by Dr. Frederick A. Cook." Also an American, Cook had supposedly reached the Pole a year earlier, in April 1908.

A long and bitter battle ensued as the counterclaims of these former colleagues were debated by the scientific establishment. The solid evidence was slim, with Peary's own expedition records proving of dubious accuracy, while Cook's records were said to have been abandoned in

△ **A frozen fracas**
A French newspaper cover from September 1909 depicts Cook and Peary fighting while in the improbable presence of penguins.

Greenland on Peary's orders. After years of investigation, America's National Geographic Society concluded it was Peary who had made it to the Pole, while Cook was discredited. But over time, doubts have crept in and it is now widely accepted that Peary got close to, but did not reach, the Pole. Both explorers relied on the expertise of local Inuits; on Peary's team, the African-American explorer Matthew Henson learned to speak Inuit.

The first proven trek over the ice had to wait until American Ralph Plaisted achieved it on a snowmobile in 1968. British explorer Sir Walter Herbert got there on foot with dog sledges a year later.

> *"We were the only pulsating creatures in a dead world of ice."*
>
> FREDERICK COOK, DIARY, 1908

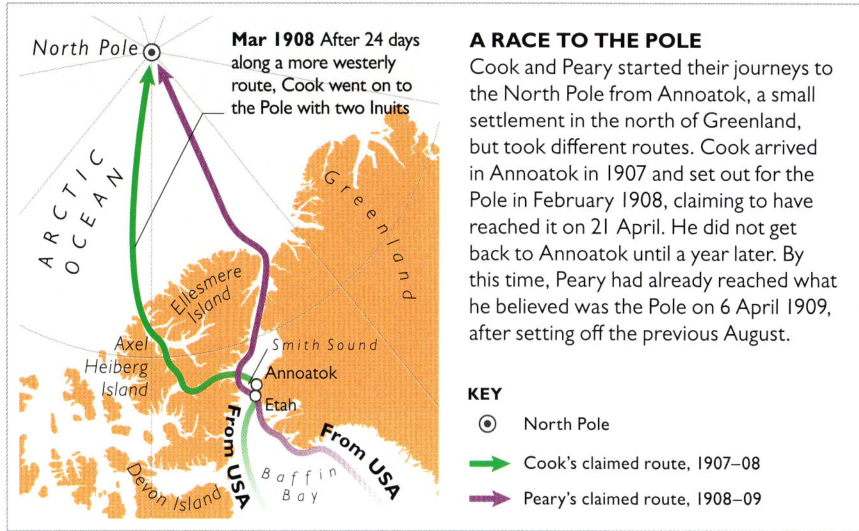

A RACE TO THE POLE
Cook and Peary started their journeys to the North Pole from Annoatok, a small settlement in the north of Greenland, but took different routes. Cook arrived in Annoatok in 1907 and set out for the Pole in February 1908, claiming to have reached it on 21 April. He did not get back to Annoatok until a year later. By this time, Peary had already reached what he believed was the Pole on 6 April 1909, after setting off the previous August.

KEY
- North Pole
- Cook's claimed route, 1907–08
- Peary's claimed route, 1908–09

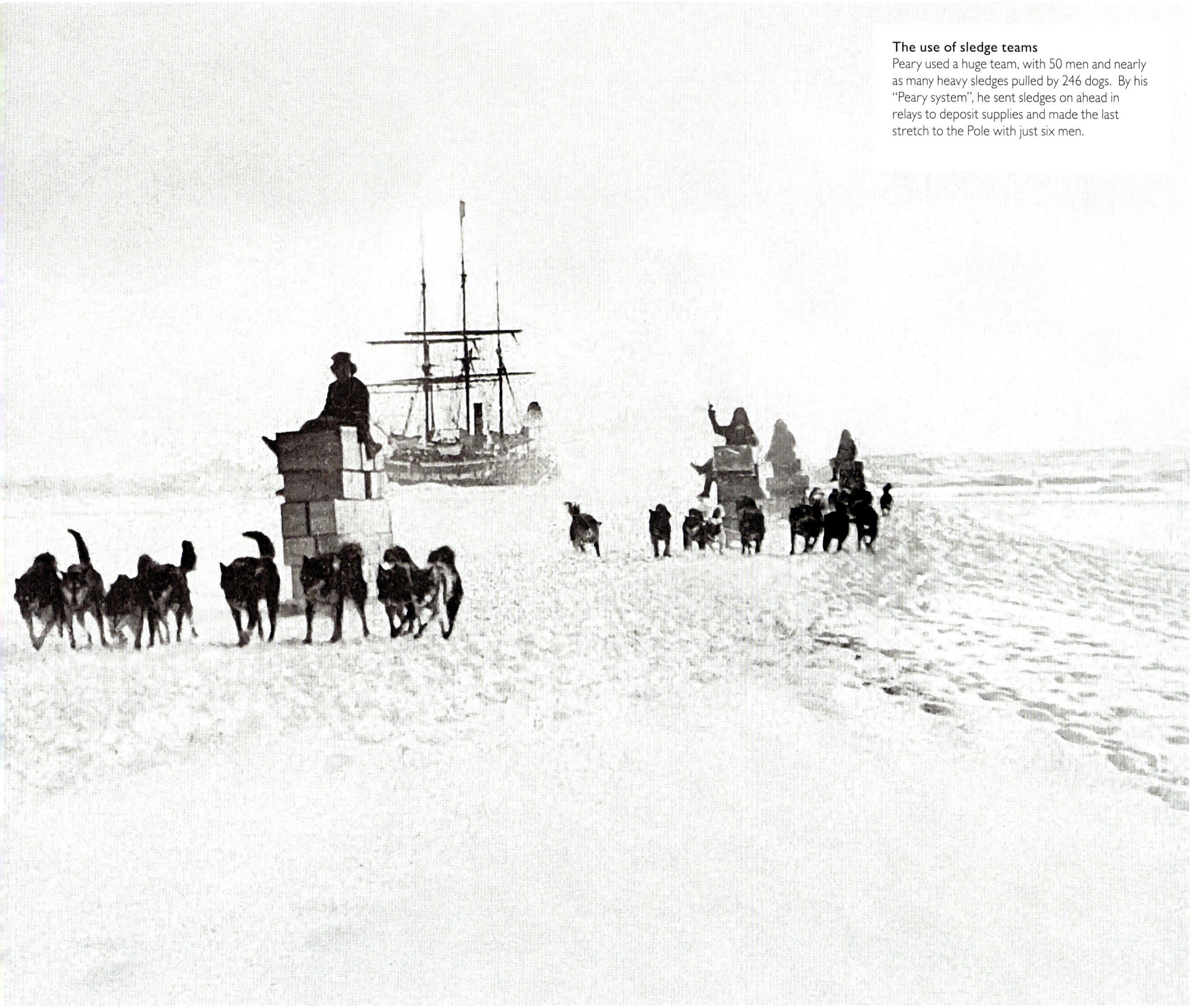

The use of sledge teams
Peary used a huge team, with 50 men and nearly as many heavy sledges pulled by 246 dogs. By his "Peary system", he sent sledges on ahead in relays to deposit supplies and made the last stretch to the Pole with just six men.

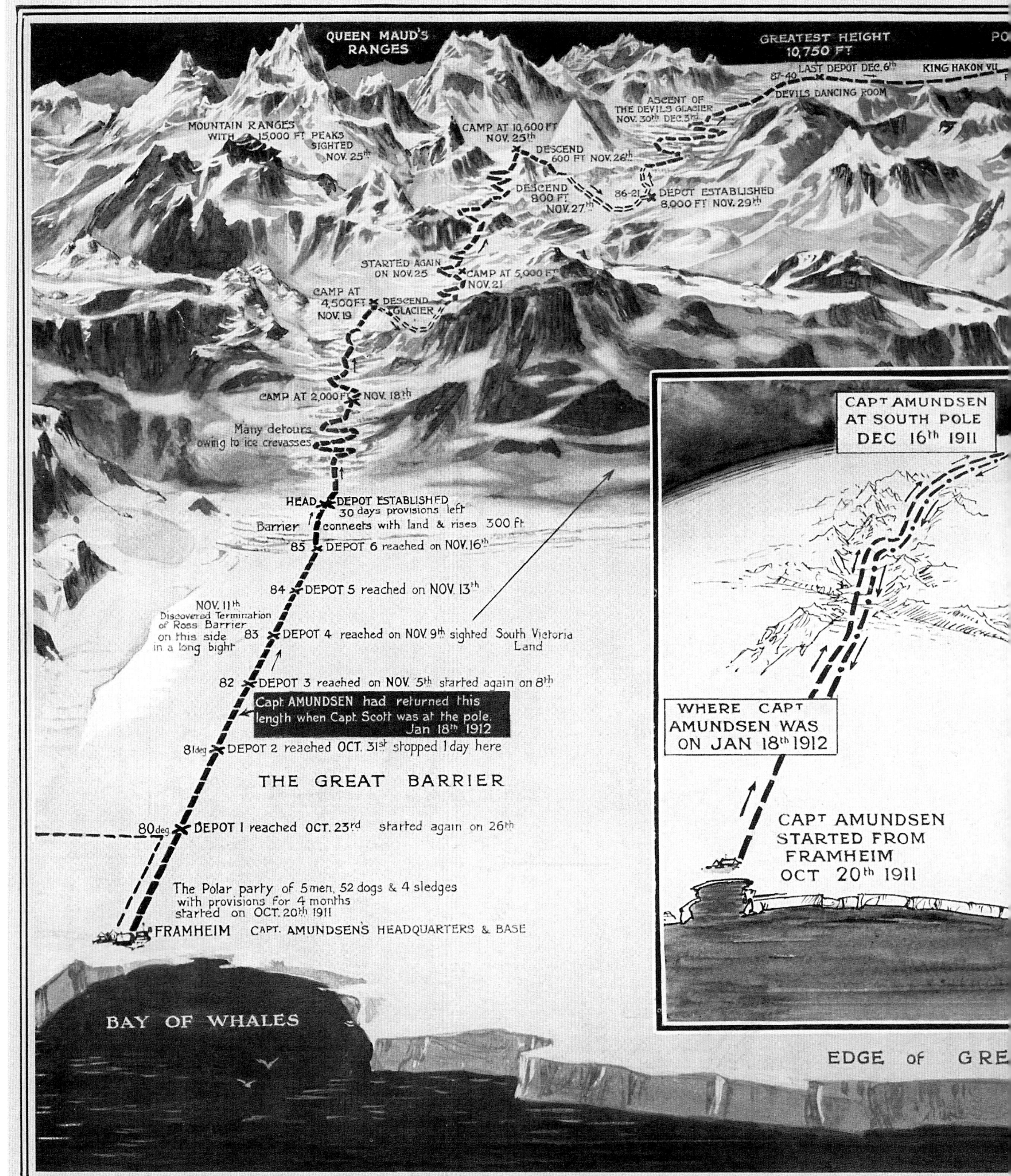
QUEEN MAUD'S RANGES
GREATEST HEIGHT 10,750 FT
LAST DEPOT DEC. 6th
KING HAKON VII
DEVILS DANCING ROOM
ASCENT OF THE DEVILS GLACIER NOV. 30th DEC. 3rd
MOUNTAIN RANGES WITH 15,000 FT PEAKS SIGHTED NOV. 25th
CAMP AT 10,600 FT NOV. 25th
DESCEND 600 FT NOV. 26th
DESCEND 800 FT NOV. 27th
DEPOT ESTABLISHED 8,000 FT NOV. 29th
STARTED AGAIN ON NOV. 25
CAMP AT 5,000 FT NOV. 21
CAMP AT 4,500 FT NOV. 19
DESCEND GLACIER
CAMP AT 2,000 FT NOV. 18th
Many detours owing to ice crevasses
HEAD DEPOT ESTABLISHED 30 days provisions left
Barrier connects with land & rises 300 ft
85 DEPOT 6 reached on NOV. 16th
84 DEPOT 5 reached on NOV. 13th
NOV. 11th Discovered Termination of Ross Barrier on this side in a long bight
83 DEPOT 4 reached on NOV. 9th sighted South Victoria Land
82 DEPOT 3 reached on NOV. 5th started again on 8th
Capt. AMUNDSEN had returned this length when Capt. Scott was at the pole. Jan 18th 1912
81deg DEPOT 2 reached OCT. 31st stopped 1 day here
THE GREAT BARRIER
80deg DEPOT 1 reached OCT. 23rd started again on 26th
The Polar party of 5 men, 52 dogs & 4 sledges with provisions for 4 months started on OCT. 20th 1911
FRAMHEIM CAPT. AMUNDSEN'S HEADQUARTERS & BASE
BAY OF WHALES
CAPT AMUNDSEN AT SOUTH POLE DEC 16th 1911
WHERE CAPT AMUNDSEN WAS ON JAN 18th 1912
CAPT AMUNDSEN STARTED FROM FRAMHEIM OCT 20th 1911
EDGE of GRE

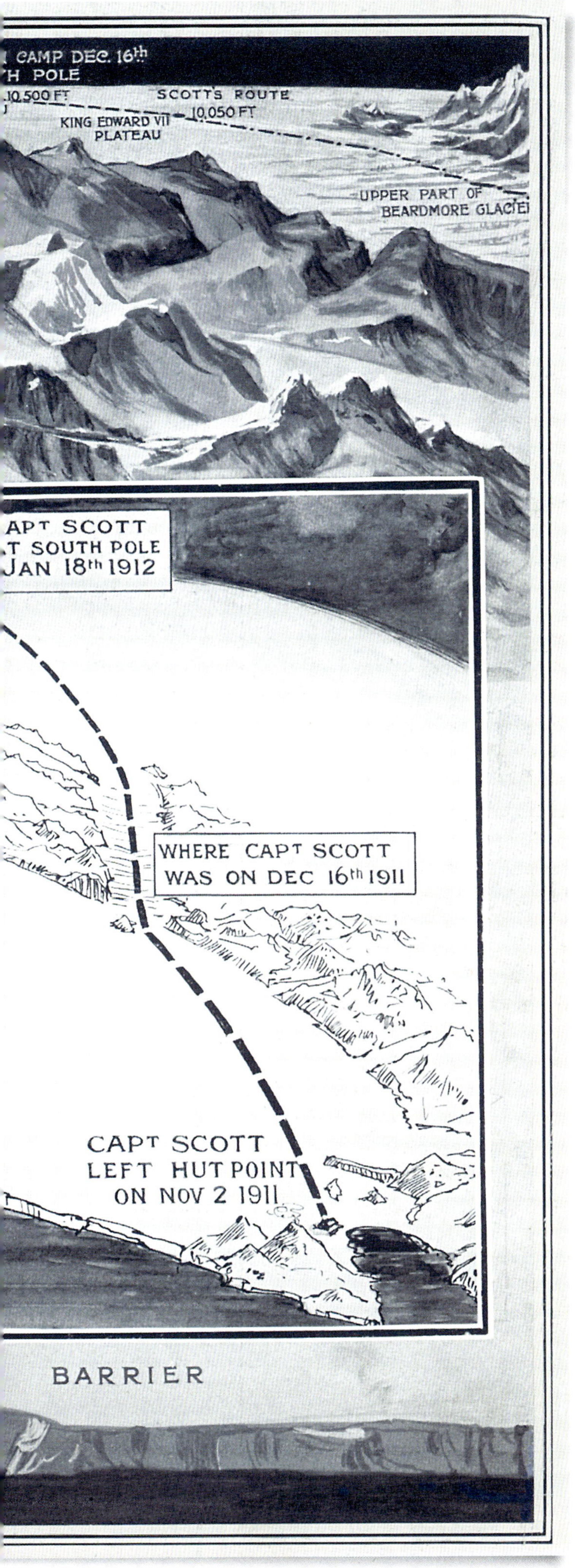

A RACE TO THE SOUTH POLE

After English explorer Robert Falcon Scott set out to be first to the South Pole in 1910, he discovered that the Norwegian Roald Amundsen had set his sights on the same destination. Scott realized he faced a race to the Pole.

Scott set sail from Cardiff, South Wales, in his ship, *Terra Nova*, on 15 June 1910. It was only when he reached Melbourne, Australia, in October that he found out about Amundsen's intentions. The Norwegian had planned to head for the North Pole, but when he learnt that Cook and Peary had already been there (see pp. 200–201), he rerouted south.

In 1911, Scott reached McMurdo Sound, while Amundsen landed at the Bay of Whales, giving him a 96.5-km (60-mile) shorter route. Amundsen's team reached the Pole in 56 days on 14 December. Scott set out from Hut Point at Cape Evans two weeks after Amundsen and trekked over tougher ground, reaching the Pole on 18 January 1912, more than a month after his rival.

Scott's disappointment at finding Amundsen's flag at the Pole turned into tragedy on the journey back. Conditions were extreme. Edgar Evans fell on a glacier and died, and Oates's feet were so badly frostbitten that he voluntarily left the tent and walked out to his death, saying, "I am just going outside and may be some time." Blizzards halted any further progress, and supplies ran out. Scott's last diary entry was from 29 March: "...the end cannot be far." Scott and his two companions' bodies were found eight months later.

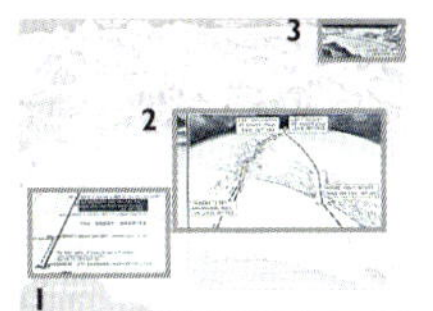

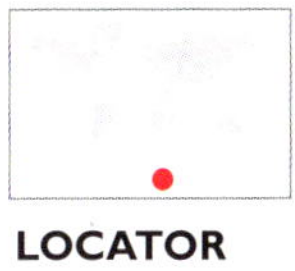

LOCATOR

KEY

1 Amundsen's route took him over the world's largest sheet of floating ice, the Great Barrier, now known as the Ross Ice Shelf.

2 When Amundsen reached the Pole, Scott was only halfway there. The Norwegian was halfway back before Scott reached the Pole.

3 On their tragic trek back, Scott and his team found the key *Glossopteris* fossil ferns in the moraines of the Beardmore Glacier.

◁ Contrasting routes
This 1913 map shows the different routes taken by Amundsen and Scott. Amundsen's shorter trek let him use dog sledges to reach the Pole on 14 December (not 16th as shown). Scott's longer route over the Beardmore Glacier meant his team had to haul their sledges by hand.

A GEOLOGICAL DISCOVERY

On 12 February 1912, as Scott and his team trudged back from the Pole, they stopped on the Beardmore Glacier for some "geologizing" (the word Scott uses in his diary). Incredibly, they collected 16 kg (35 lb) of samples and carried them on, despite the extra weight. The samples proved significant, and included a fossil of an extinct fern called *Glossopteris*. Matched with other *Glossopteris* fossils discovered in Australia, South America, India, and Africa, Scott's find supported the theory that these four continents were once part of a single prehistoric land mass.

A fossilized *Glossopteris* fern leaf

CHINESE ENVOYS IN THE WEST

In 1434, Chinese Emperor Xuande issued the Edict of Haijin, banning all trade and interaction with the rest of the world. Over the ensuing centuries, a limited form of trading was tolerated, but no foreign travel. However, in the 19th century, China began to end its isolation by sending students and diplomats overseas, to report back on the culture and customs of the wider world.

In the early 1800s, Chinese travellers were predominantly merchants or migrants, who were beginning to leave China in ever-increasing numbers due to economic pressures and the political unrest of the Taiping Rebellion. Some were also Christians, such as Lin King Chew, who travelled to New York to teach Cantonese to the Rev. Dr Cummings and other Christian missionaries going to China, and who published an account of his experiences entitled *Reminiscences of Western Travels*.

However, in 1860, after China's defeat in the Second Opium War, France and Britain forced the emperor to end the country's isolation and open up to foreign trade. In 1866, its Prince Regent, Gong, tasked the retired magistrate Binchun, together with his son, to accompany three students to Europe, and "record everything pertaining to mountains, rivers, and customs". The expedition party visited many countries, from England to Egypt, and Binchun recorded his impressions in witty poems, published in Beijing in 1868 as the best-selling *Cheng cha bi ji* (*Jottings from a Raft*). Awestruck by his first journey on a train, he describes it as running "like an unbridled horse".

The Burlingame Mission
This 1868 group portrait shows the first Chinese diplomatic delegation to the West, which was led by the American Anson Burlingame (standing centre). Formerly a diplomatic representative of the US, he became the representative of China to the world.

Offical delegations followed these early fact-finding missions. Zhang Deyi, a young student who travelled with Binchun in 1866, accompanied the Burlingame Mission to the US in 1868 (pictured above, fourth from the right), and a mission to France in 1871. His published travel journals were entitled *His Strange Tales from Over the Ocean*, and he was appointed the Qing ambassador to Britain in 1902.

In the early 1900s, Chinese women also began to venture abroad. Qian Shan Shili went to Japan with her diplomat husband in 1899, then on the Trans-Siberian Railway (right) to Russia, the Netherlands, and Italy. Her two thought-provoking travelogues note instructive differences between the array of customs she observes.

> *"When we arrived, many came to see us and take photographs of us."*
>
> BINCHUN, *JOTTINGS FROM A RAFT*, 1868

THE TRANS-SIBERIAN RAILWAY

In 1891, Russia started to build the Trans-Siberian Railway, extending 9,289 km (5,772 miles) across the country from Moscow to Vladivostok. It took 25 years to complete but became the world's longest railway. When the Trans-Mongolian Line, extending as far as Beijing, was completed in 1956, the Eurasian rail network connected China with Paris, for anyone who could afford the fare.

Exposition postcard from 1900

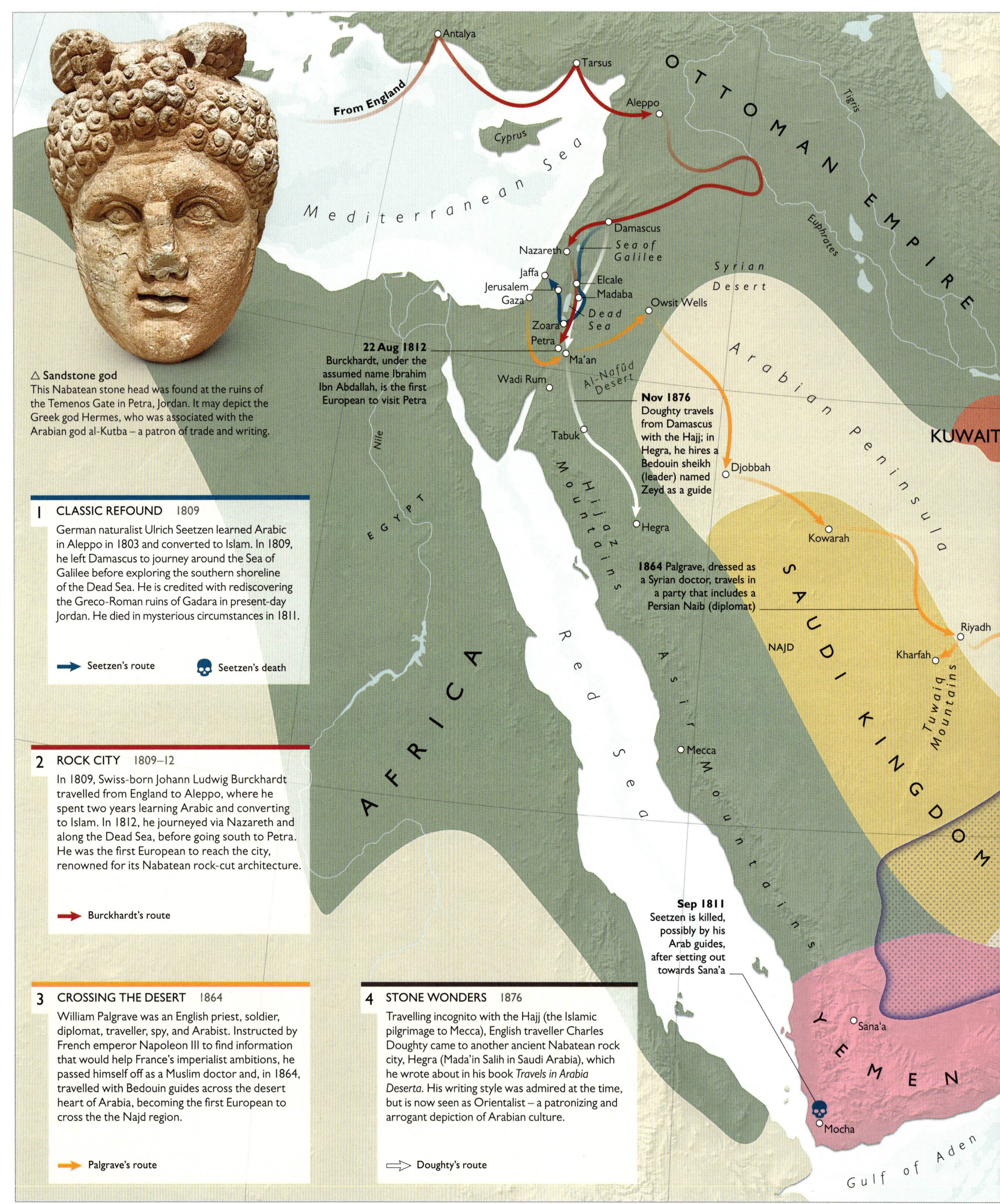

△ **Sandstone god**
This Nabatean stone head was found at the ruins of the Temenos Gate in Petra, Jordan. It may depict the Greek god Hermes, who was associated with the Arabian god al-Kutba – a patron of trade and writing.

1 CLASSIC REFOUND 1809

German naturalist Ulrich Seetzen learned Arabic in Aleppo in 1803 and converted to Islam. In 1809, he left Damascus to journey around the Sea of Galilee before exploring the southern shoreline of the Dead Sea. He is credited with rediscovering the Greco-Roman ruins of Gadara in present-day Jordan. He died in mysterious circumstances in 1811.

Seetzen's route
Seetzen's death

2 ROCK CITY 1809–12

In 1809, Swiss-born Johann Ludwig Burckhardt travelled from England to Aleppo, where he spent two years learning Arabic and converting to Islam. In 1812, he journeyed via Nazareth and along the Dead Sea, before going south to Petra. He was the first European to reach the city, renowned for its Nabatean rock-cut architecture.

Burckhardt's route

3 CROSSING THE DESERT 1864

William Palgrave was an English priest, soldier, diplomat, traveller, spy, and Arabist. Instructed by French emperor Napoleon III to find information that would help France's imperialist ambitions, he passed himself off as a Muslim doctor and, in 1864, travelled with Bedouin guides across the desert heart of Arabia, becoming the first European to cross the the Najd region.

Palgrave's route

4 STONE WONDERS 1876

Travelling incognito with the Hajj (the Islamic pilgrimage to Mecca), English traveller Charles Doughty came to another ancient Nabatean rock city, Hegra (Mada'in Salih in Saudi Arabia), which he wrote about in his book *Travels in Arabia Deserta*. His writing style was admired at the time, but is now seen as Orientalist – a patronizing and arrogant depiction of Arabian culture.

Doughty's route

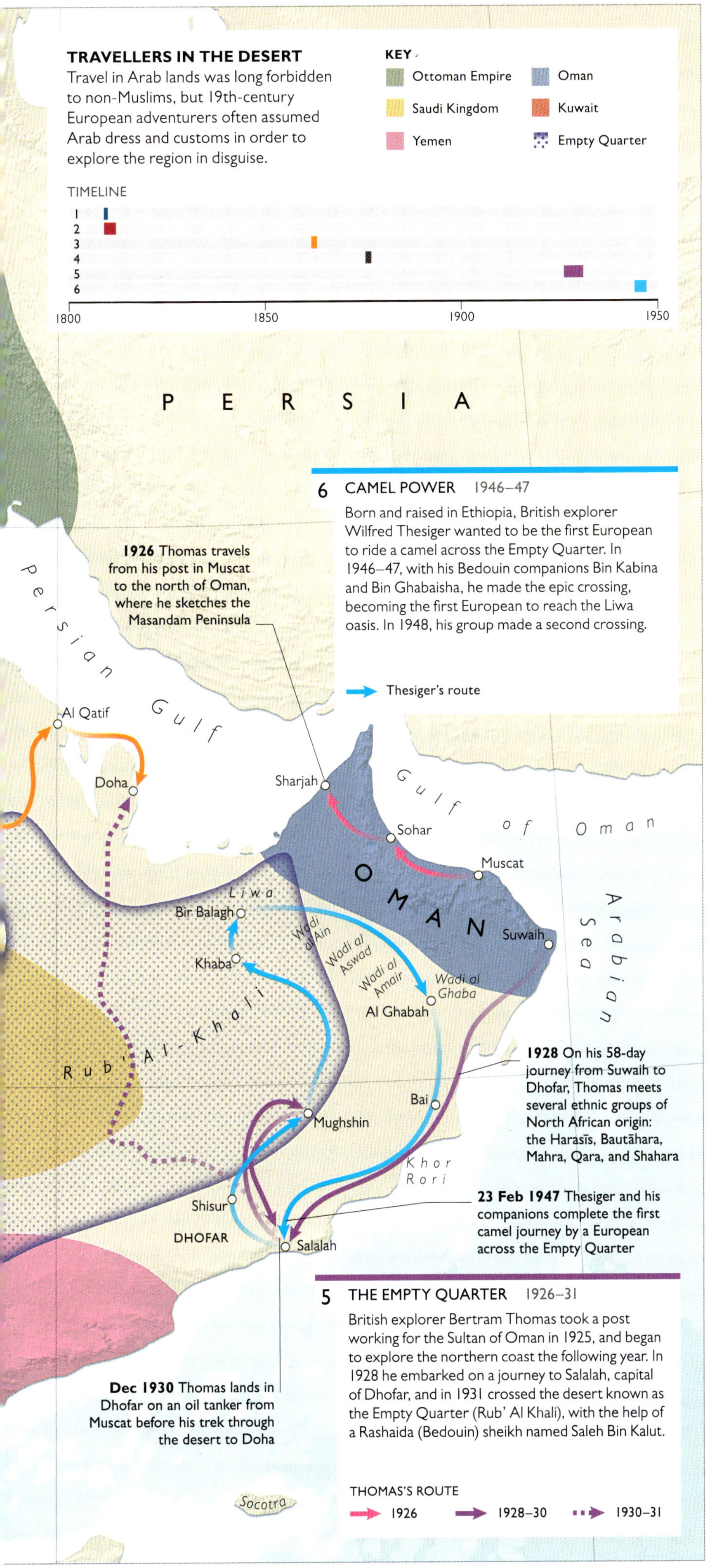

TRAVELLERS IN THE DESERT

Travel in Arab lands was long forbidden to non-Muslims, but 19th-century European adventurers often assumed Arab dress and customs in order to explore the region in disguise.

KEY

Ottoman Empire
Oman
Saudi Kingdom
Kuwait
Yemen
Empty Quarter

TIMELINE

1
2
3
4
5
6

1800 1850 1900 1950

PERSIA

6 CAMEL POWER 1946–47

Born and raised in Ethiopia, British explorer Wilfred Thesiger wanted to be the first European to ride a camel across the Empty Quarter. In 1946–47, with his Bedouin companions Bin Kabina and Bin Ghabaisha, he made the epic crossing, becoming the first European to reach the Liwa oasis. In 1948, his group made a second crossing.

Thesiger's route

1926 Thomas travels from his post in Muscat to the north of Oman, where he sketches the Masandam Peninsula

Persian Gulf
Al Qatif
Doha
Sharjah
Gulf of Oman
Sohar
Muscat
OMAN
Liwa
Bir Balagh
Suwaih
Arabian Sea
Khaba
Wadi al Ain
Wadi al Aswad
Wadi al Amair
Wadi al Ghaba
Al Ghabah
Rub' Al-Khali
Bai
Mughshin
Khor Rori
Shisur
DHOFAR
Salalah

1928 On his 58-day journey from Suwaih to Dhofar, Thomas meets several ethnic groups of North African origin: the Harasīs, Bautāhara, Mahra, Qara, and Shahara

23 Feb 1947 Thesiger and his companions complete the first camel journey by a European across the Empty Quarter

5 THE EMPTY QUARTER 1926–31

British explorer Bertram Thomas took a post working for the Sultan of Oman in 1925, and began to explore the northern coast the following year. In 1928 he embarked on a journey to Salalah, capital of Dhofar, and in 1931 crossed the desert known as the Empty Quarter (Rub' Al Khali), with the help of a Rashaida (Bedouin) sheikh named Saleh Bin Kalut.

Dec 1930 Thomas lands in Dhofar on an oil tanker from Muscat before his trek through the desert to Doha

THOMAS'S ROUTE

1926 1928–30 1930–31

Socotra

TRAVELS IN ARABIA

From the mid-1800s, Europeans explored the Arabian Peninsula, which today encompasses Saudi Arabia, Oman, Bahrain, Kuwait, Qatar, the United Arab Emirates, Yemen, and parts of Iraq and Jordan. These "Arabists" were often employed by European states with colonialist ambitions.

In 1761, King Frederick V of Denmark sent six scientists to satisfy the growing European curiosity about Arabia, as the Arabian Peninsula was known in Europe. Only the German cartographer Carsten Niebuhr returned (the others died of diseases), bringing back plant specimens, drawings, and geographical information including maps. These findings piqued the interest of other European explorers, who began to learn Arabic, adopt Arab clothing and Arabic names, and, in rare instances, convert to Islam to access places, such as the holy city of Mecca, that were forbidden to non-Muslims. Among them were German explorer Ulrich Seetzen, who visited the ruins of Gadara near the Sea of Galilee in 1809, and British writer Wilfred Thesiger, who crossed the Empty Quarter (a large expanse of desert) in 1946.

Adventurers expanded European knowledge of Arabia – returning with (often looted) artefacts, manuscripts, and samples of fauna and flora – but many were directly employed by their governments to increase European influence on the people of Arabia. The Palestinian-American academic Edward Said criticized the Arabists' "Orientalism" for the way it allowed Europeans to suggest that Arab people did not understand their own history or appreciate their ancient ruins.

"Every writer on the Orient… saw the Orient as a locale requiring Western attention…"

EDWARD SAID, *ORIENTALISM*, 1977

PETRA AND THE NABATEANS

The carved rocks of Petra in Jordan, visited by Burckhardt in 1812, are now world famous. But for 2,000 years, Nabatean culture remained unknown to Europeans. The Nabateans were originally nomadic, but their key location on ancient trading routes from Arabia to the Classical world made them wealthy. Petra was home to about 20,000 people, but now its extraordinary monuments such as the Treasury (Al-Khazneh) are all that is left.

Al-Khazneh in the 19th century

THE MODERN WORLD

DURING THE 20TH CENTURY, EXPLORERS REACHED NEARLY ALL THE REMOTE AREAS OF THE WORLD, FROM THE HIGHEST MOUNTAIN PEAKS TO THE DEEPEST OCEAN TRENCHES. OUTER SPACE BECAME THE NEW ARENA FOR EXPLORATORY MISSIONS AND, WHILE PROGRESS HAS BEEN MADE, THERE REMAINS MUCH TO DISCOVER.

△ **On the way to Everest**
Two members of the 1953 British expedition, Tom Bourdillon and Charles Evans, approach the South Col. Although the pair were unsuccessful in their attempt to summit, and turned back just short of their goal, the feat was achieved three days later by Edmund Hillary and Tenzing Norgay.

PUSHING THE BOUNDARIES

By the 20th century, media, communications, and transport connected the whole world in a dynamic way, leaving few corners untouched. Explorers set their sights on mountain peaks and ocean depths, and then looked out into space.

In the early 20th century, the world was well into its second industrial revolution, with mass-production and urbanization developing at an astonishing pace. But two world wars in the first half of the 20th century (1914–18 and 1939–45) had a devastating impact across the globe. After peace was restored in 1945, nationalist movements grew apace in almost every colony. Beginning with India and Pakistan in 1945, a succession of countries quickly obtained their independence and, by the end of 1960, three dozen new states in Asia and Africa had attained autonomy or complete independence from their European colonial rulers. The map of the world began to look remarkably different.

The Earth's extremes

During peacetime, many expeditions aimed for the remaining uncharted and extreme regions of the Earth. Gertrude Bell and Freya Stark crossed the deserts of Southwest Asia (the Middle East), Alexandra David-Néel trekked the Himalayas to reach the holy city of Lhasa in Tibet, and both Ynés Mexia and the Villas-Boas brothers led expeditions deep into the Amazonian rainforests. The first mission to scale Mount Everest was in 1922, but it was not until 31 years later that New Zealand's Edmund Hillary and Nepali-Indian Sherpa Tenzing Norgay became the first known climbers to stand at the top of the world.

The oceans, covering more than two-thirds of the Earth's surface, also remained a largely unexplored region. From the 1940s, the inventions of French naval officer Jacques Cousteau aided the study of marine life and, in 1950, he launched the research ship *Calypso*, which carried out pioneering oceanographic expeditions. In the same decade, American geologists Marie Tharp and Bruce Heezen created a groundbreaking map that showed the radically varied

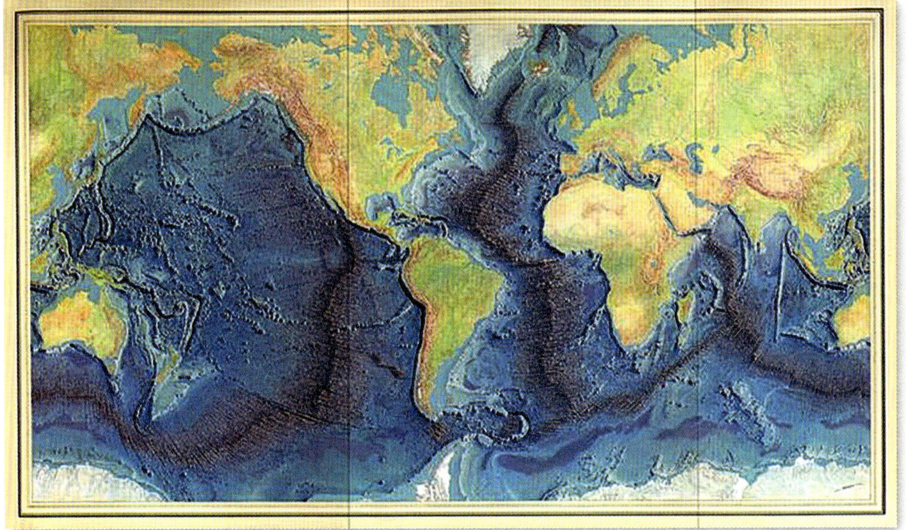

▷ **The first world map of the ocean floors**
In 1977, in collaboration with the Austrian landscape painter Heinrich Berann, American scientists Marie Tharp and Bruce Heezen published the first complete world map of the ocean floors.

THE MODERN AGE

During the 20th century, a time of political turmoil and rapid technological change, explorers reached the last remote places on Earth – both Poles, the top of Everest, the deepest depths of the ocean, and the Moon. The James Webb Space Telescope journeyed remotely throughout the solar system, and in the 21st century, new technologies like this are revolutionizing the way we explore and expanding our knowledge in unprecedented ways.

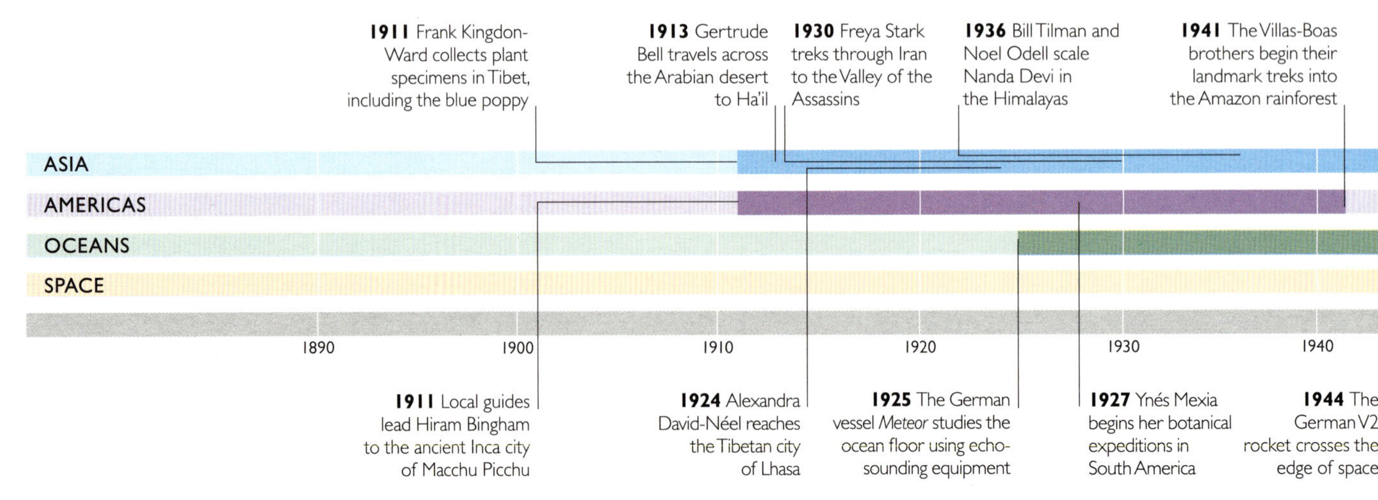

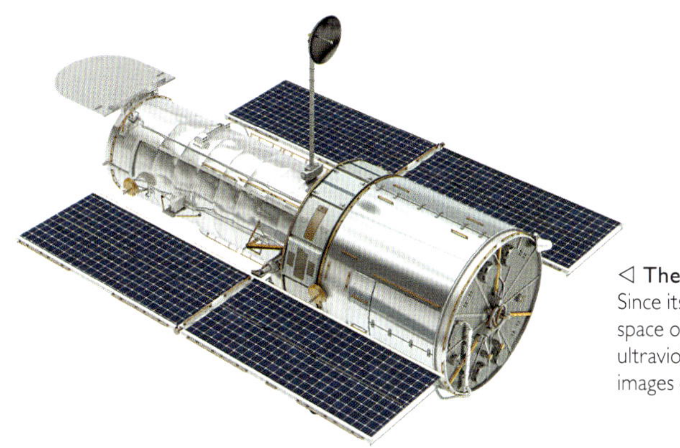

◁ **The Hubble Space Telescope**
Since its launch in 1990, Hubble has operated as a space observatory, exploring the universe in visible, ultraviolet, and infrared wavelengths, and capturing images of stars and galaxies billions of light years away.

topography of the Atlantic Ocean floor, and followed this in 1977 with a complete world map of the ocean floors. In 1960, Swiss engineer Jacques Piccard and US Navy lieutenant Don Walsh made the historic journey to the oceans' deepest point, the Mariana Trench, in the bathyscaphe *Trieste*.

Outer space

On 8 September 1944, Germany deployed the world's first long-range ballistic missile, the V2 rocket. A few months earlier, it had become the first artificial object to reach outer space when the rocket accidentally headed vertically off its launch site. From this military beginning emerged the technology to power spacecraft and satellites into space, and the USA and the Soviet Union entered a Cold War "space race". In 1957, the Soviets launched the first space satellite, *Sputnik 1*, and, in 1961, Russian Lt Yuri Gagarin became the first human to orbit Earth in *Vostok 1*. Eight years later, in 1969, the US *Apollo 11* mission landed men on the Moon.

Within a few decades, robot spacecraft visited all the planets in the Solar System and, soon after, telescopes in space peered into the distance to see galaxies formed more than 13 billion years ago, soon after the Universe began. In the 21st century, the affordability of space technology has widened the scope of human activity in space. More countries are joining the list of spacefaring nations, and commercial spaceflight is even available to private individuals – as long as they can afford the fare.

"Life... will presently stand upon this Earth as upon a footstool, and stretch out its realm amidst the stars."

H.G.WELLS, *THE OUTLINE OF HISTORY*, 1920

▷ ***Apollo 15*, August 1971**
Astronauts in the Mission Control Center in Houston, Texas, direct the United States' fourth successful Moon landings. On the screen, astronauts David R. Scott and James B. Irwin are seen performing tasks on the surface of the Moon.

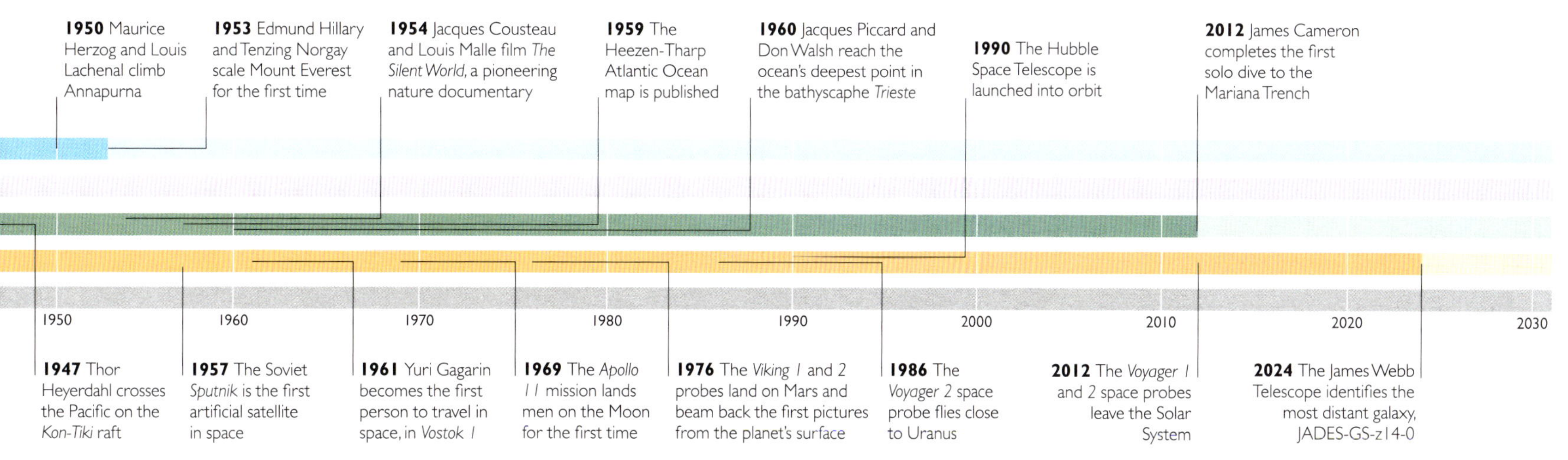

1947 Thor Heyerdahl crosses the Pacific on the *Kon-Tiki* raft
1950 Maurice Herzog and Louis Lachenal climb Annapurna
1953 Edmund Hillary and Tenzing Norgay scale Mount Everest for the first time
1954 Jacques Cousteau and Louis Malle film *The Silent World*, a pioneering nature documentary
1957 The Soviet *Sputnik* is the first artificial satellite in space
1959 The Heezen-Tharp Atlantic Ocean map is published
1960 Jacques Piccard and Don Walsh reach the ocean's deepest point in the bathyscaphe *Trieste*
1961 Yuri Gagarin becomes the first person to travel in space, in *Vostok 1*
1969 The *Apollo 11* mission lands men on the Moon for the first time
1976 The *Viking 1* and *2* probes land on Mars and beam back the first pictures from the planet's surface
1986 The *Voyager 2* space probe flies close to Uranus
1990 The Hubble Space Telescope is launched into orbit
2012 James Cameron completes the first solo dive to the Mariana Trench
2012 The *Voyager 1* and *2* space probes leave the Solar System
2024 The James Webb Telescope identifies the most distant galaxy, JADES-GS-z14-0
1950
1960
1970
1980
1990
2000
2010
2020
2030

BOTANICAL EXPLORATIONS

In the early 20th century, many intrepid plant hunters travelled the world in search of specimens that were new to scientific study. Two explorers in particular, Frank Kingdon-Ward and Ynés Mexia, went to extreme efforts to amass extraordinary botanical collections.

△ **Ynés Mexia, 1920s**
Ynés Mexia began to study botany in 1921. She collected plant specimens from all over the Americas, often solo or with a few Indigenous guides.

Starting in 1911, over a career that lasted 50 years, the British botanist Frank Kingdon-Ward went on around 24 expeditions, hunting for plants in Tibet, China, and Southeast Asia. On his first expedition, he collected 200 Tibetan species, including 22 that were new to science. Often braving extreme hardship and hazards, he brought back to Britain the first viable seeds for *Primula florindae* (giant cowslip, named after his first wife, Florinda) and the pale yellow-flowered *Rhododendron wardii*, named after him. However, his favourite discovery was the Himalayan blue poppy, which inspired his 1913 book *The Land of the Blue Poppy*, which documents his travels in Eastern Tibet.

On the other side of the world, the Mexican botanist Ynés Mexia conducted many plant-hunting expeditions between 1925 and 1938, in Mexico, Central and South America, and Alaska. Her most famous trek took her to the Amazon River, where she canoed around 4,800km (3,000 miles) from its delta to its source in the Andes. The journey took two and a half years, during which she spent three months living with the Aguaruna people and learning about their plant knowledge.

Based in California, Mexia joined the nascent environmental conservation movement and became an early campaigner for the protection of natural habitats against industrial expansion.

YNÉS MEXIA'S PLANT COLLECTION

Although Ynés Mexia didn't begin her professional career until the age of 55, in just 13 years (1925–38) she collected an astonishing 150,000 specimens, and discovered 500 new species and two new genera: *Mexianthus mexicanus* B.L. Rob. and *Spumula quadrifida*. Her collection is so vast that plant scientists are still working through it today, but she has already had 50 plants named in her honour.

Delostoma integrifolium from Ecuador, 1934

Searching for specimens in Laos, 1929
The British botanist Frank Kingdon-Ward was known for his intrepid plant hunting in the remote regions of Tibet, China, India, and Myanmar. This photograph shows his truck being rowed over a river by local people, on a platform balanced on two Indigenous wooden canoes.

A JOURNEY TO LHASA

From the mid-18th century, when all Christian missionaries were expelled from Tibet at the insistence of the Buddhist monks, the country had been largely closed to outsiders. Yet in 1924, Alexandra David-Néel embarked on a Buddhist pilgrimage in disguise and became the first European woman to reach the sacred "forbidden city" of Lhasa.

Born Louise Eugénie Alexandrine Marie David in Paris in 1868 to a French father and a Belgian mother, Alexandra David-Néel led an unconventional life from a young age. Aged 21, she converted to Buddhism, and in 1891, she journeyed to India to study Sanskrit.

In 1911, David-Néel left Paris and embarked on an extended 14-year sojourn in Asia. She spent many years in the Himalayas, close to the Tibetan border, and crossed into the country several times. While living in a cave as a Buddhist anchorite, she met a young monk called Aphur Yongden, who became her travel companion and eventually her adopted son.

In 1923, David-Néel resolved to "show what the will of a woman can achieve", undertaking a pilgrimage with Yongden from the Gobi Desert to the holy city of Lhasa. Tibet was then closed to all foreigners, except for the British, who had invaded it in 1904. To make the journey, David-Néel pretended to be Yongden's mother and smeared her face with soot to look like a beggar. Leaving behind possessions except for a compass, pistol, and purse hidden under rags, they travelled mainly on foot and at night, to avoid detection. They navigated without maps through the treacherous snow and ice of the Trans-Himalayas, and eventually reached the Tibetan capital, Lhasa, where they managed to stay in disguise for two months, until their identities were revealed and they were forced to go back to Northern India.

David-Néel returned to France with Yongden in March 1925, penniless but a national celebrity, and published an account of her travels, *My Journey to Lhasa*, in 1927.

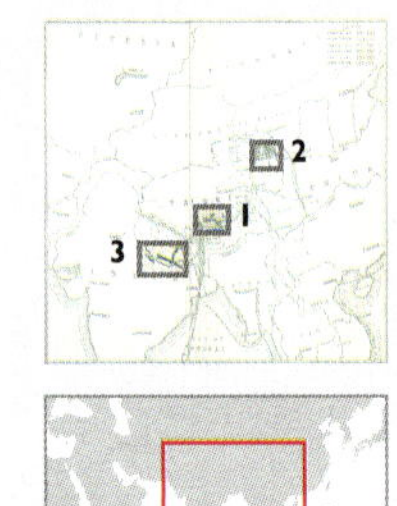

KEY

1 In February 1924, David-Néel became the first European woman to reach the Tibetan city of Lhasa.

2 David-Néel and Yongden lived at the Kumbum Monastery (1918–1921), where she translated the sacred Buddhist *Prajnaparamita* text.

3 In 1973, David-Néel's ashes and those of Yongden were scattered on the Ganges River near Benares.

"How happy I was to be there, en route for the mystery of these unexplored heights, alone in the great silence."

ALEXANDRA DAVID-NÉEL, *MY JOURNEY TO LHASA*, 1927

DESERT TRAVELS

In the early 20th century, a number of Western women, including Gertrude Bell and Freya Stark, travelled in Southwest Asia (the Middle East). Bell undertook six major expeditions (1900–12), on horseback and camelback, through the deserts of present-day Israel-Palestine, Lebanon, Syria, Jordan, Türkiye, Iraq, and Iran. In the 1930s, Stark became the first Westerner to reach a number of southern Arabian desert regions and, conducting geographical and archeological research, trekked to the heart of Iran's Valley of the Assassins.

Stark (centre) in Syria in 1928

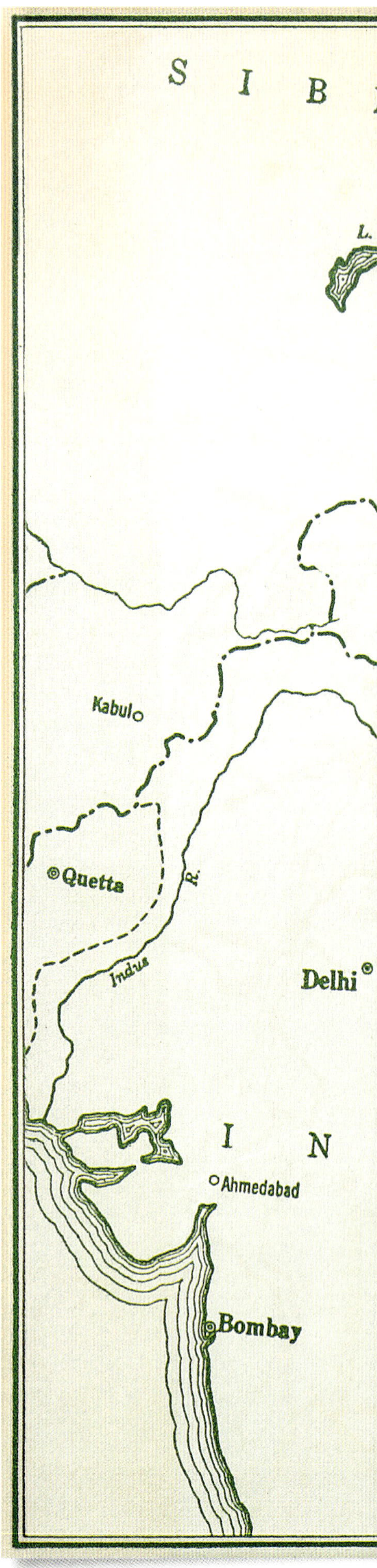

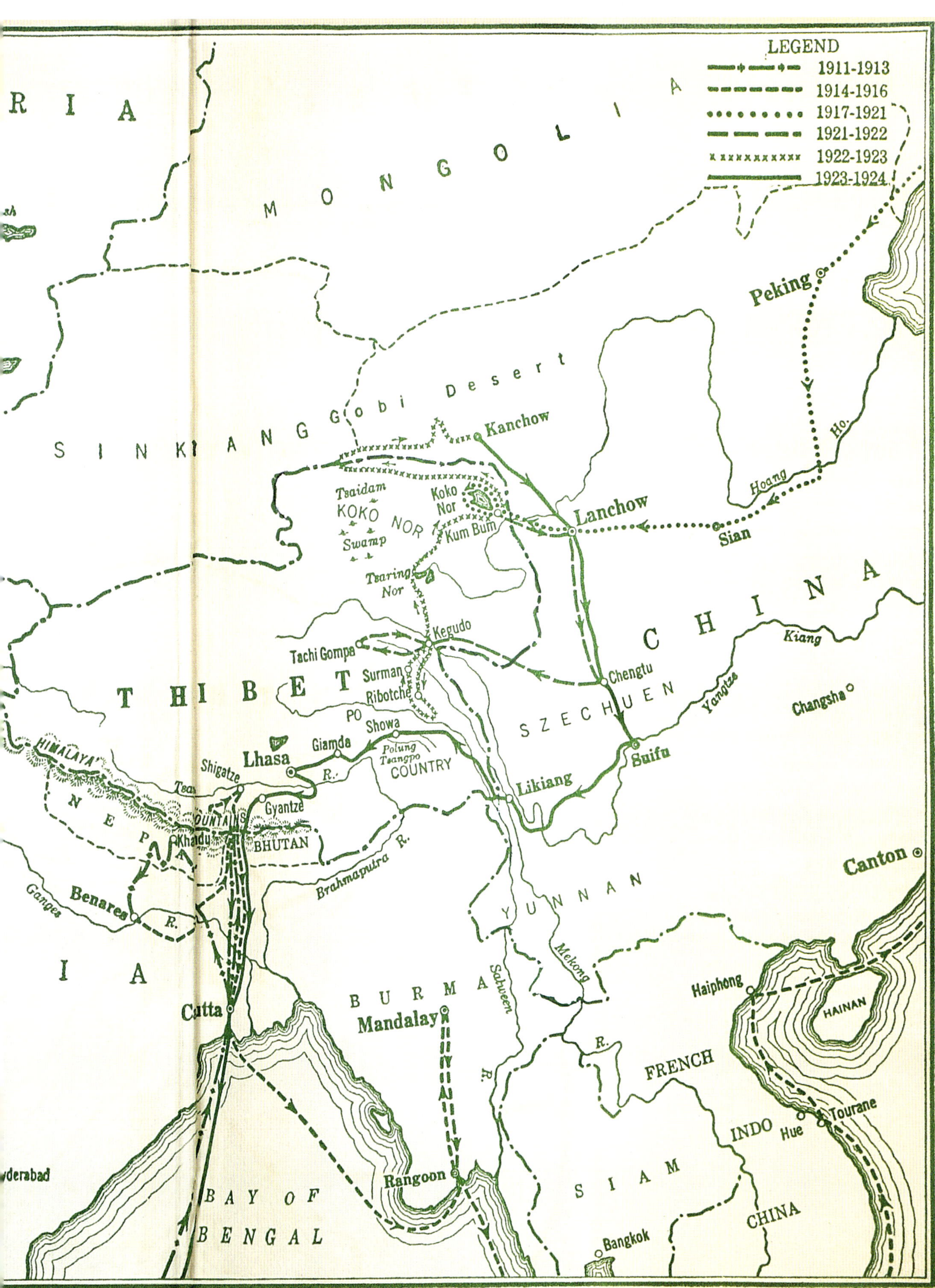

◁ **Journeys in Central Asia**
This map, included in the first edition of David-Néel's 1927 travelogue, shows the six significant journeys (1911–1924) she made into Tibet and surrounding regions in Nepal and China – culminating in her historic journey, disguised as a beggar, to the "forbidden city" of Lhasa.

A lost explorer in the Amazon
In 1951, Orlando Villas-Bôas, aided by the Kalapalos people, believed he had found the remains of the explorer, Colonel Percy Fawcett, who had vanished in the jungle in 1925. However, bone analysis disproved this and the skeleton remains a mystery.

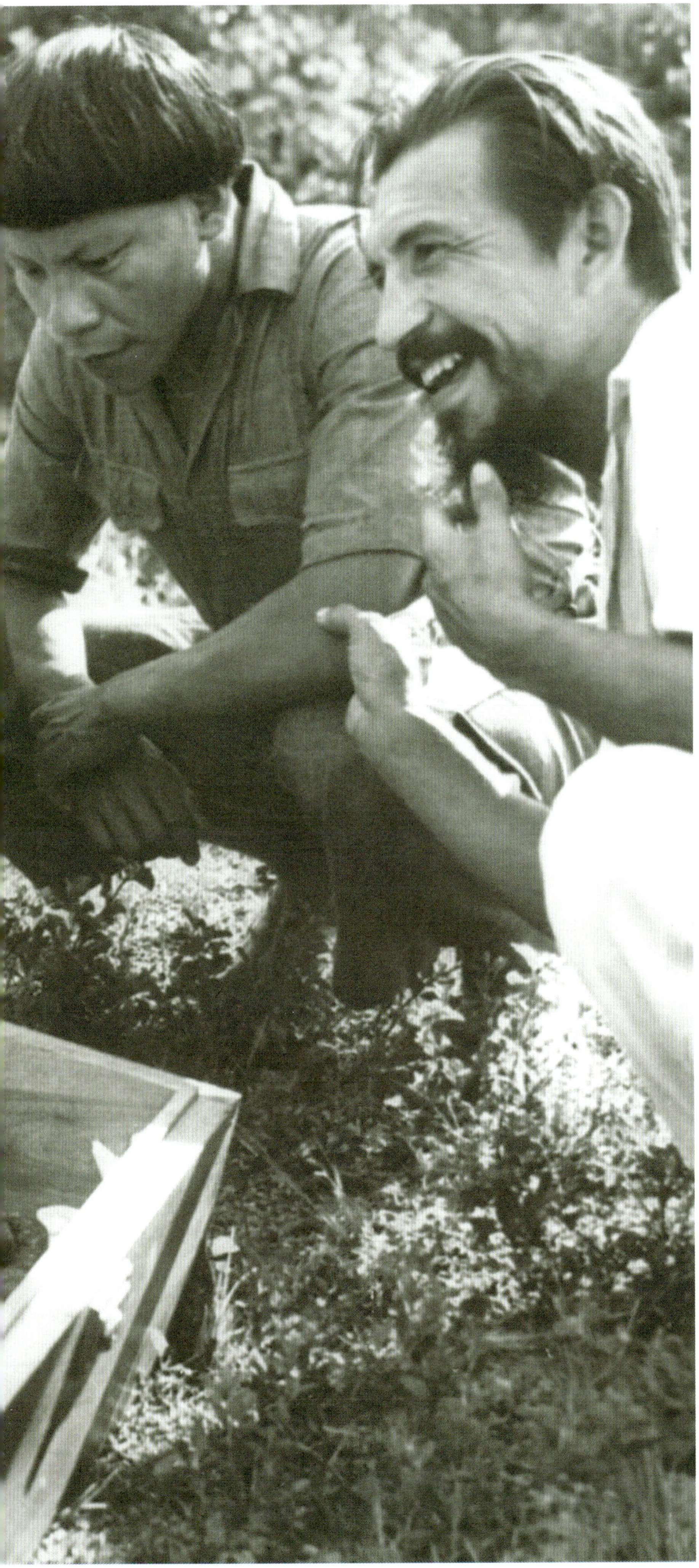

CHARTING THE RAINFOREST

In the early 20th century, the dense Amazon rainforest of central Brazil and the Mato Grosso remained largely unknown to the wider world. However, in the 1930s, the Brazilian government decided to change this.

△ **Shaman mask**
This 20th-century mask, made of feathers and mother-of-pearl shell, is used in ceremonial dances by the Tapirapé, one of the Xingu peoples.

At the start of Brazil's Estado Novo (Third Republic) in 1937, the government declared the "March to the West", an initiative to develop the country's interior. In 1941, the Roncador-Xingu expedition began a 20-year exploration of central Brazil, cutting trails, mapping rivers, and creating airstrips, roads, and three dozen towns. Among the volunteers on the expedition were three young Brazilian brothers – Orlando, Cláudio, and Leonardo Villas-Bôas. They were entranced by the beauty of the rainforest and by the 11 Indigenous groups they encountered living along the banks of the Xingu River, who had experienced no previous contact with Westerners.

The brothers saw that the Xingu peoples would have no protection from this exposure and decided to become their defenders. Forming friendships, they negotiated with them to allow the expedition to enter their lands and presented the Xingu peoples to the rest of the country as a key part of national culture. They also campaigned to protect vast areas of rainforest for the peoples living there. Orlando Villas-Bôas was well aware, however, that the "opening up" of the area would change the world of the Xingu peoples forever, as quoted in his obituary in 2002: "Each time we contact a tribe, we are contributing to the destruction of what is most pure in it."

THE XINGU UNDER THREAT

Xingu people such as the Panará have lived near the Xingu River for millennia, and have close connections with the rainforest's fragile ecosystem. Despite the efforts of the Villas-Bôas brothers and the creation of a Xingu protected reserve, their existence remains under threat due to illegal mining and deforestation.

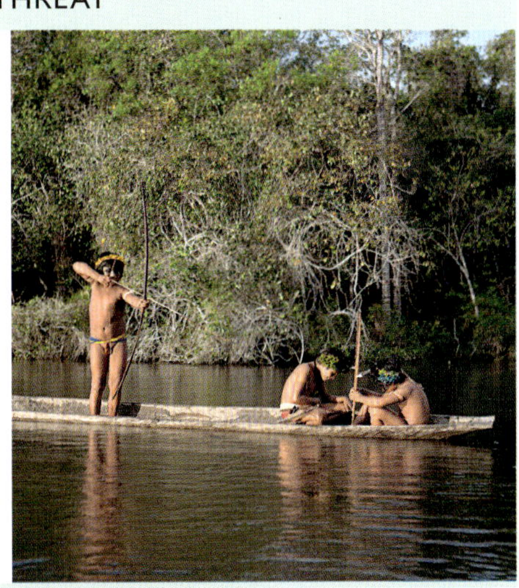

Panará children at play in 1998

ANCIENT VOYAGES REVISITED

The remarkable voyages across the Pacific Ocean made by ancient Polynesian sailors, in canoes that ranged up to 50 m (165 ft) in length, have proved inspirational to modern-day navigators. Various attempts have been made to re-enact these journeys, proving the seaworthiness of rafts and canoes, and the efficacy of Polynesian techniques of navigation and landfinding.

In 1947, Swedish adventurer Thor Heyerdahl and crew crossed the Pacific from Peru on a balsawood raft named *Kon-Tiki*. Heyerdahl had been inspired by 16th-century Spanish navigator Bartolomé Ruiz's descriptions of Indigenous people piloting balsawood rafts carrying goods along the coast of Ecuador, and proposed that people may have sailed west from the South American coast. However, although the sweet potato, native to South America, appears in eastern Polynesia and New Zealand it is most likely that the Polynesians (see pp. 24–25) made contact with South America by sailing east, as Indigenous Americans were not oceanic explorers.

Kon-Tiki's epic voyage helped to develop a new way of understanding the past, through experimental archaeology. It fuelled interest in the capabilities of ancient seafarers and inspired many imitators. The first of these, American sailor William Willis, made a solo journey on a balsawood raft in 1954, then five years later, in 1959, Czech adventurer Eduard Ingris repeated the feat on his raft, *Kantuta II*. In 1973, a new distance record was set when Spaniard Vital Alsar departed from Ecuador on three balsawood rafts, with a crew of 12 sailors, reaching the coast of Australia 161 days later.

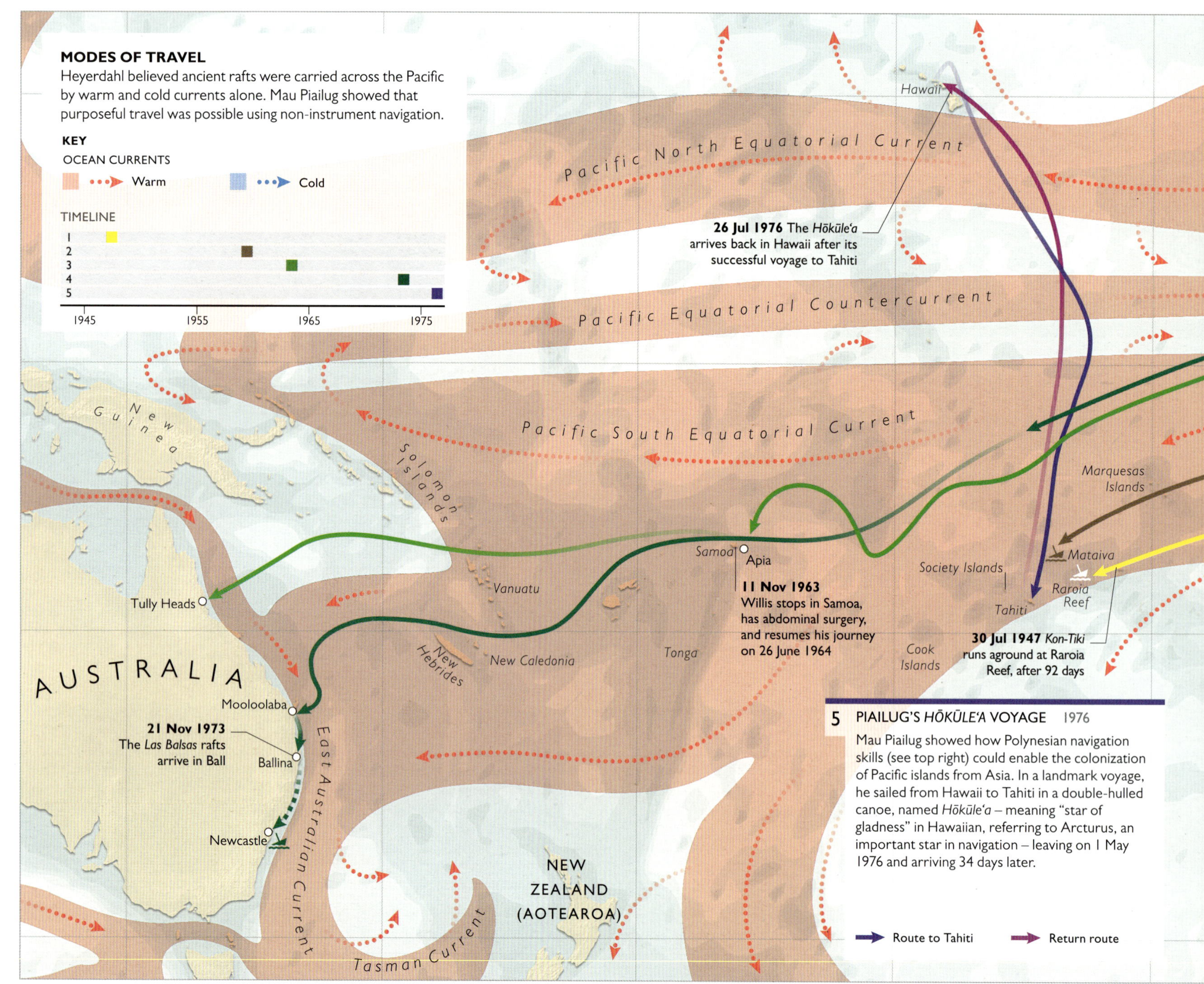

These raft-borne journeys were made by passively drifting on ocean currents from east to west, using sextants and short-wave radio for navigation. But in 1976, Micronesian navigator Mau Piailug proved purposeful journeys through the Pacific could be made using non-instrument wayfinding methods (see box). He set out in the *Hokule'a*, a reconstruction of a double-hulled Hawaiian voyaging canoe, on a voyage north to south. Measuring 19m (62ft) in length, with twin masts and a long paddle for steering, the canoe was capable of speeds of up to six knots (11 km/h, or 7 mph). Piailug navigated by using the rising and setting of the sun and stars, reading signs such as ocean swells and flight patterns of birds to detect land over the horizon, and keeping a mental record of distance travelled, speed, drift, and currents.

> *"Too many men are afraid of the sea. But I am a navigator."*
>
> MAU PIAILUG, *THE LAST NAVIGATOR*, 1987

WAYFINDING

Wayfinding is the ancient Polynesian art of navigating (see p.25) using environmental cues as guides. Navigators who know where the stars rise and set on the celestial equator can use the horizon as a "star compass", which helps them to memorize knowledge of the night sky. Wayfinding skills are passed down through the generations, but, after canoe travel in the Pacific islands was banned by European colonizers, it had declined until revived by Mau Piailug, a master navigator from the Carolinian island of Satawal (north of New Guinea).

Master navigator Piailug weaving fibres

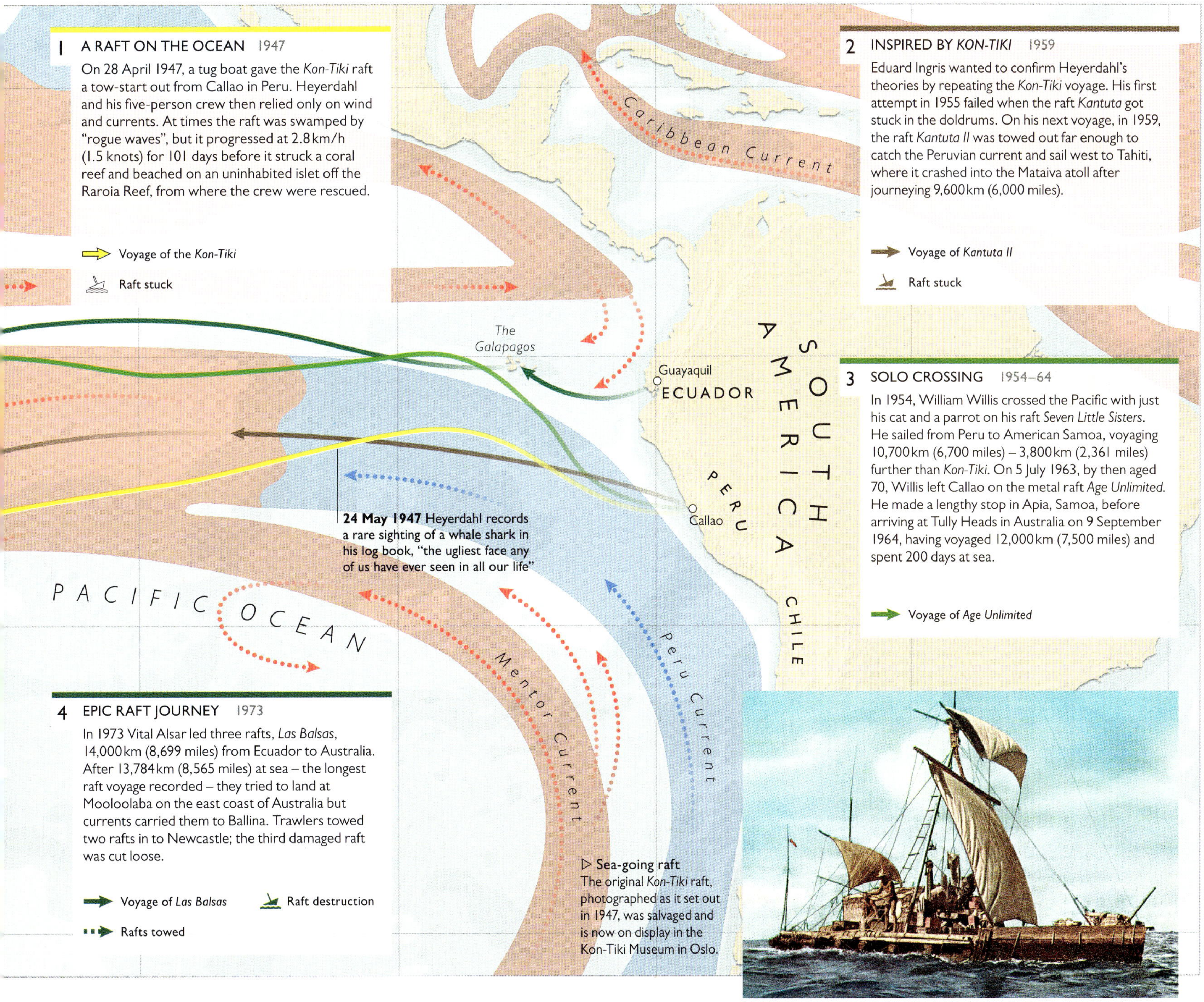

1 A RAFT ON THE OCEAN 1947

On 28 April 1947, a tug boat gave the *Kon-Tiki* raft a tow-start out from Callao in Peru. Heyerdahl and his five-person crew then relied only on wind and currents. At times the raft was swamped by "rogue waves", but it progressed at 2.8km/h (1.5 knots) for 101 days before it struck a coral reef and beached on an uninhabited islet off the Raroia Reef, from where the crew were rescued.

Voyage of the *Kon-Tiki*

Raft stuck

2 INSPIRED BY *KON-TIKI* 1959

Eduard Ingris wanted to confirm Heyerdahl's theories by repeating the *Kon-Tiki* voyage. His first attempt in 1955 failed when the raft *Kantuta* got stuck in the doldrums. On his next voyage, in 1959, the raft *Kantuta II* was towed out far enough to catch the Peruvian current and sail west to Tahiti, where it crashed into the Mataiva atoll after journeying 9,600km (6,000 miles).

Voyage of *Kantuta II*

Raft stuck

3 SOLO CROSSING 1954–64

In 1954, William Willis crossed the Pacific with just his cat and a parrot on his raft *Seven Little Sisters*. He sailed from Peru to American Samoa, voyaging 10,700km (6,700 miles) – 3,800km (2,361 miles) further than *Kon-Tiki*. On 5 July 1963, by then aged 70, Willis left Callao on the metal raft *Age Unlimited*. He made a lengthy stop in Apia, Samoa, before arriving at Tully Heads in Australia on 9 September 1964, having voyaged 12,000km (7,500 miles) and spent 200 days at sea.

Voyage of *Age Unlimited*

24 May 1947 Heyerdahl records a rare sighting of a whale shark in his log book, "the ugliest face any of us have ever seen in all our life"

4 EPIC RAFT JOURNEY 1973

In 1973 Vital Alsar led three rafts, *Las Balsas*, 14,000km (8,699 miles) from Ecuador to Australia. After 13,784km (8,565 miles) at sea – the longest raft voyage recorded – they tried to land at Mooloolaba on the east coast of Australia but currents carried them to Ballina. Trawlers towed two rafts in to Newcastle; the third damaged raft was cut loose.

Voyage of *Las Balsas*

Raft destruction

Rafts towed

▷ **Sea-going raft**
The original *Kon-Tiki* raft, photographed as it set out in 1947, was salvaged and is now on display in the Kon-Tiki Museum in Oslo.

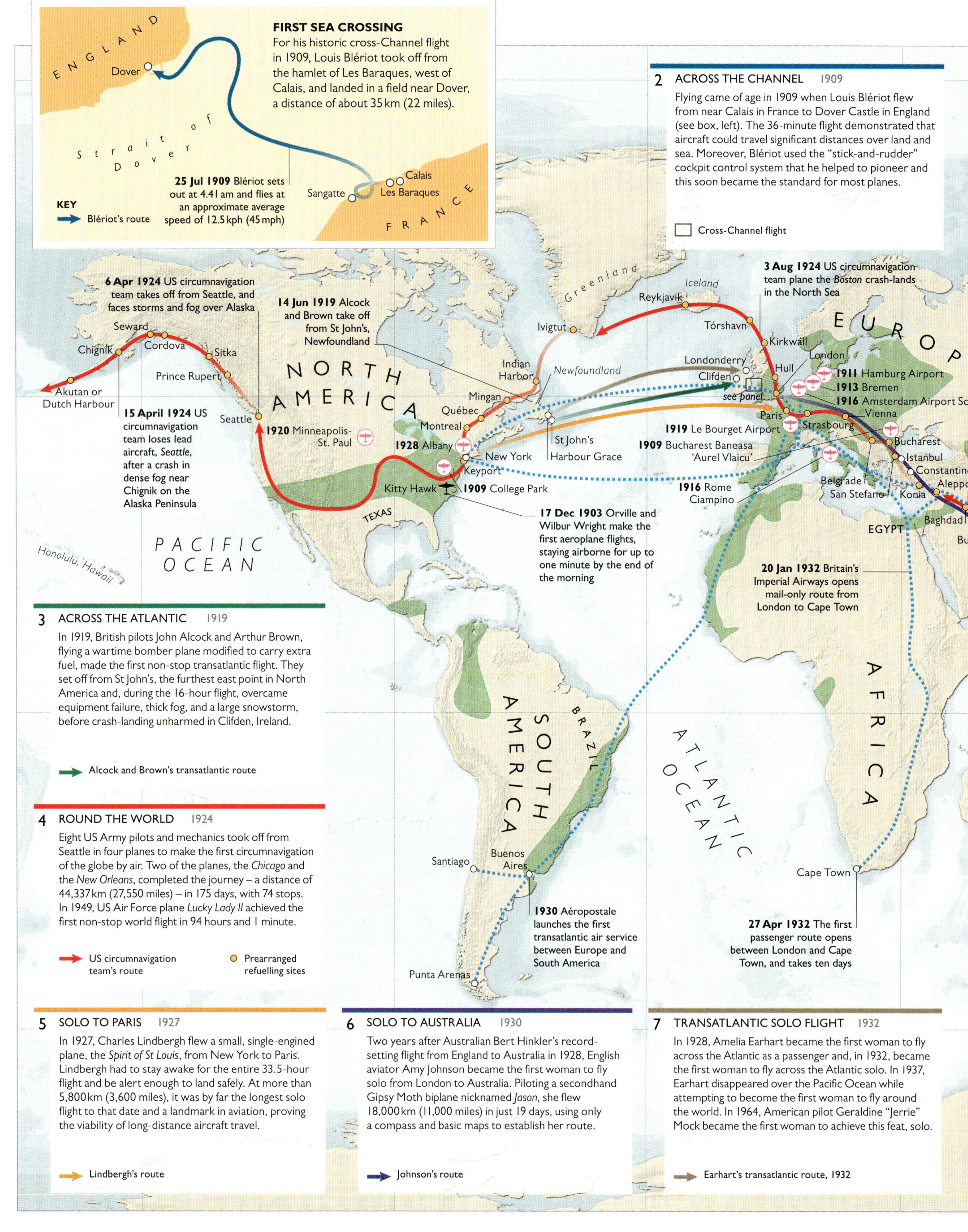

FIRST SEA CROSSING
For his historic cross-Channel flight in 1909, Louis Blériot took off from the hamlet of Les Baraques, west of Calais, and landed in a field near Dover, a distance of about 35 km (22 miles).
ENGLAND
Dover
Strait of Dover
25 Jul 1909 Blériot sets out at 4.41 am and flies at an approximate average speed of 12.5 kph (45 mph)
Sangatte
Calais
Les Baraques
FRANCE
KEY
Blériot's route
6 Apr 1924 US circumnavigation team takes off from Seattle, and faces storms and fog over Alaska
Seward
Cordova
Chignik
Sitka
Prince Rupert
Akutan or Dutch Harbour
15 April 1924 US circumnavigation team loses lead aircraft, Seattle, after a crash in dense fog near Chignik on the Alaska Peninsula
Seattle
14 Jun 1919 Alcock and Brown take off from St John's, Newfoundland
NORTH AMERICA
1920 Minneapolis-St. Paul
1928 Albany
Montreal
Québec
Mingan
Indian Harbor
Greenland
Ivigtut
Iceland
Reykjavik
Tórshavn
3 Aug 1924 US circumnavigation team plane the Boston crash-lands in the North Sea
Kirkwall
Newfoundland
Londonderry
Clifden
see panel
London
Hull
EUROPE
1911 Hamburg Airport
1913 Bremen
1916 Amsterdam Airport Schip
Paris
1919 Le Bourget Airport
Strasbourg
Vienna
Bucharest
1909 Bucharest Baneasa 'Aurel Vlaicu'
Istanbul
Constantinop
Aleppo
Belgrade
San Stefano
Konia
Baghdad
Bush
EGYPT
1916 Rome Ciampino
New York
Keyport
St John's
Harbour Grace
Kitty Hawk
1909 College Park
TEXAS
17 Dec 1903 Orville and Wilbur Wright make the first aeroplane flights, staying airborne for up to one minute by the end of the morning
PACIFIC OCEAN
Honolulu, Hawaii
20 Jan 1932 Britain's Imperial Airways opens mail-only route from London to Cape Town
AFRICA
SOUTH AMERICA
BRAZIL
ATLANTIC OCEAN
Santiago
Buenos Aires
1930 Aéropostale launches the first transatlantic air service between Europe and South America
Punta Arenas
Cape Town
27 Apr 1932 The first passenger route opens between London and Cape Town, and takes ten days

2 ACROSS THE CHANNEL 1909

Flying came of age in 1909 when Louis Blériot flew from near Calais in France to Dover Castle in England (see box, left). The 36-minute flight demonstrated that aircraft could travel significant distances over land and sea. Moreover, Blériot used the "stick-and-rudder" cockpit control system that he helped to pioneer and this soon became the standard for most planes.

Cross-Channel flight

3 ACROSS THE ATLANTIC 1919

In 1919, British pilots John Alcock and Arthur Brown, flying a wartime bomber plane modified to carry extra fuel, made the first non-stop transatlantic flight. They set off from St John's, the furthest east point in North America and, during the 16-hour flight, overcame equipment failure, thick fog, and a large snowstorm, before crash-landing unharmed in Clifden, Ireland.

Alcock and Brown's transatlantic route

4 ROUND THE WORLD 1924

Eight US Army pilots and mechanics took off from Seattle in four planes to make the first circumnavigation of the globe by air. Two of the planes, the *Chicago* and the *New Orleans*, completed the journey – a distance of 44,337 km (27,550 miles) – in 175 days, with 74 stops. In 1949, US Air Force plane *Lucky Lady II* achieved the first non-stop world flight in 94 hours and 1 minute.

US circumnavigation team's route

Prearranged refuelling sites

5 SOLO TO PARIS 1927

In 1927, Charles Lindbergh flew a small, single-engined plane, the *Spirit of St Louis*, from New York to Paris. Lindbergh had to stay awake for the entire 33.5-hour flight and be alert enough to land safely. At more than 5,800 km (3,600 miles), it was by far the longest solo flight to that date and a landmark in aviation, proving the viability of long-distance aircraft travel.

Lindbergh's route

6 SOLO TO AUSTRALIA 1930

Two years after Australian Bert Hinkler's record-setting flight from England to Australia in 1928, English aviator Amy Johnson became the first woman to fly solo from London to Australia. Piloting a secondhand Gipsy Moth biplane nicknamed *Jason*, she flew 18,000 km (11,000 miles) in just 19 days, using only a compass and basic maps to establish her route.

Johnson's route

7 TRANSATLANTIC SOLO FLIGHT 1932

In 1928, Amelia Earhart became the first woman to fly across the Atlantic as a passenger and, in 1932, became the first woman to fly across the Atlantic solo. In 1937, Earhart disappeared over the Pacific Ocean while attempting to become the first woman to fly around the world. In 1964, American pilot Geraldine "Jerrie" Mock became the first woman to achieve this feat, solo.

Earhart's transatlantic route, 1932

△ **The Wright Flyer, 1908**
In August 1908, Orville Wright travelled to Fort Myer, Virginia to demonstrate his new flying machine (Wright Model A) to the US Government and US Army.

1 FIRST FLIGHT 1903

In 1903, Wilbur and Orville Wright launched aviation with the world's first successful aeroplane, the *Flyer*, which was powered by a petrol engine linked by chain to two propellers and controlled by the Wright brothers' ingenious "wing-warp" cables. Orville made the historic first flight at Kill Devil Hills, near Kitty Hawk, North Carolina, taking off and landing safely.

Site of first flight

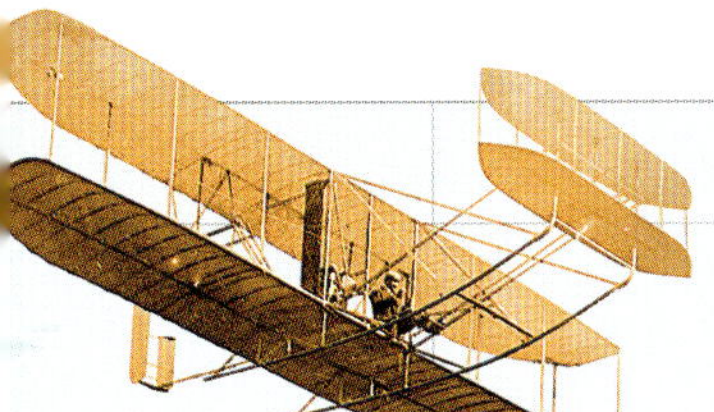

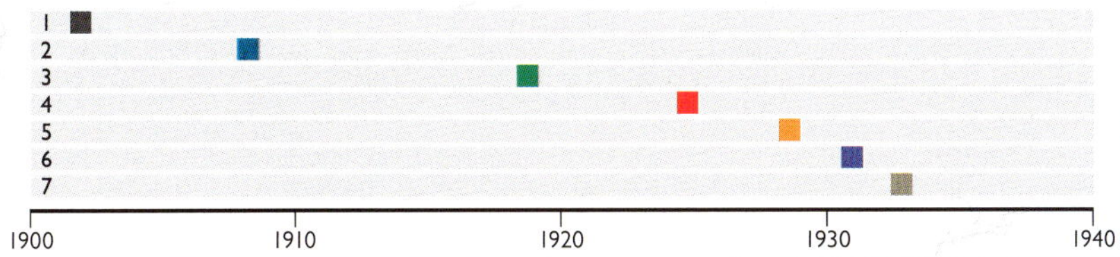

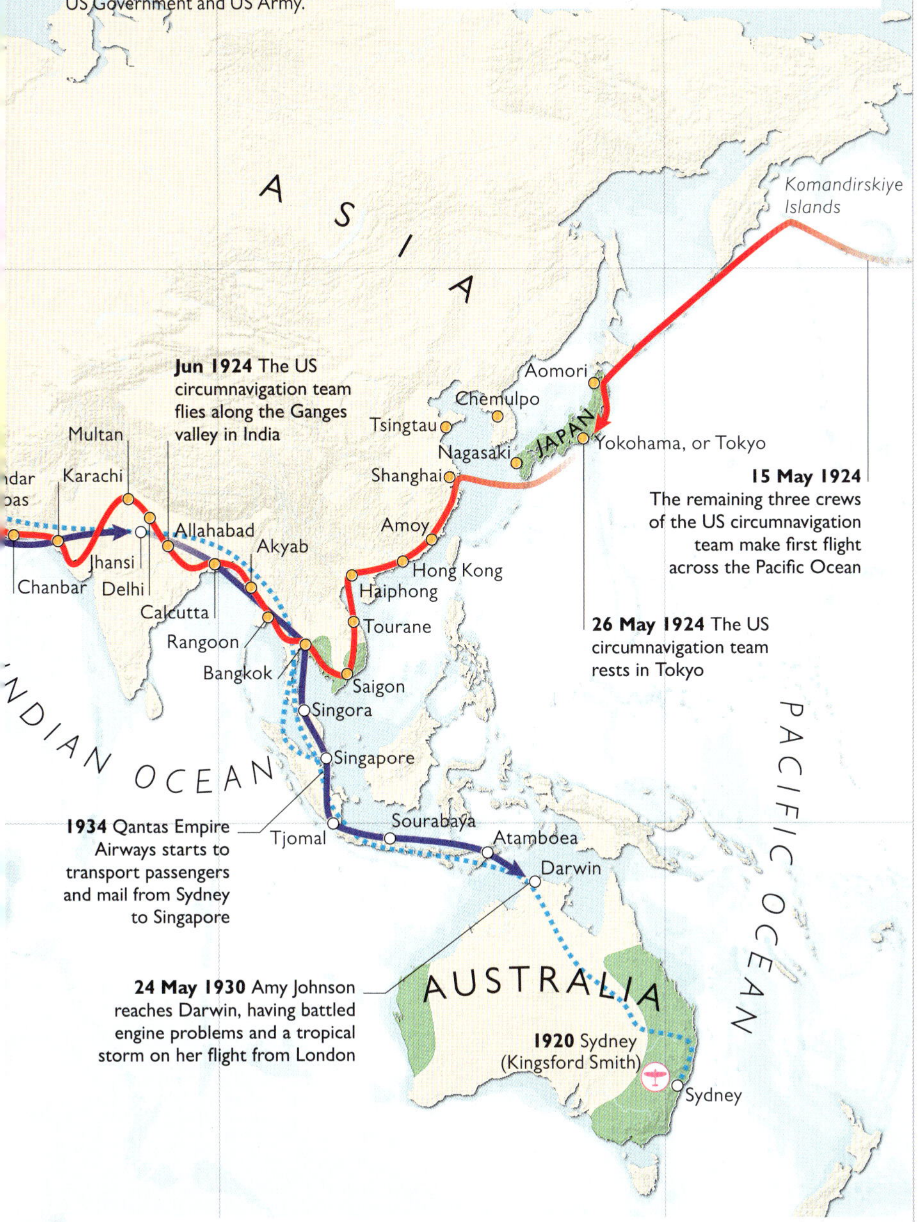

ASIA
INDIAN OCEAN
PACIFIC OCEAN
AUSTRALIA
JAPAN
Komandirskiye Islands
Aomori
Chemulpo
Tsingtau
Nagasaki
Yokohama, or Tokyo
Shanghai
Amoy
Hong Kong
Haiphong
Tourane
Saigon
Bangkok
Rangoon
Calcutta
Akyab
Allahabad
Jhansi
Delhi
Chanbar
Karachi
Multan
Singora
Singapore
Tjomal
Sourabaya
Atamboea
Darwin
Sydney
Jun 1924 The US circumnavigation team flies along the Ganges valley in India
15 May 1924 The remaining three crews of the US circumnavigation team make first flight across the Pacific Ocean
26 May 1924 The US circumnavigation team rests in Tokyo
1934 Qantas Empire Airways starts to transport passengers and mail from Sydney to Singapore
24 May 1930 Amy Johnson reaches Darwin, having battled engine problems and a tropical storm on her flight from London
1920 Sydney (Kingsford Smith)

FLYING THE WORLD
In 1919, 11 passengers boarded the first regular international passenger air service, from London to Paris. By 1925, many regions had scheduled flights. Intercontinental flights were punctuated by frequent refuelling stops.

KEY
- Areas covered by commercial air routes, 1925
- Oldest airports still operational
- Intercontinental air routes, 1930s

TIMELINE

1
2
3
4
5
6
7
1900
1910
1920
1930
1940

PIONEERS ON THE WING

In 1903, US aviation pioneers the Wright brothers made the world's first aeroplane flight. Within a few years, planes could fly long distances without stopping and soon pilots, and even commercial passengers, were flying around the world.

In the late 19th century, pioneers experimented with petrol-engine-powered flying machines and gliders. But crashes were frequent without the crucial element of control, until the Wright brothers had the idea of wing warping to roll the aircraft – twisting the wings with cables – which allowed their plane to safely bank and turn. The brothers' first flight in 1903 covered 36 m (120 ft) and lasted just 12 seconds. A year later, they achieved the first circle manoeuvre and, by 1905, were making flights of up to 39 minutes at a time. Wing warping also allowed French flyer Louis Blériot to make the first flight across the English Channel in 1909.

Technology and skills advanced enormously in World War I. In 1919, the first regular international passenger service flew between London and Paris, and John Alcock and Arthur Brown flew non-stop across the Atlantic. New records were set for distance and speed, including the first aerial navigation of the globe by a team of US Army pilots and mechanics in 1924, and the first solo non-stop Atlantic crossing by US aviator Charles Lindbergh in 1927. French pilot Raymonde de Laroche became the first licensed female pilot in 1910 and, in the early 1930s, Amy Johnson and Amelia Earhart achieved pioneering solo flights.

> *"Flying may not be all plain sailing, but the fun of it is worth the price."*
>
> AMELIA EARHART, *THE FUN OF IT*, 1932

FLYING BOATS

In the first half of the 20th century, few places in the world had airports, so airlines used flying boats, which were able to land on stretches of water. Flying boats enabled passengers to splash down anywhere from Venice and Cairo to New York and Hong Kong. Flying boats, such as the Boeing 314 Clipper operated by US airline Pan Am, were some of the largest aircraft of their age.

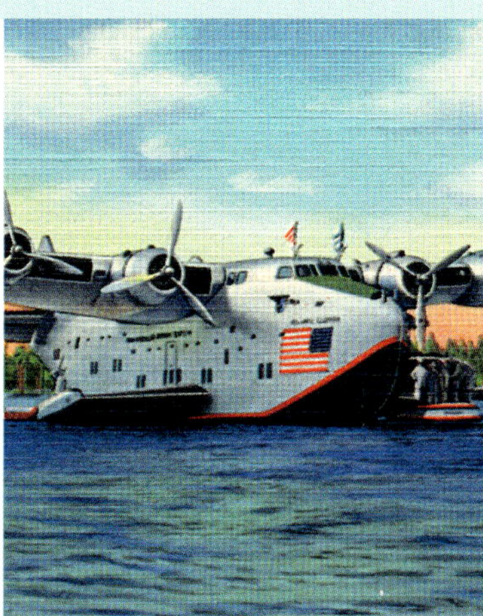

Illustration of Pan Am Clipper, 1940

DEEP-SEA EXPLORATIONS

French diver Jacques Cousteau was the first famous underwater explorer. Although he was not a trained ocean scientist, his love of the sea and his inventiveness fuelled a lifetime of exploration, starting in the early 1940s.

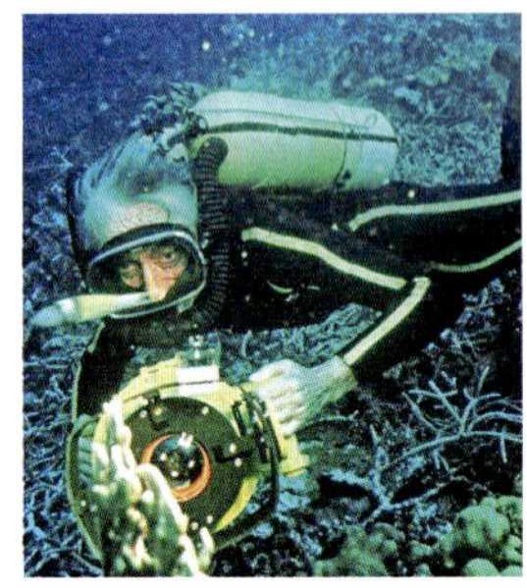

△ **Ocean explorer**
Cousteau pictured diving, aged 74, during the filming of a 1984 TV documentary series.

In 1943, Cousteau and French engineer Émile Gagnan devised the first aqualung to allow divers to swim freely underwater for long periods (this was later dubbed SCUBA after a similar device created by US diving expert Christian Lambertsen). In 1950 Cousteau converted an old Royal Navy minesweeper into a floating research ship called *Calypso*. Among its innovations was an extended underwater prow with portholes, designed for the close study and filming of aquatic life. *Calypso* also carried a small submarine (right) that could explore depths of up to 350 m (1,150 ft). On 24 November 1951 *Calypso* made its first voyage from Toulon in France to the Red Sea, between Africa and Asia, to study corals. Cousteau's book *The Silent World* and film of the same title, issued in 1953 and 1956 respectively, were based on his experiences and research into previously unknown fauna and flora. He produced many more books and films over the next 40 years.

"... man has only to sink beneath the surface and he is free."

JACQUES COUSTEAU, *TIME* MAGAZINE, 1960

UNDERWATER HABITATS

Cousteau was a pioneer of "underwater habitats", the space stations of the ocean, and created Conshelf in 1962. Others include MarineLab, which stayed underwater in a lagoon from 1985 to 2018 in Key Largo, Florida, and the Aquarius Reef Base, which has sat 19 m (62 ft) below the surface off Key Largo (right) since 1993.

Aquarius Reef Base in 2009

Going down
Jacques Cousteau stands inside the SP-350, onboard the *Calypso*, in New York Harbor in 1959. Dubbed the "diving saucer", the SP-350 was a two-man submarine created by Cousteau and engineer Jean Mollard for scientific exploration.

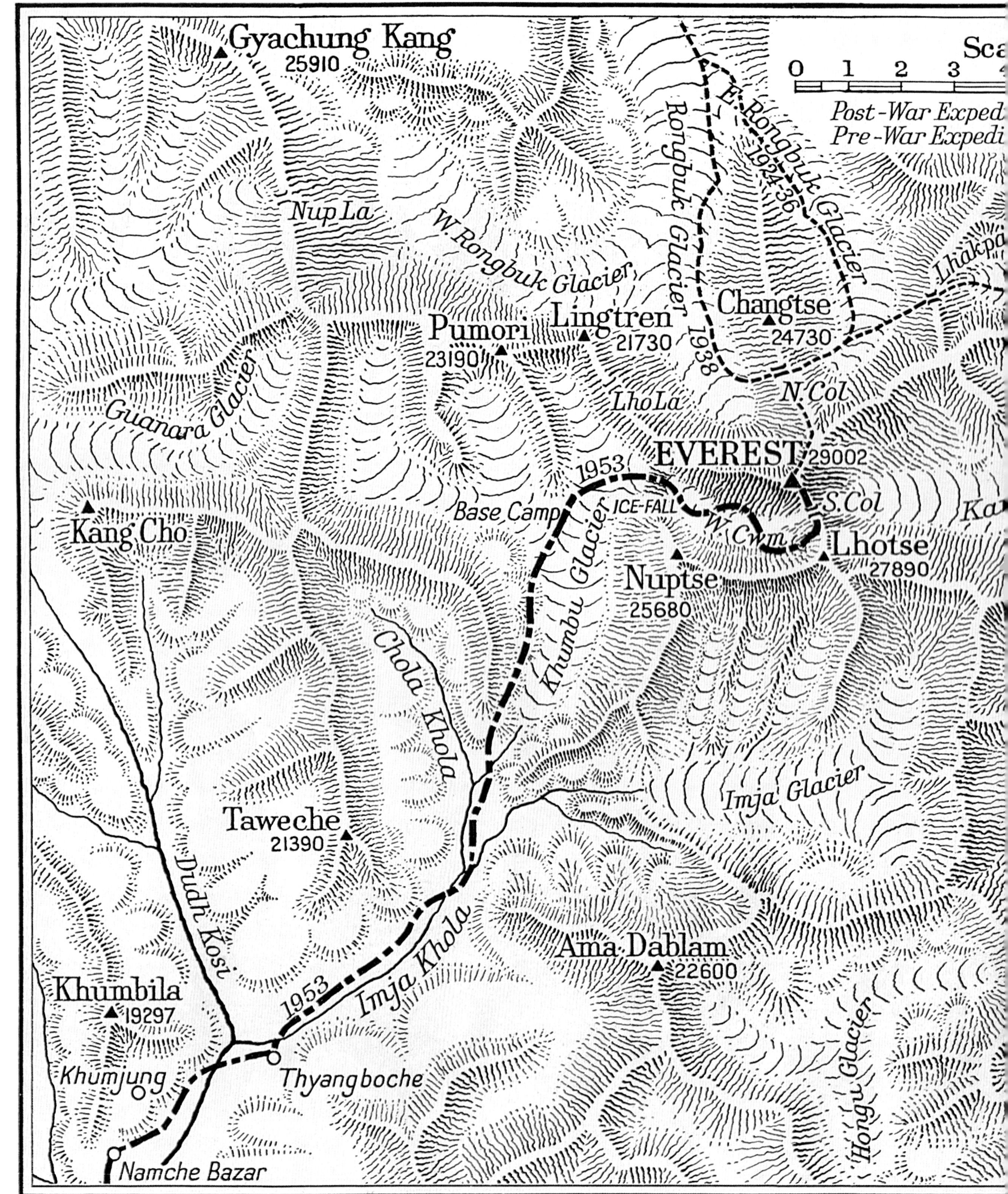
Gyachung Kang
25910
Nup La
W. Rongbuk Glacier
Rongbuk Glacier
E. Rongbuk Glacier
1924-36
1938
Changtse
24730
Lingtren
21730
Pumori
23190
Guanara Glacier
Lho La
N. Col
EVEREST
29002
1953
Base Camp
ICE-FALL
W. Cwm
S. Col
Lhotse
27890
Nuptse
25680
Kang Cho
Khumbu Glacier
Chola Khola
Imja Glacier
Taweche
21390
Dudh Kosi
Ama Dablam
22600
Imja Khola
1953
Khumbila
19297
Khumjung
Thyangboche
Hongu Glacier
Namche Bazar
Post-War Exped
Pre-War Expedi
0 1 2 3

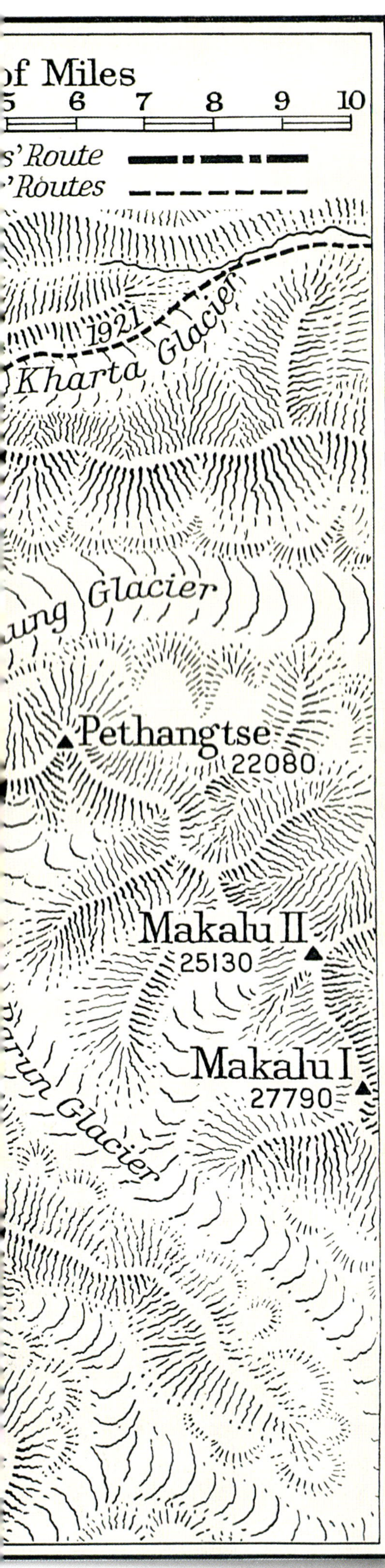

CLIMBING EVEREST

The idea of scaling Mount Everest – at 8,848.86 m (29,031.7 ft), the world's highest peak – was mooted in the 1900s by British Indian Army officers Francis Younghusband and Charles Bruce, but permission for an attempt was refused. It would be around 50 years before the first successful expedition reached the summit, in 1953.

The first organized attempt to reach Everest's summit was made by British mountaineer George Mallory. After two failed attempts in 1921 and 1922, Mallory may have reached the summit in 1924. He and his climbing partner Andrew Irvine were last sighted close to the peak. Mallory's body was recovered in 1999 at 8,155 m (26,760 ft), but Irvine's body was never found.

Three major expeditions – in 1951, 1952, and 1953 – followed, leading to a breakthrough. The final expedition was led by Nepali-Indian Sherpa Tenzing Norgay and New Zealander Edmund Hillary. After a hazardous climb, the pair pushed on to the top, reaching it at 11:30am on 29 May 1953. A British journalist who had climbed to Camp IV wired the story to *The Times* newspaper in code and the news of the first official scaling of the summit became a global sensation.

◁ Routes to Everest
Climbers reach Everest either from the north in Tibet and climb the mountain's North Col; or from the south side in Nepal, via the South Col – the route taken by Hillary and Norgay, shown here.

LOCATOR

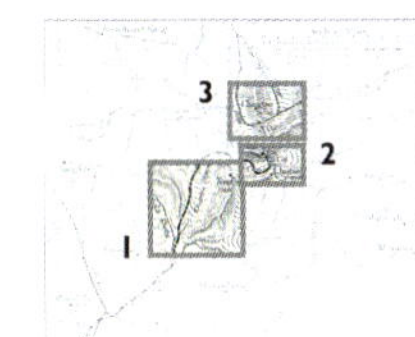

KEY

1 In 1953, Hillary and Norgay approach Everest from the southeast, via the Khumbu Glacier valley.

2 Hillary and Norgay's team then scale the South Col, setting up base camp on the peak.

3 In 1938, a British team led by Bill Tilman reaches high up the North Col, the lower peak.

Since then, over 6,000 people have scaled Everest, many more than once. In 1975, Junko Tabei from Japan became the first woman to reach the top, and in 2022, the first team comprising only Black climbers, Full Circle, reached the summit. However, success is not guaranteed, and more than 300 people have died on the mountain since the 1920s. Local Sherpas remain essential guides for those new to Everest, who rely on the Sherpas' climbing skills and mountain knowledge for survival.

> *"It is 50 to one against us but... we'll do ourselves proud."*
>
> GEORGE MALLORY'S FINAL LETTER TO HIS WIFE RUTH BEFORE DISAPPEARING, 1924

A WELL-EARNED CUP OF TEA

On the morning of 29 May 1953, Tenzing Norgay and Edmund Hillary left base camp, which was set up at 8,500 m (27,900 ft). They manoeuvred over a ridge and started climbing the perilous 17-m (55-ft) spur of icy rock now called the Hillary Step – the final section of the climb. Norgay and Hillary shook hands on the summit and took photos. A Buddhist, Norgay laid an offering of food, while Hillary left behind a crucifix. After 15 minutes, they headed back down to base camp for a cup of tea.

Norgay and Hillary pictured after their descent, 1953

SCALING THE GREAT PEAKS

By the early 20th century, the world's highest mountain ranges were some of the last unexplored regions of the Earth. Many climbers set their sights on scaling them, especially Europeans and Americans, who were often guided by local Sherpas.

△ **Benham's boots**
Gertrude Benham, the first woman to scale Mount Kilimanjaro, wore these Tibetan boots on many of her climbs worldwide.

In 1936, Bill Tilman and Noel Odell from an American–British team reached the top of Nanda Devi, a 7,816-m (25,643-ft) peak in Northern India, once thought the highest in the world. In 1950, French mountaineers Maurice Herzog and Louis Lachenal scaled the notoriously difficult Annapurna – the world's 10th-highest mountain at 8,091 m (26,545 ft) – marking the first time a mountaineer had summited a peak over 8,000 m.

There are now agreed to be 14 peaks over 8,000 m ("eight-thousanders"), all in the Himalayas or Karakoram. The first to climb them all was Italian Reinhold Messner, who did so between 1972 and 1986. With modern climbing equipment and techniques, the challenge now is to climb all 14 in as short a time as possible. In 2019, British-Nepalese climber Nirmal Purja climbed them all in just six months and six days, with the help of oxygen. On 27 July 2023, Norwegian Kristin Harila and Nepalese Tenjen Sherpa narrowed the time frame to just 92 days, although the death of porter Muhammad Hassan during their ascent of K2 (the world's second-highest mountain) led to a high-profile investigation into porters' conditions and equipment.

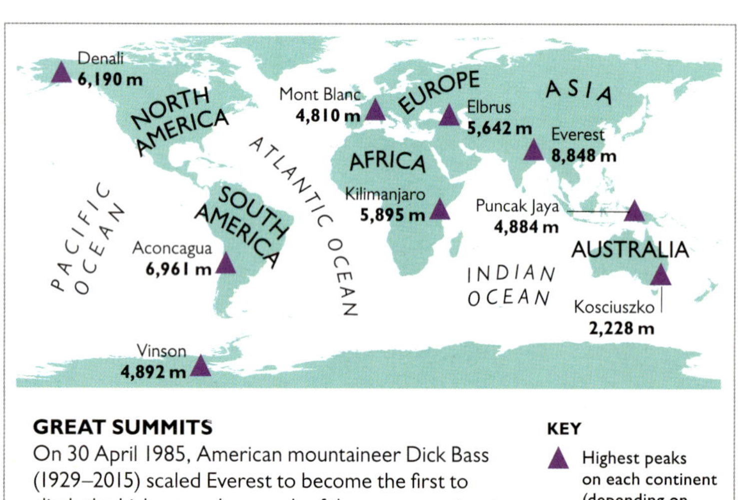

GREAT SUMMITS
On 30 April 1985, American mountaineer Dick Bass (1929–2015) scaled Everest to become the first to climb the highest peak on each of the seven continents. Since then more than 400 have succeeded, though there is debate over which peaks should be included.

KEY
▲ Highest peaks on each continent (depending on how continental boundaries are defined)

View from the top
This 1936 photograph captures British climber Bill Tilman taking in the view near the summit of Nanda Devi in India (the world's 23rd-highest peak). Tilman described the expedition in his book *The Ascent of Nanda Devi* (1937).

INTO THE DEPTHS

By 2012, 12 people had set foot on the Moon, but just three had reached the deepest point of the ocean – the trench of Challenger Deep in the Pacific Ocean. In 2019, the arrival of a new submergence vehicle allowed more to follow.

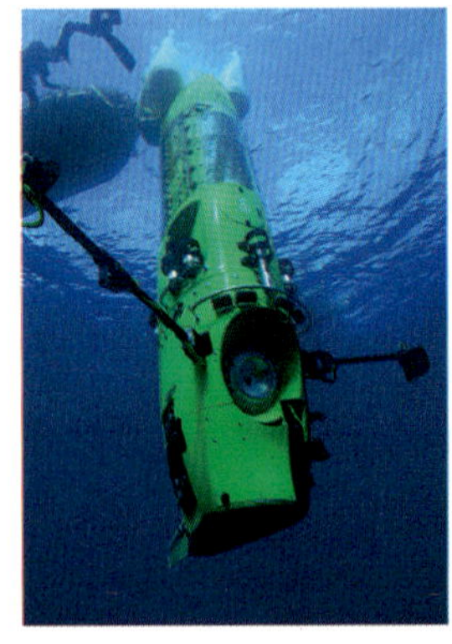

△ *Deepsea Challenger*
Cameron's vessel, made of light, superstrong glass foam, was so cramped he couldn't move his arms. This design allowed for a faster descent than the *Trieste*.

Swiss oceanographer Jacques Piccard and US Navy officer Don Walsh reached the deepest point on 23 January 1960 in the bathyscaphe *Trieste*. After a five-hour descent, the vessel touched down at a spot 10,916 m (35,814 ft) deep. The explorers spent less than 20 minutes at depth, measuring the temperature, currents, and radioactivity, and observing the marine life.

In 2012, *Titanic* film director and deep-sea explorer James Cameron descended alone in *Deepsea Challenger*. Cameron co-designed the vessel with Australian submarine designer Ron Allum. *Deepsea Challenger*'s descent took 2 hours and 37 minutes, reaching a point at which the ocean floor was 10,908 m (35,787 ft) deep. Cameron spent three hours filming and collecting samples, then took less than an hour to resurface. His films show fish surviving at these depths, despite the pressure and cold.

Piccard, Walsh, and Cameron were alone in making it to the bottom of the ocean until the arrival on the scene of the deep-submergence vehicle (DSV) *Limiting Factor* (now named *Bakunawa)* in 2019. Between 2019 and 2024, 24 more people ventured down, including retired American naval officer Victor Vescovo – the first person ever to have reached the highest point of every continent and the deepest point in all five oceans. In April 2019, Vescovo touched down at a world-record depth of 10,928 m (35,853 ft) in the trench of Challenger Deep.

DEEP-SEA FISH

In 1960, Piccard described spotting flatfish at the bottom of Challenger Deep. In 2022, *Pseudoliparis belyaevi*, a species of deep-sea snailfish, were caught in traps in Izu-Ogasawara Trench, off Japan, at 8,022 m (26,319 ft) deep, but they had died by the time that they reached the surface.

A juvenile deep-sea snailfish

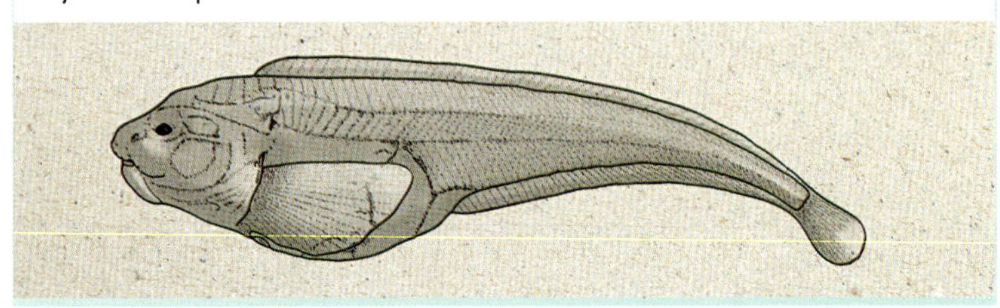

Going down
This 1960 photograph shows the *Trieste* being lowered into the water. It had two main elements: a crew cabin and a float. The bathyscaphe would descend as the float's air tanks filled with water.

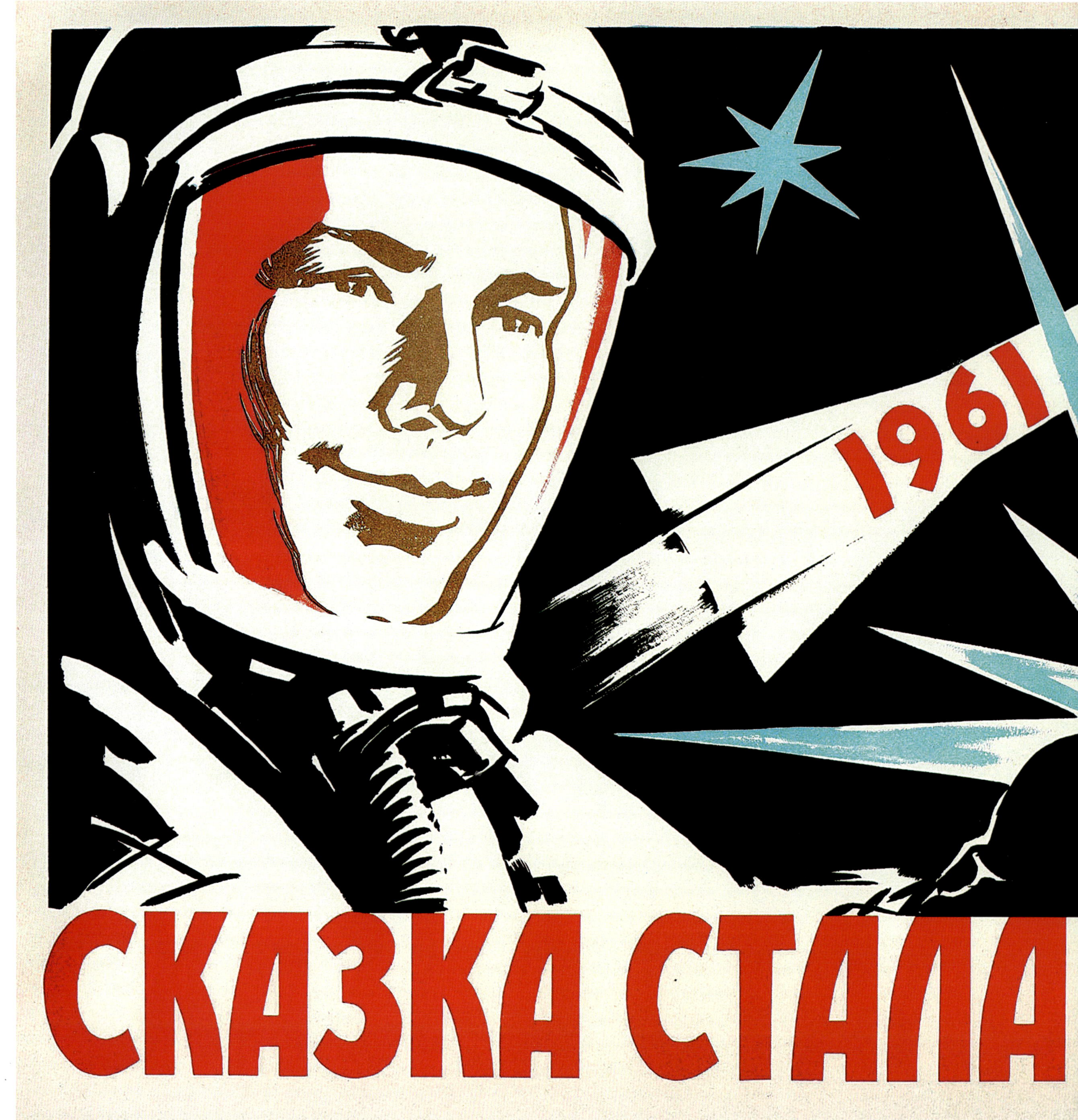

"Orbiting Earth in the spaceship, I saw how beautiful our planet is. People, let us preserve and increase this beauty, not destroy it!"

YURI GAGARIN, HANDWRITTEN NOTE ON AN AUTOGRAPH CARD, 1961

△ **Space hero**
A 1961 Soviet poster celebrates Yuri Gagarin's successful first orbit of the Earth, declaring "The fairytale has come true." The 25-year-old quickly became an international celebrity and is depicted here as Prometheus – giving fire to humankind. However, Gagarin never went into space again.

HUMANS IN SPACE

On 12 April 1961, Soviet cosmonaut Yuri Gagarin became the first human in space when he was launched into Earth's orbit by a ballistic missile in his capsule Vostok 1 for a history-making spaceflight. US astronauts repeated the feat soon afterwards, as space became the new frontier.

△ **Lift off!**
Ballistic missiles lift the Vostok 1 capsule from the Baikonur launch pad, propelling Soviet cosmonaut Yuri Gagarin towards Earth orbit on 12 April 1961.

Yuri Gagarin's spherical capsule was blasted into space at 9:07am Moscow time from the Baikonur Cosmodrome in present-day Kazakhstan. Within minutes, the single-person spacecraft was hurtling nearly 322 km (200 miles) above the Earth at almost 27,400 km per hour (300 miles per minute). The flight took 108 minutes from launch to landing. Throughout, the Vostok capsule flew gradually lower as it lost momentum and gravity hauled it in. Gagarin himself had no control over the Vostok, and computers steered the falling capsule as it hurtled earthwards. The descent was slowed by retrorocket firing, and about 6.4 km (4 miles) up, Gagarin ejected from the spacecraft and landed by parachute, although official news at the time said he had landed in his capsule.

On 5 May 1961, just 23 days after Gagarin's historic flight, American astronaut Alan Shepard – who in 1971 would walk on the Moon as commander of Apollo 14 – was launched into space. It was only a short, 15-minute suborbital flight, reaching an altitude of 187 km (116 miles), but space fever had begun. Three weeks later, on 25 May, President John F. Kennedy announced to a special session of the US Congress his intention that an American would walk on the Moon within the decade. His ambition was realized in July 1969 with the Apollo 11 mission (see pp. 230–31).

A PIONEERING COSMONAUT

The US and the Soviet Union (USSR) competed for milestones in space travel. In 1962, the USSR selected five women for space training. Originally, two women were to perform co-orbiting flights in Vostok 5 and 6 in 1963. Ultimately, however, the USSR opted for a man and a woman, and a male cosmonaut, Valeri Bykovsky, flew Vostok 5. On 16 June 1963, 26-year-old Valentina Tereshkova piloted Vostok 6, the first woman to complete a space mission.

Soviet cosmonaut Valentina Tereshkova

THE APOLLO MISSIONS

On 20 July 1969, watched on television by millions of people, two US astronauts, Neil Armstrong and Edwin "Buzz" Aldrin, stepped down from their spacecraft to become the first human beings to walk on the Moon. NASA's Apollo 11 mission would be followed by decades of lunar activity.

It took three days for Apollo 11 to reach the Moon's orbit after launching from Cape Kennedy, Florida. While Michael Collins piloted the command module *Columbia* around the Moon, Armstrong and Aldrin descended in the small lunar module *Eagle* to the Moon's surface, and Armstrong said these famous words: "That's one small step for [a] man, one giant leap for mankind." After two and a half hours of walking on the Moon and collecting 21.6 kg (47.6 lb) of lunar samples, the pair launched *Eagle* to rejoin *Columbia*. Almost three days later, *Columbia* splashed safely down on Earth.

Over the next three years, five Apollo missions and 10 astronauts journeyed to the Moon's surface. In November 1969, astronauts Charles "Pete" Conrad Jr., Alan Bean, and Richard Gordon survived two lightning strikes during liftoff in Apollo 12, and Conrad and Bean stepped on the Moon 1,426 km (886 miles) west of Apollo 11. In 1970, Apollo 13 suffered an on-board explosion 56 hours into its flight and the three-person crew got back to Earth only by using the lunar module as a lifeboat.

The Apollo programme ended in 1972, but missions to the Moon since then include Russia's probe Luna 24 in 1976 and China's robotic spaceship Chang'e 4 in 2019.

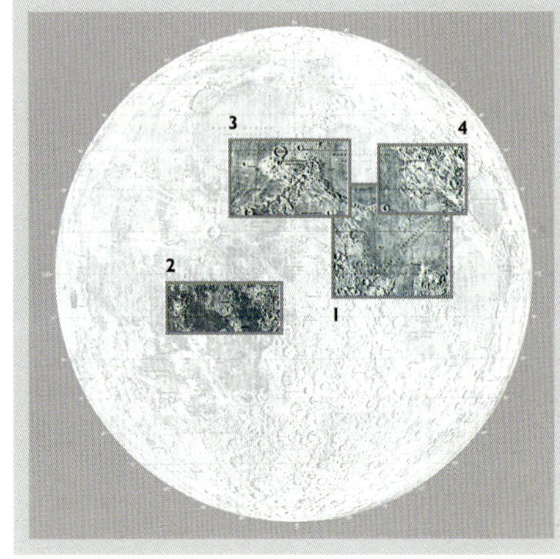

▷ **Lunar surface**
From the 17th century with the invention of telescopes, the Moon's surface has been mapped. High-powered telescopes allow it to be charted in great detail. This Lunar Reference Mosaic Map of the Moon's Earthside was first published in 1962.

▽ **Sea of Tranquility**
Before the 1969 Apollo 11 landing on the Moon's Sea of Tranquility, the uncrewed Ranger 8 and Surveyor 5 lunar probes landed here in 1965 and 1967 respectively. The photos they sent back provided crucial information.

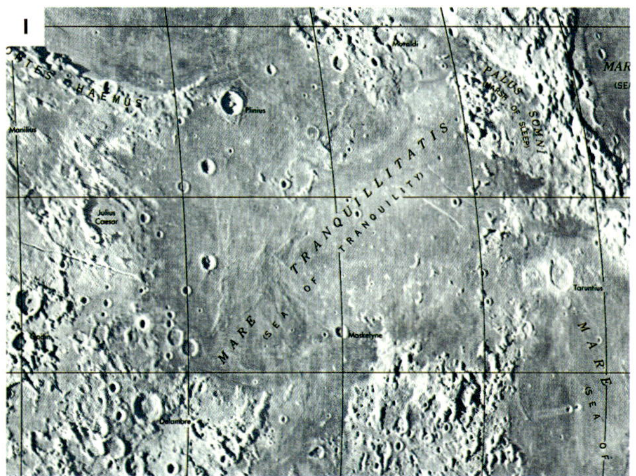

△ **Apollo 14**
In February 1971, Apollo 14 astronauts Alan Shepard and Edgar Mitchell conducted two Moon walks, or EVAs (Extravehicular activities), in the Fra Mauro highlands south of the Sea of Rains. While on one such walk, Shepard famously hit two golf balls with a makeshift club.

▽ **Apollo 15**
In July 1971, Apollo 15 touched down at the foot of the Apennine mountain range. The crew used a vehicle that came to be called the moon buggy. This craft enabled David Scott and James Irwin to travel further from the module to collect samples.

▷ **Apollo 17**
This mission, in 1972, was the last to land people on the Moon, touching down in the Taurus-Littrow highlands and valley area. The lunar samples collected by Eugene Cernan and Harrison Schmitt are still revealing new insights more than half a century after they were brought back.

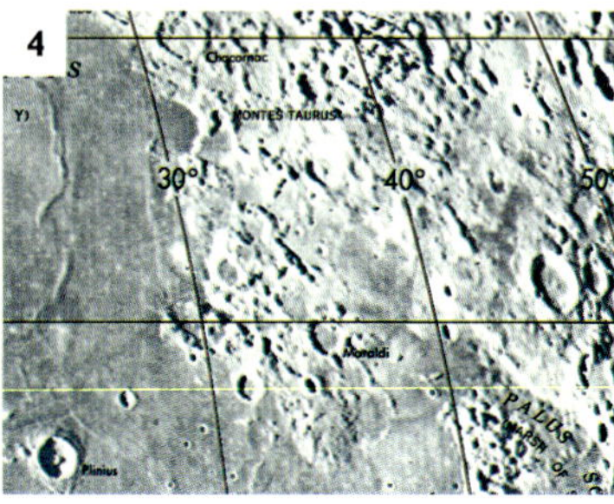

LUNAR SEAS

Mare Tranquillitatis, or the Sea of Tranquility, where the Apollo 11 lunar module landed in 1969, was named in 1651 by Italian astronomers Francesco Grimaldi and Giovanni Battista Riccioli. Maria are the large, dark patches on the Moon's surface. Early astronomers called these areas maria (Latin for "seas") because they thought they might be bodies of water. However, there are no seas on the Moon's surface. Maria are lowland plains that were created by lava flows following asteroid impacts long ago.

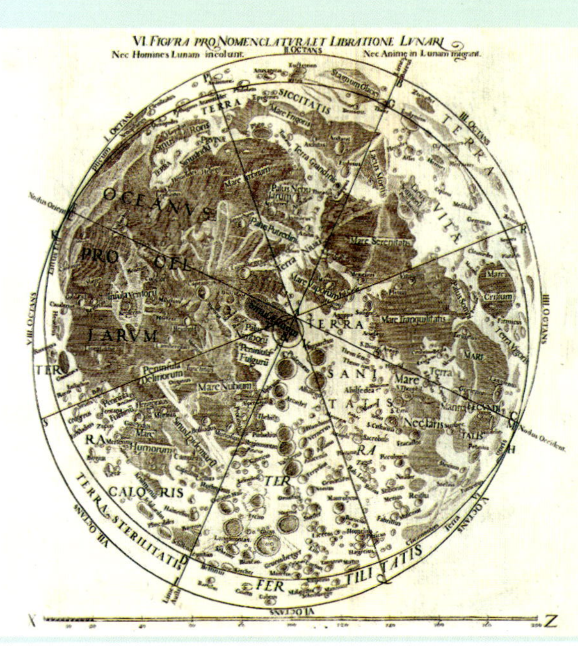

Lunar map by Giovanni Riccioli, 17th century

"... it's been a long way, but we're here."

ALAN SHEPARD, *APOLLO 14 LUNAR SURFACE JOURNAL*, 1971

◁ **Mars rover**
Landed on Mars on 18 February 2021, NASA's car-sized *Perseverance* rover is based on the successful design of the earlier *Curiosity* rover.

ROVING ON MARS

While spacecraft have flown past all of our Solar System's planets, they have landed only on Mars and, briefly, Venus. Mars has been the target of more space missions than anywhere else except the Moon, with remote-controlled robotic vehicles, called rovers, deployed to explore its surface.

In 1965, the US *Mariner 4* probe sent back the first close-up pictures of another planet when it made a flyby of Mars. In 1971, Soviet missions to land a robotic rover on the planet ended in failure: *Mars 2* crash-landed on the surface and *Mars 3* ceased transmissions 20 seconds after landing. In 1976, though, the US *Viking 1* and *Viking 2* probes landed and beamed back the first pictures from the surface. The planet Mars is the subject of intense curiosity, because it is similar enough to Earth for scientists to believe that it may harbour signs of microscopic life.

Surface of Mars
An image from *Mars Pathfinder* in 1997 shows the rocky landscape and the *Sojourner* robot vehicle next to a large rock, dubbed Yogi. Made of basalt, it indicates past volcanic activity here.

In 1997, the US *Mars Pathfinder* lander arrived on Mars complete with its *Sojourner* rover. *Sojourner* was the first wheeled vehicle to rove on a planet other than Earth. It was equipped with cameras at the front and rear, enabling it to beam back images live to Earth, as well as equipment for scientific research. Over the course of 83 sols (Martian days), or 85 Earth days, the vehicle travelled about 100 m (330 ft) over the planet surface. Finally, it lost communication with NASA, probably as a result of battery failure.

Since 1997, five missions have landed vehicles on Mars. Active from 2004, US rovers, *Spirit* and *Opportunity*, found evidence of ancient water – possibly favourable to life – before contact was lost in 2010 and 2019 respectively. China's *Zhurong* rover landed in 2021, returning images and data until May 2022, when it became inactive. US rovers *Curiosity*, which landed in 2012, and *Perseverance* (see box), which landed in 2021, are still beaming back data in 2024. On 6 August 2013, *Curiosity* played "Happy Birthday" to mark a year on Mars. This was the first time a song had been played on another planet.

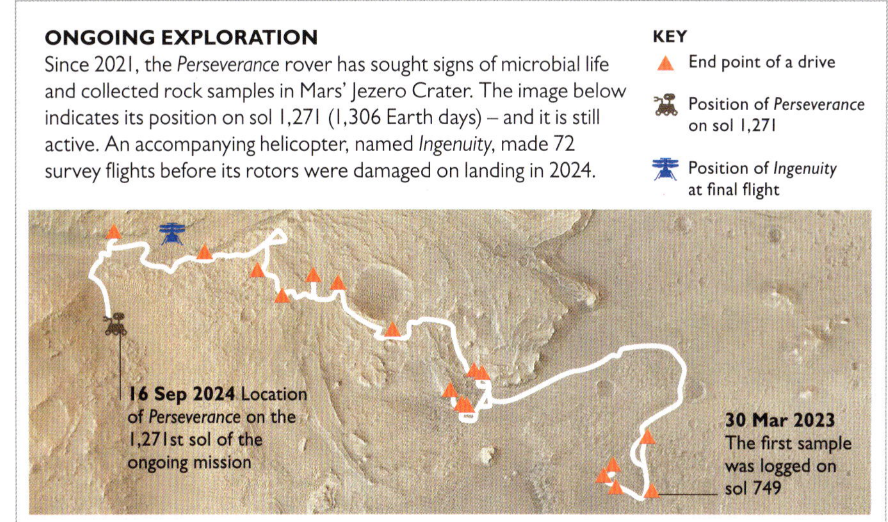

ONGOING EXPLORATION

Since 2021, the *Perseverance* rover has sought signs of microbial life and collected rock samples in Mars' Jezero Crater. The image below indicates its position on sol 1,271 (1,306 Earth days) – and it is still active. An accompanying helicopter, named *Ingenuity*, made 72 survey flights before its rotors were damaged on landing in 2024.

KEY

- End point of a drive
- Position of *Perseverance* on sol 1,271
- Position of *Ingenuity* at final flight

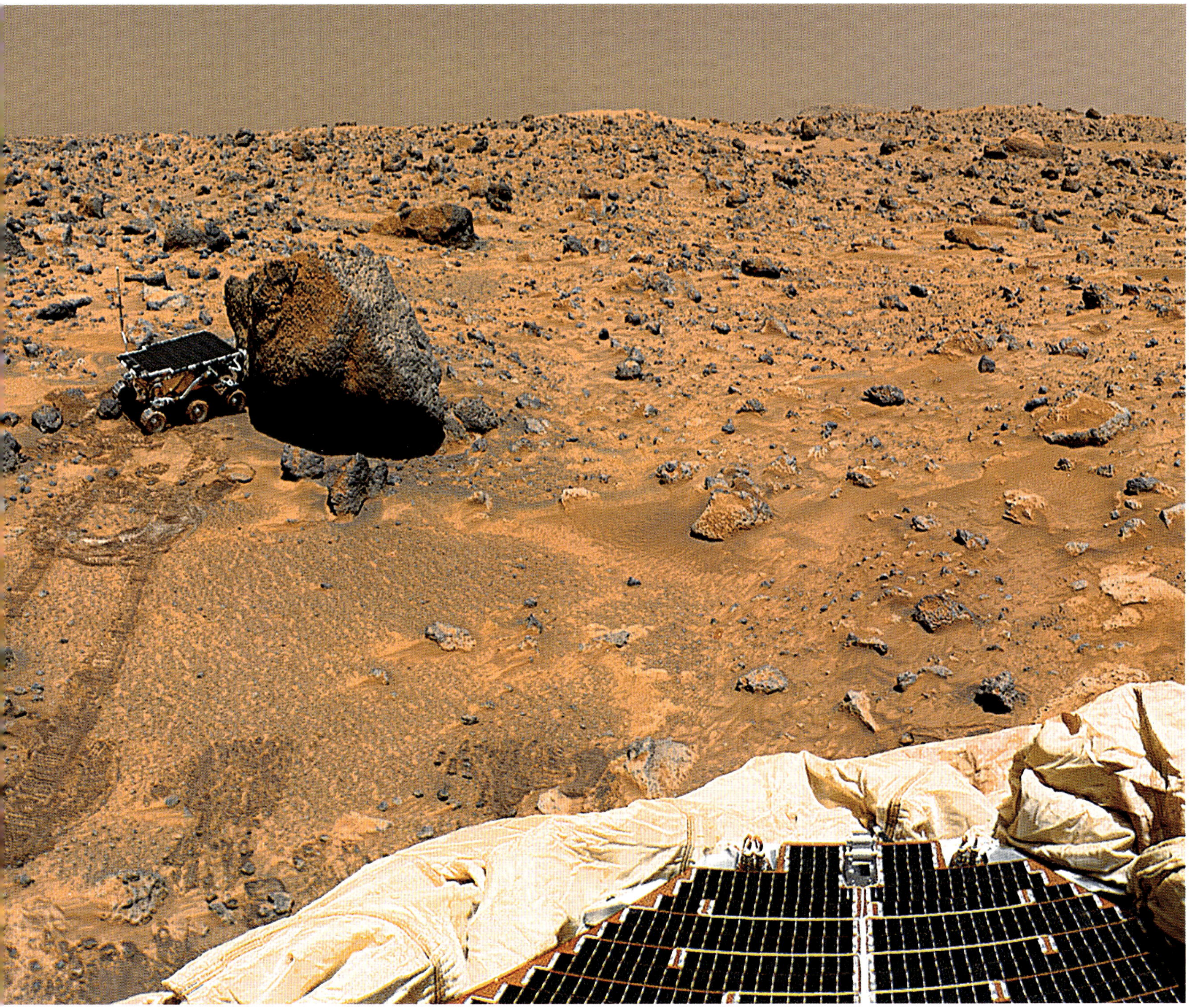

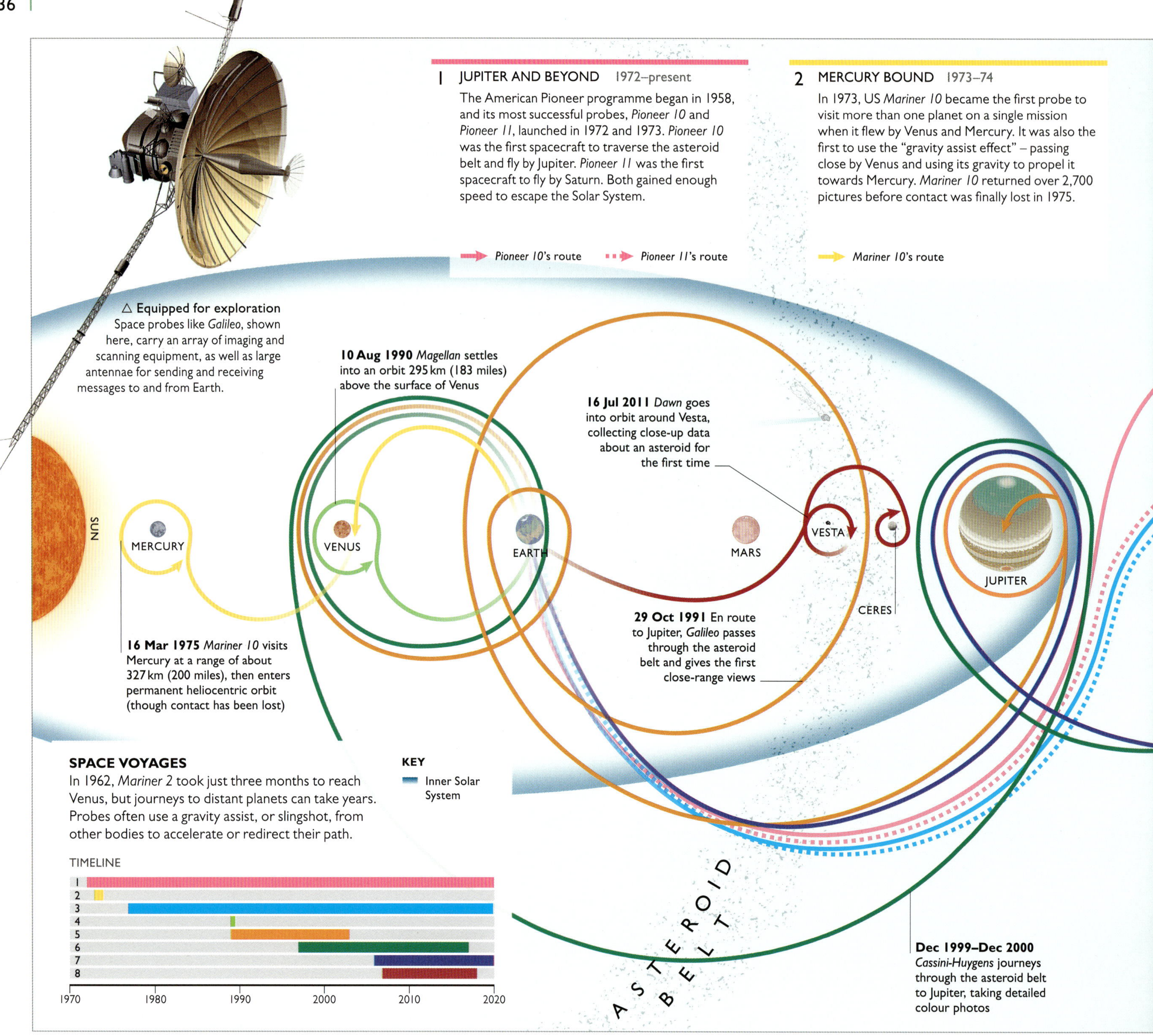

△ **Equipped for exploration** Space probes like *Galileo*, shown here, carry an array of imaging and scanning equipment, as well as large antennae for sending and receiving messages to and from Earth.

EXPLORING THE SOLAR SYSTEM

While human exploration of space stalled after the end of the US's Apollo programme in 1972, over the last 70 years, uncrewed space probes have visited all the planets and touched down on the surfaces of Mars, Venus, and several asteroids and moons. Robotic spacecraft beam back data as they journey through the Solar System, guided by onboard technology and signals sent from Earth.

The five types of space probes are atmospheric probes, interplanetary probes, orbiters, landers, and sample return missions. The first type collect data about a planet's atmosphere while interplanetary probes are designed to fly past planets, moons, and asteroids but not orbit them. Orbiters travel around a celestial body for a few years, using cameras and other instruments to study the surface. Landers detach from an orbiter to operate on the surface of a planet, such as Mars, and sample return missions collect samples from an extraterrestrial location and return them to Earth.

Since the late 1950s, some 200 space probes have been launched into the Solar System. In 1962, the American *Mariner 2* made the first successful voyage to another planet when it flew past Venus. In 1970, the Russian *Venera 7* became the first spacecraft to land on another planet when it reached the surface of Venus. In 1976, the American *Viking 1* and *Viking 2* made the first successful landings on

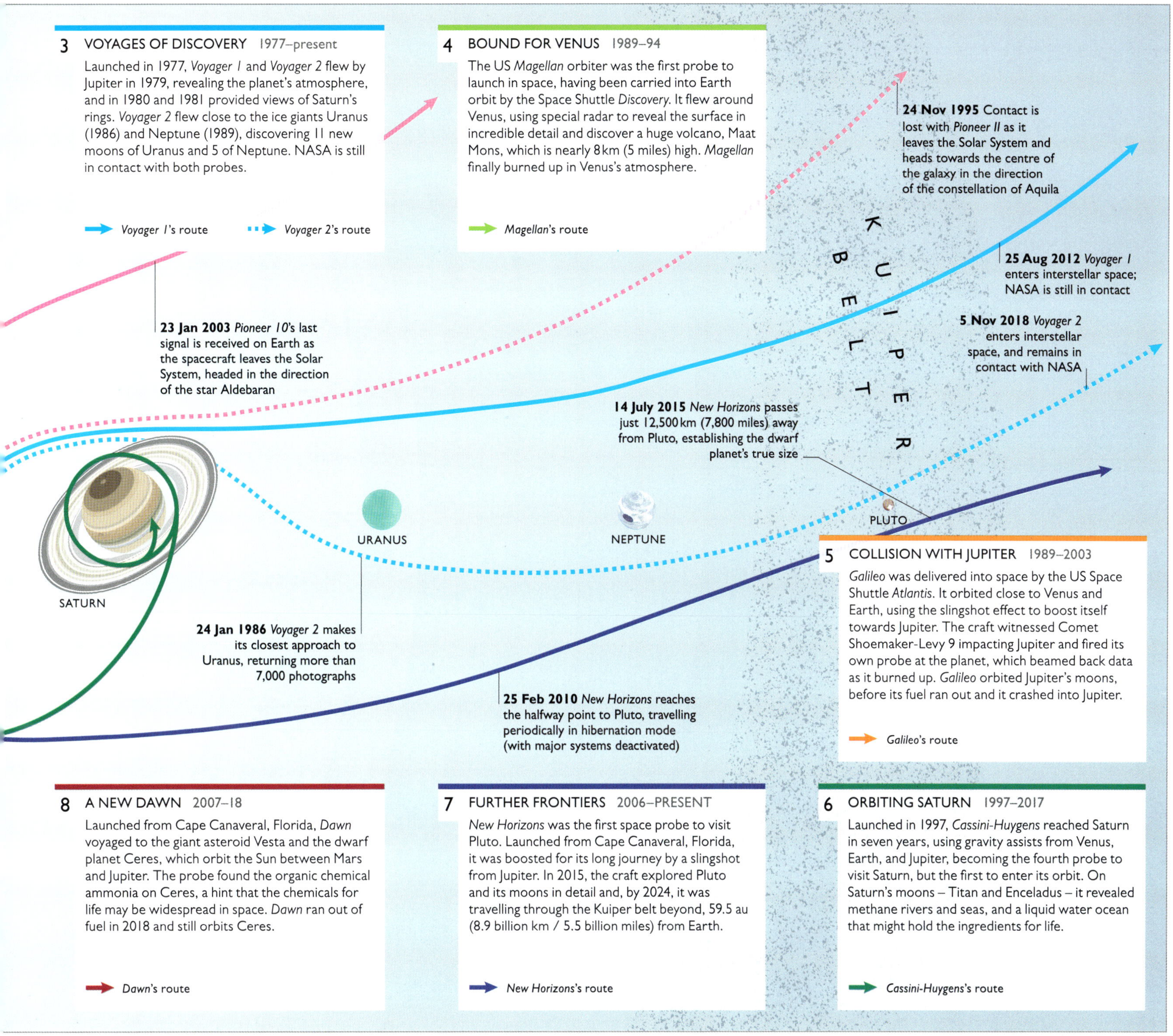

3 VOYAGES OF DISCOVERY 1977–present

Launched in 1977, *Voyager 1* and *Voyager 2* flew by Jupiter in 1979, revealing the planet's atmosphere, and in 1980 and 1981 provided views of Saturn's rings. *Voyager 2* flew close to the ice giants Uranus (1986) and Neptune (1989), discovering 11 new moons of Uranus and 5 of Neptune. NASA is still in contact with both probes.

Voyager 1's route
Voyager 2's route

4 BOUND FOR VENUS 1989–94

The US *Magellan* orbiter was the first probe to launch in space, having been carried into Earth orbit by the Space Shuttle *Discovery*. It flew around Venus, using special radar to reveal the surface in incredible detail and discover a huge volcano, Maat Mons, which is nearly 8km (5 miles) high. *Magellan* finally burned up in Venus's atmosphere.

Magellan's route

24 Nov 1995 Contact is lost with *Pioneer 11* as it leaves the Solar System and heads towards the centre of the galaxy in the direction of the constellation of Aquila

25 Aug 2012 *Voyager 1* enters interstellar space; NASA is still in contact

5 Nov 2018 *Voyager 2* enters interstellar space, and remains in contact with NASA

23 Jan 2003 *Pioneer 10*'s last signal is received on Earth as the spacecraft leaves the Solar System, headed in the direction of the star Aldebaran

KUIPER BELT

14 July 2015 *New Horizons* passes just 12,500km (7,800 miles) away from Pluto, establishing the dwarf planet's true size

SATURN

URANUS

NEPTUNE

PLUTO

24 Jan 1986 *Voyager 2* makes its closest approach to Uranus, returning more than 7,000 photographs

25 Feb 2010 *New Horizons* reaches the halfway point to Pluto, travelling periodically in hibernation mode (with major systems deactivated)

5 COLLISION WITH JUPITER 1989–2003

Galileo was delivered into space by the US Space Shuttle *Atlantis*. It orbited close to Venus and Earth, using the slingshot effect to boost itself towards Jupiter. The craft witnessed Comet Shoemaker-Levy 9 impacting Jupiter and fired its own probe at the planet, which beamed back data as it burned up. *Galileo* orbited Jupiter's moons, before its fuel ran out and it crashed into Jupiter.

Galileo's route

8 A NEW DAWN 2007–18

Launched from Cape Canaveral, Florida, *Dawn* voyaged to the giant asteroid Vesta and the dwarf planet Ceres, which orbit the Sun between Mars and Jupiter. The probe found the organic chemical ammonia on Ceres, a hint that the chemicals for life may be widespread in space. *Dawn* ran out of fuel in 2018 and still orbits Ceres.

Dawn's route

7 FURTHER FRONTIERS 2006–PRESENT

New Horizons was the first space probe to visit Pluto. Launched from Cape Canaveral, Florida, it was boosted for its long journey by a slingshot from Jupiter. In 2015, the craft explored Pluto and its moons in detail and, by 2024, it was travelling through the Kuiper belt beyond, 59.5 au (8.9 billion km / 5.5 billion miles) from Earth.

New Horizons's route

6 ORBITING SATURN 1997–2017

Launched in 1997, *Cassini-Huygens* reached Saturn in seven years, using gravity assists from Venus, Earth, and Jupiter, becoming the fourth probe to visit Saturn, but the first to enter its orbit. On Saturn's moons – Titan and Enceladus – it revealed methane rivers and seas, and a liquid water ocean that might hold the ingredients for life.

Cassini-Huygens's route

Mars, in July and September respectively. The most astonishing space flights of all are those of *Voyager 1* and *Voyager 2*. Launched in 1977, these interplanetary probes are still exploring today, heading beyond the Solar System into interstellar space. In 2024, both *Voyagers* were over 137 astronomical units (au) – equivalent to 20.4 billion km (12.7 billion miles) – from Earth, moving at a velocity of over 15.0km per second (about 34,000mph) relative to the Sun.

> *"Someday, humans will leave our cocoon in the Solar System. Voyager will have led the way."*
>
> US PHYSICIST AND FORMER ASTRONAUT JOHN GRUNSFELD, NASA NEWS CONFERENCE, 2013

A CALLING CARD FOR ALIENS

Voyager 1 and *Voyager 2* each carry a gold-plated phonograph record, bolted to their sides. These records are designed to introduce Earth to any intelligent extraterrestrial life forms that the space probes might encounter. The discs contain sounds and images of our planet, including spoken greetings in 55 ancient and modern human languages, 27 pieces of music from different eras and cultures, a variety of human and natural sounds, including laughter and animal noises, and 115 pictures of life on Earth.

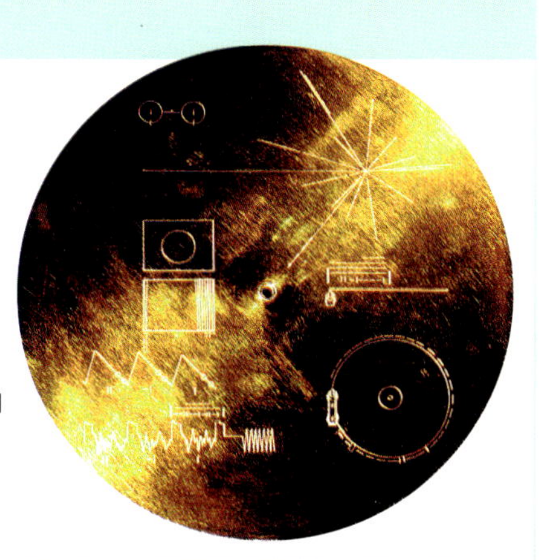

The cover of the golden record that was attached to *Voyager 1* in 1977

Star-forming nebula
In 2007, the Spitzer Space Telescope captured a view of the Eagle Nebula, located 7,000 light years from Earth. The red represents dust that has been heated by the explosion of a massive star, which also triggered the birth of new stars.

INTO DEEPEST SPACE

Since the 1960s, space probes have visited all the planets in the Solar System, and even ventured into interstellar space. But this is still a fraction of the vastness of the Universe, which can be explored only by telescopes.

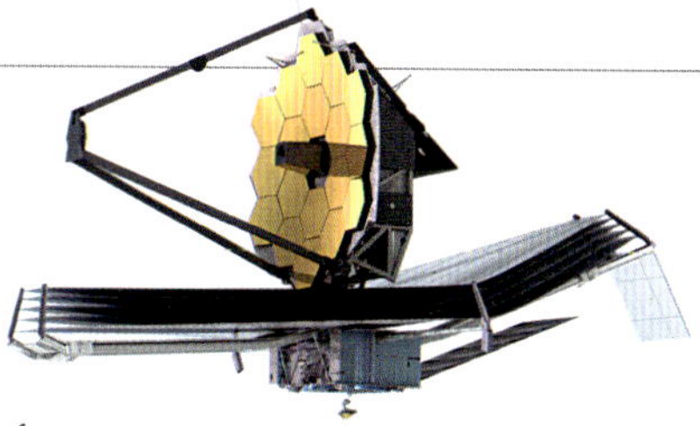

△ **Space eye**
The James Webb Space Telescope can detect objects up to 100 times fainter than the Hubble Space Telescope can.

In the early 20th century, the Universe was thought to be barely 15,000 light years (90,000 trillion miles) across. Since then, powerful telescopes have looked a million times further and discovered billions of previously unknown stars and galaxies. Astronomers have sent up space telescopes to orbit the Earth with a view undistorted by the atmosphere – most famously the Hubble Space Telescope (HST), launched in 1990. HST discovered two moons of Pluto – Nix and Hydra – and revealed that most galaxies have a black hole at the centre. In May 2009, astronauts installed the Wide Field Camera 3, or WFC3, on HST. Like an ordinary camera, WFC3 takes pictures using visible light, but also records other wavelengths. By photographing a small area, one 150th of the apparent size of a full moon, the WFC3 revealed distant galaxies 13 billion light years away, helping to date the age of the Universe to 13.7 billion years.

Meanwhile, the Spitzer Space Telescope, launched in 2003, was the first telescope to determine the presence of a planet outside of our Solar System, by measuring the dip in light from the host star when the planet crossed in front. To date, more than 5,700 "exoplanets" – planets orbiting distant stars – have been discovered, with the Kepler space telescope, launched in 2009, helping to confirm the existence of most of them. The James Webb Space Telescope, launched in 2021, is the largest and most powerful telescope in space. Orbiting the Sun 1.5 million km (1 million miles) from the Earth, it is revealing exoplanets and other aspects of distant space in even greater detail.

THE FURTHEST STAR

In 2023, the James Webb Space Telescope captured Earendel, an ancient star that existed just a billion years after the Universe began. These powerful telescopes allow us to look into the past, seeing light from long-dead stars.

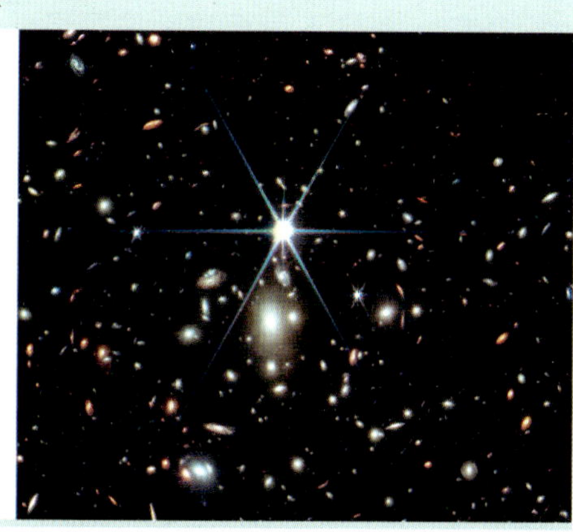

Galaxy cluster in which Earendel is found

OCEAN FRONTIERS

Until recently, the challenges of mapping underwater terrain meant that the ocean floor was charted in less detail than the Moon's surface, but developments in sonar and satellite technology are aiding our exploration and understanding of the oceans.

In World War II, the need to track enemy submarines fuelled research into sonar technology – using underwater sound waves to pinpoint objects. This led to a breakthrough in ocean exploration as sonar data could be used to map the ocean bed. In the 1950s, US geologists Marie Tharp and Bruce Heezen created the first maps of the ocean floor (see box).

Sonar is still crucial for making nautical charts and locating hazards such as shipwrecks. But using sonar technology to map is a slow process. An advanced multibeam sonar, which sends sound waves to the ocean floor and uses their echoes to find objects, takes months to chart a tiny part of the seabed.

Satellites are also expanding our oceanic knowledge. In 1995, declassified data from the US Navy's Geosat satellite showed how, with bathymetry (the study of the depths and shapes of underwater terrain), by plotting minute variations in the sea surface height, the seafloor could be mapped precisely. The gravitational pull of undersea mountains can cause water to pile up around them, swelling the sea's surface. Geosat measured the time a laser took to bounce back from the sea surface to the satellite and was able to map thousands of ocean mountains. In some instances, both depth soundings from ships and satellite altimetry were used in mapping (see map opposite). In 2023, satellite-derived bathymetry revealed 19,000 unknown undersea volcanic peaks, almost doubling the 21,000 previously mapped by sonar. In 2024, 26.1 per cent of the seabed was mapped. The United Nation's Seabed 2030 initiative plans to map the final 74 per cent by 2030.

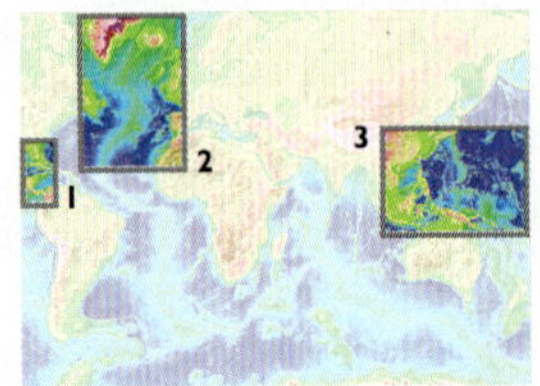

KEY

1 A deep-sea coral reef – thought to be the world's largest – stretches for 500 km (310 miles) along the US Atlantic coast.

2 The green ridge running down through the turquoise area indicates the Mid-Ocean Ridge, the world's longest mountain chain.

3 The light turquoise area in the sea indicates the Ring of Fire, a belt of underwater volcanoes around the Pacific Ocean.

MAPPING THE OCEAN FLOOR

At Columbia University in 1953, US geologists Marie Tharp and Bruce Heezen identified an underwater mountain chain that stretches across the globe – the Mid-Ocean Ridge. As women were not then allowed on research ships, Heezen went into the field while Tharp analyzed the data. Together they published the first detailed maps of the ocean floor and went on to prove the validity of the continental drift hypothesis, proposed by Alfred Wegener in 1912.

Tharp at work on a map in the 1950s

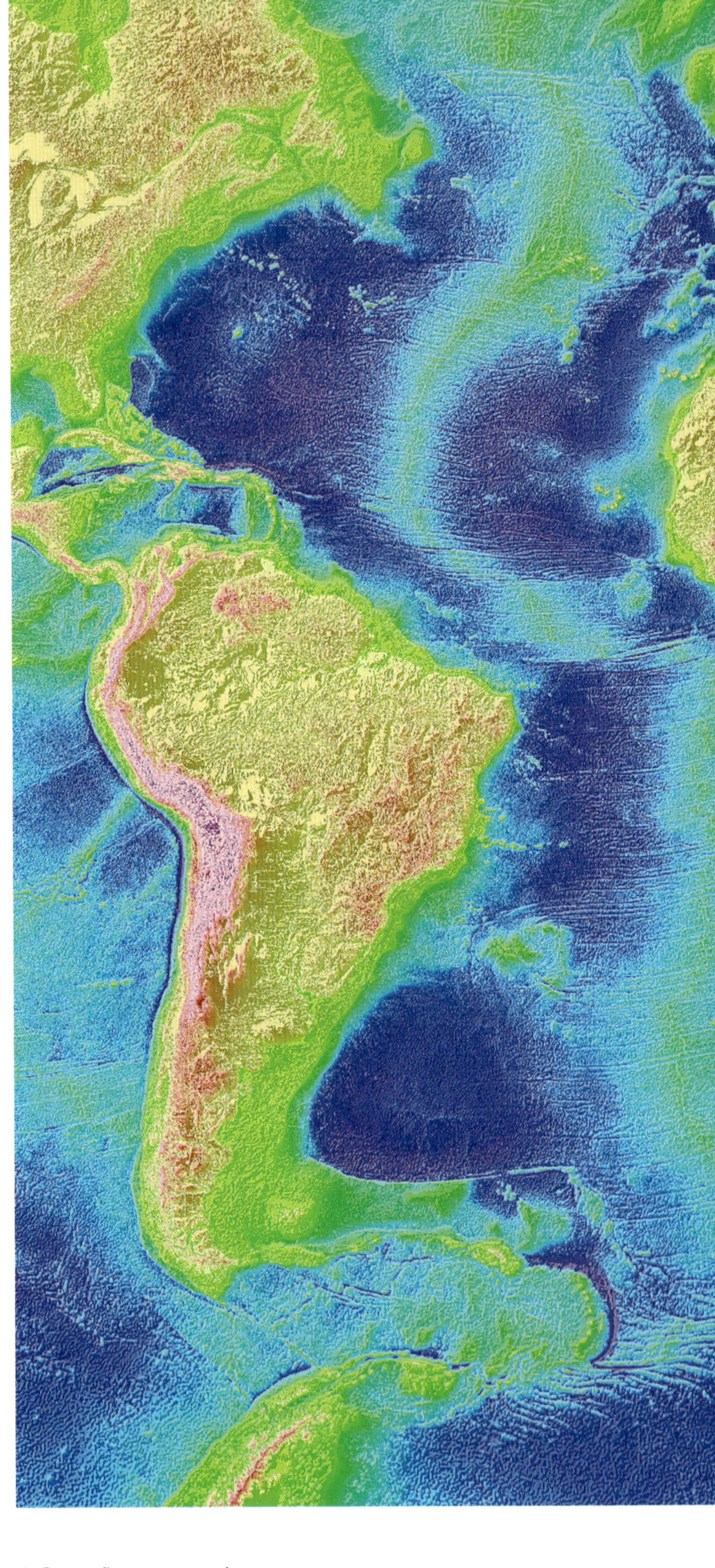

△ **Ocean floor topography**
This map, made by US geophysicists David Sandwell and Walter Smith in 1997, showed global seafloor topography in unprecendented detail. Dark blue areas relate to ocean depths, light turquoise are shallows, and green areas underwater mountains.

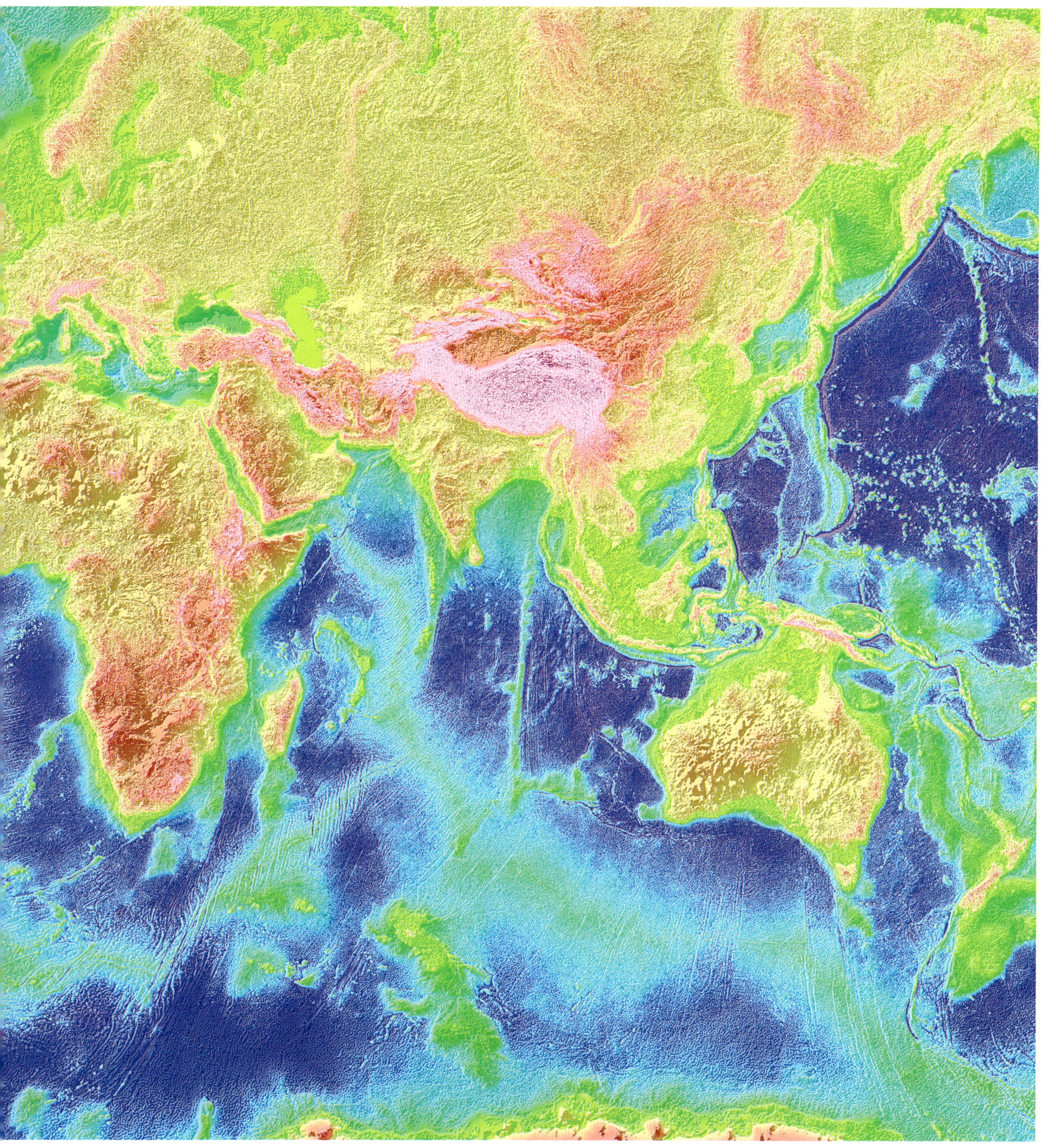

"I think our maps contributed to a revolution in geological thinking."

MARIE THARP, *LAMONT-DOHERTY EARTH OBSERVATORY*, 1999

THE NEW SPACE RACE

In the 21st century, rivalry in space is no longer a Cold War face-off between the USA and Soviet Union. More than 80 countries, as well as private enterprise, are now participating in space exploration, and collaboration exists alongside competition.

Like so many journeys of exploration across the world's surface, most missions into space are a bid to gain control over precious resources. Earth's orbit is an extremely competitive arena: it hosts technology that is vital for communication and security – satellites, space stations, weapons systems, and much more – and offers the prospect of significant strategic power. Beyond Earth's orbit, the Moon, Mars, and the asteroids have been identified as potentially rich in rare minerals used in modern technology, and the Moon may also be a source of Helium-3, which could enable the development of non-radioactive nuclear fusion power. The two major players in 21st-century space exploration, the USA and China, are prioritizing lunar missions with their respective *Artemis* and *Chang'e* projects.

As in the 20th century, exploration is also directly linked to national prestige. Russia remains in the field, and India and Japan are very active. Even countries without their own space programmes – such as Vietnam, Malaysia, Nigeria, and Sudan – are joining blocs for collaborative missions, with many sharing the view that international co-operation is crucial for the growth of our species.

> *"... a giant leap into space can be a giant leap toward peace down below."*
>
> WILLY LEY, *THE LEAP INTO SPACE*, 1957

THE MOON AND BEYOND

In 2024, China's *Chang'e 6* spacecraft became the first space probe to touch down on the far side of the Moon, returning to Earth with 1.9 kg (4.2 lb) of rock and soil samples. China is also currently working to establish an International Lunar Research Station (ILRS) near the Moon's south pole. In 2026, *Chang'e 7* aims to land on the illuminated rim of the lunar Shackleton crater and investigate the presence of water ice, with the potential to provide drinking water, oxygen, and rocket fuel.

The *Chang'e 6* capsule lands in Mongolia

THE GLOBAL SPACE RACE

Since the end of the Cold War-era space race with the collapse of the Soviet Union in 1991, space exploration has been a global arena, with new scope for competition and collaboration.

KEY

- Countries that have sent spacecraft beyond Earth's orbit
- International Space Exploration Coordination Group (ISECG) member, 2024
- Other spaceports and launch sites, 2024

TIMELINE

1, 2, 3, 4, 5, 6, 7

1952 · 1964 · 1976 · 1988 · 2000 · 2012 · 2024

CANADA

UNITED STATES

PACIFIC OCEAN

ATLANTIC

BRAZIL

Pacific Spaceport Complex

National Aeronautics and Space Administration (NASA)

Canadian Space Agency (CSA)

Midland Spaceport

Mojave Air and Space Port

Vandenberg Space Force Base

Colorado Air and Space Port

Oklahoma Air and Space Port

Mid-Atlantic Space Port

Spaceport America

Corn Ranch

Cecil Field Spaceport

Cape Canaveral

Mexican Space Agency (AEM)

SpaceX Boca Chica

Houston Spaceport

Guiana Space Centre

Alcântara Launch Centre

Brazilian Space Agency (AEB)

28 Jan 1986 In one of the worst disasters in space history, the *Challenger* shuttle breaks up over the Atlantic

31 Jan 1958 The US launches *Explorer 1* into orbit

14 May 1973 The US launches Skylab, its first space station in conjunction with the ESA

24 Dec 1979 The ESA rocket *Ariane 1* delivers its first satellite payload into orbit

22 Aug 2003 A rocket explosion kills 21 people – the deadliest space-related disaster of the 21st century

7 EUROPEAN SPACE AGENCY 1975–PRESENT

Formed in 1975 with 10 founding members, the ESA is now a collaboration of 22 European nations, with a spaceport at Kourou in Guiana. It works with the USA and, formerly, Russia, to send crews to the International Space Station, launch Earth observation, science, and communications satellites, and develop missions to Mars. The Mars Sample Return, a key collaboration with the USA, plans to collect rock and soil samples.

Spaceports and launch sites, 2024

6 INDIA 1975–PRESENT

The *Aryabhata*, India's first satellite, was launched by a Soviet Kosmos-3M rocket in 1975. In recent years, India has launched three *Chandrayaan* probes to the Moon, and one to Mars. *Chandrayaan 1*, armed with key equipment from the USA, the ESA, and Bulgaria, discovered water on the Moon in 2008. In 2023, *Chandrayaan 3* became the first space agency to land a spacecraft on the lunar south pole region, and only the fourth to land on the Moon.

Spaceports and launch sites, 2024

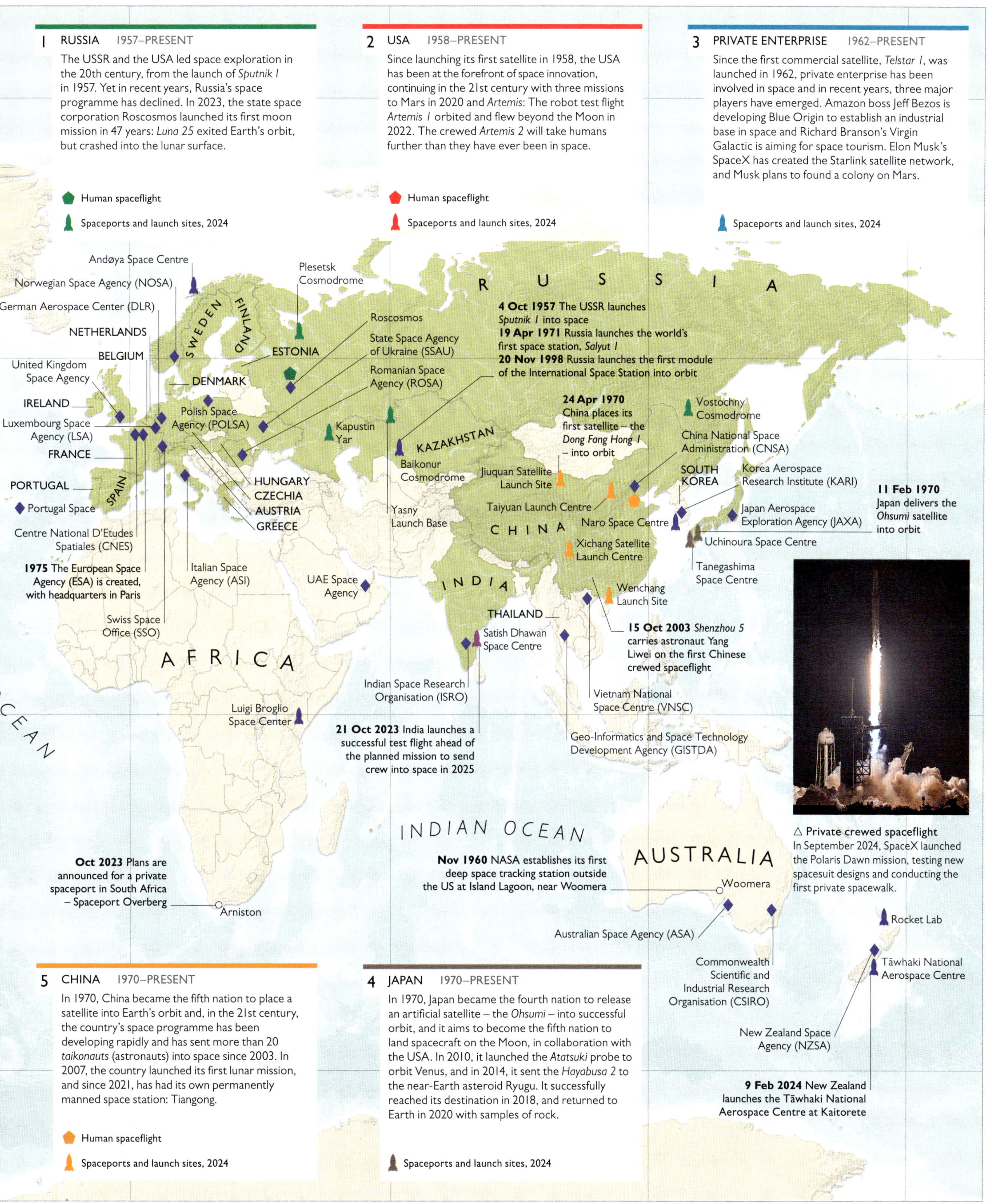

1 RUSSIA 1957–PRESENT

The USSR and the USA led space exploration in the 20th century, from the launch of *Sputnik 1* in 1957. Yet in recent years, Russia's space programme has declined. In 2023, the state space corporation Roscosmos launched its first moon mission in 47 years: *Luna 25* exited Earth's orbit, but crashed into the lunar surface.

2 USA 1958–PRESENT

Since launching its first satellite in 1958, the USA has been at the forefront of space innovation, continuing in the 21st century with three missions to Mars in 2020 and *Artemis*: The robot test flight *Artemis 1* orbited and flew beyond the Moon in 2022. The crewed *Artemis 2* will take humans further than they have ever been in space.

3 PRIVATE ENTERPRISE 1962–PRESENT

Since the first commercial satellite, *Telstar 1*, was launched in 1962, private enterprise has been involved in space and in recent years, three major players have emerged. Amazon boss Jeff Bezos is developing Blue Origin to establish an industrial base in space and Richard Branson's Virgin Galactic is aiming for space tourism. Elon Musk's SpaceX has created the Starlink satellite network, and Musk plans to found a colony on Mars.

△ **Private crewed spaceflight**
In September 2024, SpaceX launched the Polaris Dawn mission, testing new spacesuit designs and conducting the first private spacewalk.

5 CHINA 1970–PRESENT

In 1970, China became the fifth nation to place a satellite into Earth's orbit and, in the 21st century, the country's space programme has been developing rapidly and has sent more than 20 *taikonauts* (astronauts) into space since 2003. In 2007, the country launched its first lunar mission, and since 2021, has had its own permanently manned space station: Tiangong.

4 JAPAN 1970–PRESENT

In 1970, Japan became the fourth nation to release an artificial satellite – the *Ohsumi* – into successful orbit, and it aims to become the fifth nation to land spacecraft on the Moon, in collaboration with the USA. In 2010, it launched the *Atatsuki* probe to orbit Venus, and in 2014, it sent the *Hayabusa 2* to the near-Earth asteroid Ryugu. It successfully reached its destination in 2018, and returned to Earth in 2020 with samples of rock.

DIRECTORY

Mansa Abu Bakr II
Late 13th–early 14th century

A leader of the West African kingdom of Mali, Abu Bakr II is said to have led an expedition of 2,000 boats from modern-day Gambia west across the Atlantic, probably in 1311. He put his brother Kankou Moussa in charge of the kingdom while he was away. Bakr did not return from this voyage, and some researchers claim he reached South America, citing evidence such as reports by early European explorers of dark-skinned people in northeastern Brazil.

Mirza Abu Taleb Khan
1752–1806

Abu Taleb Khan was an employee of the East India Company. He travelled from India to the Middle East and Europe from 1799 to 1803, then published an account of the journey. His writings were not just accounts of his travels; they compared the culture and achievements of India with that of England and France. The other countries he visited included Malta, Turkey, Iraq, and South Africa. Indian-born but of Iranian heritage, Abu Taleb Khan was well received in London.

Edwin ("Buzz") Aldrin
1930–

As pilot of the Apollo 11 lunar lander, on 20 July 1969 Aldrin became the second person to set foot on the Moon. Three years earlier, on NASA's 1966 Gemini 12 mission, he established what was then a record by spending five and a half hours "spacewalking" in the vacuum of space. During his NASA career, Aldrin spent almost 290 hours in space. An American Air Force pilot who flew 66 combat missions from 1952 to 1953 as part of the Korean conflict, he earned a doctorate in astronautics at MIT before joining NASA in 1963. He was initially rejected by NASA because he was not a test pilot. He qualified when applicants with more than 1,000 hours of jet flying behind them became eligible.

> *"I know the sky is not the limit, because there are footprints on the Moon – and I made some of them!"*
>
> BUZZ ALDRIN, *NO DREAM IS TOO HIGH*, 2016

Alexander the Great
356–323 BCE

The king of Macedonia, Alexander led his forces to victory over the larger Persian army without losing a major battle during a decade of campaigning (334–325 BCE). Although he died at the age of just 32, Alexander forged through conquest an empire that stretched east from Greece into India and south into Egypt, encompassing approximately 5 million sq km (2 million sq miles) of land. Considered a military genius, as a teenager he was tutored by Aristotle, who gave him an interest in philosophy, medicine, and scientific investigation. His campaigns advanced the knowledge of geography and natural history.

Roald Amundsen
1872–1928

On 14 December 1911, Amundsen became the first person to reach the South Pole, beating Robert Falcon Scott by one month. Born into a Norwegian family of ship owners, he was the first to navigate Canada's Northwest Passage, a journey begun in 1903 that finally reached Alaska in 1906. In 1926, Amundsen flew over the North Pole in an airship from Norway to Alaska – the first time the Arctic Ocean had been crossed. He died in the Arctic in 1928, his plane missing and presumed crashed, during a mission to rescue the survivors of a lost airship.

Henriette d'Angeville
1794–1871

Regarded as the first female mountaineer, in 1838, d'Angeville became the first woman to successfully climb Mont Blanc, at 4,807 m (15,771 ft) the highest peak in the Alps, under her own power. (A woman named Marie Paradis reached the summit of Mont Blanc in 1808, but her companions carried her part of the way.)

Neil Armstrong
1930–2012

The commander of NASA's Apollo 11 mission, on 20 July 1969 Armstrong was the first person to set foot on the Moon. He took only one other flight into space – 1966's Gemini 8 mission, the first to successfully dock two vehicles in orbit. A test pilot and US Navy aviator who saw action during the 1950–53 Korean conflict, Armstrong joined NASA in 1962. During 1971–79, he was professor of aerospace engineering at Cincinnati University.

◁ **Buzz Aldrin**
An official NASA portrait of astronaut Aldrin taken just days before the July 1969 launch of Apollo 11, the mission on which Aldrin would become the second man to walk on the Moon.

◁ Roald Amundsen
Amundsen at the South Pole in 1911 or 1912. The cold-weather clothing required for such low temperatures included sealskin suits and anoraks sewn from army blankets.

B

William Baffin
1584–1622

A British navigator, Baffin played a key role on several 1610s voyages that searched for the Northwest Passage – a sea route that it was hoped would connect Europe to Asia without the necessity of sailing around the tip of South America. Though the Northwest Passage proved elusive then, these voyages enabled explorations of the Atlantic coasts of Canada and Greenland. Baffin later surveyed the Red Sea and Persian Gulf for the East India Company, but died during an attack on a Portuguese Persian Gulf fort. Canada's Baffin Island and Baffin Bay are named for him.

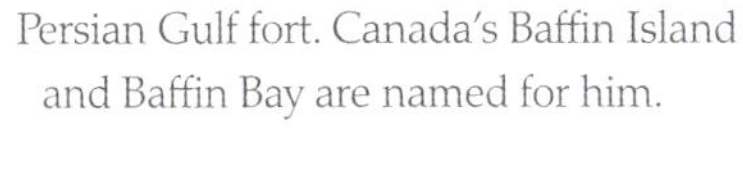

Florence Baker
1841–1916

Born Florence von Szász, Baker participated in British explorer Samuel Baker's journeys into Africa, including an 1861–65 expedition during which they became the first Europeans to reach Lake Albert. Her writings offer little detail about her role, but Samuel's accounts cite occasions on which she proved pivotal, such as stockpiling a cache of food that was later all that stood between the group and starvation. Born in Transylvania, which is now in Romania, she was orphaned, kidnapped, and sold at a white slave auction. Samuel Baker, her future husband and expedition partner, allegedly liberated her by bribing a guard after her sale, but accounts vary.

Samuel Baker
1821–93

A British explorer of east-central Africa, Baker is best known for becoming the first European to reach Lake Albert, which is located along the border of present-day Uganda and Congo, in 1864. Five years later, the khedive, or hereditary pasha, of Egypt commissioned him to lead a military expedition to the Nile River's equatorial region to suppress the slave trade there and extend Egyptian control. Baker was appointed the region's governor. The son of a successful merchant, he was knighted in 1866.

Vasco Núñez de Balboa
1475–1519

A Spanish explorer, in 1513 Balboa led an expedition across the isthmus of Panama, becoming the first European to reach the Pacific Ocean from the east. Born into a noble family, Balboa travelled to the "New World" in 1500, initially finding little success as a farmer in what is now Haiti but eventually taking charge of the Spanish colony of Darién on the Panama coast. His rival in the region, Pedro Arias Dávila, successfully plotted to have Balboa beheaded on dubious charges of treason.

Rabban Bar Sauma
c.1220–94

A monk of Turkic ethnicity who was born in China, Bar Sauma travelled west as far as France – the first person from China known to have visited Europe. A member of a Christian sect known as Nestorianism, he followed the Silk Road west from China to the Middle East on a 1275 pilgrimage to the Holy Land, failing to make it to Jerusalem due to unrest there. In 1287, Bar Sauma travelled even further west, to Constantinople, Rome, and Paris as an emissary of Arghun, a subordinate of the Great Khan. He met Pope Nicholas IV, Philip IV of France, and Edward I of England. He died in Baghdad.

Jeanne Baret
1740–1807

Baret is believed to be the first woman to circumnavigate the globe – working as an assistant to her companion, the botanist Philibert Commerçon, on a 1766 French expedition. He was the expedition's primary botanist. She was well-suited to the role, possessing a strong knowledge of the traditional medicinal uses of plants. Baret began her journey disguised as a man – French Navy rules prohibited women from sailing on its ships. When her secret came out, Commerçon and Baret left the expedition, spending years on the island of Mauritius in the Indian Ocean. Some accounts credit Baret with the voyage's discovery of a type of vine previously unknown to European science. After Commerçon's death, Baret returned to France in 1775, completing her circuit of the globe. There is no evidence that anyone noticed the historic nature of her circumnavigation at the time, though the French Navy did give her a pension.

Heinrich Barth
1821–65

A German explorer, Barth undertook extensive and arduous expeditions through North and Central Africa in the 1840s and 1850s, travelling 16,000 km (10,000 miles). His writings about his travels provided Europe with a better understanding of Africa's people and geography. Barth later served as professor of geography at the University of Berlin.

◁ Henriette d'Angeville
When d'Angeville climbed Mont Blanc in 1838 it was considered improper for a woman to wear trousers even when climbing mountains – so d'Angeville hid hers under a long woollen coat.

△ Florence Baker
Baker had survived an astonishing range of ordeals and adventures by the time this photo of her was taken in around 1870; she and her husband would soon settle into a comfortable life in England.

George Bass
1771–1803

A naval surgeon and sailor, Bass played an important role in a series of late-1790s expeditions that charted the coast of Australia. Born in England, the son of a tenant farmer, he served as ship's surgeon on voyages around England and across the Atlantic prior to his Australian travels. A skilled navigator and seaman as well as a surgeon, his name lives on in the Bass Strait, which separates Australia and Tasmania. A cargo ship on which he was travelling was lost at sea in 1803; all aboard were presumed dead.

Henry Walter Bates
1825–92

A British entomologist, in 1848, Bates travelled to Brazil's Amazon basin with the naturalist Alfred Russel Wallace. The journey helped him develop the theory of natural selection. (Wallace and Charles Darwin independently came up with similar theories, though Darwin received most of the credit.) Bates remained in the Amazon for 11 years, collecting nearly 15,000 species, 8,000 of them insects not previously known to science. After returning to London, Bates wrote about his discoveries and was assistant secretary of the Royal Geographical Society from 1864 until his death.

James P. Beckwourth
1798–c.1867

A trapper and frontiersman in the American West, Beckwourth gained fame when his memoirs were published in 1856. He worked as a guide, scout, rancher and innkeeper, as well as for the military. Born into enslavement – his mother was an enslaved Black woman, his father a white enslaver who eventually granted him his freedom – Beckwourth lived with and married into Indigenous communities including the Crow. He is credited with exploring the lowest pass through the Sierra Nevada mountains, now known as Beckwourth Pass.

◁ **Gertrude Bell**
Bell didn't just cross the Arabian Desert, she did so riding sidesaddle. Pictured here in Iraq, Bell noted that she felt more comfortable in Baghdad than in London.

"[Bennelong]... is also an example of someone who survived the clash of cultures, and still commanded respect among his people in later life."

DR KATE FULLAGAR, MACQUARIE UNIVERSITY, 2013

Gertrude Bell
1868–1926

An explorer, archaeologist, mountaineer, and diplomat in an era when none of those roles were common for women, Bell travelled extensively in Southwest Asia and Asia Minor. The Oxford-educated Bell learned to speak Arabic, Persian, and Turkish and she is believed to be the first European woman to cross Arabia. Her knowledge of the region was so highly regarded that the British government selected her to help set the borders of the nation of Iraq and choose its first ruler – she was a strong supporter of Arab self-determination. She was nicknamed the Queen of the Desert and the female Lawrence of Arabia.

Fabian Gottlieb von Bellingshausen
1778–1852

A Russian naval officer, Bellingshausen led an 1819–21 expedition that circumnavigated Antarctica. The expedition visited a number of Antarctic islands in the South Sandwich Island group, as well as possibly being the first to see the Antarctic continent, although this is disputed. Bellingshausen had previously been a member of an 1803–1806 expedition that was the first by Russia to circumnavigate the globe. Born on an island in what is now Estonia, he joined the navy at the age of 10, rising to the rank of admiral, and served as governor of the Russian port city of Kronshtadt. The Bellingshausen Sea to the west of the Antarctic Peninsula is named in his honour.

Gertrude Emily Benham
1867–1938

A mountaineer and traveller, Benham summitted more than 300 peaks during her lifetime, circumnavigated the globe seven times, and crossed Africa at least three times on foot. Born in England, she climbed mountains in Europe when young, and began her world travels in earnest only in 1903, at the age of 36, following the death of her mother. From that date, she climbed in and explored a remarkable range of locations, from the mountains of Tibet to the islands of the Pacific, her adventures interrupted only by World War I.

Benjamin of Tudela
c.1130–73

Born in the Kingdom of Navarre in northern Spain (present-day France), Benjamin, who was Jewish, spent more than 10 years travelling across Europe, Egypt, and Southwest Asia (Middle East), visiting more than 300 cities along the way. His travelogue predates that of Marco Polo by a century, and scholars generally consider it more reliable, though it is not without errors. His writings about the ruins near Mosul, in Iraq, provide an early description of the site of Nineveh.

Woollarawarre Bennelong
c.1764–1813

Bennelong was the first Aboriginal Australian to travel to Europe then return to his native country. Captured by British forces in 1789, he escaped but later revived his connections with the British

◁ **Woollarawarre Bennelong**
Here illustrated in European clothing, Bennelong opted to return to his Indigenous Australian culture in the early 19th century rather than continue living among Europeans.

on his own terms. Trust grew, and Bennelong became a go-between and interpreter between the two cultures. He sailed to England in 1792, and met King George III. Returning to Australia in 1795, he became a respected leader of his community. Bennelong Point, the site of Sydney Opera House, is named in his honour.

Vitus Bering

1681–1741

A Danish navigator, Bering led a pair of expeditions that set the stage for Russian expansion into Alaska. The first of these, on behalf of Peter the Great in 1728, determined that Russia was not connected by land to the Americas. The second, the Great Northern Expedition launched in 1733 for Peter's successor, Empress Anna, explored the Pacific coast of North America. Bering died during that arduous expedition, as did many members of his crew. His name endures in the Bering Sea and Bering Strait, both located between Russia and Alaska, as well as Bering Island, where he died and was buried in 1741.

Hiram Bingham

1875–1956

A Yale University history lecturer, Bingham shot to fame for his 1911 rediscovery, with the guidance of local farmers, of the ruins of the Inca city of Machu Picchu in present-day Peru, as well as other Inca sites. An American born in Hawaii into a family of missionaries, Bingham had previously retraced the journeys of 19th-century South American statesman Simón Bolívar through Colombia and Venezuela. He had also travelled by mule from Buenos Aires, Argentina, to the ancient Inca capital city of Cuzco in Peru. Bingham's explorations were integral to the investigation of archaeological sites in the Andes and other parts of South America. Bingham served in the US Senate from 1924–33.

17

Percentage of Russia's national income spent on Bering's 1733 expedition.

Isabella Bird Bishop

1831–1904

Advised by a doctor that travel might be beneficial for her chronically poor health, Bird became one of the most remarkable travellers of the second half of the 19th century. The daughter of an English clergyman, she visited North America, Hawaii, India, Kurdistan, the Persian Gulf, Tibet, Malaysia, Korea, Japan, and China, among many other destinations, climbing mountains and riding long distances on horseback along the way. She was the first woman to be named a fellow by the Royal Geographical Society, and published books about her adventures.

William Bligh

1754–1817

An English ship's captain, Bligh is remembered as the Captain of HMS *Bounty* against whom first mate Fletcher Christian mutinied in the South Pacific in 1789. Bligh and 18 loyal crewmen were set adrift in a 6-m (20-ft) launch, which Bligh managed to navigate more than 5,800 km (3,600 miles) to safety. In the late 1770s, he had served as sailing master on James Cook's final voyage to the South Pacific. Over the course of his career, Bligh reached more than a dozen Pacific islands previously unknown in Europe, was promoted to the rank of Vice Admiral, and served as the Governor of New South Wales from 1806 to 1808.

Lady Anne Blunt

1837–1917

An adventurer, intellectual, writer, translator, and horse enthusiast, Blunt travelled extensively in Southwest Asia (Middle East) with her husband, Wilfrid Scawen Blunt. She is thought to be the first European woman to cross the northern Arabian Desert. The Blunts' primary goal was to purchase purebred Arabian horses – most Arabian horses now descend from the animals they acquired – but she also learned to speak Arabic and wrote several books about the region. Blunt was the 15th Baroness Wentworth, the granddaughter of the poet Byron, and the daughter of famed mathematician Ada Lovelace. Blunt continued to ride, write, and travel in later life; she died during a journey to Egypt aged 80.

△ **William Bligh**
Hollywood has cast Bligh as a villain, but some historians argue he was no more tyrannical than his peers. Here he wears the medal received for his efforts in 1797's Battle of Camperdown.

Nellie Bly
1864–1922
A pioneering journalist, Bly is best remembered for completing a round-the-world trip in 72 days during 1889 and 1890 for the newspaper the *New York World*, besting the fictional Phileas Fogg's time in Jules Verne's novel *Around the World in Eighty Days*. Her other writings included exposés about the conditions in New York's asylums, jails, and sweatshops, which involved going undercover. She visited Mexico in 1886–1887 and wrote compelling reports about conditions there. Nellie Bly was a pen name – her real name was Elizabeth Cochran.

Benjamin-Louis-Eulalie de Bonneville
1796–1878
A frontiersman, US army officer, and trapper in the American West, Bonneville became famous when, in 1837, the writer Washington Irving published a popular book based on Bonneville's journals about his 1830s explorations of the Rocky Mountains in North America. Born in France, Bonneville moved to the US with his family in 1803. Both his military career and fur-trapping business endured setbacks, but by 1865 he had risen to the rank of Brigadier General.

△ Nellie Bly
Bly's round-the-world trip in 1889–90 earned her enduring fame, while her undercover investigation of conditions in an asylum led to improvements in the treatment of the mentally ill.

Louis-Antoine de Bougainville
1729–1811
A navigator and soldier, Bougainville led a French expedition around the world in 1766–69, then wrote a popular account of the journey. Earlier in his career, he had fought in the French and Indian War, and helped to found a French colony in the Falkland Islands. He later led a French fleet that opposed the English during the American Revolution. Late in his life, Bougainville was appointed to the Senate by Napoleon. Bougainville Island, the largest of the Solomon Islands, and the *Bougainvillea* plant are named after him.

Louise Arner Boyd
1887–1972
An Arctic explorer, geographer, and socialite, Boyd led pioneering scientific and mapping expeditions to Greenland in the 1930s. She also led a 1941 exploratory expedition to the region on behalf of the US government during World War II. Born the daughter of a wealthy California gold miner, Boyd lacked formal training in the sciences, but her expeditions gathered data that is still used today, including photos of Greenland's ice that track the impact of global warming. She had a final Arctic adventure in 1955 at the age of 67, flying as a passenger on a plane that passed over the North Pole.

St Brendan
c.484–578 CE
An Irish abbot, St Brendan is remembered for his legendary seven-year voyage around the Atlantic Ocean, which included the discovery of the Promised Land of the Saints. Much of his story is mythical, but the historical figure is thought to have travelled by sea to Scotland, Wales, Brittany, and possibly further.

James Bruce
1730–94
An explorer of Northern Africa and the Middle East, in 1770 Bruce reached the source of the Blue Nile and followed it to where it joined the White Nile in present-day Sudan. There is some debate about whether he was the first European explorer to accomplish this. A Scottish-born British consul based in Algiers, he travelled extensively in the region. During these trips, he came across several copies of the Book of Enoch, which had been thought lost.

Bungaree
c.1775–1830
A crewman on the HMS *Investigator*'s 1802–03 expedition to map the coast of Australia, Bungaree became the first Indigenous Australian known to have circumnavigated the country's mainland. He was recruited for the expedition by a Naval officer, Matthew Flinders, whom he had impressed on an earlier voyage with his sharp mind and bravery. Bungaree later joined an 1817 voyage to northwest Australia. He was noted for his ability to serve as an intermediary between Aboriginal people and Europeans.

△ Bungaree
Pictures often present Bungaree in military uniforms, which he wore. A popular figure, he was praised by the Europeans for his skills as a diplomat and translator as well as his witty impersonations.

▷ John Cabot
The navigator is shown here in Venetian attire – he worked in Venice before moving to England. His nationality was important to fellow Italian Giustino Menescardi, who painted this mural in 1762.

Johann Ludwig Burckhardt
1784–1817

A scholar and explorer who studied North Africa and Southwest Asia, Burckhardt was the first modern European to locate the ruins of the city of Petra in present-day Jordan. He also visited the temple of Abu Simbel in Egypt. Born to a wealthy Swiss family, Burckhardt travelled throughout the Muslim world, learning Arabic – first at Cambridge University and then in Syria and Egypt – using a Muslim alias, and wearing Muslim dress to present himself as a Muslim merchant from India, in the hope that this would improve the reception he received from local people and government officials and make him less of a target for criminals. He died in Cairo at the age of just 32, but his manuscripts helped establish his legacy.

Robert O'Hara Burke
1821–61

From 1860 to 1861, Burke led an ill-fated expedition that set out to cross Australia from south to north. The expedition achieved this goal – one member of the team reached its destination – but Burke and six others died of starvation, dehydration, and exhaustion during the trip. Born in Ireland, Burke had served as a lieutenant in the Austrian army and, following his military career, as an inspector in the police – first in Ireland, then in Australia. He had possessed no experience as an explorer when named to lead the expedition that cost him his life.

Richard Burton
1821–90

An explorer, writer, translator, soldier, scholar, and diplomat, Burton was among the first non-Muslim Europeans to complete a pilgrimage to Mecca. He also led an expedition to find the source of the Nile, though Burton was too ill to join expedition partner John Speke when he finally reached Lake Victoria, which sits between present-day Tanzania and Uganda. Burton published dozens of books about his adventures, and translated major works by other writers into English – he was said to speak 20–30 languages. He later served as a British consul in several locations, and was knighted by Queen Victoria in 1886.

John Byron
1723–86

A British admiral, Byron commanded the HMS *Dolphin* on its 1764–66 circumnavigation of the globe. In 1741, he had been a midshipman on another vessel attempting a circumnavigation, HMS *Wager*. That ship was wrecked on an island off the coast of Chile, however, and Byron didn't make it back to England until 1746. Born into a noble family – the poet Lord Byron was his grandson – Byron would earn the nickname Foul-Weather Jack for his unfortunate propensity to encounter storms while at sea. In 1769, he was named the governor of Newfoundland, Canada.

John Cabot
c.1450– c.1499

An Italian-born navigator who sailed for the English crown, in 1497 Cabot became the first modern European to reach North America, setting the stage for British colonization there. Christopher Columbus famously preceded Cabot across the Atlantic, but Columbus made landfall in the Caribbean, never setting foot in North America. Like Columbus, Cabot's goal was to reach Asia by sailing west from Europe, and like Columbus he believed he had succeeded, although the lands Cabot claimed on behalf of English king Henry VII were actually along the Atlantic coast of what is now Canada. In 1498, Cabot led an expedition of five ships and about 200 men across the Atlantic, but it failed to return. He was presumed dead in 1499. Cabot Strait, which connects Canada's Gulf of Saint Lawrence to the Atlantic, is named in his honour.

Pedro Álvares Cabral
c.1467–1520

The son of a Portuguese nobleman, in 1500 Cabral led the second European expedition to reach India by sailing around Africa's Cape of Good Hope. Cabral's route looped so far west that his expedition also landed in Brazil, probably becoming the first Europeans to do so, although it is possible that Spain's Vicente Yáñez Pinzón reached Brazil first. Cabral traded for spices along the coast of India, but storms took their toll and just four of the 13 ships returned to Portugal in 1501. It was his only expedition.

13
Number of ships on Cabral's 1500 expedition

René-Auguste Caillié
1799–c.1838

In 1828, French explorer Caillié became the first European to visit the city of Timbuktu in present-day Mali and survive the journey. A Scottish officer, Major Gordon Laing, reached Timbuktu two years prior to him, but was killed before he could return. Caillié published an account of his adventure and received a 10,000-franc prize from the Geographical Society of Paris. He had learned Arabic and Islamic ritual while living with the Brakna Moors in what is now Mauritania so that he could pose as an Arab traveller. Caillié had previously participated in an expedition that crossed Senegal's Ferlo Desert. He did no further exploring after his trip to Timbuktu.

◁ Chistopher Columbus
This 1519 portrait by Sebastiano del Piombo is said to be the definitive image of Columbus, though it's less than certain that it accurately depicts him as it was painted more than a decade after his death.

James Cameron
1954–

Best known as a filmmaker, the Canadian-born Cameron is also a pioneer of deep-sea exploration. He helped to design the submersible *Deepsea Challenger*, and in 2012 used it to become the first person to take a solo trip to the deepest known part of the ocean, the Challenger Deep section of the southern end of the Mariana Trench of the western Pacific Ocean. A number of Cameron's films have involved the sea, including *Titanic* (1997), several deep-sea documentaries, and *The Abyss* (1989).

Diogo Cão
c.1452–86

A Portuguese navigator, Cão led a pair of voyages along the Atlantic coast of Africa in the 1480s, in the hope of finding a sea route to India. He failed to achieve that goal, but did make it as far south as Cape Cross, in what is now Namibia, and was the first European to reach the mouth of the Congo River in the present-day Democratic Republic of the Congo. He marked this last achievement by installing a stone pillar to claim the land for Portugal.

Philip Carteret
c.1733–96

A British naval officer, Carteret sailed around the world twice. He served as a lieutenant on HMS *Dolphin* on its 1764–66 circumnavigation of the globe, then in 1766–69 commanded HMS *Swallow* on its circumnavigation. The *Swallow* expedition reached several Pacific islands previously unknown in Europe, including Pitcairn Island and the Carteret Islands (named in his honour). The trip was challenging – more than half of the *Swallow*'s crew died during it, and Carteret was detained by the Dutch for four months in what is now Indonesia. Carteret reached the rank of rear admiral, but was in chronically poor health following his round-the-world voyages.

Jacques Cartier
1491–1557

French navigator Cartier led three voyages to the Canadian coast and Canada's St Lawrence River between 1534 and 1542, setting the stage for French claims to the region. The Algonquins call the river the Magtogoek; other peoples called it by a variety of names. Cartier also gave Canada its name, referring to the land by the Huron-Iroquois word *kanata*, which means village. Initial attempts to establish a colony in Canada failed, and the gold and diamonds Cartier believed he had found there turned out to be worthless. He abducted several Iroquoian people, including the Haudenosaunee (Iroquois) chief Donnacona.

Evliya Çelebi
1611–c.1684

Çelebi spent more than four decades travelling extensively throughout the Ottoman Empire, which then extended into Europe, Southwest Asia, and Asia. The son of a goldsmith in Constantinople (present-day Istanbul, Türkiye) who had ties to the Ottoman court, Çelebi was employed as a scholar and entertainer, but refused any work that inhibited his travels. Some of his trips were on behalf of the government, others taken on his own account. His writings offer an insight into 17th-century Ottoman life.

Samuel de Champlain
c.1567–1635

A French explorer, Champlain made multiple voyages to North America, mapping what is now Canada's St Lawrence River (known as the Magtogoek to the Algonquin people), the Great Lakes, and surrounding areas, as well as exploring the Atlantic coast as far south as Massachusetts. Born into a seafaring family, he founded the first permanent French settlement in the Americas, on the site of what is now Quebec City. Champlain formed alliances with the Huron, Montganais, and Algonquin population, but when he introduced them to firearms it fed further bloodshed and made a long-term enemy of the Haudenosaunee (Iroquois).

40

Years Celebi spent exploring the Ottoman Empire.

Aleksei Chirikov
1703–48

A Russian official, Chirikov was second in command on Vitus Bering's expeditions to the Pacific coasts of Russia and North America. On the second of these, in 1741, Chirikov commanded the *St Paul*, one of the venture's two primary ships. The voyage made Chirikov and Bering the first explorers to reach the northwest coast of the Americas on behalf of Russia, and helped to establish Russian claims in the Americas. Chirikov's *St Paul* achieved much of the expedition's North American mapping after it was separated from the Bering-commanded *St Peter* in a storm.

Hugh Clapperton
1788–1827

Clapperton, Walter Oudney, and Dixon Denham explored West and Central Africa in the 1820s on behalf of the British government. In the process, they became the first Europeans to reach Lake Chad, located at the junction of present-day Nigeria, Niger, Chad, and Cameroon, as well as the first to provide a description of much of the region. Born in Scotland, Clapperton joined the British navy, serving in the East Indies and Canada and rising to the rank of commander. He died of dysentery near Sokoto, a city in what is now northwest Nigeria.

▽ **Hernán Cortés**
A 16th-century image depicts a meeting between the conquistador Cortés and Indigenous Mexica people. The Mexica are not in awe; it is probably a myth that they mistook the conquistador for a god.

William Clark
1770–1838
Clark was the co-leader, with Meriwether Lewis, of the most famous expedition in 19th-century US history. The Lewis and Clark expedition of 1804–06 crossed and mapped the Louisiana Territory that the US had recently purchased from France, as well as what is now the US Pacific Northwest. Clark, born into a plantation-owning family in Virginia, was an experienced soldier and frontiersman. He would go on to serve in a range of governmental roles, including governor of the Missouri Territory and superintendent of Indian affairs.

Christopher Columbus
1451–1506
Columbus's 1492 crossing of the Atlantic made the Genoese navigator the first modern European to sight the Americas and one of the most famous explorers in history. Over four transatlantic voyages on behalf of Spain, Columbus became the first European to reach South and Central America as well as the first in modern history to establish an American colony. He was granted the title Admiral of the Ocean Sea and was regarded as a talented navigator, but he failed to achieve his goal of reaching Asia. As an administrator, he was deeply flawed and at times brutal – his men enslaved, raped, and killed large numbers of Indigenous people, and Columbus ordered many acts of cruelty.

James Cook
1728–79
A British naval officer and navigator, Cook led three expeditions to the Pacific in 1768–79. Cook's voyages surveyed the coasts of New Zealand and Australia, encountered numerous Pacific islands, and searched in vain for the Northwest Passage. The English-born son of a Scottish farmhand, he had seen action in the Seven Years' War and surveyed the coast of Newfoundland, Canada. Cook died in a confrontation with locals on Hawaii caused by misunderstandings concerning a stolen boat.

▷ **Captain James Cook's telescope**
Cook used this Gregorian-type reflecting telescope on his Pacific voyages. Invented in 1661, it used concave mirrors to create magnification.

Hernán Cortés
1485–1547
A Spanish conquistador, in 1519 Cortés led an expedition that overthrew the Mexica Aztec Empire with disastrous results for the Indigenous population. He had only around 500 soldiers under his command, but enlisted support from local populations opposed to Mexica rule. It is said that as many as half of the people of the region were soon dead, with the survivors subjugated under Spanish rule. He later led an expedition to Honduras. Born to an impoverished noble family, Cortés helped Diego Velázquez conquer what is today Cuba and was mayor of Santiago. Despite the lands and titles his conquests earned him, he died in debt after returning to Spain in 1541.

Juan de la Cosa
1460–c.1510
A navigator and cartographer, de la Cosa was the owner of the *Santa María*, one of the three ships that Columbus used for his first voyage to the Americas. The *Santa María* did not survive that trip – it was shipwrecked off the coast of what is now Haiti – but de la Cosa did and visited the Americas seven times, accompanying Columbus on his first three voyages. De la Cosa died on the last of his trips, shot with a poisoned arrow in what is now Colombia when trying to colonize the land. Ten years earlier, he created the oldest known map of the Americas.

Jacques-Yves Cousteau
1910–97
An underwater explorer, filmmaker, and innovator, Cousteau was part of a team that in 1943 invented the Aqua-Lung, which enabled divers to swim underwater for extended periods. He later developed underwater living pods, two-person submarines, and underwater cameras. Born in France, Cousteau made films that won awards and popularized oceanography, while his experiments improved diving technology and informed the development of the NASA training programs.

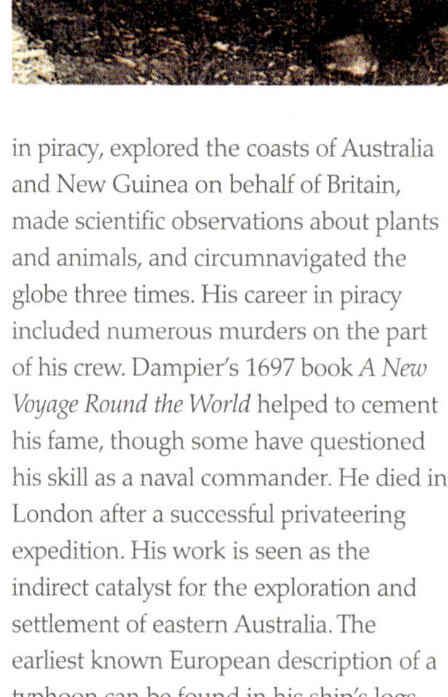

▽ Alexandra David-Néel
This image of David-Néel in Nepal probably dates to around 1912, near the outset of her 14-year journey through Asia which included the visit to Tibet that brought her fame.

Lady Elizabeth Craven
1750–1828

A playwright and writer, Craven travelled extensively across Europe in 1785–86. Her 1789 book *A Journey through the Crimea to Constantinople* detailed her adventures in the form of a series of letters. Following her separation from her first husband, the 6th Baron Craven, she wrote about the unfairness of marriage for women at the time. She later married a Prussian nobleman and they settled in London.

Peter Custis
1781–1842

Custis worked as a botanist on the 1806 Red River Expedition while still a student. That expedition crossed portions of Louisiana, Arkansas, and eastern Texas on behalf of the US government, but it was widely considered a failure because it was turned back by Spanish troops before exploring the full length of the river. The expedition did succeed in making contact with Indigenous peoples along its route, including the Caddo and Alabama-Coushatta tribes. Custis also documented many of the region's plants and animals. He returned to medical school after the journey, and later established a medical practice in New Bern, North Carolina.

D

Vasco da Gama
c.1460–1524

In 1497–99 Portuguese navigator da Gama achieved the first successful sea voyage from Europe to India, opening a lucrative new trade route. He made two further voyages to India in 1502–03 and 1524. The son of the governor of a fortress in the Portuguese province of Alentejo, da Gama was said to be brutal – during his second voyage to India, it is alleged by some late accounts that he captured a ship carrying Muslim pilgrims on their way home from Mecca, had it looted, then had the Muslims locked in the hold and burned alive. Da Gama died of malaria in 1524, during his third voyage to India.

Giovanni da Pian del Carpini
c.1180–1252

An Italian-born Franciscan friar, Carpini was sent by Pope Innocent IV to meet with the Great Khan, founder of the Mongol Empire, in 1245. Already more than 60 years old when he left, Carpini returned with a letter from the Khan in 1247. He was among the first Europeans to visit the Mongol Empire. After his return Carpini wrote a manuscript about the Mongols and was appointed archbishop of Antivari, Dalmatia (present-day Croatia).

William Dampier
1651–1715

Englishman Dampier lived an extraordinary though sometimes troubling seafaring life, during which he engaged in piracy, explored the coasts of Australia and New Guinea on behalf of Britain, made scientific observations about plants and animals, and circumnavigated the globe three times. His career in piracy included numerous murders on the part of his crew. Dampier's 1697 book *A New Voyage Round the World* helped to cement his fame, though some have questioned his skill as a naval commander. He died in London after a successful privateering expedition. His work is seen as the indirect catalyst for the exploration and settlement of eastern Australia. The earliest known European description of a typhoon can be found in his ship's logs.

Charles Darwin
1809–82

A British biologist and geologist, Darwin produced a theory that species evolve through a process of natural selection, which made him one of history's most famous scientists. The son of a doctor, he joined HMS *Beagle* as an unpaid geologist on its 1831–36 circumnavigation of the globe. The notes and observations Darwin made on that journey, plus the specimens and fossils he collected, formed the basis for his theory of evolution. During a stop in the Galapagos archipelago in 1835 he observed about 15 species of finch, each adapted to the island on which it lived.

◁ Charles Darwin
Darwin took this pocket sextant on HMS *Beagle*'s 1831–36 circumnavigation of the globe. Despite owning this navigational tool, Darwin was not the best sailor – he suffered from severe seasickness.

▷ Francis Drake
This miniature portrait was painted by Nicholas Hilliard in 1581, the year Drake was knighted by Elizabeth I following his circumnavigation of the globe.

> *"Wherever the European has trod, death seems to pursue the aboriginal."*
>
> CHARLES DARWIN, *THE VOYAGE OF THE BEAGLE*, 1839

Alexandra David-Néel
1868–1969

A world traveller, David-Néel is best known for being the first European woman to visit Lhasa, Tibet. She reached that destination in disguise in 1924 (at the time foreigners were not permitted there). Born in France, she travelled widely, visiting India multiple times, touring parts of Asia and North Africa as an opera singer, and devoting several years to slowly crossing China. She was named a Chevalier of the French Legion of Honour.

Sake Dean Mahomed
1759–1851

Dean Mahomed was among the first Indian immigrants to Europe when he arrived in the 1780s, and the first Indian to publish a book in English. In 1810, he opened England's first Indian restaurant. The venture failed, but Dean Mahomed later launched a successful bathhouse in Brighton. Prior to travelling to Europe, Dean Mahomed had served in the British East India Company's army as a trainee surgeon. He resigned and journeyed to Ireland and England in the company of the officer under whom he had served, who became a close friend.

◁ Sake Dean Mahomed
This lithograph by Thomas Mann Baynes from c.1820–25 shows Dean Mahomet in clothing he probably designed himself to wear at court. The outfit includes a silk Indian-court-style coat.

Dixon Denham
1786–1828

An explorer, soldier, and colonial administrator, Denham joined Hugh Clapperton and Walter Oudney on an 1820s expedition to West and Central Africa, which hoped to gain access to trans-Saharan trade. The group became the first Europeans to reach Lake Chad, which Denham confirmed was not the source of the Niger River. The expedition made Denham a hero – Oudney had not survived the journey and Clapperton soon returned to Africa, leaving Denham free to publish a self-serving account. He was promoted to lieutenant colonel and appointed governor of Sierra Leone, but soon died there of illness.

Bartolomeu Dias
c.1450–1500

In 1487–88, Dias led the first European voyage around the southern tip of Africa, an accomplishment that helped to open sea trade routes between Europe and Asia. It was the only expedition the Portuguese nobleman would lead, although he did join Vasco da Gama on a leg of his voyage to India, and sailed under the explorer Pedro Álvares Cabral on an expedition to Brazil. Dias died during the Cabral expedition when his ship sank in a storm off the Cape of Good Hope.

Charles Montagu Doughty
1843–1926

An Englishman who travelled widely in Arabia between 1876 and 1878, Doughty was the first European to set eyes on the ruins of Mada'in Salih in modern-day Saudi Arabia. His two-volume *Travels in Arabia Deserta* (1888) was not initially popular, but it came to be considered among the best 19th-century accounts of the region – T.E. Lawrence wrote the introduction to one edition.

Francis Drake
c.1540–96

An English admiral, Drake led the second voyage to circumnavigate the globe, in 1577–80. In the process, he became the first person to lead an expedition around the globe – Ferdinand Magellan died before completing his earlier journey. The son of a tenant farmer, Drake made raids on Spanish ships that brought great wealth to him and his sponsor, England's Queen Elizabeth I, who knighted him at the end of his round-the-world trip. He later played a major role in England's defeat of the Spanish Armada.

Paul du Chaillu
c.1831–1903

Chaillu explored Gabon from 1856 to 1859, returning to Europe with evidence of the existence of gorillas, a creature previously little known outside their region. The similarities between gorillas and humans caused a stir at the time. Chaillu's African expeditions helped to bring other species to the attention of the European scientific community as well, though his work was not always well received. Chaillu, who was mixed race, considered himself American, but there is some dispute as to whether he was born in France, the US, or on the island of Réunion. His mother may have been an enslaved person and his father was a merchant and slaveholder.

William Dunbar
1749–1810

In 1804–05, Dunbar was co-leader of one of the four expeditions commissioned by US President Thomas Jefferson to explore the Louisiana Territory that the United States had recently acquired from France. While the better-known Lewis and Clark-led expedition ventured across the northwest, Dunbar and his men remained further south, in what is now northern Louisiana and southern Arkansas. A Scottish-born scientist and plantation owner who had been living in America since 1771, Dunbar was an acquaintance of Jefferson and published articles on a range of scientific topics. After the expedition, Dunbar acted as an adviser for the 1806 Freeman-Custis Red River expedition. A prominent figure in the American agricultural community, Dunbar also invented a screw press that made the first square cotton bales.

◁ Amelia Earhart
This photo from 1936, the year before Earhart's disappearance, shows her in a prototype Stearman-Hammond Y-1, an aircraft that had its propellor behind the cockpit rather than on the plane's nose.

Henri Duveyrier
1840–1892

An explorer of North Africa, Duveyrier travelled alone into the Sahara at the age of 19. His journey lasted two and a half years and resulted in his 1864 book about the region, *Exploration of the Sahara: The Tuareg of the North*. He made several other visits to the Sahara and explored the shallow salt lakes of Algeria and Tunisia. Duveyrier learned to speak Arabic, and improved Europe's understanding of the region's geography and the Tuareg people with whom he lived for months at a time, bringing attention to their rich culture, including their poetry. He became an official of the Geographical Society of Paris, the city of his birth.

E

Gil Eanes
c.1395– c.1445

A Portuguese navigator, in 1434 Eanes became the first European to sail past Cape Bojador on the Atlantic coast of what is now Western Sahara. This helped to open the door to future Portuguese voyages to southern Africa and Asia. Rough sea had frustrated earlier attempts, including one by Eanes a year before.

Amelia Earhart
1897–1937

Earhart was an American navigator who in 1932 became the first woman to fly solo across the Atlantic, crossing from Newfoundland to Northern Ireland. She later completed the first solo flight from Hawaii to California, and the first solo flight from Los Angeles to Mexico City. Celebrated as a pioneer and nicknamed Lady Lindy, after aviator Charles Lindberg, in 1937 Earhart and her navigator, Fred Noonan, attempted to fly a plane around the world, but disappeared over the Pacific Ocean. Born in Kansas, Earhart served as a nurse's aide during World War I before discovering a passion for aviation in the 1920s.

Juan Sebastián Elcano
c.1476–1526

A Basque navigator, Sebastián Elcano completed the first circumnavigation of the globe, sailing one of Ferdinand Magellan's ships back to Spain following that explorer's death. Having joined in a failed mutiny, Elcano was not the obvious choice to take command, but other likely commanders had died and one stayed behind with a ship that needed repair, leaving him in charge of the remaining ship. Elcano was awarded a coat of arms that featured a globe and the motto *Primus circumdedisti me* ("You went round me first" in Latin). He joined a second circumnavigation in 1525, but died of malnutrition in the Pacific.

Eldad ben Mahli ha-Dani
c.851–900 CE

A Jewish traveller in ninth-century North Africa, Iberia, and Southwest Asia, Eldad claimed to have met the "Ten Lost Tribes" of Israel. For centuries, Eldad's narrative was influential in the Jewish world and Europe. Significant elements of his story are now dismissed as not factual, and some scholars believe the entire narrative should be treated as fictional. He was probably from southern Arabia.

Ennin
794–864 CE

Born in Japan, Ennin was a Buddhist monk who spent more than nine years studying and travelling in China during the Tang dynasty. He returned to Japan in 847 with hundreds of volumes of Buddhist literature and details about practices and music that furthered the religion in Japan. His teachings led to the Sammon branch of Tendai Buddhism, and his diary of his travels is among the earliest accounts of that nation by a foreigner.

▷ Leif Eriksson
This statue of Eriksson by A. Stirling Calder was a gift from the US to Iceland honouring the 1,000th anniversary of its parliament in 1932. It stands outside Hallgrímskirkja in Reykjavík.

Erik the Red (Erik Thorvaldsson)
950–c.1003 CE
A Norse explorer, Erik founded the first European settlement in Greenland, in or around 985. Born in modern-day Norway, he had moved with his family to Iceland when his father, Thorvald Asvaldsson, was exiled for manslaughter. When Erik was in turn exiled from Iceland for killing two men in around 980, he continued west, settling in Greenland and giving that island its name. He returned to Iceland a few years later and convinced an estimated 400–500 settlers to join him in the new colony, which survived into the 15th century. It is believed an expedition led by his son Leif Eriksson was the first to reach North America.

Leif Eriksson
c.970–c.1025 CE
Eriksson was a Norse explorer who is thought to have arrived in what is now eastern Canada in or around the year 1000, making him the first European to land in North America. He was probably born in Iceland, but moved to Greenland with his family in or around 985 – his father, Erik the Red, was the leader of Greenland's first colony of European settlers. Legend has it that Eriksson became chieftain of that colony soon after his Canadian trip and never returned to the land he had found. Other members of Eriksson's family did return to Canada, but hostile relations with the Indigenous population and the distances involved ended Norse attempts to establish a colony there.

Esteban (Estevanico)
c.1500–39
Born in Morocco, Esteban is generally considered to be the first person of African heritage to participate in North American exploration. Also known as Estevanico de Dorantes or Mostafa Azemmouri, he was enslaved by a member of Pánfilo de Narváez's 1527 expedition to the Gulf Coast. The journey was a disaster, involving a shipwreck and capture by the Coahuiltecan people in what is now Texas. Esteban was among the few survivors and part of a group that eventually embarked on an epic journey across what is now the southwestern US and into Mexico. Assigned to guide a 1539 Spanish expedition heading north from Mexico, he became the first non-native person to reach sections of what is now the US states of Arizona and New Mexico, before being killed, probably by the Zuni, though the reason is disputed.

Edward John Eyre
1815–1901
English-born Eyre was a sheep farmer in Australia who explored the arid lands of his adoptive country. In 1840–41, he completed a harrowing 3,200-km (2,000-mile) trek from Adelaide to Albany around the Great Australian Bight of the southwest coast. His reputation made, Eyre served as Lieutenant Governor of New Zealand, then of St Vincent, followed by terms as acting Governor of the Leeward Islands and eventually as Governor of Jamaica. In Jamaica, Eyre crushed a rebellion so cruelly that in 1866 he was sent back to England, but a grand jury declined to indict him for his actions.

F

Anna Maria Falconbridge
1769–c.1835
In the early 1790s, Falconbridge made a pair of extended visits to Sierra Leone, where her husband was employed by the Sierra Leone Company. The series of letters she wrote about her visits were the first published firsthand account by an Englishwoman in West Africa. She returned to England after her husband's death in 1792. Published in 1794 as *Narrative of Two Voyages to the River Sierra Leone during the Years 1791-1792-1793*, the book presented an unflattering view of the Sierra Leone Company.

Faxian
337–c.422 CE
A Chinese Buddhist monk, in 402 Faxian travelled by land to India. He remained there for a decade, studying Buddhist texts and visiting Buddhist shrines. He returned to China by sea in 412, making an extended stop in Sri Lanka. His account of his journey was published as *A Record of Buddhist Kingdoms*. His name is also spelled Fa Xian.

Eduard Robert Flegel
1855–86
A German explorer born in Lithuania, Flegel probed Africa's Benue River basin, becoming the first European to reach the source of that river in what is now Cameroon. He returned to the region several times, formed close bonds with the local inhabitants, and may have learned to speak Hausa, one of the local languages. His plan to establish German trade in the area was largely unsuccessful, and he died in Nigeria at the age of 30.

Matthew Flinders
1774–1814
A navigator and cartographer, Englishman Flinders circumnavigated Australia in 1801–03, mapping the coast. He was detained by the French during his return voyage to England, however, and not allowed to return there until 1810. He died barely four years later. Flinders had circumnavigated the island of Tasmania during a 1795 voyage. He helped to popularize the name of Australia for the continent, and his name now dots its landscape, including the Flinders Ranges, Flinders Bay, and Flinders Island.

▷ **Faxian**
This ninth-century painting in Mogao Caves, Gansu Province, China, probably depicts a travelling monk – the scrolls suggest an educated man. It may represent Faxian.

Georg Forster
1754–94

A German naturalist, Forster joined Captain James Cook's second circumnavigation of the globe in 1772–75 and published a popular and influential account of that expedition in 1777, *A Voyage Towards the South Pole and Round the World*. Forster's father, Johann Reinhold Forster, was the primary naturalist on that circumnavigation – Georg Forster's book was based largely on his father's journals. The younger Forster later served as a professor of natural history and a head librarian at a series of universities. His role in an effort to form a republican government in Mainz eventually forced him to leave Germany for Paris, where he died aged 39.

△ **Martin Frobisher**
Cornelius Ketel painted Frobisher's portrait shortly after the navigator's initial voyage to the "New World". He is dressed in gold, probably to suggest the wealth he expected the Americas to provide.

John Franklin
1786–1847

Franklin was an English naval officer and Arctic explorer who led three expeditions to the far north and participated in a fourth, becoming a hero even though two of his expeditions ended horribly. After commanding a ship on Captain David Buchan's unsuccessful 1818 attempt to find a sea route to the North Pole, Franklin led an 1819 overland expedition in search of the Northwest Passage. The expedition was a disaster – half of Franklin's 20 men died and there were reports of murder. Franklin led an expedition to the Arctic coast of Canada in 1825–27, but it would be his 1845 expedition, again in search of the Northwest Passage, that history would remember. Franklin and his men perished after the expedition's ships became stuck in the Arctic ice, the details of their struggles revealed only when a note was discovered in 1859. There were reports of cannibalism. While Franklin did not prove the existence of the Northwest Passage, members of his expedition may have reached Simpson Strait, which connected with the western coastal waters previously visited by Franklin.

Thomas Freeman
c.1784–1821

An Irish-born American surveyor, Freeman led the Red River Expedition, which in 1806 explored its namesake river from Louisiana to what is now Bowie County, Texas, before being turned back by Spanish troops. The men were expected to survey the river, record the flora and fauna of the region, and meet with Indigenous leaders. The expedition was considered a failure because it did not explore the full length of the river, but it did make important findings about the region and its people, plants, and animals. Freeman was later named US surveyor general and surveyed the borders between several US states.

89

Number of minutes Gagarin took to orbit the Earth

John C. Frémont
1813–90

A US Army officer, Frémont led an 1842 expedition that searched for the best route through the Rocky Mountains. He also led an 1843 expedition that explored both the Columbia River in the US northwest and the Sierra Nevada mountains in present-day California and Nevada. In 1845, Frémont led an expedition into California, where he supported a rebellion against Mexican control of the region. He was appointed military governor of California in 1847, only to be charged with mutiny and expelled from the army, though the judgment was reversed by President James K. Polk. He later served in the US Senate, as governor of the Arizona Territory, and as a Union general in the US Civil War, as well as running unsuccessfully for the Presidency.

Martin Frobisher
c.1535–94

An English navigator, Frobisher led a series of expeditions to what is today Canada in 1576–78. These voyages searched for the Northwest Passage, attempted to establish a colony, and tried to find gold. They failed to achieve any of those goals – the ore Frobisher brought back to England turned out to be worthless. Frobisher had previously looted French ships as a privateer and participated in a pair of expeditions to the Guinea coast, on Africa's Atlantic coast. Following the financial failure of his Canadian expeditions, he served under Sir Francis Drake on an expedition to the West Indies, and was knighted for his significant role in defeating the Spanish Armada in 1588. He was killed during a 1594 battle with Spanish forces on the west coast of France.

G

Yuri Gagarin
1934–68

On 12 April 1961, Gagarin became the first human to reach outer space, beating American Alan Shepard by 23 days. The Soviet cosmonaut completed a single orbit and returned to earth in less than two hours. The flight made Gagarin famous around the world and earned him honours from the Soviet government, including the title Hero of the Soviet Union. He never returned to space and died in a plane crash in 1968. Prior to entering the space programme, he had served as a pilot in the Soviet Air Force.

Aelius Gallus
c.25 BCE

A Roman prefect of Egypt during the reign of Augustus, Gallus led a military expedition to what is now Yemen, but was then called Arabia Felix. The expedition was intended to expand Roman power on the Arabian Peninsula, but it was undermined by disease, the desert environment and, reportedly, intentional deceptions by local guides. Little else is known of Gallus's life.

Giovanni da Montecorvino
1247–1328

Montecorvino (also known as John of Montecorvino) was a Franciscan missionary who founded the first Roman Catholic missions in China and India, and was appointed the first archbishop of

▽ **Maria Graham**
This portrait of Graham was probably painted in the late 1830s. At the time, her description of the 1822 earthquake in Chile led to a debate about the role of earthquakes in the formation of mountains.

present-day Beijing. His letters provide some of the earliest European accounts of those lands, and cultivated a strong presence for Catholicism in China that survived into the reign of the last Mongol khan. Born in Italy, Montecorvino served as a missionary in Armenia and Persia.

Diogo Gomes
c.1440–84

A Portuguese navigator, Gomes explored Africa's Atlantic coast on behalf of Prince Henry the Navigator. On one voyage, he sailed up the Gambia River, reaching the Gambian town now known as Kuntaur. Another expedition reached the Cape Verde Islands, though it is disputed whether he was the first European navigator to do so. Late in Gomes's life, he dictated his story to a German cartographer, creating a valuable, if not always reliable, account of early exploration along the coast of Africa.

Maria Graham
1785–1842

A British travel writer, in 1824 Graham became the first woman to have a paper published in one of the Geological Society of London's journals. Born Maria Dundas in Cumberland, she accompanied her father, a naval officer, to Bombay (present-day Mumbai) in 1808, meeting her future first husband Thomas Graham, also a naval officer, on the voyage. She remained in India for two years, later publishing a book about her time titled *Journal of a Residence in India* (1812). In 1818, Graham and her husband travelled to Italy, a trip that resulted in another travel book – *Three Months Passed in the Mountains East of Rome* – as well as a biography of the artist Nicholas Poussin, both published in 1820 after their return to London. The couple sailed for South America in 1821, where Graham explored for two years, eventually writing a pair of books about these travels. Ill-health limited her ability to travel after 1831.

James Augustus Grant
1827–92

Grant joined John Speke's 1860–63 expedition to locate the source of the Nile. A British army officer born in Scotland, Grant made significant contributions to that expedition, particularly when it came to the collection of local plants. He earned a gold medal from the Royal Geographical Society for his efforts, and later published a book about the trip, *A Walk Across Africa*. Grant had previously fought in the Sikh Wars of the 1840s, and he returned to military life after the adventure; he served in the intelligence section and reached the rank of lieutenant colonel.

H

Emperor Hadrian
76–138 CE

The emperor of Rome from 117 until his death in 138, Hadrian travelled extensively, venturing west across Europe to Britain, then south to Spain, and east to Greece and Anatolia in a tour of the Roman Empire that lasted from 121 until 125. He started a second tour in 128, visiting Arabia and the River Nile, among other destinations, and not returning to Rome until at least 132. A trip to Palestine in 134 was a military expedition. Hadrian ordered notable construction projects during his reign, including what is known as Hadrian's Wall along the northern border of Roman Britain.

Hannibal
247–c.183 BCE

A Carthaginian general, Hannibal achieved numerous victories against Roman forces and their allies during the Second Punic War (218–201 BCE), famously leading his soldiers and a team of elephants across Europe and over the Alps into Italy. Hannibal's assault on Rome ultimately ended in failure, and Carthage was eventually forced to accept unfavourable peace terms. Hannibal served as a civil magistrate for a time after the war, and continued to seek opportunities to fight Roman power. He died by suicide to avoid capture by the Romans.

Harkhuf
c.2290–70 BCE

The earliest explorer for whom travel records exist, the Egyptian Harkhuf led expeditions from that country to Nubia (today Sudan), overseeing trade missions as a court official for King Merenre and King Pepi II. Harkhuf may have previously accompanied his father on a similar journey. Born into a noble family on Elephantine, an island on the Nile River in Egypt, his expeditions may have travelled along the Nile to a location called Yam, or Iyam, most likely near the confluence of the Blue and White Niles, south of modern-day Khartoum. No records exist of any Egyptian venturing further south for the next 800 years. Details of Harkhuf's life and travels are known from inscriptions on the walls of his tomb at Aswan, Egypt.

"I will either find a way, or make one."

POSSIBLY APOCRYPHAL QUOTE WIDELY ATTRIBUTED TO HANNIBAL

▷ **Emperor Hadrian**
Hadrian (76–138 CE) was the first Roman emperor to wear a beard. His facial hair reflected his affinity with Greece, where facial hair was then viewed favourably.

▷ **Thor Heyerdahl**
In 1970, Heyerdahl crossed the Atlantic Ocean from Morocco to Barbados in the *Ra II*, a ship made from papyrus. His goal was to prove that ancient North Africans could have reached the Americas.

Sven Hedin
1865–1952
Hedin led a series of expeditions to Central Asia between 1890 and 1908, which substantially improved Europe's knowledge of the region. During these expeditions, Hedin crossed deserts and mountain ranges, followed ancient caravan routes, rediscovered the lost city of Dandan-Uilik in western China's Taklamakan Desert, and mapped and explored Tibet, among a long list of other accomplishments. The Swedish explorer's legacy was ruined by his pro-Nazi views during both world wars.

St Helena
c.248–c.328 CE
A Roman empress and the mother of Constantine the Great, Helena of Constantinople (present-day Istanbul) made a pilgrimage to the Holy Land late in her life. According to legend, there she discovered the cross on which Jesus was crucified. Believed to have been born into a Greek family of modest means in what is now Türkiye, Helena wed Roman emperor Constantius I and accepted Christianity under her son Constantine's influence. She is regarded as a saint by the Roman Catholic church.

Matthew Henson
1866–1955
Henson was a member of Robert Peary's 1909 Arctic expedition that became known as the first to reach the North Pole, though Peary's claims of having reached the Pole have been questioned. A Black American born a year after the end of slavery in the United States, Henson played a central role in Peary's numerous Arctic expeditions – he learned the Inuit language and cultivated useful skills, such as hunting and dog sled handling. He wrote a book about his experiences, *A Negro Explorer at the North Pole*. Henson later spent over 20 years working as a clerk in New York's federal customs house.

Herodotus
c.484–420 BCE
Greek historian Herodotus' account of the Greco-Persian Wars (490–479 BCE) is often cited as the first great work of historical analysis. Herodotus was probably born in a region that is now part of Türkiye but was then under Persian rule. He travelled widely in the Persian Empire, visiting locations including Egypt, Palestine, Syria, Babylonia, Libya, and Macedonia, and even venturing across the Black Sea, collecting oral histories along the way.

Thor Heyerdahl
1914–2002
A Norwegian anthropologist, Heyerdahl crossed oceans on small boats built without the use of modern tools and materials to prove that ancient people could have made similar journeys. The best-known of his adventures was in 1947, when Heyerdahl sailed more than 6,500 km (4,000 miles) across the Pacific from Peru to French Polynesia on *Kon-Tiki*, a simple balsawood raft that he had built himself. Many of his theories about transoceanic ways of travel have failed to stand the test of time, but his daring sea voyages – and books and documentaries about them – brought him fame.

Edmund Hillary
1919–2008
On May 29 1953, Hillary and Tenzing Norgay became the first people known to have reached the 8,849-m (29,032-ft) summit of Mount Everest, the highest point on Earth. In 1951, Hillary had been part of a British reconnaissance expedition of the southern flank of the mountain and that led to him being asked to join a team planning to reach the peak. Hillary led a team that used tractors to reach the South Pole in 1958, becoming the first expedition to reach it by land in more than 40 years. The son of a New Zealand beekeeper, Hillary was knighted following his Everest climb, and devoted much of his life to the welfare of Nepal's Sherpas.

Jeanne Louise Adélaïde Hommaire de Hell
1819–83
In the 19th century, Frenchwoman Hommaire de Hell travelled widely with her husband, an engineer and geographer, journeying to the Ottoman Empire, the Caucasus, Crimea, and the Caspian Steppe, as well as surrounding areas. Her accounts of their travels became popular books.

> *"I am a lucky man. I have had a dream and it has come true."*
>
> EDMUND HILLARY, 1953

◁ Henry Hudson
The 1909 Hudson-Fulton Celebration Official Commemorative Medal depicts the explorer and his sailors aboard their ship, the *Half Moon*.

Henry Hudson
c.1565–1611

An English navigator, Hudson made four unsuccessful voyages in search of a northern route to Asia. The first of those voyages, in 1607, sailed beyond the Svalbard archipelago on behalf of the Muscovy Company of London, possibly travelling further north than any previous expedition, before being turned back by ice. The second, in 1608, attempted a northeast route above Russia, but it, too, was thwarted by ice. In 1609, sailing for the Dutch East India Company, Hudson headed west, reaching North America and exploring the waterway that would be named the Hudson River in his honour. Finally, in 1610, and once again representing British interests, Hudson sailed to what is today Canada, where his ship became frozen in the ice in what is now named Hudson Bay. The ship eventually broke free, but Hudson's crew mutinied and cast him, his young son, and several others adrift; Hudson and the others were never heard from again. Hudson's voyages contributed to both English claims to Canada and Dutch claims to the area surrounding the Hudson River.

Alexander von Humboldt
1769–1859

Humboldt, a Prussian scientist, resigned from his job as a government mining engineer to explore Central and South America, eventually becoming one of the most celebrated intellectuals of the 19th century. Joined by botanist Aimé Bonpland, Humboldt achieved an astonishing range of accomplishments during the 1799–1804 trip. His successes included the discovery of a link between the water systems of the Orinoco and Amazon rivers, learning the fertilizing qualities of guano, and climbing the Ecuadorian mountain Chimborazo to a height of more than 5,790 m (19,000 ft), setting what was then a mountaineering record. Upon his return to Europe, Humboldt made major contributions in a remarkable range of scientific fields including geology, oceanography, botany, zoology, cartography, and climatology.

George Hunter
1755–1823

Hunter was co-leader of one of the expeditions commissioned by US President Thomas Jefferson in the first decade of the 19th century to explore the Louisiana Territory recently acquired by the United States. Called the Grand Expedition, it ventured across portions of northern Louisiana and southern Arkansas in 1804–05, recording details about the local plants, animals, people, and geology. The longer and better-known Lewis and Clark expedition also explored the Louisiana Territory, but much further north. A highly regarded chemist born in Scotland, Hunter had previously explored portions of Ohio and Indiana.

Hyecho
704–787 CE

A Buddhist monk from Korea, in around 721 Hyecho set out on a journey that would take him to the lands now known as China, Vietnam, Indonesia, Myanmar, India, Pakistan, Afghanistan, and Iran. When Hyecho finally turned back east, he followed the Silk Road as far as China, arriving in the town of Dunhuang in or around 727. Hyecho was not the only Buddhist to set out to see the holy sites of India, but unlike most, he continued west beyond India, and he left a journal that preserved his observations about a wide range of people and places. A partial manuscript of his journals was discovered in a Chinese cave in 1908, providing virtually all that is known about him.

Ibn Battuta
1304–68/69 or 1377

In 1325, Ibn Battuta set out from his home in Tangier, Morocco, on a pilgrimage to Mecca. He did not return for nearly three decades and travelled around 110,000 km (70,000 miles). During that time, Ibn Battuta not only made the pilgrimage multiple times, he also explored most of the Muslim world, travelling as far east as China, as far south as sub-Saharan Africa, and to Djenné in West Africa. He is often said to have travelled more extensively than anyone up to that point in history and was robbed and kidnapped several times on his journeys. Born into a family that included several Islamic judges, Ibn Battuta served that role himself in many of the locations he visited, including in India. He dictated the story of his travels to poet Ibn Juzayy at the behest of the Sultan of Morocco, resulting in a book known as *The Rihla*.

Ahmad Ibn Fadlan
c.877–c.960 CE

An Islamic scholar, in 921–22 Ibn Fadlan was part of a diplomatic mission that travelled 4,000 km (2,500 miles) north-northwest from Baghdad to the River Volga, in what is now Russia. The goal of the expedition was to teach the Volga Bulgars about Islam and construct a fortress for protection against the Khazar people. During their travels, Ibn Fadlan's mission encountered the nomadic Oghuz (Ghuzz) Turks and fair-skinned people whom Ibn Fadlan called the Rus – the latter might have been descendants of the Vikings. Ibn Fadlan's story served as the basis for the 1999 film *The 13th Warrior*.

Ibn Jubayr
1145–1217

A Muslim born in Spain, Ibn Jubayr made a pilgrimage to Mecca in 1183–85. His written description of this journey offers a valuable account of 12th-century life in Egypt, Sicily, and the Middle East. Ibn Jubayr served as secretary to the governor of Granada and took at least two other journeys east from Spain; he died during the second of these, but no written account of his later travels survives.

▷ Ibn Battuta
This statue of Ibn Battuta stands in Quanzhou, Fujian, near the Pacific coast of China, about 11,000 km (6,900 miles) from his home in Morocco.

J

Willem Janszoon
c.1570–c.1630

A Dutch navigator, in 1606, Janszoon led the first European expedition known to have set foot in Australia. His ship, the *Duyfken*, made landfall on Cape York Peninsula in the far north of Australia and charted a section of coastline. Janszoon was unaware that he had landed on a new continent – he thought he had found a new part of New Guinea. Janszoon rose to the rank of admiral and was governor of the island of Solor (now in Indonesia).

Amy Johnson
1903–41

A British aviator, Johnson became famous as the first woman to fly solo from England to Australia, completing the journey between 5 and 24 May 1930. She made several more notable flights in the early 1930s, including setting the record for fastest solo flight from London to Cape Town, South Africa, in 1936. A secretary before earning her pilot's licence in 1929, she delivered planes around Britain for the Royal Air Force during World War II. In January 1941, Johnson was forced to parachute out of an RAF plane that was low on fuel. Her chute opened, but she landed in the River Thames; her body was never recovered.

Elizabeth Justice
1703–52

A British writer, Justice's 1739 book *A Voyage to Russia* provides an outsider's look at 18th-century St Petersburg. The book featured first-hand observations about a wide spectrum of Russian society, from the poor to the nobility – even the tsarina. Justice was reduced to seeking work as a governess to the children of a British merchant in Russia because of the reckless spending of her husband.

Edmund Kennedy
1818–48

A surveyor and explorer, Kennedy was second-in-command of Thomas Mitchell's 1845–47 Australian expedition that set out to find a land route from Sydney to the Gulf of Carpentaria. He led an 1847–48 expedition that followed up on the discoveries of that earlier effort. He was killed by Aboriginal peoples during his third expedition, an 1848 exploration of Cape York Peninsula in northern Australia. Born on the island of Guernsey, he immigrated to Australia in 1840 and worked in the surveyor-general's department.

Jemima Kindersley
1741–1809

The wife of a British colonel, in 1764, Kindersley travelled by sea with her husband, Nathaniel, in India, returning to England in 1769. In 1777, the letters she wrote about her voyage were published as a book titled *Letters from the Island of Teneriffe, Brazil, the Cape of Good Hope and the East Indies*. It offered an insight into the daily life of Europeans in India, as well as locations where she made extended stops: Tenerife in the Canary Islands, Salvador in Brazil, and a Dutch colony in what is today South Africa.

Frank Kingdon-Ward
1885–1958

A British plant collector and writer, during 1909–56 Kingdon-Ward conducted more than 20 expeditions in Asia – mainly in Burma (Myanmar) – searching for plants that he hoped would be suitable for European gardens. Highly regarded as a botanist and an explorer, he wrote about his experiences and won awards from the Royal Horticultural Society and the Royal Geographical Society.

◁ **Amy Johnson**
Johnson examines a de Havilland Moth, among the most popular British civilian aircraft of the late 1920s and 1930s; she flew a Moth from England to Australia.

◁ Mary Kingsley
Kingsley seated in a canoe on the Ogowe River, Gabon, in the late 19th century. She wore the voluminous garments of Victorian Britain despite the African heat.

Mary Kingsley
1862–1900
Englishwoman Kingsley made several trips to West Africa between 1893 and 1895, spending time among local peoples and collecting specimens for the British Museum. She lived an uneventful life until 1892, when the death of her parents freed her from her 19th-century social obligations to them, allowing her to seek adventure. She wrote and lectured about Africa and was a critic of colonialism. Kingsley returned to Africa in 1900, and died there of typhoid while serving as a nurse for prisoners during the Boer War.

Johann Ludwig Krapf
1810–81
A German missionary and linguist, Kraft and fellow missionary Johannes Rebmann explored East Africa, becoming the first Europeans to view that continent's highest peaks – Rebmann saw Mount Kilimanjaro in 1848 and took Krapf to see it in 1849, when they also saw Mount Kenya – and set the stage for later exploration. Krapf is considered the first European to have devoted extended study to Swahili.

L

René-Robert Cavelier, sieur de La Salle
1643–87
A French explorer and fur trader, La Salle led a 1682 expedition down the Mississippi River to its delta, claiming the area for France and naming it Louisiana. Born into a merchant family, La Salle originally intended to become a priest, but gave that up to seek adventure. He received a land grant on the Island of Montreal from the Sulpicians, a society of priests; found success as a fur trader; and explored the region in hopes of locating a water route to Asia. La Salle died leading an ill-fated attempt to create a French colony on the Gulf of Mexico; he was killed by his own mutinous men.

Ernest-Marc-Louis Doudart de Lagrée
1823–68
Lagrée led an 1866–68 French expedition to map the Mekong River in Laos and China. It is believed that the goal was to assess whether the river could be used to connect the port of Saigon with the riches of China and upper Siam (Thailand) for the benefit of French colonial interests. A French naval captain and amateur entomologist, Lagrée suffered from poor health, and the arduous expedition took a toll – he developed fever and amoebic dysentery, and his wounds became infected. He died of an abscess of the liver in Yunnan province in southern China.

Alexander Gordon Laing
1793–1826
A major in the British army's Royal African Colonial Corps, in 1826 the Scottish-born Laing became the first European known to have visited Timbuktu in what is now Mali. He was killed only days after departing that city. Laing had previously travelled on official missions on behalf of the British colonial government to the West African lands of the Mande and Susu peoples with the goals of establishing trade and stopping the local slave trade. His account of those earlier expeditions was published in *Travels in the Timannee, Kooranko, and Soolima Countries, in Western Africa* (1825).

Richard and John Lander
1804–34 and 1806–39
The English Lander brothers were the first Europeans to map the lower course of West Africa's largest river, the Niger. Richard, who led the 1830 expedition, had participated in Hugh Clapperton's 1825 attempt to map the river, a trip that resulted in Clapperton's death from dysentery two years into the effort. Richard returned to the region in 1830 with the goal of founding a trading settlement, but was killed by locals. His brother died a few years later, probably due to a lingering illness he had contracted during the 1830 expedition.

Jean-François de Galaup, comte de Lapérouse
1741–c.1788
A French naval commander, Lapérouse was selected by King Louis XVI to lead a two-ship expedition around the world. The ships crossed the Atlantic and Pacific, surveying coastlines and studying plants and animals. On 10 March 1788 they departed Australia's Botany Bay and disappeared. The wrecks were discovered 40 years later on the Santa Cruz Islands (now the Solomon Islands). Lapérouse had earlier earned distinction for several successful military actions against the British, including a 1782 attack on Hudson Bay during the American War of Independence.

Austen Henry Layard
1817–94
Layard excavated the ancient Assyrian cities of Nimrud and Nineveh, among other archaeological discoveries. The cuneiform tablets he unearthed bettered modern understanding of Assyrian and Babylonian history. An Englishman born in Paris, Layard spent part of his childhood in Italy and worked in the London office of his uncle, a solicitor, before setting out in search of adventure in the Middle East in 1839. Layard later served as an MP, then as British ambassador to Spain and then Türkiye.

▷ Austen Henry Layard
This portrait of Layard was painted by his close friend, the symbolist artist George Frederic Watts, in or around 1852, just as Layard was beginning his career in parliament.

▽ **Meriwether Lewis**
Lewis wearing a Shoshone tippet – a garment draped over the shoulders. It was presented to him by Shoshone chief Cameahwait; Lewis called it "the most eligant [sic] piece of Indian dress I ever saw."

> *"He was the first Ambassador of America to the Court of Japan."*
>
> FORMER US PRESIDENT CALVIN COOLIDGE ON MANJIRO, 1918

Ludwig Leichhardt
1813–c.1848

A Prussian explorer and naturalist, Leichhardt began a series of expeditions across the Australian interior. The first set out from Darling Downs in eastern Australia in 1844, and its members had been given up for dead when it reached Port Essington on the northern coast in December 1845, its unexpected re-emergence launching Leichhardt to fame. In 1846 his second expedition set out to cross the continent from east to west, but was forced to turn back. He attempted that trek again in 1848, but he and his six companions disappeared, never to be seen again.

Meriwether Lewis
1774–1809

Lewis was the co-leader, with William Clark, of an 1804–06 expedition that journeyed across the Louisiana Territory, which the US had recently acquired from France, and the US Pacific Northwest. The journey became one of the most famous in United States history. Lewis had previously served as a US Army officer, and as secretary to President Thomas Jefferson. He was made governor of the Upper Louisiana Territory after the expedition, but was ill-suited to the role. He endured financial problems and died by gunshot three years after his trek; whether by suicide or murder is unclear.

Charles Lindbergh
1902–74

An American aviator, Lindbergh was propelled to global fame in 1927, when he completed the first non-stop solo transatlantic flight. An airmail pilot prior to the flight, Lindbergh took off from Roosevelt Field near New York City on 20 May, and landed at Le Bourget field near Paris 33½ hours later. He later served as a technical advisor to airlines and contributed to the development of a pump used in heart surgery. Lindbergh was back in the headlines in 1932, this time tragically – his son was kidnapped and killed. When he spoke against America joining World War II and made antisemitic comments, Lindbergh was accused of having Nazi sympathies – he even received a medal from Hermann Göring. But Lindbergh flew 50 combat missions for the US in the Pacific against the Japanese.

David Livingstone
1813–73

A Scottish physician and missionary, Livingstone spent most of his adult life exploring Africa's interior. Among his many accomplishments, he crossed the continent from coast to coast along the Zambezi River and determined that that river was not navigable along its entire length. In 1866, Livingstone set out in search of the source of the Nile. While he failed to achieve that objective, it led to one of the most famous moments in European exploration. In 1871, Livingstone was thought to have disappeared, but locating him on the shores of Lake Tanganyika, Henry Stanley famously claimed that he said: "Dr Livingstone, I presume?" Livingstone, who was unwell, declined to depart Africa and died two years later. He was buried in Westminster Abbey.

7

The percentage of Magellan's men who made it home.

Annie Londonderry
1870–1947

In 1894–95, Annie Cohen Kopchovsky, a Latvian-born woman living in the US, became the first woman to bicycle around the world, taking the name of her sponsor – Londonderry Lithia Spring Water Company – for the trip. She accomplished this alone and despite having no cycling experience; the trip seems to have been largely an effort to gain fame and funds, though a desire for adventure probably also played a role. The trip had an inauspicious start – after she biked west from New York to Chicago, the looming winter forced her to return to New York and restart, boarding a ship to Europe. Researchers have noted that much of her circumnavigation was completed on ships, not by bike.

Alexander Mackenzie
c.1764–1820

A Scottish-born fur trader, Mackenzie explored the Canadian west. In 1789, he led a small group northwest to the Arctic Ocean on what would later be named the Mackenzie River in his honour (the river is known as Dencho in the Dene language, and it has other names in other local languages). In 1793, he led an expedition that reached the Pacific Ocean in an epic Canadian journey that anticipated that of Lewis and Clark in the the following decade. He was knighted in 1802, and eventually retired to Scotland.

Ferdinand Magellan
1480–1521

A Portuguese navigator, Magellan was born into a noble family and sailed on behalf of the Portuguese crown before financial disputes saw him switch his allegiance to Spain in 1517. He led the first expedition to circumnavigate the globe, a 1519–22 voyage on behalf of Spain. The primary goal was to find a western sea route to Asia; after this was accomplished, his ships continued west to return to Europe, rather than return east. Magellan did not survive the trip – he was killed in a battle in what is now the Philippines. He had formed an alliance with the people of the island of Cebu but died attempting to help them defeat their rivals on the island of Mactan.

Susan Shelby Magoffin
1827–55

Magoffin took part in an 1846–47 journey from Missouri to New Mexico on America's iconic Santa Fe Trail, then south into Mexico. The diary she kept to record the people, plants, and animals she saw was published after her death. Born into a wealthy family in Kentucky, Magoffin was the granddaughter of Isaac Shelby, the first governor of that US state and a hero of the American War of Independence. She later returned to Missouri.

Nakahama Manjiro
1827–98

An accidental explorer, Japanese Manjiro was 14 when a fishing boat he was in on the Pacific Ocean was blown off course in a storm and ended up on Torishima Island. He was rescued by a US whaling ship and taken to Hawaii. His fellow fishermen remained in Hawaii, but Manjiro sailed on to Massachusetts with the whaling ship. He lived with the whaling ship captain's family, attended local schools, and joined the ship's crew, becoming first mate. He is regarded as the first Japanese person to live in the United States. In 1850–51, Manjiro returned to Japan where he was initially treated with suspicion in a country largely shut off from the rest of the world. His English language skills and knowledge of the West were valued when Japan was forced to open to the West following the arrival of US Commodore Matthew Perry's ships in 1853. In 1860 and 1870 Manjiro returned to the US as part of official Japanese delegations. In 1870 he also visited Europe. He served as a professor in Tokyo, but retired in 1871 after a stroke.

◁ Nakahama Manjiro
This statue of Manjiro stands in his hometown of Tosashimizu on the island of Shikoku. The town is today the sister city of Fairhaven/New Bedford, Massachusetts, where Manjiro lived in the US.

Lope Martín
Unknown–c.1564

A ship's pilot born in Portugal but of African heritage – his ancestors had been enslaved and taken to Portugal – in 1564–65 Martín completed the first round-trip voyage between the Americas and Asia. His ship was part of a four-ship Spanish fleet, but it became separated from the others in a storm and was the first to complete the trip from present-day Mexico to the Philippines and back. He was sentenced to be hanged after the pilot of another ship on that expedition accused him of abandoning the fleet. Martín escaped, and what became of him is unknown.

Robert John Le Mesurier McClure
1807–73

An Irish officer in the British navy, McClure found the Northwest Passage – a sea route through Arctic Canada connecting the north Atlantic with the north Pacific. The discovery was made during a failed 1850 attempt to locate explorer John Franklin and his men, who had been missing since 1845. McClure was not the first European to sail the Northwest Passage – his ship became trapped in ice, and he and his men completed the journey on foot and in rescue ships. McClure did not make it back to the UK until 1854. He was knighted and rose to the rank of vice admiral.

Maria Sibylla Merian
1647–1717

An entomologist and illustrator, Merian made a journey to Suriname in 1699–1701 that is sometimes cited as the first expedition to the Americas undertaken purely for the sake of scientific discovery. That journey was unfortunately cut short by malaria. Merian is also widely regarded as the first person to describe the process by which caterpillars become butterflies in true scientific detail, work she had carried out before departing on her voyage. Born in Germany, Merian was noted for her accurate illustrations of insects and plants.

▽ Maria Sibylla Merian
During the years Median spent in Suriname, she saw numerous plants and animals not previously known in Europe. It led to the publication of her book, *Metamorphosis Insectorum Surinamensium*.

Ynés Mexía
1870–1938

A botanist and explorer, Mexía travelled widely in South, Central, and North America, often alone, collecting nearly 150,000 plant specimens and recording about 500 plant species that at the time were unknown to science. Those achievements are all the more impressive considering her botanical career lasted only 13 years. Mexía spent much of her life as a social worker, and didn't discover her passion for botany until her 50s when she started her studies at the University of California, Berkley. She was born in the US, but had Mexican heritage – her father was a Mexican diplomat.

Mbarak Mombée
c.1820–85

A Bantu person born in what is today southern Tanzania, Mombée was a key member of several of the 19th century's most important East African expeditions. He was with John Speke when he reached Lake Victoria in 1858, and also helped Henry Stanley locate David Livingstone in 1871. A few years later, Mombée helped Verney Lovett Cameron become the first European to cross equatorial Africa. Previously an enslaved person in India, the multi-lingual Mombée was known for his skill as a translator as well for his work as a guide.

Moncacht-Apé
Flourished c.1700

The travels of Moncacht-Apé, an Indigenous person of the Yazoo people, were set down by a French historian of colonial Louisiana, Antoine-Simon Le Page du Pratz, in his *Histoire de la Lousiane*, published in 1758. Moncacht-Apé told Le Page that he had travelled across North America in the late 1600s from the Yazoo lands in the Mississippi River delta region north to Niagara Falls, east to the Atlantic coast, then west all the way to the Pacific, returning home two years later. If so, this pre-dates any other known journey across the continent. Due to omissions and inaccuracies in his description of North American geography, some historians are doubtful the journey happened; others say the fault may lie with Le Page's transcription.

Francisco Josué Pascasio Moreno
1852–1919

An Argentinean scientist and explorer, Moreno (also known as Perito Moreno) conducted expeditions into little-known parts of Patagonia and the Andes in the 1870s and 1880s, returning with important anthropological and geographical discoveries. His work helped to establish the border between Argentina and Chile, and his collections served as the basis of the La Plata Museum in Argentina – Moreno served as its first director. He later donated a section of the land used to create Nahuel Huapi National Park, Argentina's first national park.

Henri Mouhot
1826–61

A French explorer, Mouhout located the city of Angkor in the Cambodian jungle in 1860 – at the time it was virtually unknown outside the region. Mouhot arrived in Southeast Asia in 1858 and remained there until his death. His primary focus was zoology – Mouhot is credited with the discovery of numerous insects, including a black beetle named *Mouhotia gloriosa*, and a spider now known as *Cyphagogus mouhotii*. He died of fever in Laos, aged just 35, but the posthumous publication of his journal helped to make Angkor world-famous.

N

Naukane
c.1767–1850

An Indigenous Hawaiian, Naukane entered the fur trade and travelled extensively in the Pacific Northwest of North America and eastern Canada. Born into Hawaii's royal family, he travelled with a royal delegation to England in 1823, though the trip ended poorly – the Hawaiian king and others died of measles. As a young man he witnessed the death of explorer James Cook on a Hawaiian beach. Naukane spent his final years in the Pacific Northwest of the US.

Emperor Nero
37–68 CE

According to several Roman writers, Nero – the last emperor of Rome's first dynasty – sent a small expedition to Africa

△ **Tenzing Norgay**
Norgay on Chukhung Peak in Nepal in April 1953. He reached the summit of Everest the following month – one of the most famous feats in the history of mountaineering.

in around the year 61, by some accounts to search for the source of the River Nile. If this was the purpose of the expedition, it failed.

Adolf Erik Nordenskiöld
1832–1901
A mineralogist, geologist, and explorer, on a journey in 1878–80, Nordenskiöld became the first person to navigate the Northeast Passage – the sea route linking the Atlantic and Pacific oceans along the northern coast of Russia. During this trip his ship was stuck in ice near the Bering Strait from September 1878 until July 1879. That expedition was one of many that Nordenskiöld made to the Arctic in 1858–83, including multiple geology-focused visits to the island of Spitsbergen and an unsuccessful attempt to reach the North Pole. Born in Finland, Nordenskiöld spent his adult life mainly in Sweden, forced into exile due to his opposition to Russian rule over his homeland. He served as professor and curator of mineralogy at the Swedish State Museum.

Tenzing Norgay
c.1914–86
On 29 May 1953, Norgay and Edmund Hillary became the first people known to have reached the 8,849-m (29,032-ft) summit of Mount Everest, the highest point on Earth. Norgay, a Sherpa probably born in Nepal, took part in his first expedition aged 19, and by the time he joined Hillary's climb he had participated in more Everest expeditions than any other climber. Norgay later served as director of the Field Training Himalayan Mountaineering Institute in Darjeeling, India.

Marianne North
1830–1890
An English artist, North travelled the world in search of flowers, plants, and landscapes to paint. Between 1871 and 1885, she visited North and South America, India, Asia, and Australia, as well as numerous islands. She painted an estimated 900 plant species in their natural settings. Some of them were not previously known to science – a pitcher plant she found in Borneo is named *Nepenthes northiana* in her honour. North typically travelled alone, which was unusual for an upper-class English woman in the Victorian era – her father had served as a Member of Parliament. Many of North's paintings remain on display in the Marianne North Gallery at Kew Gardens in London.

O

Isabel Godin des Odonais
1728–92
In 1769–70, Godin travelled thousands of miles through the rainforests of South America's Amazon basin to reunite with her husband. Of the party who set out on the journey, she was the only known survivor. Born into a wealthy family of Spanish heritage, in what was then Peru but is now Ecuador, Godin had been separated from her husband for nearly 20 years – he had been prevented from returning to Peru by Spanish authorities because he was a Frenchman. The reunited couple settled in France.

Francisco de Orellana
c.1490– c.1546
The Spanish explorer Orellana led the first European journey down the Amazon River in 1541–42. That Amazon voyage wasn't planned – he was put in charge of a group sent to search for food during a larger expedition tasked with exploring the lands east of Quito, but instead followed the river all the way to the ocean. He had taken part in Francisco Pizarro's 1535 conquest of the Inca Empire in what is now Peru. In 1545, Orellana launched a second expedition of his own to explore the Amazon region. His voyage was a disaster, with Orellana among those who didn't survive.

P

Mungo Park
1771–1806
A Scottish explorer and surgeon, Park led a pair of expeditions to the Niger River region in West Africa. The son of tenant farmers, he had previously served as a medical officer on a ship that sailed to Sumatra. The first of Park's expeditions, in 1795–97, served as the basis for his book, *Travels in the Interior Districts of Africa*, which brought Park considerable fame. His second expedition, in 1805, resulted in his disappearance – it was eventually discovered that Park drowned when trying to escape an attack by local people.

St Patrick
5th century
A missionary born in Britain, St Patrick helped to bring Christianity to Ireland. His first trip to Ireland was against his will – he was captured as a teenager and enslaved there for six years. He eventually escaped, only to return later by choice on his religious mission. He is considered the patron saint of Ireland. Legend holds that St Patrick drove the snakes out of Ireland.

Robert E. Peary
1856–1920
An Arctic explorer, Peary became famous as the first to reach the North Pole, reportedly doing so on 6 April 1909. Peary's claim that he successfully reached the Pole has been questioned in recent decades, however. A US Navy officer who eventually rose to the rank of rear admiral, Peary made multiple expeditions to Greenland and elsewhere in the Arctic from 1886 onwards, and wrote a number of books about his adventures.

▷ Marianne North
This photograph shows North sketching. The artist's paintings are noted for their vivid colours. Her favourite colours included cobalt blue and rose madder – she very rarely used black paint.

▽ Ida Laura Pfeiffer
By the mid-1840s, Pfeiffer's travels had made her famous. This portrait includes a globe, symbolizing her first circumnavigation of the world.

Petachiah of Regensburg
Late 12th century

A rabbi probably from Bavaria, Petachiah made an extended journey through Poland and Russia to the Middle East and the Caucasus, in or around the 1170s. Petachiah's journals were used to write an account of his travels – *Travels of Rabbi Petachia of Ratisbon* – in the late 16th century. Little is known of him outside of this work. However, some sources have identified Petachiah as a merchant and it is believed that he was among a group of Jewish traders from Regensburg who helped to develop a trade route that extended from Mainz in present-day Germany to Kiev in present-day Ukraine.

Ida Laura Pfeiffer
1797–1858

An Austrian explorer, Pfeiffer travelled widely in the 1840s and 1850s, starting her journeying only after her children were grown up and her husband had died. She completed two trips around the world during these decades, with extended stops in the Americas and the East Indies, and wrote popular books about her adventures. Her other notable destinations included Egypt, Scandinavia, and Madagascar. An illness contracted during her trip to Madagascar claimed her life; her book about that trip – *The Last Travels of Ida Pfeiffer* – was published posthumously.

37

Number of horses included on Pizarro's 1531 expedition to Peru

Zebulon Pike
1779–1813

A US Army officer and explorer, Pike led an 1805 expedition in search of the headwaters of the Mississippi River, though the conclusion he reached about the river's source was later proved false. The following year, Pike led an expedition west into the Rocky Mountains, where he established an outpost in Colorado and tried, unsuccessfully, to climb the mountain that is now named Pikes Peak in his honour. He then headed south into what is today New Mexico, where he was escorted out of the region by Spanish forces. Pike rose to the rank of brigadier general, but died of wounds sustained while attacking the British in York, Canada, during the War of 1812.

Pius (Mau) Piailug
1932–2010

A Micronesian master navigator, Pius Piailug used virtually extinct traditional navigational techniques to guide a double-hulled canoe across more than 11,100 km (6,000 nautical miles) of open ocean from Hawaii to Tahiti in 1976. The widely acclaimed voyage helped to prove that long Pacific crossings were feasible for ancient peoples who lacked modern navigational technology. Born on Satawal in the Caroline Islands, Pius Piailug took part in a number of other voyages and mentored young navigators. His journeys are credited with spurring interest in the cultures and history of the Pacific islands.

Francisco Pizarro
c.1475–1541

A Spanish conquistador, Pizarro led a small military force that overthrew the Inca Empire and killed its emperor Atahualpa. Born the illegitimate son of a military officer and a peasant girl, he travelled to the Americas in 1502, where he initially lived in Hispaniola (modern Haiti and Dominican Republic) before joining a 1510 expedition to Colombia. He then joined Balboa's 1513 expedition that crossed Panama – the participants were the first Europeans to reach the Pacific Ocean. Pizarro served as mayor and magistrate of Panama in 1519–23, before launching a series of South American expeditions that, in 1533, toppled the Inca Empire and inaugurated centuries of Spanish colonial rule. He spent his remaining years in Peru – the final two in Lima, a city that he had founded in 1535. Pizarro was assassinated there by Spaniards loyal to his late rival, Diego de Almagro, in 1541.

△ Francisco Pizarro
Pizarro was described as prudent and discerning in temperament, belying the rash image of the man who launched an attack on the Inca Empire despite being outnumbered.

▷ **Walter Raleigh**
This miniature portrait by artist Nicholas Hilliard dates to around 1585; it's probably the earliest surviving portrait of Raleigh, who was a fashionable figure in his day.

> *"At 10 o'clock, we obtained a clearer view of the mountains of Jagga, the summit of one of which was covered by what looked like a beautiful white cloud."*
>
> JOHANNES REBMANN, DESCRIBING HIS FIRST SIGHT OF MOUNT KILIMANJARO IN HIS DIARY, 11 MAY 1848

Marco Polo
c.1254–1324
A Venetian born into a merchant family, Marco Polo accompanied his father and uncle on a 1271 journey by land to China, not returning to Venice until 1295. While in Asia, he served the court of Mongol ruler Kublai Khan in administrative roles. Later imprisoned in Genoa during its war with Venice, Polo dictated his memoirs to a fellow prisoner. The resulting book brought Polo enduring fame, although its accuracy has been questioned.

Juan Ponce de León
1460–1521
Ponce de León joined Columbus' second voyage to the Americas in 1493, landing in Hispaniola, where he served as provincial governor of eastern Hispaniola after helping to quell a 1504 rebellion. In 1508, he sailed to present-day Puerto Rico, where he amassed huge quantities of gold and was named governor. In 1513, he sailed to what he thought was an island (it was a peninsula), naming it La Florida or Land of Flowers. Ponce de León returned to Florida as its governor in 1521, but was soon killed by Indigenous Calusas when he tried to settle there.

John Wesley Powell
1834–1902
An American explorer and geologist, Powell led a pair of expeditions down the Colorado River and through the Grand Canyon, in 1869 and 1871–72. He later served as director of the US Geological Survey and of the Smithsonian Institution's Bureau of Ethnology. He accomplished all of this despite having lost his right arm below the elbow in 1862, fighting for the Union in the US Civil War. Powell returned to active duty after the injury, rising to the rank of major. Lake Powell, a massive reservoir on the Colorado River, is named in his honour.

Nikolay Przhevalsky
1839–88
A Russian geographer and explorer, Przhevalsky led numerous expeditions to Central and East Asia between 1867 and 1885. He crossed the Gobi desert and the Tian Shan mountain range, ventured into Turkestan, Tibet, and Siberia, and collected thousands of plant, insect, bird, and animal specimens. Przhevalsky's horse and Przhevalsky's gazelle are named in his honour. Przhevalsky wrote several books about his travels before dying of typhoid in what is now Kyrgyzstan.

Pytheas of Massalia
c.300 BCE
A navigator, Pytheas was the first Greek known to have visited Britain. Born in the Greek colony of Massalia on what is now France's Mediterranean coast, he sailed west to the Atlantic Ocean, then north to Brittany, England, and Scotland – and perhaps further north. He eventually turned south, travelling across the North Sea until he reached the northern coast of Europe. The account he wrote of this voyage has been lost, but details survive in references and excerpts in other works.

Q

Pedro Fernández de Quirós
c.1565–1615
A Portuguese navigator, Quirós played a major role in a pair of South Pacific voyages. He was the pilot of a 1595–96 expedition to the Solomon Islands led by Álvaro de Mendaña y Neira, and assumed command following Mendaña's death. In 1605–06, Quirós led his own expedition to the South Pacific, claiming several islands for Spain. His attempt to establish a colony on Espiritu Santo, in what is now Vanuatu, was unsuccessful, and Quirós struggled for years to convince the king of Spain to finance another Pacific voyage. Quirós eventually sailed for Peru, but he died before reaching it.

R

Walter Raleigh
c.1554–1618
Englishman Raleigh fought in the French Wars of Religion, participated in the suppression of an Irish uprising, joined his half-brother for a privateering expedition against the Spanish, and led two expeditions to South America in search of gold. He also attempted to found the first English colony in America, although he didn't travel to that colony himself. Among his other contributions, he is generally credited with making tobacco popular with the English court. Tall, handsome, and charming, Raleigh was a favourite of Queen Elizabeth I, who knighted him and appointed him captain of the Queen's Guard. Raleigh also wrote poetry. He was less popular with Elizabeth's successor, James I, who imprisoned Raleigh for treason in 1603 after he ignored the King's instruction not to attack the Spanish. He was released in 1616 but rearrested and beheaded for treason in 1618.

Johannes Rebmann
1820–76
A German missionary, Rebmann and fellow missionary Johann Ludwig Krapf explored East Africa and they were the first Europeans to see Africa's highest peaks. Rebmann saw Mount Kilimanjaro on his own in 1848, returning in 1849 when both men saw Mount Kenya for the first time. Rebmann spent nearly three decades in Africa from the 1840s, returning to Germany only in the last year of his life, when he was nearly blind. It is acknowledged that his travels were the precursor of later expeditions by British explorers, including Richard Burton, John Hanning Speke, and David Livingstone.

▷ Robert Falcon Scott
Scott writes in his diary in October 1911, seated in his corner of the Terra Nova Expedition's hut in Cape Evans, Antarctica. Photographs of Scott's wife, Kathleen, can be seen behind him.

Edith "Jackie" Ronne
1919–2009
Ronne became the first American woman to set foot on Antarctica when she served as historian for a 1947 expedition led by her husband, Finn Ronne. She and Jennie Darlington were also the first women to spend a full winter there. The Filchner-Ronne Ice Shelf is named in her honour.

Bartolomé Ruiz
c.1482–c.1532
A Spanish navigator, Ruiz served as pilot for Francisco Pizarro on some of his voyages to the Americas. While in command of one of the two ships of Pizarro's 1526–28 expedition, Ruiz was the first European to reach what is now Ecuador and make contact with the Incas. He learned that gold could be found to the south, advancing the Spanish conquest of South America.

S

Sacagawea
c.1788–1812
Sacagawea, an Indigenous Lemhi Shoshone woman, joined the Lewis and Clark expedition of 1804–06 from Fort Mandan, in present-day North Dakota, to the Pacific Ocean and back, as an interpreter. Pregnant when she and her French-Canadian trapper husband were recruited, she gave birth in 1805 yet continued onward, providing advice about plants in addition to translating for the expedition and negotiating with other peoples they encountered. Her diplomatic abilities averted trouble and ensured the mission remained peaceful.

△ Sacagawea
A one-dollar coin minted in the United States from 2000 depicts Sacagawea carrying her son. Sacagawea's image is modelled on Randy'L He-dow Teton, a modern Shoshone-Bannock woman.

Heinrich Schliemann
1822–90
German-born Schliemann made some of the most famous archaeological finds in history, excavating the ancient cities of Troy, Mycenae, and Tiryns in Greece and Anatolia. The son of a minister, his success as a businessman – at various times he traded gold, dealt in indigo dye, and was a military contractor – enabled him to retire at 36 to pursue his passion for archaeology. In 1873 he located the remains of Troy in what is now Türkiye, following a lead provided by English archaeologist Frank Calvert. Schliemann spent much of the rest of his life conducting archaeological digs and writing books about his discoveries. Critics note that some of his conclusions were incorrect, and his methods did irreparable damage to some of the sites he excavated.

Robert Falcon Scott
1868–1912
A British naval officer, Scott led a pair of expeditions to Antarctica. The second of these expeditions narrowly missed being the first to the South Pole, arriving on 18 January 1912, just 34 days after Norwegian Roald Amundsen. Scott and the four men who reached the Pole with him died during the return trip; the bodies of Scott and two others were later found in a tent just 18 km (11 miles) from a cache of supplies that could have saved them. Scott had previously led the National Antarctic Expedition of 1901–04, which travelled further south than any expedition up to that point.

Carla Serena
1820–84
In 1874, Serena embarked on a six-year solo journey that would take her to parts of the Ottoman, Russian, and Persian empires that were relatively little known in Western Europe at the time. Serena published books and numerous articles about her travels. Born in Belgium as Caroline Hartog Morgenstein, she married a Venetian merchant and began to travel in her 50s. She died in Greece.

Ernest Shackleton
1874–1922
Born in Ireland, Shackleton was a member of the 1901–04 National Antarctica Expedition, and later a 1907–09 expedition that set another "furthest point south" record. But he is best known for his third expedition to the Antarctic in 1914, which set out to cross the continent but never even reached it. In early 1915, his ship became trapped in, and then crushed by, pack ice. Shackleton and five of his men navigated a lifeboat across 1,300 km (800 miles) of Antarctic waters, then borrowed a small Chilean steamer to rescue the remaining members of the expedition, all of whom had survived the nearly two-year ordeal. Shackleton served in the British Army during World War I, then died of a heart attack during a 1921 attempt to circumnavigate Antarctica.

Qian Shan Shili
1856–1945
A Chinese writer, Shan travelled with her diplomat husband to postings in Russia, Holland, and Italy between 1903 and 1908. The accounts she wrote about these travels are often cited as the first extensive writings by a modern Chinese woman

about Europe. Earlier, Shan, who was born into a prominent family in Zhejiang province, had lived in Japan.

Jane Smart
c.1740s

Correspondence written by Smart was published in 1743 as an eight-page pamphlet titled *A Letter from a Lady at Madrass to her Friends in London*. It was the first published English-language account of life in India to be written by a woman. She lived in Madras with her husband and two daughters.

Jedediah Smith
1799–1831

An American trapper and frontiersman, in 1824 Smith helped to rediscover the "South Pass" through the Rocky Mountains in Wyoming. It became a primary route for wagon trains headed west to Oregon and California. In 1826, Smith traversed the Mojave Desert into southern California, becoming one of the first Americans of European descent to reach California by land from the east. Smith achieved a degree of financial success as a trapper. He was killed in Kansas in 1831 following an encounter with Comanche people.

Hernando de Soto
c.1496/7–1542

A Spanish explorer and enslaver based in what is now Nicaragua, de Soto joined Francisco Pizarro on his 1532 conquest of Peru. Leader of Pizarro's cavalry, de Soto was the first European to make contact with the Incan emperor Atahualpa. He returned to Spain a rich man, but soon returned to the Americas as governor of Cuba. From 1539, de Soto searched what is now the southeastern United States for treasure. The expedition engaged in repeated battles with Indigenous people, and its members became the first Europeans to see the Mississippi River.

600

Number of plants Spruce collected in Ecuador

John Hanning Speke
1827–64

A British explorer, Speke joined Richard Burton on his famous African expeditions of the 1850s. In 1858, Speke was the first European to reach Lake Victoria, the source of the White Nile (Burton was too ill to travel with him). Speke and Burton had previously become the first Europeans to reach Lake Tanganyika. Speke returned to Lake Victoria with James Grant in 1862.

Richard Spruce
1817–93

A British botanist, from 1849 Spruce spent 15 years collecting plants in Brazil, Venezuela, Ecuador, and Peru. Among other successes, he collected seeds from *Cinchona* tree species that were cultivated so their bark could be harnessed to produce quinine, a drug used to treat malaria. Spruce had previously led an expedition to the Pyrenees, in 1845–46.

> *"Difficulties are just things to overcome after all."*
>
> ERNEST SHACKLETON, *THE HEART OF THE ANTARCTIC*, 1909

▷ **John Hanning Speke**
A painting by artist James Watney Wilson, painted after 1862, depicts Speke standing before Ripon Falls on the Nile River – located close to where the White Nile flows from Lake Victoria.

◁ Freya Stark
Stark was fluent in Arabic and Persian, which enabled her to interact with the people of the Middle East better than most Europeans.

Lady Hester Lucy Stanhope
1776–1839
A British noblewoman and archaeologist, Stanhope assisted her uncle, William Pitt the Younger, when he was prime minister. In 1810, a few years after his death, she left England and never returned, travelling extensively in the Middle East. In around 1812 she was the first English woman to enter Egypt's Great Pyramid and possibly the first European woman to visit the city of Palmyra in Syria. She conducted some of the earliest modern archaeological excavations in Palestine. Eventually Stanhope settled near Mount Lebanon.

Henry Morton Stanley
1841–1904
Born in Wales, Stanley spent part of his childhood in a workhouse before sailing for America in 1859 and fighting in the US Civil War. A journalist for the *New York Herald*, in 1871 Stanley set out to locate Dr David Livingstone, an explorer who had seemingly disappeared while searching for the source of the River Nile. Stanley continued exploring Central Africa following Livingstone's death in 1873, often on behalf of King Leopold II of Belgium. In 1876, also searching for the source of the Nile, he became the first European to navigate the Congo River.

Freya Stark
1893–1993
An explorer and writer, Stark visited remote regions of Southwest Asia (Middle East), Afghanistan, Türkiye, and Central Asia, writing about her travels. Her books focus on culture, history, and everyday life. Stark usually travelled alone, at a time when that was rare for women. During World War II, she worked for the British Ministry of Information in the Middle East, creating propaganda to encourage Arabs to support the Allies. Born in Paris to an English father and mother of mixed European heritage, she spent much of her early life in Italy.

Aurel Stein
1862–1943
An archaeologist, in 1907 Stein was possibly the first European to enter the Mogao Caves in Dunhuang, China, where he found a wealth of manuscripts, art, and artefacts. On other expeditions to Central Asia, he followed the path of Alexander the Great and traced ancient trade routes that connected China to the West. Born in Hungary, he later became a British citizen.

John Lloyd Stephens
1805–52
A US diplomat posted to Central America, in 1840 Stephens and British illustrator Frederick Catherwood rediscovered Maya ruins in the jungles of what are now Belize, Mexico, and Guatemala. Trained as a lawyer but more interested in archaeology, Stephens wrote books about his travels in Europe and the Middle East before he visited Central America.

John McDouall Stuart
1815–66
A British surveyor and explorer, Stuart led six expeditions into the Australian interior. In 1860, he became possibly the first European to reach the centre of the continent. He later led the first expedition to cross the continent from south to north and return. However, his health never recovered from those challenging expeditions. Born in Scotland, he returned to Britain in 1864 but died two years later. The Stuart Highway is named in his honour – the road crosses Australia on a route similar to his final expedition.

Charles Napier Sturt
1795–1869
The son of an English judge in India, Sturt was raised in England and served in the British Army. As military secretary to the governor of New South Wales, Sturt led three expeditions into the interior of southeastern Australia with the assistance of Indigenous guides. The first two, in

> *"To awaken quite alone in a strange town is one of the pleasantest sensations in the world. You are surrounded by adventure."*
>
> FREYA STARK, *BAGHDAD SKETCHES*, 1932

1828–29 and 1829–30, explored the region's rivers. Sturt reached the Darling River and Murray River, among other achievements, but the journeys left him in poor health and nearly blind. He returned to England to recover, publishing a book about his work. He went back to Australia in 1835. His final expedition, in 1844–46, made him the first European to travel deep into Australia's interior.

T

Abel Tasman
c.1603–c.1659

A Dutch navigator sailing for the Dutch East India Company, Tasman led a 1642–43 voyage that resulted in the first European sightings of New Zealand, the Fiji Islands, Tonga, and Tasmania. Tasman named Tasmania Van Diemen's Land, but it was later renamed in his honour. The members of the expedition became the first Europeans to meet the Māori peoples. Tasman led a second voyage in the region in 1644. He later sailed to Sumatra, Siam, and Manila, then settled in Batavia.

Wilfred Thesiger
1910–2003

A British explorer and travel writer, in 1946 Thesiger crossed Saudi Arabia's Rub' Al Khali, or "Empty Quarter" in the company of Bedouin guides and spent years living among the Marsh Arabs in southern Iraq, among other adventures. He expressed greater admiration for the cultures he found there than for those in Europe. The son of a British diplomat in Ethiopia, he joined the Sudan Political Service and fought against the Italian occupation of Ethiopia during World War II. During the war, he also conducted raids behind German and Italian lines as part of Britain's Special Air Service.

Bertram Thomas
1892–1950

A British civil servant and diplomat, Thomas explored the Arabian Peninsula. In 1930–31 he became the first European known to have crossed the Rub'Al Khali, or "Empty Quarter," the world's largest continuous area of sand. Thomas wrote books about Arabia and, during World War II, served as the director of the Middle East Centre for Arab Studies.

David Thompson
1770–1857

A fur trader, surveyor, and cartographer, Thompson explored western Canada and parts of the northwestern United States. Born in Britain, in 1784 he travelled to Canada as an apprentice with the Hudson's Bay Company and defected to its rival, the North West Company, in 1797. A year later Thompson found the headwaters of the Mississippi River, and from 1807, found the source of and mapped the Columbia River. From 1817 to 1827 he surveyed the US–Canada border. His maps and journals detailed nearly 4.9 million sq km (1.9 million sq miles) of territory.

Luís Vaez de Torres
Flourished c.1605–07

A navigator who sailed for Spain, Torres commanded one of the three ships in Pedro Fernandes de Quiros' 1605–07 expedition to the South Pacific. Torres encountered the strait between New Guinea and northern Australia that would later be named the Torres Strait in his honour. Torres is generally assumed to have been Spanish.

V

Ármin Vámbéry
1832–1913

A linguist and explorer noted for his astonishing facility for languages, in 1861 Vámbéry embarked on a perilous journey from Constantinople to Samarkand, in present-day Uzbekistan. Disguised as a Sunni dervish, he travelled by donkey, camel, cart, and on foot, not returning until 1864. The trek was all the more impressive because he required a crutch to walk. He was born an impoverished Hungarian Jew in what is now Slovakia.

Willem van Ruysbroeck
c.1215–c.1295

Van Ruysbroeck (also known as William of Rubruck) was a Franciscan friar of Flemish descent who was sent by France's King Louis IX in 1253 on a diplomatic and religious mission to meet the Great Khan in Mongolia. He failed to convert the khan to Christianity and returned west, arriving in Tripoli in 1255. The account he wrote provides detail about the Mongols.

▷ **Abel Tasman**
Artist Jacob Gerritsz. Cuyp is believed to have painted this portrait of Abel Tasman in 1637, five years before Tasman led the expedition for which he would be remembered.

▽ Amerigo Vespucci
Not only did cartographer Martin Waldseemüller name South America after Amerigo Vespucci on his influential 1507 world map, he also included this portrait of Vespucci as part of it.

Francisco Vázquez de Coronado
c.1510–54

Vázquez de Coronado led the main party of a 1540–42 expedition that travelled north up the west coast of Mexico and through what are now the states of Arizona, New Mexico, Texas, Oklahoma, and Kansas. The explorer and his companions became the first Europeans to explore the American southwest, including landmarks such as the Grand Canyon, but they ultimately failed in their goal of locating gold or other riches. Born in Spain, Coronado had travelled to the Americas in 1535, and was named governor of the Spanish province of Nueva Galicia (part of present-day Mexico) in 1538, a position he resumed following the 1540–42 expedition.

Amerigo Vespucci
1454–1512

An Italian navigator, Vespucci sailed across the Atlantic at least twice. His given name, Amerigo, later provided the name of the western continents – America. Vespucci was working for the Medici family in Spain when he secured a place as an astronomer and cartographer on Alonso de Ojeda's 1499–1500 expedition, reaching the Caribbean and the Amazon River. He later joined Portuguese explorer Gonçalo Coelho's 1501–02 expedition that charted a portion of South America's Atlantic coast. Vespucci may have participated in a third voyage to the "New World", and he was among the first to conclude that the lands were a previously unknown continent and not part of Asia.

Orlando, Cláudio, and Leonardo Villas-Bôas
(1914–2002, 1916–98, and 1918–61)

The Villas-Bôas brothers explored the Brazilian interior, living among the Indigenous people for decades and advocating for their interests. They joined the 1941–60 Roncador-Xingu expedition, which made contact with numerous groups of Indigenous peoples, some of whom had never before met people of European origin. In 1961, Orlando and Cláudio played a crucial role in the creation of the 25,900 sq km (10,000 sq mile) Xingu Indigenous National Park which was designed to protect Indigenous peoples and their lands.

Alfred Russel Wallace
1823–1913

A British naturalist, Wallace developed the theory of evolution by natural selection – though Charles Darwin, who reached very similar conclusions independently, garnered more fame for this theory. Wallace's revelation was inspired in part by his 1848–52 explorations in the Amazon basin, as well as his 1854–62 expedition to the Malay Archipelago in Southeast Asia and the many specimens he collected there. Wallace had little formal education and worked as a surveyor before travelling to the Amazon.

7

Number of chapters in Xenophon's *Anabasis*, describing his travels

Samuel Wallis
1728–95

British naval officer Wallis co-led a 1766–68 two-ship expedition to circumnavigate the globe alongside Philip Carteret. Wallis visited numerous South Pacific islands, including Tahiti, but not the huge southern continent that was believed to exist and the expedition had been asked to locate. Wallis and Carteret became separated during the voyage – Wallis's ship returned to Britain in 1768, Carteret's a year later. Wallis had a long and successful naval career – he saw action in the Seven Years' War before he explored the South Pacific – and became commissioner of the British navy.

William John Wills
1834–61

Wills was second-in-command of an 1860–61 expedition that crossed Australia from south to north – the first Europeans to do so. It ended in disaster – Wills, expedition leader Robert O'Hara Burke, and other members of the group died on the return trip. Only John King made it back to the south coast. Wills was born in England and had trained as a surveyor.

Lady Mary Wortley Montagu
1689–1762

An English writer, Wortley Montagu lived in Constantinople (now Istanbul) in 1716–18 while her husband, Edward Wortley Montagu, served as ambassador to Türkiye. The 52 letters she wrote about her time in Türkiye, published after her death, are considered a classic of 18th-century travel writing. Her time in Türkiye led her to advocate for smallpox inoculation, which she saw was already used effectively in the Ottoman Empire.

Xenophon
c.431–354 BCE

Xenophon was among those selected to lead the mercenary Greek troops known as The Ten Thousand on their retreat from the Persian Empire in 401–399 BCE. A Greek historian and military strategist, his decision-making and creativity have led some to call him the greatest military tactician of his era. Xenophon's writings were held in high regard in antiquity.

Xuanzang
602–664 CE

A Buddhist monk born in China, in 629 Xuanzang embarked on a 16-year pilgrimage across the mountains and deserts of Central Asia to India, where he visited sacred sites and sought out holy texts. He eventually returned with more than 600 Sanskrit texts, some of which he translated into Chinese. Xuanzang later wrote an account of his travels.

York
c.1770–after 1815

York was an enslaved person and the only African-American member of the 1804–06 Lewis and Clark expedition to explore the Louisiana Purchase and Pacific Northwest. William Clark, the expedition's co-leader, brought York along as his "body servant". A hunter and skilled frontiersman, York was treated as more or less a full member of the group, and voted on important decisions. After the expedition, York returned to the oppression of slavery but was finally granted his freedom, probably in the early 1810s.

Z

Zhang Qian
Died 114 BCE

Zhang, a Chinese diplomat, forged new trade routes with cultures located to the west of China on behalf of the Chinese Han emperor, reaching lands previously unknown in China. On a mission in 138–125 BCE he was captured and held for ten years by the Xiongnu, a nomadic people hostile to the Han, before escaping and completing his mission. He may have carried out two more expeditions: the first, in 124 BCE, may have been blocked by hostile groups; the second, in 120 BCE, may have established trade contacts as far south as India.

Zheng He
c.1371–1433

A Chinese admiral, Zheng He made seven voyages around the rim of the Indian Ocean between 1405 and 1433, venturing as far as the Red Sea and the eastern coast of Africa. These were remarkable for the distance they covered and their scale – his first fleet is thought to have included as many as 30,000 sailors and over 60 treasure ships. Born into a Muslim family, he was captured by the Ming Army at the age of 10, castrated, then sent to serve at the Imperial Court in present-day Beijing. Zheng He was appointed Defender of Nanjing in 1425.

Zhou Daguan
c.1266–c.1346

Zhou spent 1296–97 in Angkor, the capital of the Khmer Empire, in what is now Cambodia. The written account of what this Yuan dynasty Chinese diplomat saw is among the few surviving contemporary reports of life there, aside from inscriptions carved into temple walls. Zhou was born near the southeast coast of China, but little else is known of his life.

▷ **Zheng He**
This statue of Chinese admiral Zheng He stands not in China, but in Malacca, Malaysia – a testament to the admiral's enduring impact across much of coastal Asia.

INDEX

Page numbers in **bold** refer to main entries.

B

C

D

E

L

Q

R

S

T

ACKNOWLEDGEMENTS

Dorling Kindersley would like to thank the following people for their help in the preparation of this book: Dawn Henderson for managerial reading; Diana Vowles for proofreading; Helen Peters for indexing; Priyanka Lamichhane and Michelle Harris for factchecking; Chauney Dunford and Ciara Law for editorial support; Aarushi Dhawan, Arshti Narang, and Pooja Pipil for design assistance; Manpreet Kaur and Samrajkumar S for picture research assistance.

The publisher would like to thank the following for their kind permission to reproduce their photographs:

(Key: a-above; b-below/bottom; c-centre; f-far; l-left; r-right; t-top)

2 Alamy Stock Photo: Akademie. **4 Alamy Stock Photo:** Dietmar Rauscher (tl). **Bridgeman Images:** Boucicaut Master, (fl.1390-1430) (tc). **Getty Images:** Universal Images Group / Sepia Times (tr). **5 Alamy Stock Photo:** Sam Kovak (tr); The Picture Art Collection (tl); SuperStock / Newberry Library (tc). **7 David Rumsey Map Collection, David Rumsey Map Center, Stanford Libraries. 8-9 Alamy Stock Photo:** Dietmar Rauscher (Background). **10 Alamy Stock Photo:** Phil Degginger (tl); Jon Arnold Images Ltd / Ivan Vdovin (crb). **11 Alamy Stock Photo:** History and Art Collection (tl). **Bridgeman Images:** NPL - DeA Picture Library / N. Cirani (cr). **13 Alamy Stock Photo:** The Natural History Museum (c). **Bridgeman Images:** Henri Stierlin / Bildarchiv Steffens (br). **14 Alamy Stock Photo:** Geogphotos (cla). **Getty Images:** AFP (bc). **14-15 Alamy Stock Photo:** GpPhotoStudio. **16 Alamy Stock Photo:** NPS Photo (bc). **17 Alamy Stock Photo:** The Natural History Museum (tr). **18 Getty Images:** Universal History Archive / Universal Images Group (bc). **19 Dr Stuart Bedford, Vanuatu Cultural Centre and National Library:** (br). **20 Getty Images:** Universal Images Group / Sepia Times (cla). **20-21 Alamy Stock Photo:** Lanmas. **22-23 Alamy Stock Photo:** Dietmar Rauscher. **23 The Metropolitan Museum of Art:** Rogers Fund and Edward S. Harkness Gift, 1920 (cra). **25 Alamy Stock Photo:** Album (br). **Getty Images:** Stone / Manfred Gottschalk (tc). **26 Dorling Kindersley:** Board of Trustees of the Royal Armouries / Gary Ombler (tl). **28 Alamy Stock Photo:** Heritage Images / CM Dixon / The Print Collector (tr). **28-29 Alamy Stock Photo:** Robertharding / Adam Woolfitt (b). **29 Alamy Stock Photo:** Heritage Images / CM Dixon / The Print Collector (tr). **30-31 Alamy Stock Photo:** Penta Springs Limited / Artokoloro. **32 David Coulson / TARA:** (bl). **33 Alamy Stock Photo:** imageBROKER.com GmbH & Co. KG / Raimund Franken (br). **34-35 Getty Images:** LightRocket / Wolfgang Kaehler. **35 Alamy Stock Photo:** imageBROKER.com GmbH & Co. KG / GTW (cra). **36 Alamy Stock Photo:** Heritage Image Partnership Ltd / © Fine Art Images (cla). **36-37 Alamy Stock Photo:** Adam Eastland. **38 Alamy Stock Photo:** CPA Media Pte Ltd / Pictures From History (bc). **39 Bridgeman Images:** Christie's Images (bc). **40-41 Alamy Stock Photo:** The Picture Art Collection. **42 The Metropolitan Museum of Art:** Rogers Fund, 1988 (tl). **43 Alamy Stock Photo:** Lebrecht Music & Arts (br). **44-45 Bridgeman Images:** Boucicaut Master, (fl.1390-1430) (Background). **46 Alamy Stock Photo:** Heritage Image Partnership Ltd / Ashmolean Museum of Art and Archaeology (ca); The Picture Art Collection (cl). **47 Alamy Stock Photo:** CPA Media Pte Ltd / Pictures From History (tl); CPA Media Pte Ltd / Pictures From History (cr). **48 The Metropolitan Museum of Art:** Rogers Fund, 1919 (tl). **49 Bridgeman Images:** The British Library Archive (br). **50-51 Bridgeman Images:** The British Library Archive. **52 Alamy Stock Photo:** Zev Radovan (bc). **52-53 Alamy Stock Photo:** Science History Images. **54 Alamy Stock Photo:** Historic Images (bc). **55 Bridgeman Images:** Tarker (bl). **56-57 Bridgeman Images:** Cresques, Abraham. **57 Alamy Stock Photo:** Historic Illustrations (tr). **58-59 Bridgeman Images:** Florilegius. **59 Alamy Stock Photo:** Tim Cuff (cra). **60 Getty Images / iStock:** William Dummitt (bc). **Getty Images:** Hulton Archive / Heritage Images (cla). **60-61 Alamy Stock Photo:** Manfred Gottschalk. **62 Alamy Stock Photo:** CPA Media Pte Ltd / Pictures From History (tr). **63 Alamy Stock Photo:** Wendy Kay (br). **64-65 Bridgeman Images:** Boucicaut Master, (fl.1390-1430). **65 Alamy Stock Photo:** Science History Images (cra). **66 Alamy Stock Photo:** MET / BOT (bl). **67 Bridgeman Images:** PVDE / Yahya ibn Mahmud Al-Wasiti (br). **68-69 Alamy Stock Photo:** Kingdom of Maps. **69 Bridgeman Images:** The British Library Archive (br). **70-71 Bridgeman Images:** Stefano Bianchetti. **72-73 Museo Naval de Madrid. 73 Alamy Stock Photo:** AF Fotografie (br). **74-75 Getty Images:** Universal Images Group / Sepia Times (Background). **76 Alamy Stock Photo:** NMUIM (clb); Science History Images (cra). **77 Bridgeman Images:** Christie's Images (tl); Spanish School (crb). **78 Royal Danish Library:** (bl). **79 Bridgeman Images:** The British Library Archive (br). **80 Alamy Stock Photo:** The Granger Collection (bl). **81 Museo Naval de Madrid** (br). **82-83 Bridgeman Images:** The British Library Archive / Martin Waldseemuller. **84-85 Bridgeman Images:** Flemish School, (16th century) / Josse (b). **85 Alamy Stock Photo:** Ariadne Van Zandbergen (tr). **86-87 Getty Images:** De Agostini / Dea Picture Library. **88 Alamy Stock Photo:** CPA Media Pte Ltd / Pictures From History (cl). **89 Alamy Stock Photo:** The Granger Collection (br). **90 Alamy Stock Photo:** The History Collection (cla). **90-91 Alamy Stock Photo:** Granger - Historical Picture Archive. **92 Alamy Stock Photo:** Science History Images / Photo Researchers (bl). **93 Bridgeman Images:** Brooklyn Museum / Peruvian School (br). **94-95 Alamy Stock Photo:** Album. **96 Alamy Stock Photo:** Piemags / Rmn (tl). **97 Alamy Stock Photo:** The Granger Collection (br). **98 Alamy Stock Photo:** North Wind Picture Archives (bc). **98-99 Alamy Stock Photo:** North Wind Picture Archives. **100 Bridgeman Images:** NPL - DeA Picture Library / A. De Gregorio (cla). **100-101 Getty Images:** Universal Images Group / Sepia Times. **102 Alamy Stock Photo:** World History Archive (clb). **103 Alamy Stock Photo:** Chronicle (br). **104-105 Alamy Stock Photo:** JSM Historical. **106-107 Alamy Stock Photo:** Logic Images. **107 Alamy Stock Photo:** Classic Image (br). **Bridgeman Images:** Sainsbury Centre for Visual Arts / Robert and Lisa Sainsbury Collection (cra). **108-109 Alamy Stock Photo:** Images & Stories. **109 Alamy Stock Photo:** The History Collection (br). **110-111 Alamy Stock Photo:** CPA Media Pte Ltd / Pictures From History. **111 Alamy Stock Photo:** The Natural History Museum (br). **112 Bridgeman Images:** (clb). **113 Alamy Stock Photo:** Niday Picture Library (br). **114-115 Alamy Stock Photo:** Everett Collection Inc. **115 Alamy Stock Photo:** IanDagnall Computing (br). **116-117 © The Trustees of the British Museum. All rights reserved**. **117 Alamy Stock Photo:** Chronicle (br). **Bridgeman Images. 118 Alamy Stock Photo:** WBC Art (tr). **119 Alamy Stock Photo:** Granger - Historical Picture Archive (br). **120-121 Getty Images:** Universal Images Group / Universal History Archive. **121 Getty Images:** Hulton Archive / Culture Club (br). **122-123 Bridgeman Images:** Archives Charmet. **123 Alamy Stock Photo:** Akademie (ca). **Bridgeman Images:** Georg Emanuel Opitz (br). **124-125 Alamy Stock Photo:** The Picture Art Collection. **126 Alamy Stock Photo:** Science History Images / Photo Researchers (cl). **Bridgeman Images:** Usher Gallery / Gervase Spencer (ca). **127 Alamy Stock Photo:** Vintage Travel and Advertising Archive (tr). **Bridgeman Images:** Granger / John Harrison (tl). **128 Alamy Stock Photo:** Interfoto / History (bl). **129 Alamy Stock Photo:** The History Collection (tr). **130-131 Bridgeman Images:** NPL - DeA Picture Library. **132 Alamy Stock Photo:** EDR Archives (cla). **132-133 Alamy Stock Photo:** Album. **134-135 Alamy Stock Photo:** The Natural History Museum. **135 Alamy Stock Photo:** ICP / Incamerastock (br). **136 Alamy Stock Photo:** Pictorial Press Ltd (bl). **137 Getty Images:** De Agostini Picture Library (br). **138 Alamy Stock Photo:** Vidimages (bc). **139 Alamy Stock Photo:** Pictorial Press Ltd (cra). **140-141 Alamy Stock Photo:** The Picture Art Collection. **141 Alamy Stock Photo:** History and Art Collection (cra). **Bridgeman Images:** The British Library Archive (br). **142 Alamy Stock Photo:** Niday Picture Library (bc). **142-143 Bridgeman Images:** Humboldt Innovation, Berlin / © Humboldt-Universitaet zu Berlin. **144 State Library of New South Wales:** Dixson Library (b). **144-145 Alamy Stock Photo:** Wojtkowski Cezary. **146-147 David Rumsey Map Collection, David Rumsey Map Center, Stanford Libraries. 149 Alamy Stock Photo:** The Picture Art Collection (tl). **150 Alamy**

Stock Photo: Florilegius (cla); The History Collection (bc). **150-151 Bridgeman Images:** Peter Newark American Pictures / Charles Marion Russell. **152 Getty Images:** Archive Photos / MPI (bl). **153 Alamy Stock Photo:** Heritage Image Partnership Ltd (br). **154 Alamy Stock Photo:** The Natural History Museum (bc). **Bridgeman Images:** English Photographer (cla). **154-155 Science Photo Library:** Natural History Museum, London. **156-157 Science Photo Library:** Paul D Stewart. **158 Alamy Stock Photo:** Well / BOT (bl). **159 Board of Trustees of the Science Museum:** Science Museum (cra). **160 Alamy Stock Photo:** Chroma Collection (bl). **161 Alamy Stock Photo:** Penta Springs Limited (tl). **162 Alamy Stock Photo:** Chronicle (cla); Well / BOT (bc). **162-163 Alamy Stock Photo:** Old Books Images. **164-165 Bridgeman Images:** The British Library Archive / Tupaia. **165 Alamy Stock Photo:** Historic Images (br). **166-167 Library of Congress, Washington, D.C.**. **167 Alamy Stock Photo:** M&N (br). **168 Alamy Stock Photo:** Chronicle (bc). **169 Bridgeman Images:** © Leonard de Selva / James Bruce (tr). **170 The Metropolitan Museum of Art:** Edith Perry Chapman Fund (cla). **171 Bridgeman Images:** Rene Caillie (br). **172-173 Alamy Stock Photo:** SuperStock / Newberry Library (Background). **174 Adobe Stock:** Archivist (c). **Alamy Stock Photo:** Shawshots (tl). **175 Bridgeman Images:** Christie's Images (crb); Royal Geographical Society / W Sherwill (tl). **176-177 Bridgeman Images:** Granger. **177 Alamy Stock Photo:** Pictorial Press Ltd (cra); The Granger Collection (br). **178-179 Bridgeman Images:** © Royal Geographical Society. **179 Bridgeman Images:** © National Army Museum (br). **180 Alamy Stock Photo:** History and Art Collection (bc). **181 Alamy Stock Photo:** Lakeview Images (crb). **182 Alamy Stock Photo:** BG / OLOU (cla). **182-183 Bridgeman Images:** © Archives Charmet / Louis Delaporte. **184-185 Getty Images:** De Agostini Picture Library. **185 Alamy Stock Photo:** Heritage Image Partnership Ltd / Fine Art Images (cra); History_Docu_Photo (br). **186-187 Allard Pierson, Library of the University of Amsterdam:** HB-KZL 62.04.07 / Track of H.M.S. Challenger Dec.r 1872 to May 1876 … [etc.]. Malby & Sons, ca 1900.. **187 Alamy Stock Photo:** Granger - Historical Picture Archive (br). **188 Science Photo Library:** Paul D Stewart (tl). **189 Alamy Stock Photo:** Historic Images (br). **190 Bridgeman Images:** The British Library Archive / Marc Aurel Stein (bc). **191 Bridgeman Images:** British Library (tr). **192-193 David Rumsey Map Collection, David Rumsey Map Center, Stanford Libraries. 193 Alamy Stock Photo:** GL Archive (br); Old Paper Studios (ca). **194-195 Photo Scala, Florence:** Utagawa Yoshitora. **194 Bridgeman Images:** Christie's Images / Rene Jules Lalique (br). **195 Alamy Stock Photo:** GL Archive (br). **196 Alamy Stock Photo:** 914 collection (bc); SuperStock / Newberry Library (cra). **199 Alamy Stock Photo:** Old Books Images (tl); The Picture Art Collection (br). **200 Getty Images:** Universal Images Group (tr). **200-201 Bridgeman Images:** CSU Archives / Everett Collection (b). **202-203 Alamy Stock Photo:** Chronicle. **203 Science Photo Library:** Martin Land (br). **204-205 Bridgeman Images:** Everett Collection (t). **205 Alamy Stock Photo:** Heritage Image Partnership Ltd (br). **206 Alamy Stock Photo:** Stockimo / Steven May (tl). **207 Alamy Stock Photo:** EMU history (br). **208-209 Alamy Stock Photo:** Sam Kovak (Background). **210 Alamy Stock Photo:** Auk Archive (cr); SuperStock / Sydney Morning Herald (tl). **211 Adobe Stock:** Nerthuz (tl). **Science Photo Library:** NASA (cr). **212 Bancroft Library, University of California Berkeley:** (cla). **National Museum of Natural History / Smithsonian Institution:** (b). **212-213 Getty Images:** Royal Geographical Society. **214 Alamy Stock Photo:** Royal Geographical Society / Justin Hobson (bc). **214-215 Peter Harrington**. **216-217 Arquivo Nacional - MGI**. **217 Alamy Stock Photo:** Peter Horree (cra); Ton Koene (br). **219 Alamy Stock Photo:** 914 collection (br); Photo Resource Hawaii / Franco Salmoiraghi (tr). **221 Alamy Stock Photo:** Heritage Image Partnership Ltd / Curt Teich Postcard Archives (br). **Science Photo Library:** USAF (tl). **222 Alamy Stock Photo:** Pictorial Press Ltd (cla); Stephen Frink Collection (bc). **222-223 Alamy Stock Photo:** Granger - Historical Picture Archive. **224-225 Alamy Stock Photo:** D and S Photography Archives. **225 Getty Images:** Royal Geographical Society (br). **226 The Box, Plymouth:** (cl). **226-227 Getty Images:** Royal Geographical Society. **228 Dreamstime.com:** Sorrapong Apidech (bl/Background). **SuperStock:** National Geographic / Album / Album Archivo (tl). **228-229 Bridgeman Images:** GEO Image Collection / Thomas J. Abercrombie. **230-231 Alamy Stock Photo:** Sam Kovak. **231 Alamy Stock Photo:** Science History Images (ca); Shawshots (br). **232-233 Alamy Stock Photo:** AF Fotografie. **232 Science Photo Library:** Eth Zurich (bc). **234 Alamy Stock Photo:** Stocktrek Images, Inc. (tc). **234-235 NASA:** JPL (b). **235 NASA:** ESA / DLR / FU-Berlin (cra). **236 Alamy Stock Photo:** Universal Images Group North America LLC / QAI Publishing (tl). **237 Alamy Stock Photo:** NASA / J Marshall - Tribaleye Images / Digitaleye (br). **238-239 Alamy Stock Photo:** NASA / World History Archive & ARP. **239 Alamy Stock Photo:** Alexandr Mitiuc (cra). **NASA:** NASA, ESA, CSA, STScI, J. Diego (Instituto de Física de Cantabria, Spain), J. DSilva (U. Western Australia), A. Koekemoer (STScI), J. Summers & R. Windhorst (ASU), and H. Yan (U. Missouri) (br). **240 Alamy Stock Photo:** Granger - Historical Picture Archive (bc). **240-241 Science Photo Library:** NOAA. **242 Getty Images:** Visual China Group (bc). **243 Alamy Stock Photo:** US Space Force (cr). **244-245 Library of Congress, Washington, D.C.. 246 NASA**. **247 Alamy Stock Photo:** Archivio GBB (crb). **Bridgeman Images:** Giancarlo Costa (bl). **The National Library of Norway:** (t). **248 Alamy Stock Photo:** Chronicle (bl). **249 Alamy Stock Photo:** GL Archive (crb). **National Library of Australia:** (t). **250 Alamy Stock Photo:** Hirarchivum Press (r). **Bridgeman Images:** Aunaies (clb). **251 Alamy Stock Photo:** Science History Images / Photo Researchers (tr). **252 Dreamstime.com:** Giorgio Morara (tc). **253 Alamy Stock Photo:** Album (cra). **Bridgeman Images:** Museum of New Zealand te Papa Tongarewa / Heath & Wing of London (bc). **254 Alamy Stock Photo:** Granger - Historical Picture Archive (cra). **Getty Images:** Royal Geographical Society (bl). **255 Alamy Stock Photo:** Album (tc). **Wellcome Collection:** (bl). **256 Dreamstime.com:** Brian Logan (br). **Library of Congress, Washington, D.C.. 257 Alamy Stock Photo:** CPA Media Pte Ltd / Pictures From History (br). **258 Alamy Stock Photo:** Science History Images / Photo Researchers (clb). **259 Alamy Stock Photo:** Heritage Image Partnership Ltd / Ashmolean Museum of Art and Archaeology (cla); IanDagnall Computing (br). **260 Getty Images:** Ullstein Bild (tr). **261 akg-images:** Yvan Travert (br). **Getty Images:** Archive Photos / The New York Historical Society (tc). **262 Getty Images:** Bettmann (l). **263 Alamy Stock Photo:** Chronicle (tl). **Mary Evans Picture Library:** © The Pictures Now Image Collection (br). **264 Alamy Stock Photo:** Science History Images (cla). **265 Alamy Stock Photo:** Heritage Image Partnership Ltd / Fine Art Images (cra). **Shutterstock.com:** KPG-Payless2 (c). **266 Getty Images:** Royal Geographical Society (r). **267 © the Board of Trustees of the Royal Botanic Gardens, Kew:** (br). **268 Alamy Stock Photo:** Universal Art Archive (cla). **SuperStock:** Iberfoto Archivo (crb). **269 Alamy Stock Photo:** Incamerastock / ICP (tr). **270 Alamy Stock Photo:** Independent Picture Service / Todd Strand (bl). **Library of Congress, Washington, D.C.. 271 Alamy Stock Photo:** Photo12 / Archives Snark (r). **272 Alamy Stock Photo:** Artepics (t). **273 Alamy Stock Photo:** Uber Bilder (b). **274 Getty Images:** Universal Images Group / Picturenow (t). **275 Alamy Stock Photo:** VanVang (br)

Endpaper images: *Front & Back:* **Alamy Stock Photo:** Pola's Archives

DK LONDON

Senior Editor Alison Sturgeon
Senior Art Editor Gadi Farfour
Project Art Editor Katie Cavanagh
Editors Simon Beecroft, Claire Cross, Alethea Doran, Rob Dimery, Abigail Mitchell, Vicki Murrell, Fiona Plowman
Production Editor Rob Dunn
Senior Production Controller Rachel Ng
Managing Editor Gareth Jones
Managing Art Editor Luke Griffin
Art Director Maxine Pedliham
Design Director Phil Ormerod
Publishing Director Georgina Dee
Managing Director Liz Gough

DK DELHI

Senior Editor Anita Kakar
Senior Art Editor Anjali Sachar
Project Art Editor Sonakshi Singh
Editors Saumya Agarwal, Aashirwad Jain
Art Editors Debjyoti Mukherjee, Mitravinda V K
Senior Picture Researcher Deepak Negi
Senior Jacket Designer Suhita Dharamjit
Senior Jackets Coordinator Priyanka Sharma Saddi
DTP Coordinator Tarun Sharma
Senior DTP Designer Harish Aggarwal
DTP Designers Mohd Rizwan, Rajdeep Singh
Deputy Managing Art Editor Vaibhav Rastogi
Senior Managing Editor Rohan Sinha
Managing Art Editor Sudakshina Basu
Picture Research Manager Virien Chopra
Pre-production Manager Balwant Singh
Production Manager Pankaj Sharma
Creative Head Malavika Talukder

LOVELL JOHNS

Cartography and research Clare Varney
Project management Louisa Keyworth

TOUCAN BOOKS LTD.

Text Mike Robbins
Picture Researcher Sharon Southren
Editor Dorothy Stannard
Senior Editor Julie Brooke
Senior Designer Dave Jones
Editorial Director Ellen Dupont

First published in Great Britain in 2025 by
Dorling Kindersley Limited,
20 Vauxhall Bridge Road, London SW1V 2SA

The authorised representative in the EEA is
Dorling Kindersley Verlag GmbH. Arnulfstr. 124, 80636 Munich, Germany

10 9 8 7 6 5 4 3 2 1
001–341850–Apr/2025

A CIP catalogue record for this book is available from the British Library.
ISBN: 9780-2-4168-2791

Printed and bound in the UAE

www.dk.com

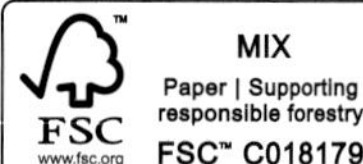

This book was made with Forest Stewardship Council™ certified paper – one small step in DK's commitment to a sustainable future.
Learn more at www.dk.com/uk/information/sustainability

Polus Arcticus
Circulus Arcticus
AMERICA MEXICANA
Tropicus Cancri
Circulus Aequinoctialis
OCEANUS PERUVIANUS
Tropicus Capricorni
EL MAR PACIFICO
TERRA AUSTRALIS
MAGALLANICA
Circulus Antarcticus
Polus Antarcticus
MEXICANA
PERUANA
La Bermuda
Insula Salomonis
Hanc continentem Australem nonnulli Magellanicam regionem ab eius inuentore nuncupant.